T0392538

Microbial Syntrophy-mediated Eco-enterprising

Developments in Applied Microbiology and Biotechnology

Microbial Syntrophy-mediated Eco-enterprising

Edited by

Raghvendra Pratap Singh
Azoth Biotech Private Limited, Noida, Uttar Pradesh, India

Geetanjali Manchanda
DAV University, Jalandhar, Punjab, India

Kaushik Bhattacharjee
Indian Institute of Technology Guwahati, Guwahati, Assam, India

Hovik Panosyan
Yerevan State University, Yerevan, Armenia

Academic Press is an imprint of Elsevier
125 London Wall, London EC2Y 5AS, United Kingdom
525 B Street, Suite 1650, San Diego, CA 92101, United States
50 Hampshire Street, 5th Floor, Cambridge, MA 02139, United States
The Boulevard, Langford Lane, Kidlington, Oxford OX5 1GB, United Kingdom

Library of Congress Cataloging-in-Publication Data
A catalog record for this book is available from the Library of Congress

British Library Cataloguing-in-Publication Data
A catalogue record for this book is available from the British Library

ISBN 978-0-323-99900-7

For information on all Academic Press publications
visit our website at https://www.elsevier.com/books-and-journals

Publisher: Stacy Masucci
Senior Acquisitions Editor: Linda Versteeg-Buschman
Editorial Project Manager: Bernadine A. Miralles
Production Project Manager: Punithavathy Govindaradjane
Cover Designer: Miles Hitchen

Typeset by STRAIVE, India

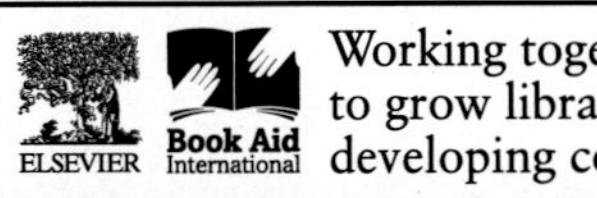

Contents

Contributors

Azza A.M. Abd El-Aal
Soil Microbiology Department, Soils, Water and Environment Research Institute, Agricultural Research Center, Giza, Egypt

Abdul Salam Abdul Raheem
Environmental Biotechnology Program, Life Sciences Department, College of Graduate Studies, Arabian Gulf University, Manama, Bahrain

Mian Nabeel Anwar
Department of Civil and Environmental Engineering, University of Alberta, Edmonton, AB, Canada

Archana
Department of Endocrinology, Sanjay Gandhi Post Graduate Institute of Medical Sciences, Lucknow, Uttar Pradesh, India

Gayane Ye. Avetisova
Laboratory of Strain-Producers of BAS and Biosynthesis, Scientific and Production Center "Armbiotechnology" of the National Academy of Sciences of Armenia; Institute of Pharmacy, Yerevan State University, Yerevan, Armenia

Emre Babur
Department of Soil Science and Ecology, Faculty of Forestry, Kahramanmaras Sutcu Imam University, Kahramanmaraş, Turkey

Dawoud Bahzad
Petroleum Research Center, Kuwait Institute for Scientific Research, Kuwait City, Kuwait

Dhritiksha M. Baria
Department of Microbiology and Biotechnology, University School of Science, Gujarat University, Ahmedabad, Gujarat, India

Martin L. Battaglia
Department of Animal Science, Cornell University, Ithaca, NY, United States

Hitarth B. Bhatt
Department of Microbiology, Faculty of Science, Atmiya University, Rajkot, Gujarat, India

Siddharth Shankar Bhatt
Uttaranchal University, Dehradun, Uttarakhand, India

Edison Dausacker Bidone
Geoscience (Geochemistry) Pos-Graduation Program, Chemistry Institute, Fluminense Federal University—UFF, Niteroi, RJ, Brazil

Qianru Chen
Marine Agriculture Research Center, Tobacco Research Institute of Chinese Academy of Agricultural Sciences, Qingdao, China

Vanessa de Almeida Moreira
Geoscience (Geochemistry) Pos-Graduation Program, Chemistry Institute, Fluminense Federal University—UFF, Niteroi, RJ, Brazil

Angelo Cezar Borges de Carvalho
Geoscience (Geochemistry) Pos-Graduation Program, Chemistry Institute, Fluminense Federal University—UFF, Niteroi, RJ, Brazil

Turgay Dindaroğlu
Department of Soil Science and Ecology, Faculty of Forestry, Kahramanmaras Sutcu Imam University, Kahramanmaraş, Turkey

Laura Drummond
Microbial Biotechnology, DECHEMA Research Institute, Frankfurt, Germany

Luiz Francisco Fontana
Laboratory of Micropaleontology, Rio de Janeiro State Federal University—UNIRIO, Rio de Janeiro, RJ, Brazil

Naila Ilyas
Marine Agriculture Research Center, Tobacco Research Institute of Chinese Academy of Agricultural Sciences, Qingdao, China

Wael A. Ismail
Environmental Biotechnology Program, Life Sciences Department, College of Graduate Studies, Arabian Gulf University, Manama, Bahrain

S.K. Jayasekara
Microbiology and Soil Ecosystems Research Project, National Institute of Fundamental Studies, Kandy, Sri Lanka

Changliang Jing
Marine Agriculture Research Center, Tobacco Research Institute of Chinese Academy of Agricultural Sciences, Qingdao, China

Alok Kalra
Division of Crop Protection and Production, CSIR-Central Institute of Medicinal and Aromatic Plants, Lucknow, Uttar Pradesh, India

Sashi Kant
Department of Immunology and Microbiology, University of Colorado School of Medicine, Anschutz Medical Campus, Aurora, CO, United States

Masuma Khawary
Microbial Pathogenesis and Microbiome Lab, Department of Microbiology, Central University of Rajasthan, Ajmer, Rajasthan, India

Bin Li
ChinaTobacco Chengdu Industrial Co., Ltd., Chengdu, China

Yiqiang Li
Marine Agriculture Research Center, Tobacco Research Institute of Chinese Academy of Agricultural Sciences, Qingdao, China

Jose V. Lopez
Halmos College of Natural Sciences and Oceanography, Nova Southeastern University, Dania Beach, FL, United States

Lusine H. Melkonyan
Laboratory of Strain-Producers of BAS and Biosynthesis, Scientific and Production Center "Armbiotechnology" of the National Academy of Sciences of Armenia; Institute of Pharmacy, Yerevan State University, Yerevan, Armenia

Mohit Mishra
Amity University, Raipur, Chhattisgarh, India

Pooja Misra
Crop Protection Division, CSIR-Central Institute of Medicinal and Aromatic Plants, Lucknow, Uttar Pradesh, India

Juliana Ribeiro Nascimento
Geoscience (Geochemistry) Pos-Graduation Program, Chemistry Institute, Fluminense Federal University—UFF, Niteroi, RJ, Brazil

Ekrem Ozlu
Great Lakes Bioenergy Research Center, W.K. Kellogg Biological Station, Michigan State University, Hickory Corners, MI, United States

Ani M. Paloyan
Laboratory of Protein Technologies, Scientific and Production Center "Armbiotechnology" of the National Academy of Sciences of Armenia, Yerevan, Armenia

Saurabh Pandey
Department of Biochemistry, School of Chemical and Life Sciences, Jamia Hamdard, New Delhi, India

R.R. Ratnayake
Microbiology and Soil Ecosystems Research Project, National Institute of Fundamental Studies, Kandy, Sri Lanka

Raunak
Microbial Pathogenesis and Microbiome Lab, Department of Microbiology, Central University of Rajasthan, Ajmer, Rajasthan, India

Vikram H. Raval
Department of Microbiology and Biotechnology, University School of Science, Gujarat University, Ahmedabad, Gujarat, India

Rana Roy
Department of Agroforestry & Environmental Science, Sylhet Agricultural University, Sylhet, Bangladesh

Elisamara Sabadini-Santos
Geoscience (Geochemistry) Pos-Graduation Program, Chemistry Institute, Fluminense Federal University—UFF, Niteroi, RJ, Brazil

Jana Sedlakova-Kadukova
Department of Ecochemistry and Radioecology, Faculty of Natural Sciences, University of Ss. Cyril and Methodius in Trnava, Trnava, Slovakia

Mahmoud F. Seleiman
Plant Production Department, College of Food and Agriculture Sciences, King Saud University, Riyadh, Saudi Arabia; Department of Crop Sciences, Faculty of Agriculture, Menoufia University, Shibin El-Kom, Egypt

Ana Elisa Fonseca Silveira
Geoscience (Geochemistry) Pos-Graduation Program, Chemistry Institute, Fluminense Federal University—UFF, Niteroi, RJ, Brazil

Akanksha Singh
Division of Crop Protection and Production, CSIR-Central Institute of Medicinal and Aromatic Plants, Lucknow, Uttar Pradesh, India

Satya P. Singh
UGC-CAS Department of Biosciences, Saurashtra University, Rajkot, Gujarat, India

Sucheta Singh
Molecular Biology and Biotechnology Division, CSIR-National Botanical Research Institute, Lucknow, Uttar Pradesh, India

Suman Singh
Department of Botany, University of Lucknow, Lucknow, Uttar Pradesh, India

Atul Kumar Srivastava
Research and Development Department, Uttaranchal University, Dehradun, Uttarakhand, India

Deeksha Tripathi
Microbial Pathogenesis and Microbiome Lab, Department of Microbiology, Central University of Rajasthan, Ajmer, Rajasthan, India

Takshashila Tripathi
Department of Neuroscience, Physiology and Pharmacology, University College London, London, United Kingdom

Shikha Uniyal
Research and Development Department, Uttaranchal University, Dehradun, Uttarakhand, India

Ömer Suha Uslu
Field Crops Department, Agriculture Faculty, Kahramanmaras Sutcu Imam University, Kahramanmaraş, Turkey

Yingjie Yang
Marine Agriculture Research Center, Tobacco Research Institute of Chinese Academy of Agricultural Sciences, Qingdao, China

Ping Zou
Marine Agriculture Research Center, Tobacco Research Institute of Chinese Academy of Agricultural Sciences, Qingdao, China

About the editors

Dr. Raghvendra Pratap Singh is an eminent scientist in microbial biotechnology. He received his PhD from Gurukula Kangri University, India. His research relates to the area of ecology of myxobacteria, plant–microbe interaction, and microbial genomics. He is a member of the Aquatic Biodiversity Society and has a certification from the Food and Drug Association of India. He has received several awards and grants from various scientific agencies and societies, such as the Young Scientist Award from ABA, 2017, the SERB-DST grant, the Chinese Postdoctoral Grant, the DBT travel grant, etc. He has participated in several national and international scientific meetings and conferences, such as FEMS, 2013 in Italy, Myxo, 2016 in Switzerland, AMI, PTPB, 2014, etc.

Dr. Singh has actively contributed 51 research articles, 2 patents, 9 book chapters, 3 general articles, etc., in microbiology and biotechnology journals. He has deposited more than 450 gene sequences, 7 whole genome sequences, and 3 metagenomes at NCBI. He has authored four books, one of which is published by the Indian Council of Agricultural Research, Government of India.

Dr. Geetanjali Manchanda is head of the Botany Department at DAV University, Jalandhar, India. She received her MSc degree from Delhi University and her PhD from Panjab University, Chandigarh. She has extensively worked on plant–microbe interactions in stressed and contaminated environments, with special focus on mycorrhizae for the fortification of various crops.

She has received prestigious research grants from DST, India, and IFS, Sweden. She has contributed prolifically to the scientific community by publishing research papers and book chapters. She had recently authored three books one of which is on the use of omics technology for microbiology, published by ICAR, India.

Dr. Kaushik Bhattacharjee received his PhD in microbiology from the North-Eastern Hill University of India, by conducting research into the interdisciplinary field of microbial diversity in extreme environments, pharmaceutical microbiology, and medicinal chemistry. His postdoctoral training was at the Department of Botany, North-Eastern Hill University of India and at IASST, Guwahati, India.

He has so far contributed over 20 publications in journals of high repute and has published about 6 book chapters. He also serves as an editorial board member and invited journal reviewer for many highly reputed journals with publishing groups such as Springer Nature, Elsevier, PLoS, and Taylor & Francis. He was presented with the Outstanding Reviewer Award from Elsevier for the year 2018. He also serves as a certified mentor at Publons Academy, Clarivate Analytics, USA. His broad area of research interests includes environmental microbiology and pharmaceutical microbiology.

Professor Hovik Panosyan graduated in biology from Yerevan State University (YSU) in 1999. He received his PhD in microbiology, from the Institute of Botany, NAS, Armenia, in 2003. He has been a faculty member at YSU since 2002 and was promoted to associate professor in 2011. His main area of research is microbial ecology and biology of extremophilic microbes. He has been awarded numerous research fellowships and awards, including FEBS Short-Term Fellowship (2009 and 2004), FEMS Research Fellowship (2009), NFSAT (2011), and DAAD (2013), and has participated in international research with partners in the United States, Europe, and Asia. He is currently the coordinator and leader of international research and educational programs, as well as the ISME ambassador for Armenia.

He has had work experience at the University of Bergen, Norway, LMU Munich, Germany, the University of Nevada, USA, and the Institute of Biomolecular Chemistry Naples, Italy. He is actively engaged in studying the microbial community of extreme environments, such as terrestrial geothermal springs, alkali-saline soils, subterranean salt deposits, and copper and molybdenum mines in Armenia, based on culture-dependent and molecular techniques. He has published 4 books, 25 book chapters, and more than 60 research papers in peer-reviewed journals.

CHAPTER

1

Anabaena-azollae, significance and agriculture application: A case study for symbiotic cyanobacterium

Azza A.M. Abd El-Aal

Soil Microbiology Department, Soils, Water and Environment Research Institute, Agricultural Research Center, Giza, Egypt

Abstract

Anabaena azollae is a heterocystous filamentous nitrogen-fixing cyanobacterium that is naturally growing symbiotically in specialized leaf cavities of a small eukaryotic water fern *Azolla pinnata*. It is well documented that *Anabaena azollae* has been successfully grown and propagated freely in synthetic media like BG-110 with pleasant biomass.

A. azollae is considered a promising natural biosource for agricultural, medicinal, and industrial applications. Where the cyanobacterium biomass and/or extract greatly enhanced the physical and chemical properties of soil texture. It is also well-known to produce several biologically active substances against a wide array of plant-infecting pathogens. Moreover, it is considered a good phycoremediator of industrial wastewater.

To the best of our knowledge, this is the first review of *A. azollae*, a case study for its characteristics, significance with a special focus on its agricultural applications.

The commercial production and exploitation of *A. azollae*-derived materials with interesting properties such as fungicides, bactericides, nematicides, insecticides, biofertilizers, as well as its phycoremediation ability were highlighted in detail in this review.

Numerous growths promoting substances like indole acetic acid (IAA), gibberellic acid (GA), bioactives such as fatty acids, polysaccharides, phenolic compounds were extracted from *A. azollae*, and reported to have in vitro and in vivo microbicidal effects. Additionally, the high nitrogenase activity of *A. azollae* has been accepted as a pioneer indicator for its biofertilization ability. Strikingly, the induced dehydrogenase activity and its polysaccharides excretions were reported to increase soil fertility by increasing the soil microbial communities. Thus, *A. azollae* could provide multiple benefits to the agricultural sector and can be considered as a promising and safe bio-inoculant in recent trends of organic farming.

Keywords: *Anabaena-azollae*, *Azolla pinnata*, Bio-fertilizer, Antifungal activity, Antinematode activity, Bioinsecticidal activity, Bioremediation

1 Introduction

Bio-fertilizer was recognized as an option in sustainable agriculture to increase soil fertility and crop manufacturing. Because of their potential role in food safety and sustainable crop production, the exploitation of beneficial microbes as biofertilizers has become of paramount importance in the agricultural sector (Itelima et al., 2018). Furthermore, biofertilizers such as nitrogen fixers, cyanobacteria, bacteria, and aquatic Azolla fern are becoming increasingly important in sustainable agriculture, where

Microbial Syntrophy-mediated Eco-enterprising. https://doi.org/10.1016/B978-0-323-99900-7.00006-7

different complementary combinations of microbial inoculants are required for the management of significant nutrients such as nitrogen and phosphorus (Brahmaprakash and Sahu, 2012).

Bio-fertilizer inoculation is now regarded to restrict and minimize the use of mineral fertilizers and is an efficient instrument for soil development in less polluted settings, Reduction of agricultural expenses, maximization of crop yields owing to the availability of nutrients, and development of substances (Al-Erwy et al., 2016). Azolla, a free-floating aquatic fern with a dichotomous branch, is naturally accessible in India's tropical belt. The exposed dorsal lobe has a particular cavity comprising its symbiotic partner, *Anabaena azollae*, a blue-green alga.

Azolla's abundant development not only makes the combined nitrogen added to the ecosystem helpful but can also provide green manure (Rkyadav et al., 2014).

Biofertilizers maintain a rich soil environment with all sorts of micro- and macro-nutrients via nitrogen fixation, phosphate, and potassium solubilization or mineralization. The release of plant growth regulating biologically active substances including phytohormones, such as auxin, Gibberellins, and cytokinin's, antibiotic production, and organic soil biodegradation (Divya and Ram, 2018).

Azolla also produces many other components that enhance the overall soil fertility by enhancing the amount of nutrients available to plants, as the organic matter in the soil increases, and the soil structure improves (Nevine and El-Shahat, 2018).

Purushottam and Jiban (2015) was noted that the soil's biological health owing to the implementation of Azolla resulted in improved mineralization and a consequent boost in the soil's microbial status.

The use of Azolla fern and cyanobacteria as a bio-fertilizer is advocated to minimize the dependency on chemical fertilizer. Azolla supplements rice nitrogen by setting atmospheric nitrogen in the soil for plant growth, crop manufacturing, and soil fertility improvements (Bharati et al., 2016).

The impacts of the use of magnetite, diatoms, and some biofertilizers (azolla and cyanobacteria) on the development yield and quality of Valencia orange cultivated under saline soil circumstances in El Bustan County, Egypt, was the best combination for achieving the highest total yield (Hoda et al., 2013).

The effect on the development of cyanobacteria, Azolla, bacteria with biostraw as biofertilizers or coupled with urea, rice output showed several advantages over chemical fertilizers and enhanced saline soil fertility (Abd El-Aal et al., 2013).

Arafa and Abd El-Aal (2013) additionally, the organic fertilizer humic acid gave the highest outcomes for all soil characteristics using the combination of *Spirulina platensis* and *Azolla pinnata*. It also saves on using bio-organic fertilizers for colored cotton without influencing the characteristics of the colored fiber.

Algae are a big and varied group of microorganisms that can perform photosynthesis as they capture sunlight energy. *Anabaena azalea* plays a significant role in farming where it is used as a biofertilizer and soil stabilizer. *Anabaena azallea* is a symbiotically related photoautotrophic cyanobacterium with the tiny eukaryotic water fern *Azolla pinnata*, in specialized leaf cavities of the fern under natural conditions. It is a multicellular organism with two separate, interdependent kinds of cells. Which are "vegetative" cells and heterocyst's; to fix atmospheric nitrogen. Cyanobacteria *Anabaena azallea* from the symbiosis has also been separated and cultivated as a bioagent suppressor against *Fusarium oxysporum* and *Alternaria alternata* separately of the fern. This antifungal activity may be attributed to the presence of bioactive compounds identified Filters such as phenolic compounds, saponins, and alkaloids that act as natural defense mechanisms against pathogenic fungi in cyanobacterial culture (Abd El-Aal, 2013). Mohamed et al. (2015) discovered that *Anabaena azollae* played a major role in enhancing the pelargonium potential and antibiotic studies have shown promising control results; *Fusarium oxysporum* and *Rhizoctonia solani* by various microorganisms, such as *Pseudomonas fluorescens* and/or extracts of either single or combined *Pleurotus columbinus* or *Anabaena azollae*. The combination of *Anabaena azollae* extract, *Spirulina platensis* and *Pleurotus columbinus* was the best therapy to achieve high output in the manufacturing of herbal and oil.

In addition, the 25% reduction in the recommended bioagent treatment dose was higher than the recommended N fertilizer dose in plant yield and oil quality and quantity production (Mohsen et al., 2015).

Therefore, this review is regarded as the first of its kind to highlight and summarize the potential for using *Anabaena azollae* as the biological control of certain agricultural diseases; Integrated pest management, and plant growth improvement to promote sustainable agricultural technology.

2 *Anabaena azollae* description isolated from *Azolla pinnata*

The Azolla cyanobiont is classified taxonomically in phylum Cyanophyta, Order-Nostocales, Family-Nostocaceae. It was first named Nostoc and then renamed *Anabaena azollae*.

Abd El-Aal (2013) revealed the morphological characteristics of *Anabaena azollae* isolated from *Azolla pinnata* and demonstrated that it can be cultured free in BG11 medium (Figs. 1–4).

FIG. 1

Azolla pinnata.

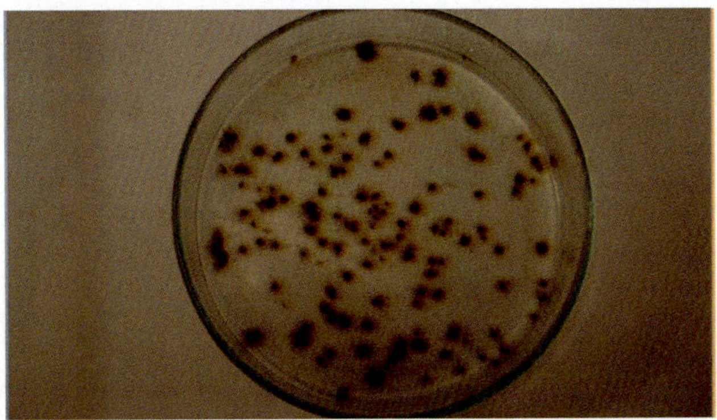

FIG. 2

Anabaena azollae colonies isolated from *Azolla pinnata*.

FIG. 3

Growth of *Anabaena azollae* free in BG11 medium.

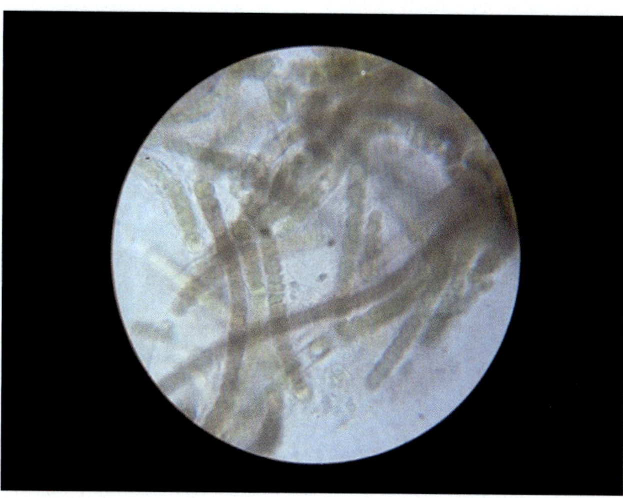

FIG. 4

Cells in *Anabaena azollae* grown in BG11 medium.

3 Use of *Anabaena azollae* in the farming sector

3.1 Professional biostimulation on plant growth

Biostimulants are products that reduce the need for fertilizers and increase plant growth, resistance to water and abiotic stresses. In small concentrations, these substances are efficient, favoring the good performance of the plant's vital processes, and allowing high yields and good quality products. In addition, biostimulants applied to plants enhance nutrition efficiency, abiotic stress tolerance, and/or plant quality traits, regardless of its nutrient contents. Several researches have been developed in order to evaluate the biostimulants in improving plant development subjected to stresses, saline environment, and development of seedlings, among others. Furthermore, various raw materials have been used in biostimulant compositions, such as humic acids, hormones, algae extract, and plant growth-promoting bacteria (Yakhin et al., 2017).

Cyanobacteria excrete a great number of substances that influence plant growth and development. These microorganisms have been reported to benefit plants by producing growth-promoting regulators (the nature of which is said to resemble gibberellin and auxin), vitamins, amino acids, polypeptides, antibacterial and antifungal substances that exert phytopathogen biocontrol and polymers, especially exopolysaccharides, that improve soil structure and exoenzyme activity (Oluwaseyi et al., 2017). In three cyanobacterial strains, plant growth promoters are estimated (*Anabaena azollae* Strasburger, *Spirulina platensis* Geitler, and *Nostoc muscorum* C. Agardh in the presence of potassium nitrate, sodium chloride, and tryptophan, with different concentrations. The highest concentrations of IAA and GA3 were determined and applied to maize that was planted in soil with EC (5.4). During the maize cultivation period, three times cyanobacterial therapies are applied. Mineral fertilization has been implemented in compliance with the Egyptian Ministry of Agriculture guidelines, with 100% to control, and 75% to all other treatments. Final results showed that the highest yield was obtained through the treatment of maize (*Spirulina platensis* with tryptophan, *Anabaena azollae* with both potassium nitrate and tryptophan).

These results were nearly three times greater than the yield of the control. For yield improvement, it is recommended to use the selected strains, especially in sandy and saline soils. Applying different cyanobacterial strains to maize grown in a soil affected by relatively high EC, treated with different additives, such stress was found to be overcome if cyanobacteria were applied during the optimum period of IAA and GA3 production (Al Awamri et al., 2018).

3.2 *Anabaena azollae* as biofertilizer

Azolla biofertilizer may be a promising approach to achieve better N use efficiency (NUE) in paddy rice fields due to its great potential for biological N fixation (BNF) (Bharati et al., 2016). One of the Hopeful biofertilizers for a variety of crops, including rice (Joshi et al., 2012), wheat (Babu et al., 2015), taro (Petruccelli et al., 2015), and soybean (Sholkamy et al., 2015) is *Azolla anabaena.* When used in a rotating rice-wheat cropping method, Azolla is useful for wheat (Gaind and Singh, 2015).

Manipulation of some cyanobacteria like *Anabaena azollae* or *Spirulina platensis* or white rot fungi like *Pleurotus columbinus* under 50%, 75%, and 100% of N fertilizer on *Pelargonium graveolens* L. were tested to study the growth, herb yield, essential oil % and essential oil yield, and its components. Geranium seedlings were soaking with these bio-agents then plants were sprayed with the bio-agent suspension five times after sowing. Results indicated that inoculation with these bio-agents caused a significant increase in plant height, a number of branches/plant, herb fresh and dry weights per plant (g) and per fed (ton), essential oil percentage in the herb, and essential oil yield per plant (cc) and per fad (L) as well as Total carbohydrates (%). The highest increase in these parameters was obtained when plants were treated with *Anabaena azollae*(A)+*Spirulina platensis*(B)+*Pleurotus columbinus* (C) in the presence of 75% of N fertilizer. The lowest mean of all parameters in two cuts for both seasons was obtained in the plants which fertilized by 50% of N. The highest percentages of Geranyl formate, linalool, citronellol, and geraniol in essential oil were recorded with the same treatment. In addition, the highest total carbohydrates percentages and phenol content were recorded in herb of treated plants with this treatment.

There is no doubt that the use of different bio-agent treatment of *Anabeana Azollae, spirulina platensis,* and *pleurotus columinus* in different combinations and application of 75% N- fertilizer on Pelargonium plant led to an increase in plant growth as: plant height, a number of branches/plant and herb fresh and dry weights per plant and per fed. As well as essential oil yield per plant and per fed (Mohsen et al., 2015).

The role of *Azolla pinnata*, *Anabaena azolla*, *Pleurotus columbinus*, and *Azotobacter* sp. in the presence of urea (46.5% N) as a source of nitrogen fertilizer on the growth and yield of wheat on sandy soil. It was found that the total count of bacteria, fungi, azotobacter, and algae at different treatments were higher than those of other treatments especially with the treatment of Mix only which gave the highest values of total bacteria count. Also, for different microbes such as fungal, algae and N2-fixing bacteria count it was noticed that treatment (Mix of biofertilizer+75% of the recommended dose of nitrogen) gave the more optimum results for different types of microbes count. The results have showed that the highest values for IAA production were at (mix + urea 50%) and (mix + urea 75%), while (mix only) and (mix + urea 75%) gave mostly higher N2-ase activity at 120 days of incubation compared to other treatments. Also, straw and grain yields were significantly increased with (mix + 75%) of the recommended dose of nitrogen fertilizer. Also it was noticed that treatments of the mix of biofertilizer have a pivotal role in increasing N, P, and K uptake in straw and grain yields.

The use of microorganisms in plant production especially in cereal crops can improve growth and yield components, lower use of mineral fertilizers, and higher microbiological activity of soil.

So, the application of different kinds of biofertilizers enriched soil fertility and so it is helpful to improve the soil properties such as organic matter content,

As well as, macronutrients uptake (N, P, and K) in the wheat cropping system, which is reflected on the yield and its components. Hence, it is imperative to popularize the use of biofertilizers, which is a low-cost input technology.

to reduce the dependence on inorganic fertilizers and contribute to a pollution-free atmosphere, which is the need of the day (Taha et al., 2017).

3.3 *Anabaena azollae* biological control

3.3.1 Antifungal activity

Anabaena azollae can be used as a bioagent suppressor for *Fusarium oxysporum* (Fig. 5) and *Alternaria alternates* (Fig. 6) pathogenic fungal members.

In this respect, Different cyanobacterial strains are known to produce intracellular and extracellular metabolites with various biological activities, including antibacterial and antifungal effects (Mohamed et al., 2011). It has been reported that several attempts to recombine isolated *Anabaena azollae* with cyanobacterium-free Azolla are unsuccessful (John and Jeff, 2002).

Some fungal diseases may attack geranium plants (*Pelargonium graveolens* L. Herit.) causing a drop in harvested crop and deterioration in oil yield, as root rot and wilt syndromes. Antibiosis studies

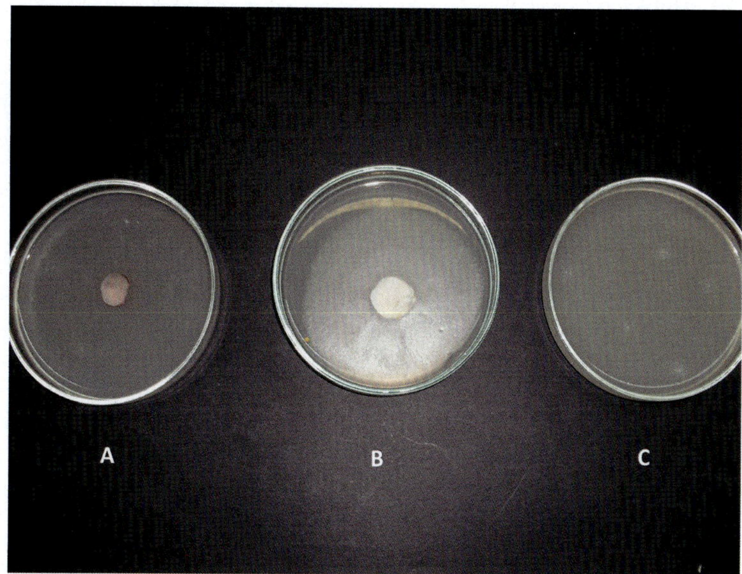

FIG. 5

Antibiosis among *Anabaena azollae* and *Fusarium oxysporum*. (A) *Anabaena azollae*+ *Fusarium oxysporum*. (B) (Control) *Fusarium oxysporum*. (C) (Control) *Anabaena azollae*.

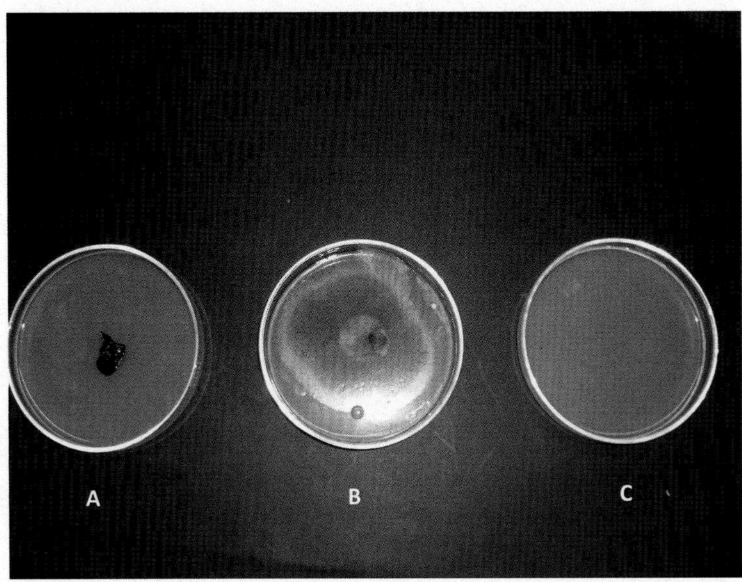

FIG. 6

Antagonistic activity of *A. azollae* toward (A) *Anabaena azollae+ Alternaria alternat.* (B) (Control) *Alternaria alternat.* (C) (Control) *Anabaena azollae.*

showed promising results in control of; *Fusarium oxysporum* and *Rhizoctonia solani*, by different microorganisms such as *Pseudomonas fluorescens* and/or the extracts of either *Pleurotus columbinus* or *A. azollae* individually or in combination at the second cut of the second season.

The mixture of Pseudomonas and the extract of both Pleurotus and Anabaena increased geranium plant length. Due to either Fusarium or Rhizoctonia disease(s), the fresh and dry weights of plants increased in the second cut in the second season. *A. azollae* extract treatment was the superior management of diseases of either Fusarium or Rhizoctonia as demonstrated by fresh geranium crop yield.

And the yield of second season pelargonium oil, in the second cut. Highly significant rises in fresh yields of *A. azollae*, *Pseudomonas* sp. interaction therapy. And then *pleurotus* sp. Compared to the control.

The influence of bio-agents on pelargonium productivity caused by stress from Fusarium or Rhizoctonia. Treatments with bio-agents played a major role in avoiding the disease and increasing fresh and dry weights. Any of the three used microbes raised the fresh weight about four times compared to control, while being doubled in the case of dry weight for the second cut (Mohamed et al., 2015).

3.3.2 Antinematode activity

Root-knot nematodes, Meloidogyne spp., are among the most dangerous nematodes in agriculture, causing a worldwide total loss of US$ 100 billion (Entsar, 2018).

Symptoms of nematode infection are root gall formation, resulting in reduced growth, nutrients, and water absorption, wilting increase, and mineral deficiency, resulting in weak and low-yielding plants (Ping et al., 2014).

Chemical nematicides have been found to be an effective measure for controlling nematodes but have a toxic residual effect on the environment, especially on non-target organisms and human health. Moreover, the use of chemical nematicides in organic farming is banned. Therefore, the creation of alternative, healthy environmental strategies for controlling nematodes is urgent (Anastasiadis et al., 2008). Research on nematode control over the last decades has concentrated on suggesting strategies to inhibit egg hatch, degradation of hatching factor or metabolite production (Fereidoun and Abolfazl, 2020). One of the biological control practices recently attempted is the study of the nematocidal capacity of cyanobacterial culture filtrates, which parasitize plant-parasitic nematodes (Nikoletta et al., 2020).

Cyanobacteria that excrete a large number of substances have been reported to benefit plants by producing growth-promoting regulators (PGPRs), vitamins, amino acids, polypeptides, phytopathogenic antibacterial and antifungal substances, and polymers, in particular exopolysaccharides, that improve soil structure and exoenzyme activity (Zaccaro et al., 2001).

In the productivity greenhouse experiment, the combination of mixing five algal culture filtrates of *Spirulina platensis*, *Oscillatoria* sp., *Anabaena oryzae*, *Nostoc muscorum*, and *Phormedium fragile*, with *A. pinnata* aqueous extract filtrate and compost extract achieved the highest reduction in the number of the 2nd stage juveniles in soil, the numbers of galls, developmental stages, females, egg masses, Egg numbers/egg mass in roots of cucumber plants comparing with the individual treatment and the non-treated control. In addition, all combinations significantly improved the fresh weight of roots and shoots and increased the plant yield (Shawky et al., 2009).

Two algal cultures, *Spirulina platensis*, *Anabaena azollae*, *Azolla pinnata* and *Pleurotus columbinus* were also used in the control of root-knot nematode in addition to olive mill waste, *Meloidogyne javanica* in banana was monitored under both laboratory and commercial greenhouse conditions. Laboratory experiment revealed that high juvenile mortality percentage occurred during all the exposure periods of all treatments, the best results were after 72h exposure. *Spirulina platensis* followed by *Anabaena azollae*, *Azolla pinnata*, *Pleurotus columbinus,* and olive watery extract significantly increased juveniles mortality up to 70% after 72h at the highest concentration of 1:10 (85.2, 81.4, 79.9, 73.5 and 71.7%, respectively).

In the productivity greenhouse experiment, the combination of culture filtrates of *Spirulina platensis*, *Anabaena azollae*, *Azolla pinnata*, *Pleurotus columbinus*, and olive mill wastewater achieved the highest reduction in the number of total nematodes in both soil and roots, also in numbers of galls. In addition, all combinations significantly increased the crop yield of banana plants comparing with the individual treatment and the control.

Combined treatments significantly increased the regulation activities of CO_2, dehydrogenase, and nitrogenase. It may be advised that the use of biological control agents in bananas against root-knot nematode is preferred in order to reduce the inputs of chemical nematicides (Shawky et al., 2014).

3.3.3 Bioinsecticidal activity

Pesticides kill beneficial predators, parasites, and pathogens as well as pests, and can result in secondary pest outbreaks or rapid resurgence of pests that were previously suppressed. Biological control is the use of non-chemical and environmentally friendly methods of controlling insect pests and diseases through the action of natural control agents. Biological pest control is the use of pathogens, predators, and parasitoids to kill pests by reducing their populations or eliminating them entirely from our farms, gardens, and forests, thereby increasing productivity and safety. Microbial antagonists have been used to control pests and diseases in recent years. A common example is *Bacillus thuringiensis*, which is toxic to many insect species. Furthermore, Entomopathogenic nematodes from the families

Steinernematidae and Heterorhabditidae have been used to suppress pest insect populations in a variety of agroecosystems, and in several cases, their positive effects on crop yield have been demonstrated (Nafiu and Mustapha, 2014).

With a wide variety of host plants and world distribution, the two-spotted spider mite, *Tetranychus urticae* Koch, is one of the most common mite pest species (Rabie et al., 2018). *T. urticae* is an important one in global distribution. Its phytophagous nature, high reproductive potential, and short life cycle rapid resistance development too many acaricides often after a few applications.

On the other hand, the great reliance on Chemical pesticides had its serious drawbacks, manifested in resistance problems and high residue levels in food products (fruits, vegetables, grains, and seeds) that may hinder its marketing (Adekunle et al., 2019). Phytoseiid mites are important biological control agents because of their well-known capacity to suppress pest mite populations, mainly tetranychids in diverse cropping system (James et al., 2013). Two mite predators of the family Phytoseiidae have been found in association with the two-spotted spider mite on cucumber and pepper fields in Egypt. *Neoseiulus barkeri* (Hughes) (Acari: Phytoseiidae) is an oligophagous predatory mite. *T. negevi* Swirski and Amitai (Acari: Phytoseiidae) is the agriculturally important predator of spider and eriophyid mites (Momen, 2010).

Significant biocontrol agents are many species of predatory mites and the species *Phyto seiulus macropilis* and *Neoseiulus californicus* are used to monitor the two-spotted spider mite *Tetranychuschus Urticae*, a major worldwide agricultural pest (Morgana et al., 2020).

For optimal biological mite management, it is important to know if the tested compounds have adverse undesirable effects on the predatory mite. The toxicity effects of some saves compounds (Manure Tea (M.Tea), Manure neem Tea (M.N.Tea) and Boric acid) or some microorganisms (*Spirulina platensis, A. azollae, Paenibacillus polymyxa, Pleurotus columbinus*) comparing with recommended compound "Abamectin" were tested against the adult females of two-spotted spider mite, *Tetranychus urticae* Koch and adult females of its predatory mite, *T. negevi* Swirski and Amitai under laboratory conditions. The mortality percentages of *T. urticae* and toxicity of *T. negevi Swirski* and Amitai were recorded after 24, 48, and 72h from treatment, which were increased gradually with increasing the exposure time of the tested compounds. Abamectin gave highly percent reduction and toxicity against the adult females of *T.urticae* and its predatory mite, *T. negevi* so came in the first category. M.Tea, M.N.Tea, and Boric acid came in the second category with moderate mortality percentages against *T. urticae* and toxicity of *T. negevi*. The microorganisms "*Paenibacillus polymyxa, Anabaena azollae, Pleurotus columbinus*, and *Spirulina platensis*" came in the last category. Generally, the adult female of *T. negevi* was more tolerant comparing with adult females of *T. urticae*.

Laboratory observations reported that no adult females of *T. negevi* deposited eggs due to the effectiveness of Abamectin and M.Tea. The gross fecundity (the number of eggs laid per female) of the adult females of *T. negevi* reached 31.8, 22.3, 12.5, 14.2, 7.8 and 5.1 eggs when treated with *Anabaena azollae, Spirulina platensis*, Boric acid, *Pleurotus columbinus, Paenibacillus polymyxa* and M.N.Tea, respectively. Comparing with control which the fecundity reached 58.2 eggs (Nour El-Deen et al., 2015).

4 *Anabaena azollae* bioremediation

Rapid growth in population and industrialization has resulted in disposing of various harmful compounds into the environment. The major sources of environmental contamination include industries, agrochemicals, mining activities, and waste disposals (Navarro and Vincenzo, 2019). Treating industrial effluent is very important as it may contain heavy metals, for instance, arsenic, cadmium, mercury, chromium, cobalt, nickel, zinc, lead, and copper, which are major pollutants of freshwater reservoirs Due to their non-biodegradable, poisonous and persistent nature (Arezoo et al., 2017). These contaminants affect the human health and surroundings (Florence et al., 2015).

Biological treatment has given preference over chemical treatment because the chemical materials only react with a small number of waste materials and heavy metals, and large portions of waste material remain unaffected.

However, chemical materials are very costly and produce a large quantity of chemical sludge. Thus, biological methods are appropriate to treat wastewater by producing dense biomass produced from organic matter, which is easily removed by sedimentation. Microbes also feed on dissolved organic matter resulting in small sludge production as compared to the chemical treatment (Samer, 2015).

Olive oil extraction is a process that is conducted by mechanical procedures in olive mills. During this process, large amounts of liquid effluents and solid residues are produced, with a high organic load, the nature of which depends on the technology of the extraction process and the system employed (Adnan and Ghaida, 2020).

Seven strains of algal species including *Anabaena azollae* were tested for degradation of phenolic compounds, a decrease of COD, decolorization of olive mill wastewater (OMWW), and biomass produced. In addition to recording the change in pH, E.C, Carbohydrate, Indole acetic acid, and gibberellins.

Degradation of phenolic compounds in 20 and 10% olive mill wastewater (OMWW) treated by *Anabaena azollae* was 19.02, 28.37% respectively after 30 days and reduction of COD in 20 and 10% (OMWW) treated by *Anabaena azollae* was 25.0, 30.54% respectively after 30 days decolorization of (OMWW) of 20 and 10% concentrations treated with *Anabaena azollae* was 9.9 and 21.5% after 30 days increased the carbohydrate content from o.56 g/L to 1.11 g/L by *Anabaena azollae* after 10 days of incubation but after 20 days, the value of carbohydrates began to decrease reached to 0.99 after 30 days in the concentration of 20% (OMWW). While in the concentration of 10% decrease total carbohydrate from 1.16 g/L to 0.51 g/L after 10 days of incubation by *Anabaena azollae* and decrease to 0.44 g/L after 30 days of incubation (Rokia, 2017).

5 Conclusion

In conclusion, this review indicated the morphological characteristics of *Anabaena azollae* isolated from *Azolla pinnata* and it can be free cultured in BG11 medium, as well as, it can be used, as a bio-agent suppressor for the pathogenic fungal members' *Fusarium oxysporum* and *Alternaria alternate*. The results also showed that the extract of *Anabaena azollae* was the best. Played an important role in increased yielding potentials and disease control in pelargonium plantations.

It could be recommended that application of *Anabaena azollae* as bio-control agents and plant growth-promoting substances to increase plant nutrients availability and has a nematicidal effect to control root-knot nematode. And Its toxicity effects on the two-spotted spider mite, *T. urticae* Koch, and its predatory mite, *T. negevi* Swirski and Amitai. Also the reduction of 25% of N recommended dose with bio-agent treatment was more superior to a recommended dose of N fertilizer in the yield of Pelargonium plant and in quality and quantity of oil production. The beneficial effect is to reduce the use of chemical fertilizers, as well as to improve macronutrients uptake (N, P, and K) in wheat crop, which is reflected on the yield and its components and enhanced microbiological activity, and produce various biologically active substances like growth-promoting substances as Indol Acetic Acid (IAA), Gibberellic acid (GA), (Nitrogenase, dehydrogenase) enzymes and carbohydrate excretion which help soil aggregation and again may result in soil fertility for sustainable crop production. Also, it recommended that using treatments (*Anabaena azollae* with KNO3) and (*Anabaena azollae* with tryptophan), each separately; can improve the yield of Maize three times more than control, while treatment (*Anabaena azollae* with NaCl) could improve the yield of Maize two times more than control. And the possibility of the transformation of the components of the olive mill wastewater from high toxicity components to less toxic taking advantage of the wastewater in fertilization and plant irrigation for the containing of elements and nutrients to the plant.

Acknowledgments

The authors wish to thank Soils, Water and Environment Research Institute (SWERI), Agriculture Research Center, Giza, Egypt for the facilities provided to conduct this research work.

References

Abd El-Aal, A.A.M., 2013. Characterization of *Anabaena azollae* isolated from *Azolla pinnata*. Egypt. J. Agric. Res. 91 (3). https://doi.org/10.21608/ejar.2013.164847.

Abd El-Aal, A.A.M., Elsherif, M.H., Shehata, H.S.H., El-Shahat, R.M., 2013. Efficiency use of nitrogen, biofertilizers and composted biostraw on rice production under the saline soil. J. Appl. Sci. Res. 9 (3), 1604–1611.

Adekunle, W.A., Michael, J.B., Mark, D.L., Laura, C.L., Fang, Z., Doug, B.W., 2019. Physiological resistance alters behavioral response of *Tetranychus urticae* to acaricides. Sci. Rep. 9, 19308.

Adnan, K., Ghaida, A.-R., 2020. Sustainable environmental management and valorization options for olive mill byproducts in the Middle East and North Africa (MENA) region. Processes 8, 671.

Al Awamri, A., Reda, M.E.S., Azza, A.E.-A.M., Eman, M.M., 2018. Production of algal growth promoters and studying their effects on maize crop. Egypt. J. Phys. 19.

Al-Erwy, A.S., Abdulmoneam, A.-T., Sameera, O.B., 2016. Effect of chemical, organic and bio fertilizers on photosynthetic pigments, carbohydrates and minerals of wheat (*Triticum aestivum*. L) irrigated with sea water. Int. J. Adv. Res. Biol. Sci. 3 (2), 296–310.

Anastasiadis, I.A., Giannakou, I.O., Prophetou-Athanasiadou, D.A., Gowen, S.R., 2008. The combined effect of the application of a biocontrol agent *Paecilomyces lilacinus*, with various practices for the control of root-knot nematodes. Crop Prot. 27, 352–361.

Arafa, A.S., Abd El-Aal, A.A.M., 2013. Evaluation of the technological properties of the bio-organic colored cotton. J. Agron. 12, 78–85.

Arezoo, A., Ahmed, A., Mashallah, R., Meisam, A., 2017. Removal of heavy metals from industrial wastewaters: a review. ChemBioEng 4, 37–59.

Babu, S., Prasanna, R., Bidyarani, N., Singh, R., 2015. Analyzing the colonization of inoculated cyanobacteria in wheat plants using biochemical and molecular tools. J. Appl. Phycol. 27, 327–338.

Bharati, K., Ashok, K.P., Santosh, R.M., 2016. Aquatic *microphylla Azolla*: a perspective paradigm for sustainable agriculture, environment and global climate change. Environ. Sci. Pollut. Res. 23, 4358–4369.

Brahmaprakash, G.P., Sahu, P.K., 2012. Biofertilizers for sustainability. J. Indian Inst. Sci. 92 (1), 37–62.

Divya, V., Ram, H.M., 2018. Role of biofertilizers in integrated plant nutrient system (ipns). Int. Res. J. Nat. Appl. Sci. ISSN: (2349-4077).

Entsar, H.T., 2018. Abundance and distribution of plant parasitic nematodes associated with some different plant hosts. Egypt Acad. J. Biol. Sci. 10 (2), 99–109.

Fereidoun, F., Abolfazl, H., 2020. Recent advances in the development of environmentally benign treatments to control root-knot nematodes. Front. Plant Sci. 22. https://doi.org/10.3389/fpls.2020.01125.

Florence, C., Julien, C., Roseline, B., Valérie, B., Pierre, B., Martine, R., 2015. Soil contamination and human health: a major challenge for global soil security. In: Global Soil Security Symposium, College Station, United States, pp. 275–295.

Gaind, S., Singh, Y.V., 2015. Soil organic phosphorus fractions in response to long-term fertilization with composted manures under rice wheat cropping system. J. Plant Nutr. 39, 1336–1347. https://doi.org/10.1080/01904167.1086795.

Hoda, M.M., Faten, A.A.-K., Azza, A.M.A.-E.A., 2013. Effect of magnetite and some biofertilizer application on growth and yield of valencia orange trees under El–Bustan condition. Nat. Sci. 11 (6), 46–61.

Itelima, J.U., Bang, W.J., Onyimba, I.A., Sila, M.D., Egbere, O.J., 2018. Bio-fertilizers as key player in enhancing soil fertility and crop productivity: a review. Direct Res. J. Agric. Food Sci. 6 (3), 73–83.

James, A.M., Gilberto, J.D.M., Nazer, F.S., 2013. Revision of the lifestyles of phytoseiid mites (Acari: Phytoseiidae) and implications for biological control strategies. Syst. Appl. Acarol. 18 (4), 297–320.

John, C.M., Jeff, E., 2002. Regulation of cellular differentiation in filamentous cyanobacteria in free-living and plant associated symbiotic growth states. Microbiol. Mol. Biol. Rev. 66, 94–121.

Joshi, L.N., Rana, A., Shivay, Y.S., 2012. Evaluating the potential of rhizo-cyanobacteria as inoculants for rice and wheat. J. Agric. Technol. 8, 157–171.

Mohamed, E.-A.H.O., Mostafa, M.E.-S., Metwally, A.M., El-Wahab, A.I.A., Mona, M.I., 2011. Antagonistic activity of some Fungi and Cyanobacteria species against *Rhizoctonia solani*. Int. J. Plant Pathol. 2 (3), 101–114.

Mohamed, T.N.M., Ismail, M.A., El-Aal, M.H.A., Azza, A.M., Heba, S.S.H., 2015. Biological control of Pelargonium graveolens diseases and impacts on oil contents and essential crop parameters. Egypt. J. Appl. Sci. 30 (5), 118–144.

Mohsen, M.A.M., Ismail, H., Kasem, H.M.A., Azza, A.E.-A.M., 2015. Bio-influence of some microorganisms on *pelargonium graveolensl*. Plant Egypt. J. Biotechnol. 49, 106.

Momen, F.M., 2010. Ntra- and interspecific predation by *neoseiulus barkeri* and t. negevi (acari:phytoseiidae) on different life stages: predation rates and effects on reproduction and juvenile development. Acarina 18 (1), 81–88.

Morgana, M.F., Angelo, P., Pedro, H.M., Eraldo, L., Arne, J., 2020. Compatibility of two predator species for biological control of the two-spotted spider mite. Exp. Appl. Acarol. 80, 409–422.

Nafiu, B.S., Mustapha, S., 2014. Fundamentals of biological control of pests. Jib Rev. 1, 2349–2724.

Navarro, F., Vincenzo, T., 2019. Waste mismanagement in developing countries: a review of global issues. Int. J. Environ. Res. Public Health 16 (6), 1060.

Nevine, M.T., El-Shahat, R.M., 2018. Influence of Azolla, some blue green algae strains and humic acid on soil, growth, productivity, fruit quality and storability of "Canino" apricot cultivar grown under clay loamy soil. J. Plant Prot. 8 (1), 1–11.

Nikoletta, N., Zbigniew, A., Maria, D., Nikolaos, M., 2020. Review, nematicidal amendments and soil remediation. Plan. Theory 9, 429.

Nour El-Deen, M.A., Ismail, H., Sholla, M.E.S., Abou-zeid, M.Y., Azza, A.M., 2015. Effectiveness of some microorganisms, biogas manure and boric acid on *Tetranychus urticae* Koch and *Typhlodromus negevi Swirski* and Amitai under laboratory conditions. Middle East J. Appl. Sci. 5 (2), 445–452.

Oluwaseyi, S.O., Bernard, R.G., Babalola, O., 2017. Mechanisms of action of plant growth promoting bacteria. World J. Microbiol. Biotechnol. 33 (11), 197.

Petruccelli, R., Bati, C.B., Carlozzi, P., Padovani, G., Vignozzi, N., Bartolini, G., 2015. Use of Azolla as a growing medium component in the nursery production of olive trees. Int. J. Basic Appl. Sci. 4, 333–339.

Ping, L., Richard, F.D., Robert, C.K., Marc, W.V.I., Harald, S., 2014. Physiological effects of *Meloidogyne incognita* infection on cotton genotypes with differing levels of resistance in the greenhouse. J. Nematol. 46 (4), 352–359.

Purushottam, S., Jiban, s., 2015. Improving soil fertility through Azolla application in low land rice: a review. Aja 2 (2), 35–39.

Rabie, E.A.E., Basha, A.E., Mostafa, E.M., El Mageed, A.E.A., 2018. Seasonal abundance of the two spotted spider mite, *Tetranychus urticae* koch on four cotton cultivars at dakahlia governorate, Egypt. Zagazig J. Agric. Res. 45, 5.

Rkyadav, G.A., Singh, Y.V., Singh, P.K., 2014. Advancements in the utilization of Azolla-Anabaena system in relation to sustainable agricultural practices. Proc. Indian Natl. Sci. Acad. 2, 301–316.

Rokia, Y., 2017. Bioremediation of Olive Mill Wastewater for using in Biofertilization. Master thesis Agric. Sci. (Agric. Microbiology), Fac. of Agric., Ain Shams Univ.

Samer, M., 2015. Biological and chemical wastewater treatment processes. In: Wastewater Treatment Engineering. Intech Open, London, https://doi.org/10.5772/61250.

Shawky, S.M., El-Aal, A., Azza, A.M., Ahlam, M.A.-G., 2014. Comparative efficacy of some algal species, Azolla, Pleurotus and olive mill in controlling root-knot nematode on a banana. Egypt. J. Agronematol. 13 (2), 23–39.

Shawky, S.M., Soha, S.M.M., Azza, A.M., 2009. Efficacy of some algal, Azolla and compost extract in controlling root-knot nematode and its reflection on cucumber. Bull. Fac. Agric. 60, 443–459.

Sholkamy, E.N., El-Komy, H.M., Ali, H.M., 2015. Enhancement of soybean (*Glycine max* L.) growth by biofertilizers of *Nostoc muscorum* and *Nostoc rivulare*. Pak. J. Bot. 47, 1199–1204.

Taha, A.A., El-Zehery, T.M., Azza, A.E.-A., Thanaa, E.-k., 2017. Comparative of some microorganisms on Sandy soil fertility, and wheat productivity. J. Soil Sci. Agric. Eng. 8 (5), 203–208.

Yakhin, O.I., Aleksandr, A.L., Ildus, A.Y., Patrick, H.B., 2017. Biostimulants in plant science: a global perspective. Front. Plant Sci. 7, 2049.

Zaccaro, M.C., Salazar, C., Zulpa, D.G., Cans, S.C.d.M.M., Stella, A.M., 2001. Lead toxicity in cyanobacterial prophyrin metabolism. Environ. Toxicol. Water Qual. 16, 61–67.

CHAPTER

The bioremediation of agricultural soils polluted with pesticides

2

S.K. Jayasekara and R.R. Ratnayake

Microbiology and Soil Ecosystems Research Project, National Institute of Fundamental Studies, Kandy, Sri Lanka

Abstract

The use of pesticides in agriculture is dramatically increased. They are indispensable elements in protecting the crops from pests, weeds, and diseases while enhancing the harvest. Imprudent high-intensity application of pesticides in agriculture for prolonged durations has imposed serious crises viz., barren soils, polluted waterways, biodiversity obliteration, severe health concerns, etc. Ambitious efforts are being made to reverse these impacts. Among various technologies available today for remediation of pesticide-contaminated soils, bioremediation seems to be environmentally safe and cost-effective. The present chapter provides an overview of pesticides, their role in agriculture, and the negative impacts of over-application. The chapter mainly discusses the potential use of microorganisms in rectifying pesticide-contaminated agroecosystems and the influence of biotic and abiotic factors on the process. Finally, the chapter provides an update of recent research outputs which monitored and evaluated the efficiency of the technique, its advantages and disadvantages, and indications for future studies.

Keywords: Agricultural soils, Agrochemicals, Bioaugmentation, Bioattenuation, Soil pollution, Soil bioremediation, Pesticides, Microbial bioremediation, Indigenous microorganisms

1 Introduction

Many countries in the world where the main role in their economy is played by agriculture are currently facing an alarming situation due to various adverse impacts imposed on the environment and public health by pesticides. A recent study conducted on estimating the global scale risk of agrochemical-related pollution revealed that 64% of agricultural land around the world is under the threat of getting polluted by these chemicals (Tang et al., 2021). Contamination of groundwater resources, soil microbial population imbalances, and accumulation in food chains are among the fundamental impacts of extensive pesticide over-use in agriculture which attributes to long-term deleterious impacts such as environmental pollution, biodiversity destruction, global warming, and severe risks on human and animal health.

With the ever-growing global population, the requirement of a sufficient food supply to meet the increasing demand is unavoidable. In this challenging process of securing food for all, the use of agrochemicals has become a part and parcel of enhancing the crop yield quality and quantity (Sharma et al., 2019). However, the adverse effects emanated by the extensive over-use of these chemicals emphasize the importance of their limited use and the significance of discovering efficient, ecofriendly techniques to remediate the contaminated sites (Singh et al., 2017).

Microbial Syntrophy-mediated Eco-enterprising. https://doi.org/10.1016/B978-0-323-99900-7.00007-9

Bioremediation is one such emerging technology which is currently recognized as an environmentally friendly approach to clean the agrochemical contaminated soil. The technique utilizes biological systems to decontaminate organic pollutants in the environment into harmless or less concentrated forms (Vishwakarma et al., 2020; Zouboulis et al., 2019). The current chapter mainly discusses the applications of bioremediation in rectifying pesticide-associated pollution of agricultural soils with a special focus on the potential use of various microorganisms in the process.

2 An overview of pesticides; their role in agriculture and the adverse impacts associated with the over-use

2.1 Overview of pesticides and their roles in agriculture

The invention of agriculture dates back to about 10,000 years ago into the era of Mesopotamian civilization (Kislev et al., 2004). Since then the losses of harvest due to infectious diseases, animals or insect attacks, and competitive growth of other plants became a great incentive to find efficient modes of preventing the losses. From ancient times, attention was paid to develop highly effective and time-saving techniques to protect the food crops (Abubakar et al., 2020). Subsequently, the elimination of pests and diseases using chemical compounds emerged. The earliest documented application of a chemical compound as a pesticide is the use of sulfur about 4500 years ago by Sumerians. It was the dawn of highly sophisticated agrochemicals available for sale in today's markets.

In 2014, the Food and Agricultural Organization of the United Nations in collaboration with the World Health Organization endorsed "The International Code of Conduct on Pesticide Management". This particular document provided overall guidance to everybody from pesticide manufacturers to the common general public on safe handling of pesticides from production to disposal. The same document broadly defined a pest as "any species, strain or biotype of plant, animal or pathogenic agent injurious to plants and plant products, materials or environments, vectors of parasites or pathogens of human and animal disease and animals causing public health issues". The document further described pesticides to be "any substance, or a mixture of substances composed of chemical or biological ingredients intended for repelling, destroying or controlling any pest, or regulating plant growth".

Currently, pesticides are categorized into several groups based on various criteria. Their toxicity levels, organisms inhibited, chemical composition, mode of entry into the host, working mechanism, sources of origin (synthetic chemical pesticides or biopesticides) are among the major principles of classifying pesticides (Akashe et al., 2018). Pesticides are divided into primary categories on the basis of pest organism they inhibit and cover a wide range of chemicals used for multiple purposes such as insecticides, herbicides (weedicide), fungicides, rodenticides, nematicides, biocides, algicides, acaricides, molluscicides, etc. (Pandey et al., 2018; Brandt et al., 2017; Rani et al., 2017; Yadav and Devi, 2017; Gavrilescu, 2005). The major chemical classes of organic pesticides are comprised of organochlorines, organophosphates, organometallic compounds acetamides, carbamates, triazoles and triazines, neonicotinoids, and pyrethroids (Gilden et al., 2010) while organochlorines are extremely toxic as they are nonbiodegradable and have a tendency of bioaccumulation (Ortiz-Hernandez et al., 2013). There are also heavy metal-containing inorganic pesticides, viz., lead arsenate, chromated copper arsenate, copper acetoarsenite, borax, and boric acid complexes, etc. (Tarla et al., 2020).

Herbicides are a broad group of phytotoxic pesticides that are used to kill unwanted plants in agricultural fields. They are grasses or weeds with a higher growth rate that compete with crop plants for nutrients, water, space for growth and act as bearers of pests (Gupta, 2011). Other pesticides target macro and microorganisms that are attracted to the fields such as insects, worms or nematodes, rodents, bacteria, fungi, algae, etc. that compromise the growth and yield of desired crops. In addition, "biopesticides" are extracted from biological systems such as animals, plants, and microorganisms that include microbial pesticides, plant-incorporated protectants, and biochemical pesticides.

2.2 Adverse impacts of pesticides over-use in agriculture

Soil is the basis of agriculture that renders space for the growth of food crops (Ashraf et al., 2014). Although the large-scale application of pesticides is currently an integral part of boosting crop yields, their long-term unsustainable addition into the soil contributes greatly to soil pollution. When these agrochemicals are added into the soils, they are not completely utilized for the intended purpose. Instead, around 99% become persistent as contaminants and distribute in the agricultural soils in a multiphasic manner. For instance, they may accumulate in the fields by attachment with the soil particles, dissolving in soil water, or by suspension in the soil atmosphere (Mishra et al., 2021; Boopathy, 2000). The problem intensifies when used in high quantities because their natural degradation becomes arduous. As a result, the amounts that persist in the soil drastically increase leading to pollution of soil and water, damage the microflora, microfauna, and hinder the mineral absorption of plants (Van der Werf, 1996).

The serious environmental concerns associated with the widespread application of pesticides initiate the deterioration of soil health. The soil functions, soil quality, soil productivity, and soil biodiversity are extremely threatened by extensive agrochemical usage. The indigenous microbial communities in soil that carry out essential agricultural-ecosystem processes are directly exposed to the deposited agrochemicals. Those chemicals interact with soil microorganisms and change their biochemical and physiological behavior leading to pesticides associated inhibition of microorganisms (Singh and Walker, 2006). Ultimately, the soil becomes infertile because the main soil microbial functions, viz., increasing the bioavailability of nutrients by nutrient cycling (nitrogen fixation, phosphate solubilization), production of plant growth stimulation hormones, plant debris or cellulose degradation, inhibition of plant pathogens, get highly restricted (Hayat et al., 2010). Another common risk linked with pesticide usage is the development of resistance in the targeted pests and weeds that make them uncontrollable and continue to destroy the crops (De Bon et al., 2014).

Moreover, the displacement of pesticides from agricultural fields through wind and rain creates additional agrochemical-enriched sites such as natural surface water bodies and groundwater resources. They act as extra exposure routes of agrochemicals for humans and animals (Pérez and Eugenio, 2018). Especially, the runoff of pesticides into the lakes and coastal waters consequently leads to oxygen depletion, eutrophication, kills of fish and other aquatic organisms, stinking water, and release of offensive odors into the environment while degrading water quality (Chen et al., 2017). Nitrates from agricultural runoff or leachate are recognized as the most common groundwater pollutant that contaminates drinking water (The United Nations World Water Development Report, 2013). Thereby, pollution of potable water has currently become a major concern of pesticides over-use (Rasmussen et al., 2015). Eventually, these chemical compounds end up entering food chains in the ecosystem in a persistent manner and become toxic to non-target species including humans, flora and fauna, and soil microbial communities (Lozowicka et al., 2016).

The emergence of health hazards on humans due to injudicious use of pesticides and other persistent organic pollutants in agriculture is drastically increased with time. Acute pesticide poisoning (APP) is a phenomenon that might cause chronic health issues in humans (Boedeker et al., 2020). The 48-h occupational or unintentional exposures to pesticides may develop certain respiratory, neurotoxic, cardiovascular, endocrine, gastrointestinal, nephrotoxic disorders, and allergic reactions in humans. These conditions may be fatal or non-fatal (Henao and Arbelaez, 2002). In addition, the bioaccumulation of agrochemicals in living beings causes malfunctions of the endocrine and reproductive systems (Vos et al., 2000). Another well-known example is methemoglobinemia (blue-baby syndrome), a deadly illness among infants which is caused by polluted water. Humans, especially farmers, when exposed to lower doses of agrochemicals for an extended period, experience reduced immunity and intelligence, cancer, and allergic or non-allergic asthma (Margni et al., 2002; Hoppin et al., 2009; Yadav et al., 2015; Alavanja et al., 2003).

It is evident that the adverse impacts associated with improper application of pesticides are extremely diverse and complicated. Currently, the problem is severe that urgent steps must be followed to minimize the intensity of the impacts. Most of the farmers are not aware of the safe handling of pesticides and the importance of using personal protective equipment. In addition, their lack of knowledge on hidden threats behind the intensive use of harmful chemicals further complicates the situation. Therefore, conducting awareness programs to make them knowledgeable is important. Moreover, developing ecofriendly alternatives such as biopesticides and publicizing them might play a significant role in keeping adverse impacts of pesticides over-use to a minimum level.

3 Conventional methods of remediating agricultural soils polluted with pesticides

Excavation of polluted land and transporting its soil into a landfilling site is one of the conventional methods of remediating soils contaminated with various pesticides. Another approach is a containment of the highly polluted areas on-site by setting boundaries to the area and covering up (Zouboulis and Moussas, 2011). However, these methods are not capable of providing a sustainable solution for the issue. Instead, they may add more pressure to the prevailing situation. For instance, landfilling creates another polluted scrap of land. In addition, excavation and transportation of contaminated soils demand a large amount of capital and labor investment. On-site containment of pollutants could worsen the problem as it is not helpful in the complete removal of pollutants from the site. Therefore, these methods are neither ecofriendly nor cost-effective.

Furthermore, incineration of pollutants at elevated temperature, dechlorination, and UV radiation are some of the physical and chemical methods utilized for destroying soil pollutants or for their conversion into less/or non-toxic forms (Zouboulis and Moussas, 2011). Unfortunately, these methods are far more complicated and not user-friendly or environmentally friendly. As a consequence, novel technologies that apply biological methods of detoxifying pollutants emerged.

4 Bioremediation of agricultural soils polluted with pesticides

The long-term environmental and public health concerns that occurred in the aftermath of irresponsible anthropogenic activities are usually dreadful and not easily reversible. Cleaning up pesticides polluted agricultural soils also shares the same severity. Although an ideal pesticide should have the ability to cause rapid destruction of the target pest and easily degrade into non-toxic substances, most of the synthetic compounds that are currently utilized are not readily degradable (Doolotkeldieva et al., 2018). However, untiring efforts have been made by scientists to overcome this problem through inventing ecofriendly and low-cost modes of removing environmental pollutants. Bioremediation is one such emerging technology that carries out the decontamination of polluted sites including agricultural lands using biological systems namely, green plants, microorganisms, or their enzymes, in order to restore the original condition of the site (Glazer and Nikaido, 1995).

The bioremediation process aims to bring considerable decrease in the agrochemical persistence in the environment (Velázquez-Fernández et al., 2012; Lal et al., 2010). It involves a series of biologically mediated reactions that modify the chemical structure of the particular xenobiotic compound into harmless states including carbon dioxide, water, minerals, and non-toxic biomass (Vaish and Pathak, 2020; Azubuike et al., 2016). Among the available bioremediation techniques, microbial bioremediation is considered to be more effective, inexpensive as well as ecofriendly. Moreover, there are in situ and ex situ approaches to accomplishing bioremediation (Ying, 2018). In in situ bioremediation, pesticides are biologically degraded onsite under natural environmental conditions using bioattenuation, biostimulation, or bioaugmentation whereas ex situ bioremediation needs excavation and movement of contaminated soils to another site for treatment such as using bioreactors or composting (Megharaj et al., 2011).

4.1 Microbial bioremediation of agricultural soils contaminated with pesticides

The diversity of microorganisms is incredible that the majority of them contribute greatly to ecologically important functions. Among all, the maintenance of environmental balance rests crucially upon microbial activities involved in nutrient cycling between soil, water, living organisms, and the atmosphere (Tortora et al., 2010). Microbial role in scavenging pollutants, a part of the nutrient cycling process, is known for decades. Usually, bacteria, fungi, protists, and other microbial types in any soil, including agricultural soils, are constantly employed in a process of organic matter degradation (Speight, 2018). Although the soil-inhabited microbial population degrades various soil pollutants to fulfill their energy and carbon needs, they are not capable of efficient removal when the pollutants are added on large scale. Biotechnological approaches such as bioremediation could be utilized to enhance these microbial activities (Yang et al., 2020). Microbial bioremediation is a process that uses different microbial species to clean up the pollutants in soil. Bacteria, archaea, and fungi are considered to be typical key bioremediating agents (Strong and Burgess, 2008).

4.1.1 In situ bioremediation

In situ methods are mostly preferred over ex situ methods for bioremediation. It is usually comprised of three different strategies that the use of particular technology is decided by various factors, such as site conditions, types of the microorganisms present and their specific functions, quality and quantity of the chemical pollutant present in the site (Ying, 2018). These strategies could be explained as the underlined fundamental theories applied in microbial bioremediation technologies.

4.1.1.1 Bioattenuation

This particular remediation technique is also called natural attenuation which is a process that completely depends on the natural microbial degradation. There is no anthropogenic involvement in the process while its productivity is determined by the natural microbial metabolic potential in degradation, removal, alternation, immobilization, or detoxification of various pesticides molecules (Abatenh et al., 2017). In this process, soil microorganisms have to adapt and interact with the polluted environment. The degradation process could be either anaerobic or aerobic depending on the type of microorganisms present. However, the process is less advantageous in terms of time consumption due to the lack of pertinent microbial species in sufficient population densities. If the agrochemical type used is a persistent organic pollutant, the duration required for remediation is extended. In addition, continuous screening or assessment of pollutants removal is needed to confirm the efficient performance of bioattenuation.

4.1.1.2 Biostimulation

This process brings in human activities for stimulating the natural microbial pesticides degradation. The intentional addition of water, fertilizers, nutrients (growth supplements and trace metals), electron donors, and electron acceptors into the polluted agricultural soil stimulates the indigenous microbial activities (Owsianiak et al., 2010). A stepwise approach is usually recommended to stimulate naturally existing bacterial and fungal communities. The initial step is the addition of fertilizers, growth supplements, and trace minerals to stimulate microbial growth. Second, the adjustment of environmental parameters viz., pH, temperature, moisture, redox conditions, and oxygen of soil in favor of microbial communities is done (Kumar et al., 2011; Adams et al., 2015; Hussain et al., 2009). The fortification of soil with these physicochemical parameters is extremely important to accelerate the microbial metabolism of agrochemicals in the soil. One of the key determinants of a successful biostimulation process is the availability of properly balanced carbon, nitrogen, and phosphorus ratio (C: N: P ratio) in the targeted agricultural site. These are macronutrients necessary for microbial growth (Wolicka et al., 2009; Madhavi and Mohini, 2012).

Bioventing is also an in situ bioremediation method which uses a biostimulation strategy. The pesticide degrading aerobic microbial growth is stimulated by flushing oxygen through the contaminated soil (Hinchee, 1993).

4.1.1.3 Bioaugmentation

During bioaugmentation, individual microorganisms or syntrophic microbial consortia with a confirmed potential of degrading pesticides are added into the polluted agricultural soils. Choosing the correct microbial strains for the process is extremely important. The selected species must be able to degrade the maximum amount of pollutants to a minimum toxic level within a limited time period. However, a thorough knowledge on the population dynamics of exotic and indigenous microorganisms that reside in the same habitat is extremely important to get a favorable outcome of the process (Cameotra and Singh, 2008).

The bioaugmentation process aims to expand the remediation capacity of existing microbial populations in the site with the assistance of microbial supplements. The preparation of microbial inocula for bioaugmentation is done using general microbiological techniques and advanced genetic engineering techniques. After the isolation of microorganisms from the contaminated agricultural fields, they can be enriched in pesticides containing broth media to increase the population of pesticide degrading microorganisms. If the site is contaminated with a specific pesticide, the enrichment can be done using that particular agrochemical in the medium.

Recently, during bioaugmentation, genetically modified microorganisms (GMMs) with enhanced pesticides degrading capacity are added to the polluted agricultural sites. Using GMMs as bioremediators might be faster in completely removing pollutants than the soil inhabited indigenous microbial species

(Sayler and Ripp, 2000). However, their competitive growth with the natural microbial population in the soil acts as a limitation of using GMMs in the process. Moreover, their release into the agroecosystems is quite challenging due to the insufficiency of knowledge on their interactive adaptability to the fluctuations of nutrients, water, temperature, and other biotic and abiotic factors in the targeted environment. Therefore, the addition of GMMs must be done under continuous surveillance.

By genetic modification, the rates of existing pesticides degradation pathways of microorganisms could be enhanced. Moreover, easily culturable microorganisms with fast growth rates could be incorporated with genes responsible for degrading recalcitrant pesticides molecules.

Furthermore, bacterial plasmids which carry specific genes encoding essential enzymes for the catabolism of pesticides and related molecules can be transferred into indigenous bacterial species. Therefore, GMMs can be used effectively for bioremediation although further research is needed for broadening the applications (Kulshreshtha, 2013).

4.1.2 Ex situ bioremediation

4.1.2.1 Composting

There are multiple strategies of using composting for decontamination of agricultural soil viz., on-site direct composting of polluted agricultural soils, the addition of compost into the soil as a fertilizer/nutrient supplement or surfactant to stimulate the indigenous microbial growth, and addition of compost as a source of pollutant degrading microbes (bioaugmentation). Composting uses an aerobic, thermophilic microbial treatment process in which the contaminated soil is piled up, covered and periodically aerated to facilitate microbial activities. It is also an ecofriendly, less expensive method of bioremediating agricultural soil pollutants. Moreover, adding chemical fertilizers into the soils could be replaced by compost. As an organic nutritional supplement, compost can improve the soil quality, microbial balance, and subsequent crop growth (Lim et al., 2016; Xia Guo et al., 2019: Chen et al., 2015). With proper management, composting can be practiced as an in situ bioremediation technique.

4.1.2.2 Land farming

In land farming, pesticides-containing soil is strewn over a large plot of land which is allocated for scavenging a certain pesticide. Spreading polluted soil over a large surface area tends to dilute the pesticides present and the aerobic indigenous microbial activity targets the biodegradable proportions of these pesticides (Castelo-Grande et al., 2010). Periodic soil tillage is carried out to improve the aeration and facilitate the complete degradation of pesticide molecules. Mainly, nutrient amendment and proper irrigation could stimulate the indigenous microbial functions to enhance land farming-mediated bioremediation. This process can be conducted in situ as well.

4.1.2.3 Slurry phase biological treatment

The excavated soil is mixed with nutrients and water inside a series of tanks. The contents mixed into aqueous slurry are then provided with proper reaction temperature, oxygen, and pH in order to facilitate aerobic microbial pollutant degradation (Gavrilescu, 2005).

4.1.2.4 Biopiles

Biopiles is an ex situ bioventing technique that the polluted soil that must be excavated and transported to another site. Soil is then piled up in heaps and aerated to facilitate aerobic microbial activities. It is also a combination of composting and land farming. In this strategy, bioremediation is accelerated by providing optimal temperature, pH, moisture, and essential nutrients for the rapid growth of indigenous microbial species (Gavrilescu, 2005).

All these strategies indicate the enormous potential of using microorganisms in remediating polluted soil ecosystems including agricultural soils. However, the microbial species involved as well as the physicochemical conditions of the contaminated site should work conjointly to attain the full productivity of the processes.

4.2 Factors affecting the microbial bioremediation

Microorganisms show great potential in degrading diverse groups of organic pollutants due to their metabolic mechanisms and inherent adaptability to environmental fluctuations. Therefore, successful bioremediation demands a healthy population of microorganisms in the environment and favorable environmental parameters that facilitate their growth. The biotic factors that influence the bioremediation process are the microorganisms involved and their nutritional requirements while abiotic factors are environmental factors.

4.2.1 Biotic factors affecting the microbial bioremediation

The biotic factors that have a severe impact on the bioremediation process simply include the richness of microbial population and their functional diversity (enzyme production, toxic metabolite production, tendency of mutation and horizontal gene transfer, microbial interactions such as competition for limited carbon sources, symbiosis, antagonism, predation, etc.) in the contaminated soil and other flora and fauna which directly influence the productivity of the bioremediation.

The biochemical activities of microorganisms such as bacteria fungi, protists, and algae are vital for the remediation of pesticides. The potential of microorganisms to carry out bioremediation by degradation, removal, alteration, immobilization, or detoxification of environmental pollutants depends on their enzymes-mediated biochemical pathways (Abatenh et al., 2017). Microorganisms use pesticides and related compounds as carbon and energy sources for their rapid growth in the natural environment. The breaking down of particular pesticides into their less toxic molecular states is usually achieved by the interactive action of microbial communities instead of individual microorganisms.

The microbial species suitable for the bioremediation process must exhibit a set of unique characteristics. Most importantly, they must be capable of adapting to constant fluctuations in the environmental conditions such as temperature, pH, moisture, and aeration while being a resistant genotype for the particular pollutant considered. Moreover, they must have potential in metal-processing in the contaminated site (Stelting et al., 2010). In addition, microbial strategies of adapting to harsh conditions in the environment include making changes in the growth rate, gene expression, physiological or enzymatic activities, and undergoing changes in intimate or symbiotic associations with other organisms. Furthermore, they synthesize bioactive compounds, participate in biofilm formation, and produce biosurfactants when they are exposed to extremes in temperature, salinity, or depletion of micronutrients (Mangwani et al., 2014).

The fundamental mechanism of microbial bioremediation of pesticides is the ability of microorganisms to break down xenobiotic compounds by producing a wide range of enzymes. Microbial oxidoreductases, oxygenases, monooxygenases, dioxygenases, laccases, manganese peroxidases, lignin peroxidases, microbial versatile peroxidases, and other extracellular hydrolytic enzymes viz., amylases, proteases, lipases, DNases, pullulanases, and xylanases are important groups of enzymes in catalyzing this process (Karigar and Rao, 2011). The type of microorganisms involved in the bioremediation process is also an extremely important rate-limiting biotic factor of the process.

4.2.1.1 Bacteria in bioremediation

Bacteria have been the most prominent group of bioremediating microorganisms that have been applied so far. Bacterial genera included in gamma-proteobacteria (*Pseudomonas*, *Aerobacter*, *Acinetobacter*, *Moraxella*, and *Plesiomonas*), beta-proteobacteria (*Burkholderia*, *Neisseria*), alpha-proteobacteria (*Sphingomonas*), Actinobacteria (*Micrococcus*), and Flavobacteria (*Flavobacterium*) are recognized as efficient pesticides bioremediating bacterial groups (Mamta and Khursheed, 2015). According to a recent study conducted on bacterial endosulfan pesticide degradation, five bacterial genera including *Klebsiella*, *Acinetobacter*, *Alcaligenes*, *Flavobacterium*, and *Bacillus*, were found to be endosulfan degraders (Kafilzadeh et al., 2015). It proves that different bacterial genera participate in co-metabolizing the same pesticides. Moreover, enzymatic mineralization of certain pesticides by bacterial species needs both aerobic and anaerobic conditions (Langenhoff et al., 2002). For instance, organochlorines which are considered as a major highly persistent group of agrochemicals need anaerobic conditions for dechlorination and aerobic conditions for breaking down organic or aromatic constituents of the compound (Baczynski et al., 2010). In addition, some anaerobic bacteria including *Dehalobium chlorocoercia* DF1 and *Dehalococcoides mccartyi* have been successfully utilized for the bioremediation of polychlorinated biphenyls (Payne et al., 2013).

4.2.1.2 Fungi in bioremediation

The potential of fungi to degrade pesticides including different insecticides, fungicides, and herbicides has been well-recognized. Fungi which show a saprotrophic nutritional mode are highly efficient in degrading the soil pollutants such as agrochemicals. They use diverse enzymatic pathways for hydrolyzing organic compounds. In addition, fungi transform agrochemicals into nontoxic molecules by making slight alternations in their molecular structure and release them into the soil for subsequent easy degradation by other soil microflora (Hai et al., 2012). When choosing fungal isolates for bioremediation of agricultural soils with pesticides and related compounds, these characteristics could be considered as a baseline.

The process which utilizes fungi for bioremediation is called "Mycoremediation". There are previous reports of using filamentous fungi and members of the fungal class basidiomycetes (e.g., white-rot fungi and brown-rot fungi) for this purpose. *Aspergillus flavus*, *A. niger*, and *Trichoderma harzianum* have been found capable of degrading chlorpyrifos and endosulfan (George et al., 2014). The efficiency of white-rot fungi; *Lentinus subnudus* and *Trametes hirsute* to degrade pesticides and herbicides such as Dichloro diphenyl trichloroethane (DDT) and endosulfan has been previously reported (Nwachukwu and Osuji, 2007; Singh and Singh, 2016; Singh and Kuhad, 1999). Moreover, *Pleurotus ostreatus*, *Trametes versicolor*, *Lentinula edodes*, and *Bjerkandera adusta* are some other fungal species that are capable of carrying out bioremediation of pesticides and herbicides contaminated soils (Singh, 2006).

The commonly used herbicides include atrazine, metolachlor, clodinafop propargyl (CF), 2, 4-dichlorophenoxyacetic acid (2, 4-D), diuron, paraquat, and glyphosate (GP) (Pandey et al., 2018). *Mucor genevensis*, *Phoma glomerata*, *Chrysosporium pannorum*, *Aspergillus penicillioides*, *A. niger*, and *Fusarium oxysporum* are useful in degrading 2, 4-D and its derivatives (Vroumsia et al., 2005). Diuron is used in cotton cultivations. *Rhizoctonia solani*, *Pestalotiopsis versicolor*, *Sporothrix cyanescens* and *Cunninghamella echinulata* have shown higher potential in breaking down diuron (Ellegaard-Jensen et al., 2013). Many fungi including *Aspergillus*, *Rhizopus*, *Fusarium*, *Penicillium*, *Trichoderma*, and *Phanerochaete* are proficient to degrade atrazine which is a selective herbicide (Mougin et al., 1994).

Most of the above fungi are recognized as lignocellulolytic fungi. They are considered the most promising fungi in bioremediation applications as they are capable of producing unique sets of extracellular enzyme complexes to degrade recalcitrant compounds such as lignin (Anastasi et al., 2013). In a recent study conducted by the authors (Jayasekara et al., 2020), several filamentous fungi; *Trichoderma* spp., *Aspergillus* spp., and *Penicillium* spp., and 18 basidiomycetes isolates including *Earliella scabrosa*, *Trametes hirsuta*, *Schizophyllum commune*, *Annulohypoxylon stygium*, *Lentinus sajor-caju* isolates recorded to be highly lignocellulolytic with higher total cellulase, lignin peroxidase, manganese peroxidase, and laccase enzyme activities. These oxidative enzymes target cellulose polymer and highly recalcitrant lignin in allochthonous organic matter. Therefore, these fungal isolates may have the potential of degrading various environmental pollutants including pesticides.

The authors further studied the potential use of lignocellulolytic microorganisms in the degradation of sugarcane bagasse which is a highly recalcitrant abundant cellulosic biomass in many countries in the world. A combination of *E. scabrosa* and *Aspergillus niger* was found to be highly efficient in degrading this recalcitrant biomass (Madusanka et al., 2019). Therefore, adding lignocellulosic biomass such as straw, decaying litter, bagasse, sawdust, etc., and lignocellulolytic fungi into the agricultural soils polluted with pesticides could be a promising way of simultaneous cleaning up of soil pollutants and adding nutrients for the plants and beneficial microbial growth.

4.2.1.3 Synthetic microbial communities in bioremediation of pesticides

Although microbial bioremediation is practiced as a promising method of neutralizing the agrochemicals in soil, incomplete degradation of target pesticide molecules is recognized as a drawback in the process. On the other hand, the resulted by-products might be greater in toxicity than the parent molecule creating a severe impact on living beings. The development of synthetic microbial communities is focused on overcoming these issues. In a synthetic microbial population, two or more defined microbial populations are assembled in a well-characterized and controlled environment while retaining their key metabolic functions. Therefore, a synthetic community with a known, simple, and easily controllable microbial structure could be capable of efficiently carrying out targeted agrochemical degradation. On the contrary, natural monocultures or microbial communities might contain various unknown microbial species with numerous unknown functions. Another advantage of using a synthetic microbial population is their resistance to fluctuating environmental conditions and adverse interactions imposed by invasive species such as antagonism, predation, etc. (Brenner et al., 2008; De Roy et al. 2014).

Designing a cooperative and stable microbial community which is focused on the successful completion of an expected biochemical function is a sequential process. The community must be initiated using harmless microorganisms. Next, it is allowed to degrade, detoxify or transform toxic waste material into target end products. Then the overall bioremediation process of the microbial community is optimized by continuous monitoring of its growth and function (Liu et al., 2017). Several synthetic microbial consortia are reported to be efficient in degrading certain pesticides and herbicides. For instance, a three-member consortium of two fungal strains and a bacterium (*Mortierella* LEJ702, *Variovorax* SRS16, and *Arthrobacter globiformis* D47) was found to be breaking down Diuron (Ellegaard-Jensen et al., 2013). A fungal-bacterial consortium consisted of *Mortierella* sp. LEJ702 and *Aminobacter* sp. MSH1 was found to be successfully degrading 2, 6-dichlorobenzamide (Šašek et al., 1993). Therefore, using synthetic microbial consortia could contribute in an extremely efficient way to bioremediation of pesticides contaminated agricultural soils. However, the process requires the stable initial establishment of the community and continuous routine monitoring of its progress.

4.2.2 Abiotic factors affecting the microbial bioremediation

The abiotic factors that determine the process output are mainly environmental conditions in the polluted site such as pH, water holding capacity, soil texture, permeability, soil temperature, nutrient status, dissolved oxygen content, presence of electron donors and acceptors, contaminant load, and the nature of the pollutants accumulated in the site (chemical structure and composition, hydrophobicity, hydrophilicity, toxicity, biodegradability, etc.) (Mohan et al., 2006; Wang et al., 2007; Sihag et al., 2014).

4.2.2.1 Characteristics of the pesticide contaminant in the soil

The structure of the pesticide molecule and its concentration in the soil directly impact the microbial bioremediation process. Especially, the structure determines the chemical and physical features of the pesticide which eventually decide its overall microbial biodegradability. For instance, the pesticide molecules composed of phenyl rings with non-polar alkyl or halogen groups are highly resistant to microbial activities as they are insoluble in water. The solubility of a contaminant in an aqueous medium is also playing a major role in bioremediation steps. When a pesticide molecule contains hydroxyl or carboxylic groups attached to phenyl rings, they are highly bioavailable for microbial degradation as they are polar molecules with hydrophilic nature (Cork and Krueger, 1991; Chowdhury et al., 2008).

Pesticides concentration in the soil ecosystem is an important rate-limiting factor of a microbial bioremediation reaction. Higher pesticides concentration in soil inhibits the microorganisms and their enzymes leading to a reduced bioremediation rate. However, the presence of appropriate pesticides concentration in soil is essential for the induction of microbial pesticides degrading enzymes (Adams et al., 2015; Prakash and Suseela, 2000).

4.2.2.2 Nutrient availability in the soil for microbial growth

Microbial bioremediation is highly dependent on the availability of macro and micronutrients for microbial growth. Carbon, nitrogen, phosphorus, potassium, and calcium are essential nutrients for microbial cell growth and development. Although excess concentrations in soil cause microbial toxicity, metal ions are indispensable in microbial metabolism. The function of essential enzymes depends on the presence or absence of metal ions (Mg^{2+}, Co^{2+}, Mn^{2+}, etc.) in the soil. Especially, trace elements such as Fe, Zn, Cu, Mn, Mo, and Ni are also important due to their role in metalloenzymes. These metals play an important role in microbial cell membrane stability, nucleic acid structure, function and metabolism, and protein synthesis (Dedyukhina and Eroshin, 1991). Therefore, the application of bioremediation strategies into the soil with a lack of essential nutrients for microbial growth is not a better approach to remediating the pesticide contaminants.

The addition of soil amendments like compost, manure, and decaying plant debris is a better option for enriching the polluted soil with favorable nutrients as well as desired microbial populations. This is not a novelty to traditional agriculture. For instance, Sri Lankan farmers still practice this method to augment their arable lands including paddy fields and home gardens. They usually add cattle dung, paddy straw, sawdust, rice husks, and *Gliricidia sepium* leaves into the soil to improve the nutrients in the soil.

4.2.2.3 pH of the soil

An optimal pH of about 6–8 in range must be maintained in the pesticides contaminated soils in order to induce the bioremediation process (Adams et al., 2015). When the soil pH goes down the bioremediation activities of fungi and acidophilic bacteria are enhanced while other microbial activities are suppressed (Stapleton et al., 1998). Soil pH affects the pesticides solubility and bioavailability for soil microorganisms. Maintenance of a proper soil pH could therefore accelerate the microbial bioremediation process (Racke et al., 1997).

4.2.2.4 Temperature of the soil

Temperature is a crucial factor which determines the rate of any biochemical reaction. Soil temperature level starting from 30°C to about 40°C, which also falls under the mesophilic category of microbial growth temperature requirements, enhances and optimizes the microbial bioremediation of pesticides. Moreover, the molecular structure of the pesticide molecules is altered by the temperature fluctuations in the soil and at higher temperatures, increases their solubility (Margesin and Schinner, 2001; Topp et al., 1997). Maintaining the soil temperature at which the microbial activities are optimal is important to carry out an efficient and optimal bioremediation process.

4.2.2.5 Oxygen availability in the soil

The level of oxygen in the pesticide-contaminated soil determines the predominant microbial species inhabiting the soil based on their oxygen requirement. Accordingly, the bioremediation process could be aerobic or anaerobic. The complete removal of certain pesticides needs both aerobic and anaerobic conditions in the environment. Some pesticides are completely removed from the environment when the soil is anaerobic. As an example, rapid removal of herbicides atrazine and trifluralin could be observed under anaerobic conditions. The steps of the bioremediation of a compound may be altered according to the availability of oxygen. For instance, the initial steps of biodegrading polycyclic aromatic hydrocarbons (PAH) need oxygen for the oxidation of benzene rings by monooxygenase and dioxygenase enzymes activities (Sihag et al., 2014). The anaerobic oxidation reaction needs ferrous, nitrate, and sulfate ions as electron acceptors. In addition, aerobic reactions are more eco-friendlier than the anaerobic remediation processes. For instance, the anaerobic conditions tend to release greenhouse gases, like methane, into the atmosphere thereby leading to global warming (Bamforth and Singleton, 2005). Apparently, providing proper aeration for in situ or ex situ bioremediation processes is more advantageous in terms of achieving an ecofriendly pollutants remediation process.

4.2.2.6 Moisture content of the soil

Moisture or water is an essential element in biochemical reactions. Water acts as the reaction medium which dissolves and disperses all the reactants. Water-soluble pesticides are easily contacted with the hydrolytic microbial enzymes in the environment. Therefore, wet soil is capable of accelerating the bioremediation process than dry soil. However, plenty of water or water-clogging conditions in the agricultural fields could adversely impact oxygen diffusion in soils (Chowdhury et al., 2008).

4.2.2.7 Organic matter content in the soil

Agricultural practices heavily disturb the natural microbial diversity, population density, and decomposing reactions in the soil (Adriaens and Hickey, 1993). Crop harvesting also removes a considerable share of organic matter from agricultural soils. Therefore, the organic matter content in agricultural soil is lower than that of natural surface soil. When the soil is contaminated with pesticides and related agrochemicals, indigenous microbial activities get disturbed and result in subsequent organic matter decrease. However, soil organic matter may interfere in microbial pesticides removal processes by facilitating the adsorption of pesticides molecules to organic matter. Although the availability of organic matter in the soil diversely impacts the microbial pesticide bioremediation process, the initial presence of organic matter is essential for the development of active pesticides biodegradable, indigenous microbial population (Perucci et al., 2000).

In view of the above-mentioned facts, the complete removal of an agrochemical from the polluted agricultural site using an effective microbial bioremediation technology demands the proper environmental conditions and a group of most suitable microorganisms. In other words, an efficient bioremediation process is always an output of a strong correlation between biotic and abiotic factors in the polluted agricultural site. Most importantly, it is a complex interrelationship developed by the co-existence of biodegradable pollutants, microbial population capable of degrading those pollutants, and the environmental factors facilitating the process. However, it is obvious that the setting up of all the required factors in perfect order to enhance the effectiveness of the process is extremely complicated. It must be accomplished by continuous monitoring and gradual optimization of the process parameters.

5 Research on bioremediation of agricultural soils polluted with pesticides

5.1 Current areas of research on microbial bioremediation

The topic "bioremediation" is very popular among researchers. A number of researches have already been conducted on different areas related to bioremediation. The current chapter focuses on research activities carried out on microbial bioremediation of pesticides. Isolation and identification of pesticides degrading individual microorganisms and microbial consortia and evaluating their pesticides bioremediation potential are one of the main study topics in bioremediation research. Several admirable works have been conducted so far in this area of research (Pan et al., 2016; Leitao, 2009; Aust, 1995). Carrillo-Pérez et al. (2004) isolated several bacterial species of *Pseudomonas*, *Neisseria*, *Moraxella*, and *Acinetobacter* that completely degrade DDT. Recently, pesticides are developed in a composite manner, in particular, to improve their functional spectrum. For instance, Velpar K is a hybrid herbicide composed of hexazinone and diuron which is used in sugarcane cultivation. Degradation of such complex compounds using efficient microbial consortia has also been investigated (Ramos and Yoshioka, 2012).

Currently, researches focus on genetic engineering approaches of developing genetically modified, efficient microorganisms and microbial consortia for bioaugmentation of pesticides and related residues. There are reports on using genetically modified bacteria in the bioremediation of pollutants (Hrywna et al., 1999; Menn et al., 2008). Research on microbial growth kinetics in pesticides contaminated soils, patterns of metabolism, and genetic diversity have earned enormous interest from researchers (Wirsching et al., 2020). Moreover, a number of researches have been conducted to determine the impact of certain physicochemical factors such as incubation temperature, pH, and nitrogenic source on the bioremediation process (Odukkathil, and Vasudevan, 2013).

Studies on the evaluation of the efficacy of bioremediation as a pollutant removal technology have also earned immense importance. For instance, a recent study observed that bioaugmentation and biostimulation are effective in the removal of polyaromatic hydrocarbons containing pollutants. However, the efficiency of the two techniques varied on the molecular weight of the pollutant (Sun et al., 2012). A study conducted to evaluate the pesticides removal potential of composting found that immature compost has very lower levels of organic pesticide pollutants than in the initial substrates utilized in the composting process suggesting the apparent removal of those xenobiotic compounds during composting (Vogtmann and Fricke, 1992; Strom, 1998).

Studying the impact of pesticides on soil microbiological factors is also quite important in establishing a stable bioremediation strategy. The output of these studies could provide momentous insights about the impact of pesticide molecules on different microorganisms and thereby could modify the composition of microbial types utilized in the remediation strategies. For instance, a field study conducted to observe the influence of herbicides 2, 4-D, and glyphosate on microbial biomass revealed that the herbicides reduced the microbial biomass in the soil (Wardle and Parkinson, 1992).

Microbial biomass which is composed of live microorganisms is an integral proportion of soil organic matter content. The reduction of microbial biomass indicates the decline of live microbial count in the soil. Another investigation showed the impact of imazethapyr herbicide on microbial biomass (Perucci and Scarponi, 1994). In the laboratory-scale experiment, the soil was treated with the herbicide at the recommended rate for the field crop, 10-fold of recommended rate, and 100-fold of the recommended rate. Although the recommended rate did not impact on soil microbial biomass carbon content, over-application of the herbicide drastically decreased the soil microbial biomass carbon contents. This observation suggests that pesticide over-use above the recommended rate could adversely affect the microbial populations in soil.

Soil respiration (production of carbon dioxide) measurements are effective indicators of soil health. When the soil is polluted with pesticides, their inhibitory effects on the soil microbial population would make fluctuations in the release of CO_2 from soil. A study conducted using a series of the isopropyl amine salt of glyphosate (47, 94, 140, and 234 $\mu g\, g^{-1}$ soil) inoculated soil showed that the glyphosate is capable of vitalizing the microbial functions in soil (Haney et al., 2000). An increase in C and N mineralization further confirmed the rapid microbial removal of glyphosate. Therefore, designing laboratory and field trials to understand the effect of different pesticides on soil microbial parameters is extremely important to confirm their inhibitory effects against microbial population. Then the bioremediation strategies could be altered to meet the utmost efficiency.

Many researchers have emphasized the importance of scaling up the laboratory experiments up to on-site field experiments because; some theories which show the potential of practical application in the laboratory may exhibit inefficiency at the field trials. Bioremediation strategies which perform well in the field trials could be considered as efficient pollutant removal processes. Many field trials have been currently conducted on remediating agrochemicals (Antonious, 2012; Pussemier et al., 1998).

Above mentioned examples provide only a mere glimpse of the various researches that have been conducted so far on bioremediation. Bioremediation is a complex microbial process and its complexity drastically increases with the excessive addition of agrochemicals with higher functional and structural diversity. Therefore, research activities on the subject should be broadened purposefully.

5.2 Novel trends in microbial bioremediation technology

Bioremediation technologies are rapidly evolving. The modifications added into the basic process target optimizing it. This section aims to summarize the emerging trends of bioremediation process that lead to technological advancements in the overall process. Modification of microorganisms and designing site-specific bioremediation schemes with process alternations are key areas of attention in terms of adding novelties to the process (Vishwakarma et al., 2020).

The use of proficient microbial consortia is an essential component of a progressing bioremediation technique. However, adding microbial cells into the contaminated site has certain limitations such as rapid cell viability loss and unequal dispersal in the field. Instead of adding a microbial consortium into the soil, the current trend is to incorporate plasmids that harbor genes encoding for specific pesticides hydrolyzing enzymes. These plasmids are uptaken by a competent indigenous microbial population in the field through conjugation. Microbial cell division increases the number of plasmid copies in the field. This is a novel approach in bioremediation that replaces conventional cell-mediated bioaugmentation (Garbisu et al., 2017).

The bioavailability of pesticides is a dire necessity of microbial bioremediation. The use of biosurfactants; surface-active compounds are introduced as an ecofriendly approach to enhancing hydrophilicity of pesticides molecules. Instead of using synthetic chemicals like Tween 80, surfactants of plant or microbial origin are more beneficial for this process due to their biodegradability and lower toxicity. Humic acid is a plant-based surfactant (Roy et al., 1997). Some microbial-derived surfactants viz., lipopolysaccharides, rhamnolipids, glycolipids, phospholipids, etc., are extracted from *Pseudomonas*, *Bacillus*, yeasts, and yeast-like fungi including *Starmerella bombicola* (Naughton et al., 2019).

Rhizoremediation is recognized as an efficient approach to treating pesticides contaminated agricultural soils. The process basically involves the mutual interactions between the plant and the plant's rhizosphere-inhabiting microorganisms. Microbes increase in their population density by feeding on plant root exudates. The enhancement of microbial activities in the vicinity of plant roots facilitates the microbial breakdown of pollutants (Saravanan et al., 2019).

Using microbial biofilms for pollutant removal and improving soil health is another novelty in agriculture. Some researchers in Sri Lanka currently have proved that deteriorated agricultural soils could be reinstated using microbial biofilms developed under laboratory conditions (Seneviratne et al., 2011; Seneviratne and Kulasooriya, 2013). They have observed that the direct application of bio-filmed biofertilisers (BFBFs) to depleted agricultural soils could promote better yield while restoring the soil shortly after the application. Currently, they have upgraded their BFBFs to the commercial level targeting various crops including tea, rice, maize, and vegetables.

Another most important trend that emerged in current bioremediation technology is the application of "–omics" for understanding the microbial agroecosystems functions. This is a collective of technologies which includes genomics, proteomics, metabolomics, transcriptomics, glycomics, lypomics, systems biology, bioinformatics, and computational biology tools for studying the behavior of microorganisms at the molecular level. These techniques enable the exploration of the overall biochemical activities of microorganisms through unraveling important information viz., specific genes, proteins, factors affecting cellular metabolism and catabolism, nucleic acids, etc. (Aardema and MacGregor, 2002).

Especially, recently developed metagenomics approaches enable the culture-independent assessment of microbial populations for application in bioremediation. Enrichment of microbial target genes using stable isotope probing (SIP) technology, metagenome extraction, and library construction, using function-based and sequence-based metagenomics approaches for accessing microbial biodegradative genes, and microbial community profiling using direct sequencing are recognized as sequential steps of applying metagenomics in bioremediation (Pushpanathan et al., 2014). Incorporation of novel technologies into the conventional bioremediation strategies, under a well-planned upstream process, continuous monitoring, and downstream process would contribute greatly to the overall optimization of microbial bioremediation.

5.3 Advantages and disadvantages of microbial bioremediation process

5.3.1 Advantages of microbial bioremediation process

It is compulsory to understand the positive and negative impacts associated with the bioremediation process and enacting strategies to reduce the drawbacks. One of the main advantages of bioremediation is its ecofriendly mechanism of action. Bioremediation is carried out using an appropriate microbial population for degrading the pesticide pollutants into simpler nontoxic forms such as water and carbon dioxide. Thus, the emission of toxic byproducts during the process is minimized.

Bioremediation is economically feasible and cost-effective which could be practiced on-site. Therefore, wasting labor costs on polluted soil excavation and transportation into the ex situ soil treatment plants is not required. It has minimal remediating site destruction as the microbial activities carried out on-site are less harmful. On the contrary, the use of conventional methods such as landfills demands large plots of extra lands that are eventually get polluted due to the dumping of pollutants.

Bioremediation process permanently removes the toxic pesticides contaminants from the agricultural soils. Therefore, the risk of long-term persistence of pesticides in the food chains is restricted. It will reduce the emergence of severe health concerns in the future.

5.3.2 Disadvantages of microbial bioremediation process

Bioremediation technology has certain disadvantages. Particularly, the process is limited to the removal of biodegradable agrochemicals. Other non-biodegradable pesticides including chlorinated compounds could remain intact in the soils that will accumulate in food chains. In addition, microbial pesticides metabolism also may produce poisonous compounds. For instance, under anaerobic conditions in the agricultural field, greenhouse gasses such as methane and carbon dioxide emission into the atmosphere may occur which will account for intense global warming in the future.

There are various types of synthetic agrochemicals that are used in modern agricultural practices. Those artificial compounds might have complex chemical structures that the bioremediation end products may be extremely hazardous and highly persistent in the natural environment. A common example is the production of vinyl chloride by the microbial remediation of trichloroethylene; this vinyl chloride is a carcinogenic compound.

Moreover, bioremediation is a time-consuming process. Before starting the cleaning up of the polluted site, a sequential evaluation and optimization of the process must be assessed under laboratory conditions. Especially, for bioaugmentation, a suitable microbial consortium must be developed under intensive laboratory procedures. Before the addition of the consortium into polluted agricultural soil, they must be assessed for their physicochemical growth characteristics because the effectiveness of the whole bioremediation process is determined by the efficient growth of the appropriate microbial population and the environmental parameters of the site (Zouboulis and Moussas, 2011).

5.4 Indications for future research

As previously mentioned, a more productive bioremediation process depends on the microbial capacity to degrade pollutants. Currently, the phenomenon called "aging" of pesticides changes the bioavailability of pollutants with the time that the microorganisms are being challenged due to their lack of adaptability to the aging process. Therefore, studies must be conducted to understand the methods of making these pesticides and related chemical molecules bioavailable for microbial activities (De Lorenzo et al., 2018).

Microorganisms, although they are primitive in the organization, their metabolic pathways are extremely complicated and not fully explained in modern science. Having an in-depth understanding of the metabolic pathways of indigenous and exogenous microorganisms employed in bioremediation is essential for designing an efficient bioremediation process (Dvořák et al., 2017). Therefore, an extensive number of research is required for unraveling the underlying mechanisms of microbial metabolisms related to bioremediation. A combined approach of modern molecular biology and genetics, microbiology, and biochemistry could be followed to achieve this goal.

Most of the bioremediation technologies are based on the hydrolytic potential of diverse groups of microbial enzymes. Having an in-detailed understanding of the pesticides degrading enzymes including their molecular structure, functional properties, rate-limiting factors, and enzyme inhibitory factors is essential especially in optimizing the bioremediation process. Therefore, microbial enzymatic studies can be further extended to unfold the theories behind the enzymatic activities (Vishwakarma et al., 2020).

Adding different types of pesticides into the same agricultural field might magnify the complexity of bioremediation process due to the lack of a clear view of interactions in-between pesticides. If any chemical reaction occurs among the pesticides molecules, it could produce unidentified end products insensitive to microbial activities. An end product which is naturally susceptible to microbial attack might be more beneficial for the bioremediating process. However, this information must be confirmed with further research. Another concern on end products released after bioremediation of pesticides is their toxicity. It is compulsory to guarantee that the bioremediation does not pollute the agricultural soils by releasing toxic end products. Therefore, it is necessary to evaluate the characteristics of end products and if found toxic, strategies must be established to remove them. In order to do that the downstream process of bioremediation must be comprised of proper methods of neutralizing these chemicals. Research activities on these particular concepts might be intensely beneficial for organizing a well-intended bioremediation process.

Moreover, microorganisms may develop genetic modifications during the process. These genetic modifications being unpredictable and not intended for the productivity of the bioremediation process may cause serious adverse impacts on humans, animals, and plants. Therefore, research activities must be designed to understand the post-bioremediation impacts on microbial population and remedies for rectifying them if their prevalence in the environment is not ecofriendly.

Furthermore, discovering greener alternatives for agrochemicals could also be a more desirable approach to reducing pesticides usage and consequent mitigation of the adverse impacts associated with their over application.

6 Conclusion

The application of pesticides is an integral part of modern agriculture. The excessive use of pesticides could impose adverse impacts on the natural environment and its living beings. Bioremediation is emerged as a promising technique for rectifying pesticide-associated pollution of agricultural soils. However, the beneficial output of the process basically relies upon the biodegradability of the pesticide molecules, availability of an efficient microbial population with the potential of degrading targeted pesticides as well as the environmental conditions in favor of the process. In addition, progress monitoring criteria should be established to periodically confirm that the aforementioned three factors are in optimal interconnection to achieve effective bioremediation. However, the application of bioremediation techniques without sufficient knowledge about the behavior of pesticides in the field, their interactions in-between and the environment, microbial population utilized, mechanisms of enzymatic functions on pesticides, bioremediation-mediated fate of pesticides, and influence of environmental parameters on the process, factors limiting the bioremediation and the nature of end products formed, could make the bioremediation process a failure. Therefore, it is necessary to design further research to fully understanding the working mechanisms of bioremediation. Bioremediation technology has currently reached new ventures by means of advances in scientific disciplines such as microbiology, biochemistry, molecular biology, genetic engineering, and analytical instrumentation. Development of genetically engineered microorganisms with enhanced pesticides degradation potential for bioaugmentation, application of metagenomics approaches for studying the beneficial microbial communities in agroecosystems, and inventing analytical instruments for studying the process parameters are some of the recently evolved technological advancements in bioremediation. Yet, it is essential to develop novel bioremediation technologies specifically designed for removing pesticide pollutants in agricultural ecosystems. Another important step is properly optimizing the effective laboratory-scale experiments on bioremediation up to the field-scale application. Microbial bioremediation is already accepted as an ecofriendly and cost-effective strategy for efficiently cleaning up the pesticide pollutants in agricultural soils. Focusing on discovering novelties to improve the prevailing technologies while minimizing the process-associated drawbacks would assist in making bioremediation a greener approach to reversing pesticides-associated soil pollution.

References

Aardema, M.J., MacGregor, J.T., 2002. Toxicology and genetic toxicology in the new era of "toxicogenomics": impact of "-omics" technologies. Mutat. Res. 499 (1), 13–25. https://doi.org/10.1016/s0027-5107(01)00292-5.

Abatenh, E., Gizaw, B., Tsegaye, Z., Wassie, M., 2017. The role of microorganisms in bioremediation – a review. Open J. Environ. Biol. 2 (1), 038–046. https://doi.org/10.17352/ojeb.000007.

Abubakar, Y., Tijjani, H., Egbuna, C., Adetunji, C.O., Kala, S., Kryeziu, T.L., et al., 2020. Pesticides, history, and classification. In: Egbuna, C., Sawicka, B. (Eds.), Natural Remedies for Pest, Disease and Weed Control. Elsevier Inc, pp. 29–42, https://doi.org/10.1016/B978-0-12-819304-4.00003-8.

Adams, G.O., Fufeyin, P.T., Okoro, S.E., Ehinomen, I., 2015. Bioremediation, biostimulation and bioaugmention: a review. Int. J. Environ. Bioremediat. Biodegrad. 3, 28–39. https://goo.gl/9XY7ni.

Adriaens, P., Hickey, W.J., 1993. In: Stone, D.L. (Ed.), Biotechnology for the Treatment of Hazardous Waste. Lewis Publishers, Ann Arbor, MI, United States, pp. 97–120.

Akashe, M.M., Pawade, U.V., Nikam, A.V., 2018. Classification of pesticides: a review. Int. J. Res. Ayurveda Pharm. 9 (4), 144–150. https://doi.org/10.7897/2277-4343.094131.

Alavanja, M.C., Samanic, C., Dosemeci, M., Lubin, J., Tarone, R., Lynch, C.F., et al., 2003. Use of agricultural pesticides and prostate cancer risk in the agricultural health study cohort. Am. J. Epidemiol. 157 (9), 800–814. https://doi.org/10.1093/aje/kwg040.

Anastasi, A., Tigini, V., Varese, G.C., 2013. The bioremediation potential of different ecophysiological groups of fungi. In: Goltapeh, E.M., Danesh, Y.R., Varma, A. (Eds.), Fungi as Bioremediators. Springer Berlin Heidelberg, Berlin, pp. 29–49, https://doi.org/10.1007/978-3-642-33811.

Antonious, G.F., 2012. On-farm bioremediation of dimethazone and trifluralin residues in runoff water from an agricultural field. J. Environ. Sci. Health B 47, 608–621. https://doi.org/10.1080/03601234.2012.668454.

Ashraf, M.A., Maah, M.J., Yusoff, I., 2014. Soil contamination, risk assessment and remediation. In: Hernandez-Soriano, M.C. (Ed.), Environmental Risk Assessment of Soil Contamination. IntechOpen, Rijeka, Croatia, https://doi.org/10.5772/57287.

Aust, S.D., 1995. Mechanisms of degradation by white rot fungi. Environ. Health Perspect. 5, 59–61. https://doi.org/10.1289/ehp.95103s459.

Azubuike, C.C., Chikere, C.B., Okpokwasili, G.C., 2016. Bioremediation techniques-classification based on site of application: principles, advantages, limitations and prospects. World J. Microbiol. Biotechnol. 32 (11), 180. https://doi.org/10.1007/s11274-016-2137-x.

Baczynski, T.P., Pleissner, D., Grotenhuis, T., 2010. Anaerobic biodegradation of organochlorine pesticides in contaminated soil—significance of temperature and availability. Chemosphere 78 (1), 22–28. https://doi.org/10.1016/j.chemosphere.2009.09.058.

Bamforth, S.M., Singleton, I., 2005. Bioremediation of polycyclic aromatic hydrocarbons: current knowledge and future directions. J. Chem. Technol. Biotechnol. 80 (7), 723–736. https://doi.org/10.1002/jctb.1276.

Boedeker, W., Watts, M., Clausing, P., Marquez, E., 2020. The global distribution of acute unintentional pesticide poisoning: estimations based on a systematic review. BMC Public Health 20, 1875. https://doi.org/10.1186/s12889-020-09939-0.

Boopathy, R., 2000. Factors limiting bioremediation technologies. Bioresour. Technol. 74 (1), 63–67. https://doi.org/10.1016/s0960-8524(99)00144-3.

Brandt, M.J., Johnson, K.M., Elphinston, A.J., Ratnayaka, D.D., 2017. Chemistry, microbiology and biology of water. In: Twort's Water Supply, pp. 235–321, https://doi.org/10.1016/b978-0-08-100025-0.00007-7.

Brenner, K., You, L., Arnold, F.H., 2008. Engineering microbial consortia: a new frontier in synthetic biology. Trends Biotechnol. 26, 483–489. https://doi.org/10.3389/fmicb.2012.00203.

Cameotra, S.S., Singh, P., 2008. Bioremediation of oil sludge using crude biosurfactants. Int. Biodeter. Biodegr. 62 (3), 274–280. https://doi.org/10.1016/j.ibiod.2007.11.009.

Carrillo-Pérez, E., Ruiz-Manriquez, A., Yeomans-Reina, H., 2004. Isolation, identification and evaluation of a mixed culture of microorganisms with capability to degrade DDT. Rev. Int. Contam. Ambient. 20 (2), 69–75. https://www.scopus.com/inward/record.uri?partnerID=HzOxMe3b&scp=11144303047&origin=inward.

Castelo-Grande, T., Augusto, P.A., Monteiro, P., Estevez, A.M., Barosa, D., 2010. Remediation of soils contaminated with pesticides: a review. Int. J. Environ. Anal. Chem. 90, 438–467. https://doi.org/10.1080/03067310903374152.

Chen, M., Xu, P., Zeng, G., Yang, C., Huang, D., Zhang, J., 2015. Bioremediation of soils contaminated with polycyclic aromatic hydrocarbons, petroleum, pesticides, chlorophenols and heavy metals by composting: applications, microbes and future research needs. Biotechnol. Adv. 33 (6), 745–755. https://doi.org/10.1016/j.biotechadv.2015.05.003.

Chen, R., Deng, M., He, X., Hou, J., 2017. Enhancing nitrate removal from freshwater pond by regulating carbon/nitrogen ratio. Front. Microbiol. 8 (1712). https://doi.org/10.3389/fmicb.2017.01712.

Chowdhury, A., Pradhan, S., Saha, M., Sanyal, N., 2008. Impact of pesticides on soil microbiological parameters and possible bioremediation strategies. Indian J. Microbiol. 48, 114–127. https://doi.org/10.1007/s12088-008-0011-8.

Cork, D.J., Krueger, J.P., 1991. Microbial transformations of herbicides and pesticides. Adv. Appl. Microbiol. 36, 1–66. https://doi.org/10.1016/S0065-2164(08)70450-7.

De Bon, H., Huat, J., Parrot, L., Sinzogan, A., Martin, T., Malezieux, E., et al., 2014. Pesticide risks from fruit and vegetable pest management by small farmers in sub-Saharan Africa. A review. Agron. Sustain. Dev. 34 (4), 723–736. https://doi.org/10.1007/s13593-014-0216-7.

De Lorenzo, V., Prather, K.L., Chen, G.Q., O'Day, E., von Kameke, C., Oyarzún, D.A., et al., 2018. The power of synthetic biology for bioproduction, remediation and pollution control: the UN's Sustainable Development Goals will inevitably require the application of molecular biology and biotechnology on a global scale. EMBO Rep. 19 (4). https://doi.org/10.15252/embr.201745658.

De Roy, K., Marzorati, M., Van den Abbeele, P., Van de Wiele, T., Boon, N., 2014. Synthetic microbial ecosystems: an exciting tool to understand and apply microbial communities. Environ. Microbiol. 16, 1472–1481. https://doi.org/10.1111/1462-2920.12343.

Dedyukhina, E.G., Eroshin, V.K., 1991. Essential metal ions in the control of microbial metabolism. Process Biochem. 26 (1), 31–37. https://doi.org/10.1016/0032-9592(91)80005-a.

Doolotkeldieva, T., Konurbaeva, M., Bobusheva, S., 2018. Microbial communities in pesticide-contaminated soils in Kyrgyzstan and bioremediation possibilities. Environ. Sci. Pollut. Res. 25, 31848–31862. https://doi.org/10.1007/s11356-017-0048-5.

Dvořák, P., Nikel, P.I., Damborský, J., de Lorenzo, V., 2017. Bioremediation 3.0: engineering pollutant-removing bacteria in the times of systemic biology. Biotechnol. Adv. 35 (7), 845–866. https://doi.org/10.1016/j.biotechadv.2017.08.001.

Ellegaard-Jensen, L., Amand, J., Kragelund, B.B., Johnsen, A.H., Rosendahl, S., 2013. Strains of the soil fungus *Mortierella* show different degradation potential for the phenylurea herbicide diuron. Biodegradation 24, 765–774. https://doi.org/10.1007/s10532-013-9624-7.

Garbisu, C., Garaiyurrebaso, O., Epelde, L., Grohmann, E., Alkorta, I., 2017. Plasmidmediated bioaugmentation for the bioremediation of contaminated soils. Front. Microbiol. 8, 1966. https://doi.org/10.3389/fmicb.2017.01966.

Gavrilescu, M., 2005. Fate of pesticides in the environment and its bioremediation. Eng. Life Sci. 5 (6), 497–526. https://doi.org/10.1002/elsc.200520098.

George, N., Chauhan, P.S., Sondhim, S., Saini, S., Puri, N., Gupta, N., 2014. Biodegradation and analytical methods for detection of organophosphorus pesticide: chlorpyrifos. Int. J. Pure Appl. Sci. Technol. 20, 79–94. http://www.ijopaasat.in.

Gilden, R.C., Huffling, K., Sattler, B., 2010. Pesticides and health risks. J. Obstet. Gynecol. Neonatal Nurs. 39 (1), 103–110. https://doi.org/10.1111/j.1552-6909.2009.01092.x.

Glazer, A.N., Nikaido, H. (Eds.), 1995. Microbial Biotechnology: Fundamentals of Applied Microbiology. Freeman, New York, NY.

Gupta, P.K., 2011. Herbicides and fungicides. In: Gupta, R.C. (Ed.), Reproductive and Developmental Toxicology. Academic Press, pp. 503–521, https://doi.org/10.1016/B978-0-12-382032-7.10039-6.

Hai, F.I., Modin, O., Yamamoto, K., Fukushi, K., Nakajima, F., Nghiem, L.D., 2012. Pesticide removal by a mixed culture of bacteria and white-rot fungi. J. Taiwan Inst. Chem. Eng. 43 (3), 459–462. https://doi.org/10.1016/j.jtice.2011.11.002.

Haney, R., Senseman, S., Hons, F., Zuberer, D., 2000. Effect of glyphosate on soil microbial activity and biomass. Weed Sci. 48 (1), 89–93. https://doi.org/10.1614/0043-1745(2000)048[0089:EOGOSM]2.0.CO;2.

Hayat, R., Ali, S., Amara, U., Khalid, R., Ahmed, I., 2010. Soil beneficial bacteria and their role in plant growth promotion: a review. Ann. Microbiol. 60, 579–598. https://doi.org/10.1007/s13213-010-0117-1.

Henao, S., Arbelaez, M.P., 2002. Epidemiological situation of acute pesticide poisoning in the Central American Isthmus, 1992-2000. Epidemiol. Bull. 23 (3), 5–9. 12608345.

Hinchee, R.E., 1993. Bioventing of petroleum hydrocarbons. In: Mathew, J.E. (Ed.), Handbook of Bioremediation. CRC Press, Boca Raton, FL, https://doi.org/10.1201/9780203712764.

Hoppin, J.A., Umbach, D.M., London, S.J., Henneberger, P.K., Kullman, G.J., Coble, J., et al., 2009. Pesticide use and adult-onset asthma among male farmers in the agricultural health study. Eur. Respir. J. 34 (6), 1296–1303. https://doi.org/10.1183/09031936.00005509.

Hrywna, Y., Tsoi, T.V., Maltseva, O.V., Quensen, J.F., Tiedje, J.M., 1999. Construction and characterization of two recombinant bacteria that grow on ortho- and para-substituted chlorobiphenyls. Appl. Environ. Microbiol. 65 (5), 2163–2169. https://doi.org/10.1128/AEM.65.5.2163-2169.1999.

Hussain, S., Siddique, T., Arshad, M., Saleem, M., 2009. Bioremediation and phytoremediation of pesticides: recent advances. Crit. Rev. Environ. Sci. Technol. 39 (10), 843–907. https://doi.org/10.1080/10643380801910090.

Jayasekara, S.K., Kathirgamanathan, M., Ratnayake, R.R., 2020. Isolation, identification and study of the potential applications of tropical fungi in Lignocellulolysis. In: Chaurasia, P.K., Bharati, S.L. (Eds.), Research Advances in the Fungal World: Culture, Isolation, Identification, Classification, Characterization, Properties and Kinetics. Nova Science Publishers, Inc, New York, USA.

Kafilzadeh, F., Ebrahimnezhad, M., Tahery, Y., 2015. Isolation and identification of endosulfan-degrading bacteria and evaluation of their bioremediation in Kor River, Iran. Osong. Public Health Res. Perspect. 6 (1), 39–46. https://doi.org/10.1016/j.phrp.2014.12.003.

Karigar, C.S., Rao, S.S., 2011. Role of microbial enzymes in the bioremediation of pollutants: a review. Enzyme Res. https://doi.org/10.4061/2011/805187.

Kislev, M.E., Weiss, E., Hartmann, A., 2004. Impetus for sowing and the beginning of agriculture: ground collecting of wild cereals. Proc. Natl. Acad. Sci. U. S. A. 101 (9), 2692–2694. https://doi.org/10.1073/pnas.0308739101.

Kulshreshtha, S., 2013. Genetically engineered microorganisms: a problem solving approach for bioremediation. J. Bioremed. Biodegr. 4, 1–2. https://goo.gl/JkdMBV.

Kumar, A., Bisht, B.S., Joshi, V.D., Dhewa, T., 2011. Review on bioremediation of polluted environment: a management tool. Int. J. Environ. Sci. 1, 1079–1093. https://goo.gl/P6Xeqc.

Lal, R., Pandey, G., Sharma, P., Kumari, K., Malhotra, S., Pandey, R., et al., 2010. Biochemistry of microbial degradation of hexachlorocyclohexane and prospects for bioremediation. Microbiol. Mol. Biol. Rev. 74 (1), 58–80. https://doi.org/10.1128/MMBR.00029-09.

Langenhoff, A.A.M., Staps, J.J.M., Pijls, C., Alphenaar, A., Zwiep, G., Rijnaarts, H.H.M., 2002. Intrinsic and stimulated in situ biodegradation of Hexachlorocyclohexane (HCH). Water Air Soil Pollut. Focus 2, 171–181. https://doi.org/10.1023/A:1019943410568.

Leitao, A.L., 2009. Potential of *Penicillium* species in the bioremediation field. Int. J. Environ. Res. Public Health 6 (4), 1393–1417. https://doi.org/10.3390/ijerph6041393.

Lim, S.L., Lee, L.H., Wu, T.Y., 2016. Sustainability of using composting and vermicomposting technologies for organic solid waste biotransformation: recent overview, greenhouse gases emissions and economic analysis. J. Clean. Prod. 111 (Part A), 262–278. https://doi.org/10.1016/j.jclepro.2015.08.083.

Liu, S.H, Zeng, G.M., Niu, Q.Y., Liu, Y., Zhou, L., Jiang, L.H., Tan, X.F., Xu, P., Zhang, C., Cheng, M., 2017. Bioremediation mechanisms of combined pollution of PAHs and heavy metals by bacteria and fungi: a mini review. Bioresour. Technol. 224, 25–33.

Lozowicka, B., Jankowska, M., Hrynko, I., Kaczynski, P., 2016. Removal of 16 pesticide residues from strawberries by washing with tap and ozone water, ultrasonic cleaning and boiling. Environ. Monit. Assess. 188 (1), 1–19.

Madhavi, G.N., Mohini, D.D., 2012. Review paper on—parameters affecting bioremediation. Int. J. Life Sci. Pharma Res. 2, 77–80. https://goo.gl/tBP2C6.

Madusanka, T.Y.G., Jayasekara, S.K., Ratnayake, R.R., Seneweera, S., Rupasinghe, C.P., 2019. Biological pre-treatment of sugarcane bagasse for cellulosic ethanol production. In: Proceedings Young Graduates' Forum. International Symposium on Agriculture & Environment. Faculty of Agriculture, University of Ruhuna, Mathara, Sri Lanka, p. 21.

Mamta, R.J.R., Khursheed, A.W., 2015. Bioremediation of pesticides under the influence of bacteria and fungi. In: Singh, S., Srivastava, K. (Eds.), Handbook of Research on Uncovering New Methods for Ecosystem Management through Bioremediation. IGI Global, pp. 51–72, https://doi.org/10.4018/978-1-4666-8682-3.ch003.

Mangwani, N., Shukla, S.K., Rao, T.S., Das, S., 2014. Calcium mediated modulation of *Pseudomonas mendocina* NR802 biofilm influences the phenanthrene degradation. Colloids Surf. B. Biointerfaces 114, 301–309. https://doi.org/10.1016/j.colsurfb.2013.10.003.

Margesin, R., Schinner, F., 2001. Biodegradation and bioremediation of hydrocarbons in extreme environments. Appl. Microbiol. Biotechnol. 56, 650–663. https://doi.org/10.1007/s002530100701.

Margni, M., Rossier, D., Crettaz, P., Jolliet, O., 2002. Life cycle impact assessment of pesticides on human health and ecosystems. Agric. Ecosyst. Environ. 93 (1–3), 379–392. https://doi.org/10.1016/s0167-8809(01)00336-x.

Megharaj, M., Ramakrishnan, B., Venkateswarlu, K., Sethunathan, N., Naidu, R., 2011. Bioremediation approaches for organic pollutants: a critical perspective. Environ. Int. 37 (8), 1362–1375. https://doi.org/10.1016/j.envint.2011.06.003.

Menn, F.M., Easter, J.P., Sayler, G.S., 2008. Genetically engineered microorganisms and bioremediation. In: Rehm, H.J., Reed, G. (Eds.), Biotechnology: Environmental Processes. Wiley-VCH Verlag GmbH, Weinheim, Germany, https://doi.org/10.1002/9783527620951.ch21.

Mishra, M., Singh, S.K., Kumar, A., 2021. Environmental factors affecting the bioremediation potential of microbes. In: Microbe Mediated Remediation of Environmental Contaminants, pp. 47–58, https://doi.org/10.1016/B978-0-12-821199-1.00005-5.

Mohan, S.V., Kisa, T., Ohkuma, T., Kanaly, R.A., Shimizu, Y., 2006. Bioremediation technologies for treatment of PAH-contaminated soil and strategies to enhance process efficiency. Rev. Environ. Sci. Biotechnol. 5, 347–374. https://doi.org/10.1007/s11157-006-0004-1.

Mougin, C., Laugero, C., Asther, M., Dubroca, J., Frasse, P., Asther, M., 1994. Biotransformation of the herbicide atrazine by the white rot fungus Phanerochaete chrysosporium. Appl. Environ. Microbiol. 60 (2), 705–708. https://doi.org/10.1128/AEM.60.2.705-708.1994.

Naughton, P., Marchant, R., Naughton, V., Banat, I., 2019. Microbial biosurfactants: current trends and applications in agricultural and biomedical industries. J. Appl. Microbiol. 127 (1), 12–28. https://doi.org/10.1111/jam.14243.

Nwachukwu, E.O., Osuji, J.O., 2007. Bioremedial degradation of some herbicides by indigenous white rot fungus, *Lentinus subnudus*. J. Plant Sci. 2 (6), 619–624. https://doi.org/10.3923/jps.2007.619.624.

Odukkathil, G., Vasudevan, N., 2013. Toxicity and bioremediation of pesticides in agricultural soil. Rev. Environ. Sci. Biotechnol. 12, 421–444. https://doi.org/10.1007/s11157-013-9320-4.

Ortiz-Hernandez, L., Sanchez-Salinas, E., Dantan-Gonzalez, E., Castrejon-Godinez, M.L., 2013. Pesticide biodegradation: mechanisms, genetics, and strategies to enhance the process. In: Chamy, R. (Ed.), Biodegradation-Life of Science. IntechOpen, Rijeka, Croatia, pp. 251–287, https://doi.org/10.5772/56098.

Owsianiak, M., Dechesne, A., Binning, P.J., Chambon, J.C., Sørensen, S.R., Smets, B.F., 2010. Evaluation of bioaugmentation with entrapped degrading cells as a soil remediation technology. Environ. Sci. Technol. 44 (19), 7622–7627.

Pan, X., Lin, D., Zheng, Y., Zhang, Q., Yin, Y., Cai, L., et al., 2016. Biodegradation of DDT by *Stenotrophomonas* sp. DDT-1: characterization and genome functional analysis. Sci. Rep. 6. https://doi.org/10.1038/srep21332, 21332.

Pandey, C., Prabha, D., Negi, Y.K., 2018. Mycoremediation of common agricultural pesticides. In: Prasad, R. (Ed.), Mycoremediation and Environmental Sustainability. Springer, Cham, pp. 155–179, https://doi.org/10.1007/978-3-319-77386-5_6.

Payne, R.B., Fagervold, S.K., May, H.D., Sowers, K.R., 2013. Remediation of polychlorinated biphenyl impacted sediment by concurrent bioaugmentation with anaerobic halorespiring and aerobic degrading bacteria. Environ. Sci. Technol. 47, 3807–3815. https://doi.org/10.1021/es304372t.

Pérez, A.P., Eugenio, N.R., 2018. Status of Local Soil Contamination in Europe: Revision of the Indicator "Progress in the Management Contaminated Sites in Europe". Publications Office of the European Union, Luxembourg, https://doi.org/10.2760/093804.

Perucci, P., Dumontet, S., Bufo, S.A., Mazzatura, A., Casucci, C., 2000. Effects of organic amendments and herbicide treatment on soil microbial biomass. Biol. Fertil. Soils 32, 17–23. https://doi.org/10.1007/s003740000207.

Perucci, P., Scarponi, L., 1994. Effects of the herbicide imazethapyr on soil microbial biomass and various soil enzyme activities. Biol. Fertil. Soils 17, 237–240. https://doi.org/10.1007/BF00336329.

Prakash, N.B., Suseela, D.L., 2000. Persistence of butachlor in soils under different moisture regime. J. Indian Soc. Soil Sci 48, 249–256.

Pushpanathan, M., Jayashree, S., Gunasekaran, P., Rajendhran, J., 2014. Microbial bioremediation. In: Microbial Biodegradation and Bioremediation, pp. 407–419, https://doi.org/10.1016/b978-0-12-800021-2.00017-0.

Pussemier, L., Goux, S., Elsen, Y.V., Mariage, Q., 1998. Biofilters for on farm clean-up of pesticide wastes. Med. Fac. Landbouww. Univ. Gent. 63, 11–27.

Racke, K.D., Skidmore, M.W., Hamilton, D.J., Unsworth, J.B., Miyamoto, J., Cohen, S.Z., 1997. Pesticide fate in tropical soils. Pure Appl. Chem. 69, 1349–1371.

Ramos, M.A.G., Yoshioka, S.A., 2012. Bioremediation of herbicide Velpar K® in vitro in aqueous solution with application of EM-4 (effective microorganisms). Braz. Arch. Biol. Technol. 55 (1), 145–149. https://doi.org/10.1590/S1516-89132012000100018.

Rani, M., Shanker, U., Jassal, V., 2017. Recent strategies for removal and degradation of persistent and toxic organochlorine pesticides using nanoparticles: a review. J. Environ. Manage. 190, 208–222.

Rasmussen, J.J., Wiberg-Larsen, P., Baattrup-Pedersen, A., Cedergreen, N., McKnight, U.S., Kreuger, J., et al., 2015. The legacy of pesticide pollution: an overlooked factor in current risk assessments of freshwater systems. Water Res. 84, 25–32. https://doi.org/10.1016/j.watres.2015.07.021.

Roy, D., Kommalapati, R.R., Mandava, S.S., Valsaraj, K.T., Constant, W.D., 1997. Soil washing potential of a natural surfactant. Environ. Sci. Technol. 31 (3), 670–675. https://doi.org/10.1021/es960181y.

Saravanan, A., Jeevanantham, S., Narayanan, V.A., Kumar, P.S., Yaashikaa, P., Muthu, C.M., 2019. Rhizoremediation—a promising tool for the removal of soil contaminants: a review. J. Environ. Chem. Eng. https://doi.org/10.1016/j.jece.2019.103543, 103543.

Šašek, V., Volfová, O., Erbanová, P., Vyas, B.R.M., Matucha, M., 1993. Degradation of PCBs by white rot fungi, methylotrophic and hydrocarbon utilizing yeasts and bacteria. Biotechnol. Lett. 15, 521–526. https://doi.org/10.1007/BF00129330.

Sayler, G.S., Ripp, S., 2000. Field applications of genetically engineered microorganisms for bioremediation processes. Curr. Opin. Biotechnol. 11, 286–289. https://goo.gl/hd9ZLX.

Seneviratne, G., Jayasekara, A.P.D.A., De Silva, M.S.D.L., Abeysekera, U.P., 2011. Developed microbial biofilms can restore deteriorated conventional agricultural soils. Soil Biol. Biochem. 43 (5), 1059–1062. https://doi.org/10.1016/j.soilbio.2011.01.026.

Seneviratne, G., Kulasooriya, S.A., 2013. Reinstating soil microbial diversity in agroecosystems: the need of the hour for sustainability and health. Agric. Ecosyst. Environ. 164, 181–182. https://doi.org/10.1016/j.agee.2012.10.002.

Sharma, A., Kumar, V., Shahzad, B., Tanveer, M., Sidhu, G.P.S., Handa, N., et al., 2019. Worldwide pesticide usage and its impacts on ecosystem. SN Appl. Sci. 1, 1446. https://doi.org/10.1007/s42452-019-1485-1.

Sihag, S., Pathak, H., Jaroli, D.P., 2014. Factors affecting the rate of biodegradation of polyaromatic hydrocarbons. Int. J. Pure Appl. Biosci. 2, 185–202.

Singh, R.P., Manchanda, G., Li, Z.F., Rai, AR, 2017. Insight of proteomics and genomics in environmental bioremediation. In: Bhakta, J.N. (Ed.), Handbook of research on inventive bioremediation techniques, first ed. vol. 1. IGI Global, USA, doi:10.4018/978-1-5225-2325-3.

Singh, B., Singh, K., 2016. Microbial degradation of herbicides. Crit. Rev. Microbiol. 42 (2), 245–261. https://doi.org/10.3109/1040841X.2014.929564.

Singh, B.K., Kuhad, R.C., 1999. Biodegradation of lindane (γ-hexachlorocyclohexane) by the white rot fungus *Trametes hirsute*. Lett. Appl. Microbiol. 28 (3), 238–241. https://doi.org/10.1046/j.1365-2672.1999.00508.x.

Singh, B.K., Walker, A., 2006. Microbial degradation of organophosphorus compounds. FEMS Microbiol. Rev. 30 (3), 428–471. https://doi.org/10.1111/j.1574-6976.2006.00018.x.

Singh, H., 2006. Mycoremediation: Fungal Bioremediation. Wiley, London, pp. 181–215.

Speight, J.G., 2018. Mechanisms of transformation. In: Reaction Mechanisms in Environmental Engineering, pp. 337–384, https://doi.org/10.1016/B978-0-12-804422-3.00010-9.

Stapleton, R.D., Savage, D.C., Sayler, G.S., Stacey, G., 1998. Biodegradation of aromatic hydrocarbons in an extremely acidic environment. Appl. Environ. Microbiol. 64 (11), 4180–4184. https://doi.org/10.1128/AEM.64.11.4180-4184.1998.

Stelting, S., Burns, R.G., Sunna, A., Visnovsky, G., Bunt, C., 2010. Immobilization of *Pseudomonas* sp. strain ADP: a stable inoculant for the bioremediation of atrazine. In: 19th World Congress of Soil Science, Soil Solutions for a Changing World, Brisbane, Australia.

Strom, P.F., 1998. Evaluating pesticide residues in yard trimmings compost. Biocycle 39 (11), 80.

Strong, P.J., Burgess, J.E., 2008. Treatment methods for wine-related and distillery wastewaters: a review. Biorem. J. 12 (2), 70–87. https://doi.org/10.1080/10889860802060063.

Sun, G.-D., Xu, Y., Jin, J.-H., Zhong, Z.-P., Liu, Y., et al., 2012. Pilot scale ex situ bioremediation of heavily PAHs-contaminated soil by indigenous microorganisms and bioaugmentation by a PAHs-degrading and bioemulsifier-producing strain. J. Hazard. Mater. 233, 72–78. https://doi.org/10.1016/j.jhazmat.2012.06.060.

Tang, F.H.M., Lenzen, M., McBratney, A., Maggi, F., 2021. Risk of pesticide pollution at the global scale. Nat. Geosci. 14, 206–210. https://doi.org/10.1038/s41561-021-00712-5.

Tarla, D.N., Erickson, L.E., Hettiarachchi, G.M., Amadi, S.I., Galkaduwa, M., Davis, L.C., et al., 2020. Phytoremediation and bioremediation of pesticide-contaminated soil. Appl. Sci. 10, 1217. https://doi.org/10.3390/app10041217.

The United Nations World Water Development Report, 2013. United Nations World Water Assessment Programme (WWAP). United Nations Educational, Scientific and Cultural Organization, Paris.

Topp, E., Vallayes, T., Soulas, G., 1997. Pesticides: microbial degradation and effects on microorganisms. In: Van Elsas, J.D., Trevors, J.T., Wellington, E.M.H. (Eds.), Modern Soil Microbiology. Mercel Dekker, Inc., New York, USA, pp. 547–575.

Tortora, G.J., Funke, B.R., Case, C.L., 2010. Microbiology: An Introduction, tenth ed. Pearson Education, Inc., Publishing, San Francisco, CA.

Vaish, S., Pathak, B., 2020. Bio nano technological approaches for degradation and decolorization of dye by mangrove plants. In: Patra, J.K., Mishra, R.R., Thatoi, H. (Eds.), Biotechnological Utilization of Mangrove Resources. Academic Press, pp. 399–412, https://doi.org/10.1016/B978-0-12-819532-1.00019-6.

Van der Werf, H.M.G., 1996. Assessing the impact of pesticides on the environment. Agric. Ecosyst. Environ. 60 (2–3), 81–96. https://doi.org/10.1016/S0167-8809(96)01096-1.

Velázquez-Fernández, J.B., Martínez-Rizo, A.B., Ramírez-Sandoval, M., Domínguez-Ojeda, D., 2012. Biodegradation and bioremediation of organic pesticides. In: Soundararajan, R.P. (Ed.), Pesticides—Recent Trends in Pesticide Residue Assay. InTech, https://doi.org/10.5772/48631.

Vishwakarma, G.S., Bhattacharjee, G., Gohil, N., Singh, V., 2020. Current status, challenges and future of bioremediation. In: Bioremediation of Pollutants, pp. 403–415, https://doi.org/10.1016/b978-0-12-819025-8.00020-x.

Vogtmann, H., Fricke, K., 1992. Organic chemicals in compost: how relevant are they for use of it. In: Jackson, D.V., Merilott, J.M., L'Hermite, P. (Eds.), Composting and Compost Quality Assurance Criteria. Commission of the European Communities, Luxembourg, pp. 227–236.

Vos, J.G., Dybing, E., Greim, H.A., Ladefoged, O., Lambre, C., Tarazona, J.V., et al., 2000. Health effects of endocrine disrupting chemicals on wildlife, with special reference to the European situation. Crit. Rev. Toxicol. 30 (1), 71–133. https://doi.org/10.1080/10408440091159176.

Vroumsia, T., Steiman, R., Seigle-Murandi, F., Benoit-Guyod, J.L., 2005. Fungal bioconversion of 2, 4-dichlorophenoxyacetic acid (2, 4-D) and 2, 4-dichlorophenol (2, 4-DCP). Chemosphere 60 (10), 1471–1480. https://doi.org/10.1016/j.chemosphere.2004.11.102.

Wang, L., Barrington, S., Kim, J., 2007. Biodegradation of pentyl amine and aniline from petrochemical wastewater. J. Environ. Manage. 83 (2), 191–197. https://doi.org/10.1016/j.jenvman.2006.02.009.

Wardle, D.A., Parkinson, D., 1992. Influence of the herbicides, 2, 4-D and glyphosate on soil microbial biomass and activity: a field experiment. Soil Biol. Biochem. 24, 185–186.

Wirsching, J., Pagel, H., Ditterich, F., Uksa, M., Werneburg, M., Zwiener, C., Berner, D., Kandeler, E., Poll, C., 2020. Biodegradation of pesticides at the limit: kinetics and microbial substrate use at low concentrations. Front. Microbiol. 11. https://doi.org/10.3389/fmicb.2020.02107.

Wolicka, D., Suszek, A., Borkowski, A., Bielecka, A., 2009. Application of aerobic microorganisms in bioremediation in situ of soil contaminated by petroleum products. Bioresour. Technol. 100 (13), 3221–3227. https://doi.org/10.1016/j.biortech.2009.02.020.

Xia Guo, X., Tao Liu, H., Biao Wu, S., 2019. Humic substances developed during organic waste composting: formation mechanisms, structural properties, and agronomic functions. Sci. Total Environ. 662, 501–510. https://doi.org/10.1016/j.scitotenv.2019.01.137.

Yadav, I.C., Devi, N.L., 2017. Pesticides classification and its impact on human and environment. In: Kumar, A., Singhal, J.C., Techato, K., Molina, L.T., Singh, N., Kumar, P., et al. (Eds.), Environmental Science and Engineering. Studium Press LLC, USA, pp. 140–158.

Yadav, I.C., Devi, N.L., Syed, J.H., Cheng, Z., Li, J., Zhang, G., et al., 2015. Current status of persistent organic pesticides residues in air, water, and soil, and their possible effect on neighboring countries: a comprehensive review of India. Sci. Total Environ. 511, 123–137. https://doi.org/10.1016/j.scitotenv.2014.12.041.

Yang, Y., Singh, R.P., Song, D., Chen, Q., Zheng, X., Zhang, C., Zhang, M., Li, Y., 2020. Synergistic effect of *Pseudomonas putida* II-2 and *Achromobacter* sp. QC36 for the effective biodegradation of the herbicide quinclorac. Ecotoxicol. Environ. Safety 188 (188), 109826. https://doi.org/10.1016/j.ecoenv.2019.109826.

Ying, G.G., 2018. Remediation and mitigation strategies. In: Maestroni, B., Cannavan, A. (Eds.), Integrated Analytical Approaches for Pesticide Management. Academic press, pp. 207–217, https://doi.org/10.1016/B978-0-12-816155-5.00014-2.

Zouboulis, A.I., Moussas, P.A., 2011. Groundwater and soil pollution: bioremediation. In: Niriagu, J.O. (Ed.), Encyclopedia of Environmental Health, pp. 1037–1044, https://doi.org/10.1016/B978-0-444-52272-6.00035-0.

Zouboulis, A.I., Moussas, P.A., Psaltou, S.G., 2019. Groundwater and soil pollution: bioremediation. In: Niriyagu, J. (Ed.), Encyclopedia of Environmental Health. Elsevier Inc., USA, pp. 369–381, https://doi.org/10.1016/b978-0-12-409548-9.11246-1.

CHAPTER

3

Multifunctional properties of polysaccharides produced by halophilic bacteria and their new applications in biotechnology

Hitarth B. Bhatt[a], Dhritiksha M. Baria[b], Vikram H. Raval[b], and Satya P. Singh[c]

[a]*Department of Microbiology, Faculty of Science, Atmiya University, Rajkot, Gujarat, India,*
[b]*Department of Microbiology and Biotechnology, University School of Science, Gujarat University, Ahmedabad, Gujarat, India,*
[c]*UGC-CAS Department of Biosciences, Saurashtra University, Rajkot, Gujarat, India*

Abstract

Halophilic bacteria produce exopolysaccharides (EPS) as part of their defense strategy against extreme conditions of salt concentrations, pH, low nutrient availability, and sometimes high temperatures prevalent in wide-ranging saline habitats of salt lakes, marine salterns, deep-sea, saline desert, and the sea ice in polar regions. The halophilic microorganisms have developed unique metabolic processes and defense mechanisms for their survival in extreme conditions. One of such strategies includes the ability to produce novel exopolysaccharides. EPS from halophilic bacteria have focused attention due to their unique structural, physiological, and rheological properties. EPS are significant in terms of their diverse composition, size, and structure. It is, therefore, interesting to explore and identify such a wide diversity of the properties and functions of the EPS and their potential applications in food, medicine, cosmetics, textile, pharmaceuticals, agriculture, paint, petroleum, and wastewater treatment. This chapter has focused on the production, purification, and functional properties of the EPS from halophilic microorganisms. Besides, strategies to improve the production and multi-functional properties and potential applications of the EPS are also illustrated.

Keywords: Exopolysaccharide (EPS), Halophilic bacteria, Microbial adaptations, Multifunction properties, Structure-function relationship, EPS applications

1 Introduction

The saline environments raise a series of challenges to microorganisms, such as high salinity and osmotic pressure, exposure to UV radiation in intertidal ponds, low nutrient availability and water activities, low temperatures and high pressure in saline deserts, deep sea, and other saline habitats (Bhatt et al., 2018; Raval et al., 2018). These habitats harbor highly diverse microorganisms and thus represent

Microbial Syntrophy-mediated Eco-enterprising. https://doi.org/10.1016/B978-0-323-99900-7.00014-6

a rich source of high potential and useful biomolecules (Bhatt and Singh, 2020; Bhatt et al., 2017; Gan et al., 2020a; Nichols et al., 2005a; Pandey et al., 2012; Patel et al., 2006; Poli et al., 2010; Purohit and Singh, 2013).

Over the past decade, there have been increasing interests in the development of natural polymers useful in various industries and applications (Singha, 2012; Wang et al., 2019; Rana and Upadhyay, 2020). Polysaccharides are produced from plants and are generally used as food additives (Delbarre-Ladrat et al., 2014; Poli et al., 2011). However, attention has now shifted to exopolysaccharides (EPS) of microbial origin. EPS are synthesized by the polymerization of sugar residues (Bragadeeswaran et al., 2011). The EPS molecules are highly diverse with varying molecular weights (10–1000 kDa) and chain length. These are either homopolysaccharides, consisting of only one type of monosaccharide, or heteropolysaccharides, constituted by multiple monosaccharides (Nwodo et al., 2012). The EPS composition thus significantly varies with respect to molecular size, chemical composition, and physicochemical properties.

The production of EPS is one of the adaptive strategies of halophilic bacteria as it protects the cells by forming a coating around them. Compared with the EPS of the microorganisms from wastewater treatment plants, which are rarely stressed and have no food limitation, the EPS molecules of marine microorganisms are significant with higher carbohydrate content (Tansel, 2018). EPS display high heterogeneity in their structure and function (Yadav et al., 2011). The variations are mainly due to the number and types of functional groups attached to the EPS (Nichols et al., 2005a). Thus, EPS offers a wide range of physiological functions such as retention of nutrients to prevent starvation, biofilm formation, establishing symbiotic relationships, protection against desiccation, and presence of toxins or antibiotics in their surroundings (Jones, 2012; Singha, 2012; Vu et al., 2009). The diverse properties and functions of the EPS have opened a wide arena of numerous industrial applications in food, cosmetics, textile, pharmaceuticals, agriculture, paint, petroleum, and wastewater treatment (Casillo et al., 2018; Cojoc et al., 2009; Poli et al., 2018; Radchenkova et al., 2018; Sahana and Rekha, 2020; Wang et al., 2019).

In this chapter, we describe the production, purification, characterization techniques, and multifunctional properties of the EPSs produced by halophilic bacteria. Further, the structure-function relationship, promising engineering strategies for higher yield or modified molecular structure are also discussed. Further, the recent advances in the potential applications of EPSs produced by halophilic bacteria are also presented.

2 Eco-physiological role

A number of biogeochemical processes occur and are regulated in microorganisms in the microenvironment. These processes are confined in the "bacterial film", an extracellular matrix of polymeric constituents (Van Colen et al., 2014). EPSs play a crucial role in cellular physiology, including biofilm formation and flocculation, associated with the accumulation of bacterial cells, cell adherence, and water holding capacity to avoid drying (Tian, 2008). The adhesion of cells to the biotic and abiotic surfaces leads to the formation of biofilm. The process can be categorized into two stages: the first includes reversible sorption due to hydrophobicity and intermolecular forces, while in the second stage, the polymeric substance allows the cells to adhere and grow onto the surface.

The eco-physiological roles of EPS depend on the environmental niche of the microorganisms and are often responsible for the defense against both biotic and abiotic stresses (Quesada et al., 2004). In halophilic bacteria, the EPS plays a vital role in combating conditions of high salinity (Llamas et al., 2012).

Exopolymers play an essential role in bacterial adhesion, absorbing nutrients from flowing water to ensure its availability for bacteria. It acts as a protective barrier against predators and toxic chemicals, and helps in water retention and metal binding (Onbasli and Aslim, 2008; Poli et al., 2011) (Fig. 1). The biofilm matrix is significant in ecological dynamics, with the formation of a micro-environment comprising biological and chemical processes (Poli et al., 2011). EPSs play an effective role in the uptake and degradation of hydrocarbon in bacteria (Caruso et al., 2018, 2019).

The stabilization and architecture of ion-gated channels or metal cation pumps are mediated through these exopolysaccharides. These channels form small surface particles, allowing them to sink to the bottom and adhere to sediments or substratum. The strategies in halophilic bacteria to remain stable under the extremities of salt, temperature and pH make them highly attractive thus marine microbes offer new biocatalysts and other products with high value (Raval et al., 2013; Singh et al., 2013).

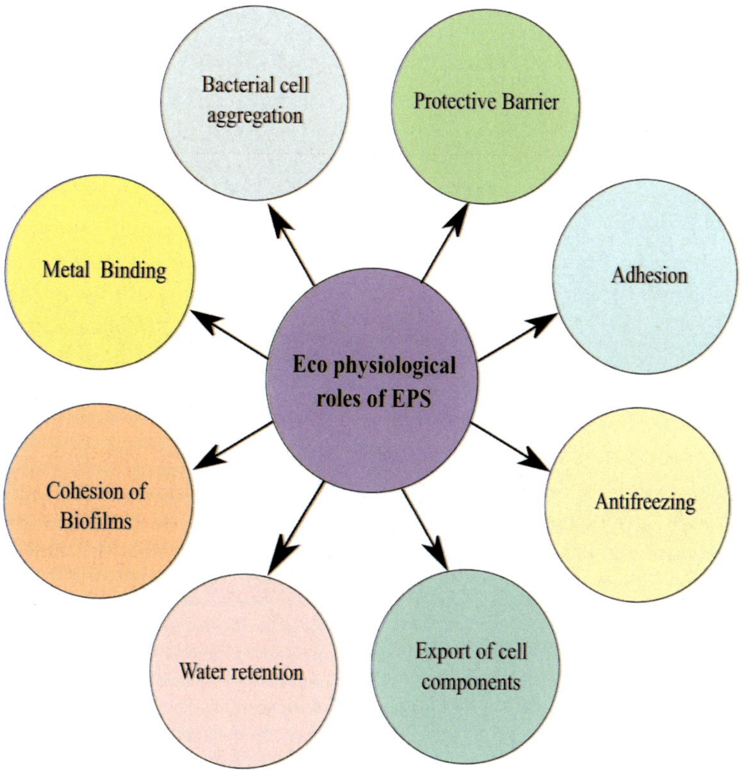

FIG. 1

Schematic illustration of eco-physiological roles of EPS.

Halophilic bacteria from saline habitats have been investigated over the last few years shows potential biomolecules (Dodia et al., 2006, 2008a, b; Joshi et al., 2008; Nowlan et al., 2006; Pandey and Singh, 2012; Purohit and Singh, 2013). Halophilic bacteria are involved in the synthesis of a large amount of protective extracellular biopolymers, with the typical occurrence of proteins and exopolysaccharides (More et al., 2014).

3 EPS producing halophilic bacteria

EPS producing halophilic bacteria have been isolated from sediments of salt lakes and saline soils from salterns. EPS with high sulfur content with the ability to form stable emulsion has been reported in *Halomonas anticariensis* strain FP35, *Halomonas maura* strain S-30, *Halomonas eurihalina* strain F2–7, *Halomonas ventosae* strain A112 (Poli et al., 2010). *Salipiger mucosus* A3T, a halophilic species isolated from the Spanish Mediterranean seaboard produced an EPS with enhanced fucose content during the exponential phase (Llamas et al., 2010).

A halotolerant bacteria *Bacillus licheniformis* PASS26 isolated from Ribandar saltern of Goa, India was investigated for EPS production (Insulkar et al., 2018). Another halophilic, non-pathogenic marine bacterium, *Cobetia marina* strain DSMZ 4741 isolated using Zobell medium synthesizes a novel K-antigen-like EPS named L_6. The potential bioactivity of the unusual structure of EPS is yet to be elucidated (Lelchat et al., 2015). In yet another instance, *Halolactibacillus miurensis* SEEN MKU3 from the soil of Salt and Marine Chemicals salt pan in Tuticorin, Southeast coast of India was reported to produce EPS using Zobell marine agar medium supplemented with 10% NaCl (Arun et al., 2017).

A novel marine bacterium, *Zunongwangia profunda* SM-A87 with a high potential of EPS production (8.90 g/L) has been described from southern Okinawa Trough deep-sea sediment (Qin et al., 2010). Another novel halophilic bacterium, *Halomonas smyrnensis*, with the ability to produce levan was isolated from saltern regions of Turkey. *Halomonas alkaliantarctica* produced a fructo-glucan based polymer with enhanced viscosity at high NaCl concentrations and low pH (Poli et al., 2010).

A moderate halophile *Virgibacillus salarius* from the bottom sediments of Wadi-El-Natron, lake in Egypt has been reported (Gomaa and Yousef, 2020). While the production of enzymes such as amylases and proteases has been reported in *Virgibacillus* genus (Cosa et al., 2011), the EPS production in these organisms is being reported for the first time (Gomaa and Yousef, 2020).

Exopolysaccharide producing halophilic luminescent bacteria are isolated from Mandovi and Zuari estuarine network connecting the coastal regions of Goa have been described. The halophilic bio-luminescent strain *Vibrio harveyi* VB23, showed potent exopolymer secretion with high viscosity emulsification capacity (Bramhachari and Dubey, 2006).

The moderate halophilic bacterium, *Chromohalobacter canadensis* 28 isolated from Pomorie salterns, Burgas Bay, Black Sea, Bulgaria has been studied for EPS production. The analysis of functional properties revealed that this is the first halophilic bacterium reported to synthesize a polymer containing a polyglutamic acid (PGA) fraction (Radchenkova et al., 2018). A haloalkaliphilic *Bacillus* sp. I-450 isolated from is reported from the heavily polluted tidal mudflats of the Korean Yellow Sea, Inchon city with an ability to synthesize exopolysaccharide during the late exponential phase.

4 Production and fermentation strategies of EPS

The rheological properties and yields of exopolysaccharides depend upon certain critical parameters correlated with fermentation strategies such as the utilization of carbon and/or nitrogen sources, temperature, pH, mineral ions, agitation, and consumption of oxygen. The cultural and nutritional conditions affect both the quality and yield of the polymer. However, the degree of branching in EPS, number of branching units and molecular mass can be physiologically controlled by various parameters (Suresh Kumar et al., 2007). The limitation of nutrients in marine bacteria such as phosphorus, sulfur, nitrogen, and phosphorus can lead to enhance in EPS synthesis (Sutherland, 1982). Similarly, the EPS production may also be enhanced under osmotic stress, varying salinity, pH, and temperatures along with carbon sources (Nicolaus et al., 2010) (Table 1).

An EPS producing halophilic bacterium was isolated from the Lake Qarun, Fayoum Province, Egypt using modified S-G medium (Sehgal and Gibbons) supplemented with 2% *w/v* sucrose. The halophilic bacterial strain, EG1HP4QL produced an EPS with a yield of 5.9 g/L at 35 °C (pH 8.0). The EPS produced from this strain had significant activity against sunflower oil ($44.7 \pm 0.5\%$), o-xylene ($64.0 \pm 1\%$), and kerosene ($65.7 \pm 0.8\%$). This strain facilitated bioremediation of oil-contaminated regions. EPS produced by this strain has shown promising application in the cleaning of saline regions polluted with oil hydrocarbons (Ibrahim et al., 2020).

Virgibacillus salarius BM02, a moderately halophilic bacterium isolated from Wadi El-Natron, Egypt was described as a high EPS producing bacterium. One factor at a time (OFAT) followed by Box-Behnken experimental design used to study the significant effect of nutrients such as nitrogen, carbon, pH on the polymer production and physio-chemical properties, for instance, total sugars/protein ratio (12.56) and intrinsic viscosity ($0.13\,\mathrm{dL\,g^{-1}}$) were evaluated. The EPS production was $3.53 \pm 0.33\,\mathrm{g\,L^{-1}}$ in sucrose and peptone as nitrogen source. It is well recognized that carbon and nitrogen sources play a critical role in EPS production. The impact, however, may vary with various bacterial strains (Malick et al., 2017). Further, the acidic and alkaline pH leads to the reduction in the EPS yield while neutral pH (6.5–7.0), promotes polymer production (Gomaa and Yousef, 2020).

Metal ions and precursors are essential for the growth and synthesis of EPS. Elements such as potassium, phosphorus, and other cations play a crucial role in the metabolism, regulation of carbohydrates, and lipid uptake. Besides, magnesium and potassium salts serve to act as a cofactor for some enzymes and are also required in many transport processes along with carbohydrate metabolism (Survase et al., 2007a, b). The addition of precursors; amino acids and nucleotide phosphate sugars, enhances EPS production. For example, the use of the sugar nucleotides UDPG and UMP, and amino acids L-lysine, involved in scleroglucan production is well documented (Survase et al., 2007b).

Response surface methodology (RSM) is a widely used technique in the optimization of fermentation parameters (Banik et al., 2007). It is an efficient method for analyzing multiple factors, since experimental trials are obligatory as compared to classical one-factor analysis methods. Furthermore, the collaboration of multiple factors allows to determine specific points leading to EPS hyperproduction. Until now RSM has been reported significant approach for media composition optimization using different microbial species (Imran et al., 2016; Ragavan and Das, 2019; Suryawanshi et al., 2019; Wang et al., 2017a, b).

Table 1 Overview of fermentation parameters for EPS production by halophilic bacteria.

Microorganisms	Carbon Sources	Optimum NaCl concentration (%)	pH	Medium	Temperature	EPS (g/L)	References
Halomonas eurihalina	Glucose	7.5	7.2	MY medium	32	2.8	Hussain et al. (2020)
Halomonas maura S-30	Glucose	2.5	7.0	MY medium	32	3.8	Arias et al. (2003)
Halomonas ventosae Al-16	Glucose	7.5	7.2	MY medium	32	0.30	Mata et al. (2006)
Halomonas anticariensis	Glucose	7.5	7.2	MY medium	32	0.4	Mata et al. (2006)
Halomonas almeriensis	Glucose	7.5	7.0	MY medium	32	1.7	Llamas et al. (2012) and Martínez-Checa et al. (2005)
Halomonas stenophilia	Glucose	5	7.2	MY medium	32	3.89	Amjres et al. (2011)
Halomonas smyrnensis AAD6	Sucrose	5	7.0	CD medium	25	1.8	Poli et al. (2009) and Sarilmiser et al. (2015)
Halomonas Xianhensis SUR308	Glucose	2.5	7.0	MY medium	32	7.87	Biswas and Paul (2017)
Virgibacillus salarius	Sucrose	8	7.0	HS medium	40	5.87	Gomaa and Yousef (2020)
Halobacillus strain EG1HP4QL	Sucrose	5	8.0	SG medium	40	5.9	Ibrahim et al. (2020)
Cobetia marina DSMZ4741	Glucose	3	7.6	ZM medium	20	0.4	Lelchat et al. (2015)
Chromohalobacter canadensis 28	Lactose	15	8.5	Basal medium	30	0.31	Radchenkova et al. (2018)
Halolactibaci-llus miurensis SEEN MKU3	Glucose	7.5	8	CD medium	32	2.5	Arun et al. (2017)
Vibrio alginolyticus CNCM I-4994	Glucose	3	7.2	CD medium	25	Not mentioned	Drouillard et al. (2015)
Salipiger mucosus A3^{T}	Glucose	2.5	7.0	MY medium	32	1.35	Llamas et al. (2010)

MY medium, malt extract-yeast extract medium; SG medium, Sehgal and Gibbons medium; CD medium, chemically defined medium; HS medium, Hestrin-Schramm agar.

5 Extraction and purification of EPS

The exopolymer is extracted from the production medium by separating the cell biomass using the appropriate centrifugation technique. The EPS in the cell-free medium is then precipitated using solvents like methanol, acetone, or ethanol (Grivaud-Le Du et al., 2017). In a subsequent step, pellets are dialyzed against ultrapure water and lyophilized to obtain the partially fractionated EPS. Since exopolymers remain embedded in the biofilm matrix, its extraction and further recovery is a challenging process. The contamination of intracellular molecules and debris is often attached and extracted along with exopolysaccharide fractions after cell disruption (Di Donato et al., 2016). It is essential to first dissociate capsular EPS from the cells. The method for dissociation depends on the nature of the association between the polysaccharide and cells (Fig. 2). Generally, heat treatment at 60 °C in saline solution is used for the detachment of capsular EPS from the cells (Mende et al., 2013; Notararigo et al., 2013) or the food matrix, e.g., kefir grains (Enikeev, 2012). The alternate methods include boiling of cell suspension in water for 15 min, followed by sonication or prolonged agitation in 0.05 M EDTA or 0.5% phenol. In certain cases, autoclaving is employed but it may lead to undesirable broth viscosity (Wang et al., 2014).

EPS synthesis was investigated in a moderately halophilic bacterium *Gracilibacillus* sp. SCU50 by centrifugation of the broth at 30 °C and 200 rpm for 72 h. For the further extraction of EPS centrifugation, the process was carried out at 12,000 × *g* and 4 °C for 20 min to separate cells from the culture. Subsequently, in the supernatant, three volumes of chilled absolute ethanol were added to obtain the carbohydrates. The mixture was incubated overnight at 4 °C. The precipitates again dissolved in deionized water and treated with Sevag reagent (chloroform: *n*-butanol = 4:1, v/v) and 4% (w/v) trichloroacetic acid to eliminate the free proteins. The precipitates were dialyzed (M_w cut-off: 8000–14,000 Da) against ultrapure water for 48 h. The EPS thus obtained was further purified by anion exchange chromatography using Q-Sepharose Fast Flow column (Gan et al., 2020a). The column eluate is collected and analyzed for total carbohydrate content by the phenol sulfuric acid method (Dubois et al., 1956). The fractions were then pooled, dialyzed, and lyophilized to obtain a dry product. Sephacryl S-400 HR in gel-filtration chromatography is used for purification (Gan et al., 2020a).

The purification steps also include heating at high temperature (90–95 °C) to denature enzymes that hydrolyze the polysaccharide. Generally, lyophilization of the purified EPS is preferred for long-term storage and further structural characterization and biochemical analysis (Di Donato et al., 2018).

6 Quantification and characterization of the EPSs

The primary characterization of a polysaccharide includes the monosaccharide composition, the anomeric configurations, the ring size, sequence of the monosaccharides, the presence and the positions of the appended groups, and the glycosylation sites. Further, to evaluate the potential applications of exopolysaccharides, it is essential to study their composition and structure. Analytical methods to characterize polysaccharides include chemical and spectroscopic techniques. Glycosyl analysis establishes the type and the relative amount of the monosaccharides present in the polymer. Therefore, the polysaccharide is subjected to acid hydrolysis by sulfuric acid, hydrochloric acid or trichloroacetic acid. The acid hydrolysis cleaves glycosidic linkages of the polymers to release monosaccharides units (Rana and Upadhyay, 2020). The released monomers are reduced to form sugar derivatives which are separated by

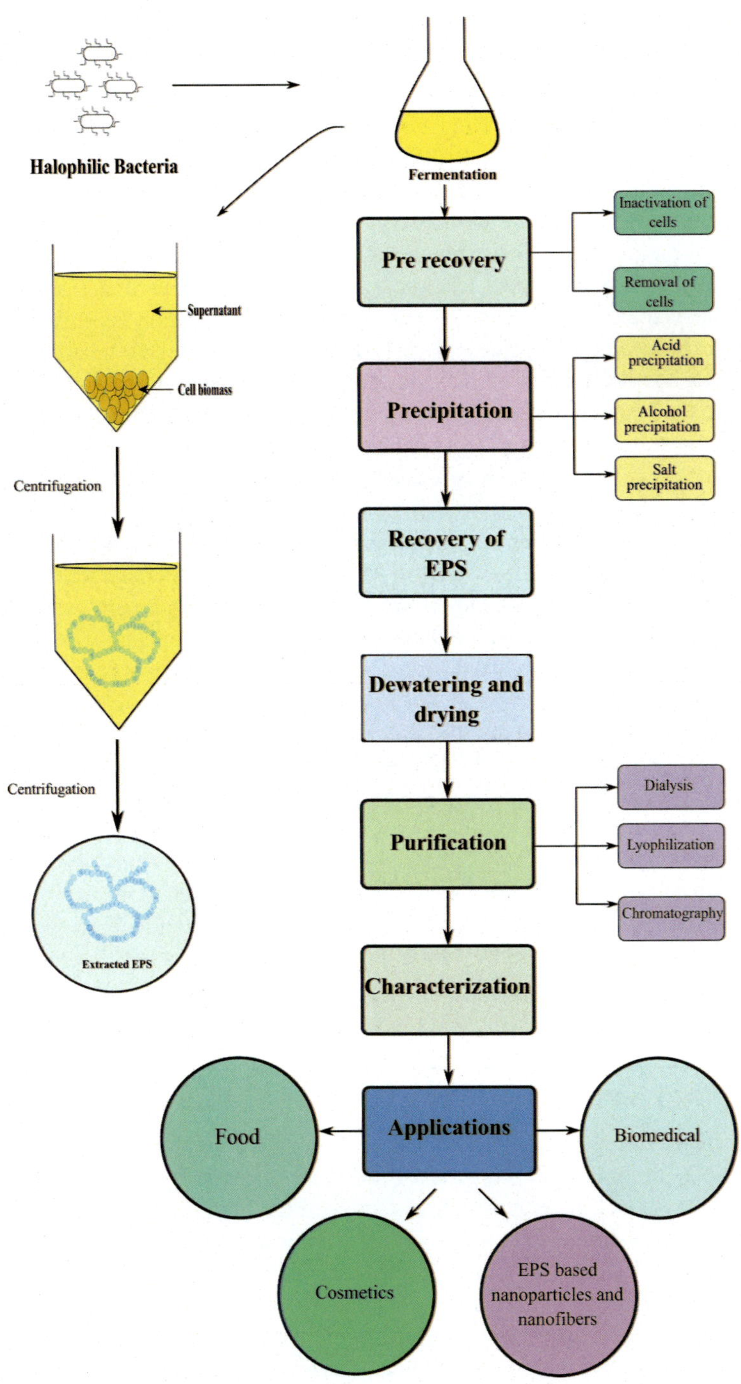

FIG. 2

Schematic flow diagram for the extraction, purification, and assessment of the bioactivities of EPS.

gas chromatography coupled to mass spectrometry (GC-MS) and identified by their relative retention times compared with the standards sugars (Casillo et al., 2018; Schmid et al., 2015; Zhang et al., 2015).

For the determination of absolute D or L configurations, the solvolysis is performed with an optical chiral pure alcohol, such as 2-(+)-butanol or 2-(+)-octanol. The diastereoisomeric butyl or octyl-glycosides obtained can be acetylated and analyzed by GC–MS and the configuration is deduced by comparison with authentic standards (Casillo et al., 2018). High performance anion exchange chromatography (HPAEC) with pulsed amperometric detection (PAD) are alternative methods to gas chromatography-mass spectrometry (GC–MS) to study the linkage and composition of exopolysaccharide. The main advantage of this technique over GC–MS is that it does not require sugar derivatization steps (Whitney and Howell, 2013).

Fourier transform infrared (FTIR) is used to detect types of functional groups and characterization of bonds in exopolysaccharides. FT-IR reveals the types of sugar rings, the presence of amino sugars, uronic acids, and sulfate groups. The technique focuses on bonds which vibrate at characteristic frequencies (Finore et al., 2014). FTIR spectrum provides molecular fingerprinting information for organic, polymeric substances, and inorganic materials. The FTIR data of unknown material is compared with libraries of known standards (Finore et al., 2014).

NMR is another useful tool for the structural determination of polysaccharides. NMR Spectroscopy is used to analyze glycosidic linkages, ring configuration, and anomeric configuration of the monomeric units. While 2D NMR such as Nuclear Overhauser effect spectroscopy (NOESY), Correlation Spectroscopy (COSY), and Total Correlation Spectroscopy (TOCSY) are major techniques used for analyzing the chemical composition of exopolysaccharides (Poli et al., 2018; Rana and Upadhyay, 2020). The role of NMR in carbohydrate structural elucidation has been illustrated in different literature (Agrawal, 1992; Duus et al., 2000; Larsen and Engelsen, 2015). Size Exclusion Chromatography coupled with multi-laser light scattering (SEC-MALLS) is an advanced technique to determine the molecular weight of exopolysaccharides (Willis et al., 2013). Moreover, other techniques, such as X-ray diffraction (XRD), differential scanning calorimetry (DSC), and thermal gravimetric analysis (TGA) are also used to study the exopolysaccharide structure (Hubbard et al., 2012; Zhang et al., 2015).

7 Chemical composition of EPS

Based on the monomeric composition, exopolysaccharides are categorized mainly in two types, i.e., homopolysaccharides such as pullulan, curdlan, dextran, scleroglucan, and cellulose; and heteropolysaccharides, viz. hyaluronic acid and xanthan (Patel et al., 2010). These are also associated with various other organic components like amino acids, organic acids, and inorganic components like phosphates and sulfates (Poli et al., 2011). The majority of the homopolysaccharides are neutral while most of the heteropolysaccharides are polyanionic (Czaczyk and Wojciechowska, 2003). Homopolysaccharides are grouped into different categories, such as fructans, α-D-glucans, β-D-glucans, etc., although, heteropolysaccharides comprised repeating units of L-rhamnose, D-galactose, D-glucose, etc. (Patel and Prajapat, 2013). Linear homopolysaccharides (cellulose and curdlan) have 1,4-β-glycosidic and 1,3-β-glucan linkage, respectively. The usual components of exopolysaccharides are hexoses, pentoses, uronic acids, and amino sugars (Parikh and Madamwar, 2006) (Table 2).

The high-molecular-weight fraction of the exopolysaccharide has the prevalence of carbohydrates (~98%), while the protein content did not exceed more than 2%. The EPS consists of two polymers;

Table 2 Properties of exopolysaccharides obtained from halophilic bacteria.

Microorganisms	Monosaccharide composition	Molecular weight	Functional properties	Properties and activities	References
Alteromonas hispanica F32	Glc:Man:Rha:Xyl 18:63:7:12 sulfate and phosphate group	1.9×10^7 Da	Low toxicity, higher biodegradability, high selectivity	Emulsifying activity	Mata et al. (2008)
Alteromonas macleodii	Gal:Glc:Rha:GlcA:GalA:Man:Fuc 5.9:2.6:2.5:2.0:1.9:1.4:1.0	1.10×10^6 Da	–	Not tested	Le Costaouëc et al. (2012)
Hahella chejuensis strain 96 CJ10356	Glc:Gal 1:6.8 Traces of ribose and xylose	2.2×10^3 Da	Low toxicity, high specific activity	Emulsifying activity	Yim et al. (2004) and Casillo et al. (2018)
Salipiger mucosus A3^T	Glc:Man:Gal:Fuc 1.5:2.5:2.5:1	2.5×10^5 Da	High jellifying activity at acidic pH pseudoplasticity	Metal binding capacity, emulsifying activity	Llamas et al. (2010)
Bacillus licheniformis strain I-450	Glc:Gal:Fr	2.2×10^6 Da	Highly soluble in water	Gel forming, shear thinning property	Kumar et al. (2004)
Bacillus licheniformis strain B3–15	Man:Glc	6.0×10^5 Da	Highly stable, Degradation temperature 220 °C	Antiviral activity, immunostimulant	Maugeri et al. (2002)
Bacillus licheniformis strain PASS26	Fru:Glc:Gal:Man:Gal ac	5.6×10^4 Da	Highly stable, Degradation temperature 350 °C	Wound healing efficiency, antitumor activity, emulsifying activity	Insulkar et al. (2018)
Chromohalobac-ter canadensis 28	GlcN:Glc:Rha:Xyl:unknown sugar 36.7:32.3:25.4:1.7:3.9	Above 1000 kDa	High swelling behavior	Foaming capacity, antioxidant activity, high stabilizing, and emulsifying activity	Radchenkova et al. (2018)
Kocuria rosea ZJUQH	Glucose	5.659×10^3 Da	–	Not tested	Gu et al. (2017)
Vibrio alginolyticus CNCM I-4994	Galacturonic acid/*N*-acetylglucosamine (3/1, by relative ratio)	1.16×10^6 Da	–	Not tested	Drouillard et al. (2015)

Aphanothece halophytica GR02	EPS1: Ara:Rha:Fuc:Man:Glc:Gal (Trace:0.06:0.05:0.08:1:0.75) and GlcA 3.58% of polysaccharide dry weight EPS 2 Ara:Fuc:Man:Glc (1:2.08:1.57:2.87) and GlcA 15.78% of polysaccharide dry weight	EPS 2-above 2000 kDa	–	Metal binding capacity, gelling property	Li et al. (2001)
Halomonas smyrnensis AAD6^T	Fructose	Above 1000 kDa	Degradation temperature 253 °C	Biocompatibility Bioflocculating activity	Sarilmiser and Oner (2014), Sam et al. (2011), and Poli et al. (2009)
Halomonas maura S-30	Man:Gal:Glc:GlcA 34.8:14:29.3:21.9	4.7×10^6 Da	High viscosity Pseudoplasticity	Biosorption of heavy metal	Arias et al. (2003)
Halomonas almeriensis	EPS 1: Man:Glc:Rha (72:27.5:0.5) EPS 2: Man:Glc (70:30)	6.3×10^6 Da	Low viscosity High-pseudoplasticity	Bio-detoxifier	Llamas et al. (2012)

a neural EPS fraction comprised mannose and glucose (~3:1) and another negatively charged EPS fraction consisting of mannose. Both the fractions also contain traces of galactose, galactosamine, and glucosamine (Ibrahim et al., 2020). Only limited reports are available on the structural characterization of the EPSs from *Halobacillus* and *Bacillus* genera. These microorganisms are characterized by the production of neutral EPSs of high molecular weights, mainly composed of glucose (Glc), galactose (Gal), mannose (Man), and fructose (Fru). The EPS produced by haloalkaliphilic bacterium *Bacillus* sp. I-450 has high water-holding and flocculating activities. The EPS derived from this bacterium predominately consist of Fru, Glc, and Gal (Kumar et al., 2004). *Bacillus licheniformis* B3–15 produced EPS of mannose when grown on kerosene as the sole carbon source (Maugeri et al., 2002). *Halobacillus* sp. EG1HP4QL produces glucomannan with high emulsifying activity and even a slight change in the EPS structure leads to changes in physiochemical properties (Ibrahim et al., 2020). The structural analysis involved the determination of the glycosidic bonds sequence and polysaccharide conformation or distribution of monomeric components using different spectroscopic and chromatographic methods (Di Donato et al., 2018).

A moderate halophilic bacterium *Alteromonas hispanica* F32 isolated from hypersaline habitat in Spain produced EPS in MY medium supplemented with 7.5% (w/v) of sea salts. The molecular weight was determined as 1.9×10^7 Da, a value for this kind of microorganisms. The EPS comprised Man, Glc, Xyl, and Rha, in addition to phosphate groups and sulfate groups (Mata et al., 2008). Further, *Halomonas* sp. AAD6 (JCM 15723) releases EPS, levan in the growth medium. The polymer levan consisted of residue β-(2→6)-fructo furanosyl with antioxidant and anticancer potential (Abdel-Fattah et al., 2012). Since the production of levan is rather expensive, the large-scale production of levan at a large scale using low-costs substrates appears highly significant (Küçükaşik et al., 2011).

8 Structure-function relationship

The microbial communities play a significant role in biogeochemical cycles in marine habitats. Since the exopolysaccharides significantly contribute to biogeochemical cycling in the marine environment, a deep insight into its ecological function is required. Chemical characterization of these polymers through advanced techniques will aid to find structure-function relationships and eventually their applications. Most of the exopolysaccharides showing emulsifying properties include deoxyhexoses, uronic acids, and fatty acids in their composition. For a good emulsifying agent, the inclusion of a protein moiety is very important for its commercial viability (BeMiller, 2007; Gutierrez et al., 2007). The presence of proteins has been observed for many exopolysaccharides with good emulsifying activity (Satpute et al., 2010).

EPS produced by marine bacteria generally contains higher levels of uronic acids, notably D-glucuronic and D-galacturonic acid, and the presence of acidic substituents, such as phosphates, sulfates, and pyruvates (Kennedy and Sutherland, 1987). This makes EPS highly polyanionic (negatively charged), which may be attributed to different anionic groups (e.g., COO^-, SO_4^-) (Kennedy and Sutherland, 1987). Cations such as Na^+ and Ca^{2+} abundant in marine ecosystems are attracted with the anionic groups of the EPS allowing the formation of cross-linking among EPS chains and therefore microscopic gels (Chin et al., 1998). Such gels are refractory to degradation by bacteria and may constitute a network of nutrient "hot spots" (Azam and Malfatti, 2007).

The polyanionic nature of EPS from halophilic bacteria serves important ecological functions in marine systems. A key role of polyanionic EPS is in its potential role in controlling essential trace metal soluble iron (Fe^{3+}) bioavailability. Many reports have suggested that single anionic residues, such as glucuronic and galacturonic acids (Hassler and Schoemann, 2009; Hassler et al., 2011a), and marine bacterial EPS containing high levels of uronic acids (Gutierrez et al., 2008; Hassler et al., 2011a), can effectively bind Fe^{3+}. Therefore, EPS with a high content of uronic acids and/or negatively charged groups contribute to the biogeochemical iron cycle in marine environments. Further, there are some reports on the marine bacterial EPS binding with toxic metal ions such as Cu, Pb, Ni, Cd, Cr, Al, and Ur (Bhaskar and Bhosle, 2006; Gutierrez et al., 2008; Iyer et al., 2005).

A large fraction of the EPS produced by bacteria in the ocean is of glycoprotein composition (Suja et al., 2019). The amino acid and peptide of these glycoprotein biopolymers confer amphiphilic features to these macromolecules (Gutierrez et al., 2009). This certainly explains their ability to interact with hydrophobic compounds, such as oil hydrocarbons.

Sulfates are not commonly found in microbial EPSs, although they are reported to be present in EPSs produced by halophilic bacteria (Llamas et al., 2010; Mata et al., 2006, 2008). Sulfated EPSs are of great potential in medicine since they have a number of bioactive properties: antiproliferative, antiviral, anticoagulant, antiangiogenic, etc. (Arena et al., 2009; Matou et al., 2005; Yim et al., 2005). Fucose and fucose-rich oligosaccharides can be used in biocosmetics, the food industry, and in medicine (Vanhooren and Vandamme, 1999). Further, the presence of acetyl groups in EPS confers hydrophobicity, contributing to its emulsifying property (Ashtaputre and Shah, 1995). *Halomonas* produces a sulfated exopolysaccharide, mauran, having potential biotechnological applications (Arias et al., 2003). In addition, this polymer showed a thixotropic and pseudo-plastic behavior, with unchanged viscosity in the presence of high salt concentrations or extreme pH. The chelating properties of mauran also offer its usefulness in bio-remediation (Arias et al., 2003).

9 Improved EPS with the modified properties: Engineering strategies

Traditional strategies for the improvement in the bacterial EPS include strain selection and optimization of the cultivation conditions. However, such strategies can bring the desired changes only marginally due to the physiological constraints of the microorganisms. Therefore, among the newer approaches, metabolic engineering either by manipulation of the concerned genes or by altering the pathways affecting gene expression and enzyme activity appears significant (Yang, 2011).

With the complete view of the EPS producing bacterial genome, it is possible to select appropriate strategies to improve the properties and release of the polymer. The modification of the expression of a single gene or groups of genes can be used to increase the EPS yield (Bajaj et al., 2007). *Halomonas smyrnensis* AAD6T synthesizes levan as EPS involving levansucrase enzyme. The levansucrase could be overexpressed by boric acid as a stimulator through the quorum sensing-based signaling (Sarilmiser et al., 2015). Further, the gene encoding phosphocarrier protein of the phosphoenolpyruvate sugar phosphotransferase system (PTS) for fructose uptake was knocked out in *Halomonas smyrnensis* AAD6T. The mutant strain displayed nearly threefold higher efficiency of the levan production compared with the wild-type strain (Aydin et al., 2018). Efforts on the genome reconstruction of the halophilic *Halomonas smyrniensis* strain AAD6 predicted the stimulatory effect of mannitol on levan biosynthesis, which was later experimentally confirmed (Ates et al., 2011, 2013).

10 Multifunctional properties of EPS

10.1 Cryoprotection

The temperature for seawater at the salinity of 35 g/L approaches − 1.8 °C. During the ice formation, salts are excluded from the ice and concentrate in brine pockets. In the sea ice brine, salinity may vary from nearly zero to > 150 g/L at the ice–seawater interface (Staley and Gosink, 1999). Microorganisms in this environment have to survive fluctuations in temperature, salinity, nutrient concentrations, and light. Microbial communities in the poles are dominated by diatoms and bacteria mainly inhabit 10–20 cm layer of sea ice, where nutrients and light are still available (Nichols et al., 2005a; Staley and Gosink, 1999). The secretion of EPS molecules, which modify the surrounding conditions of the cells, provides a protective environment and thus allows the survival of the cells.

The cryoprotective activities of EPSs from halophilic bacteria are reported. A mucoid *Pseudomonas* sp. ID1 produces EPS composed mainly of glucose, galactose, and fucose (Carrión et al., 2015). The impact of EPS on wild-type strain was compared with a mutant variant lacking EPS production. The wild-type strain displayed higher viability after freezing at − 20 and − 80 °C as compared to the mutant strain that establishes the cryoprotective action of the EPS.

EPS not only provides cryoprotection to the native strain, but it also protects non-EPS producing organisms, such as *Escherichia coli*. A new EPS from *Pseudoalteromonas* sp. SM20310 was reported with a different structure compared to other known EPS of marine bacterial origin (Liu et al., 2013). The EPS enhanced cell survival after freeze–thaw cycles and could also considerably improve high-salinity tolerance of strain. Supplementation of EPS into *E. coli* cultures allowed an 18-fold increase in the cell survival after three freeze–thaw cycles (Liu et al., 2013). Thus, the extracellular release of the polymer also protects microorganisms which do not produce them. Besides, the carbohydrates generated from the EPS may serve as a source of organic compounds and nutrients (Riedel et al., 2006), affecting the biogeochemical cycling of the nutrients (Vancoppenolle et al., 2013).

10.2 Facilitates biochemical interactions

The exopolymeric substance forms microzone surrounding the cells and thus connecting bacteria and surrounding cells and tissues (Decho, 1990; Logan and Hunt, 1987). Exopolymer creates a hydrated layer, which provides a buffer zone against sudden changes in the environmental conditions (Dudman, 1977; Nichols et al., 2005a; Wang et al., 2019). This stable environment may help localization of secreted exoenzymes essential in the cycling of both organic and inorganic material in the marine environment (Decho, 1990; Liu et al., 2020). The hydrated exopolymer matrix maintains the exoenzyme activity in the proximity of the cell, thereby facilitating the cellular uptake of small molecules for metabolic conversion to energy and biomass (Decho and Herndl, 1995). Such environments may also promote symbiotic relationships between bacteria and other organisms. The microzones around cells also facilitate the transfer of nutrients from one organism to another (Paerl, 1976). Members of the several halophilic genera produce polysaccharide-containing exopolymers that potentially benefit the survival of other marine organisms by facilitating attachment to surfaces (Casillo et al., 2018; Holmström and Kjelleberg, 1999; Satpute et al., 2010).

10.3 Temperature stability

Temperature of the outer environment greatly influences physiochemical properties of EPS. The thermal stability of the EPS adds to its versatility for commercial applications. The majority of the microbial exopolysaccharides are unstable at 200 °C and beyond (Cojoc et al., 2009). However, EPS from *Paenibacillus* is reported to be stable at as high as 200 °C (Pooja and Chandra, 2009). Moreover, the EPSs from halophilic bacteria, *Rhodobacter johrii* CDR-SL 7 Cii and *Sphingobium yanoikuyae* BBL01 are reported stable even beyond 200 °C (Bhatia et al., 2021; Sran et al., 2019).

10.4 Emulsifying property

Emulsifiers of microbial origin have attracted attention due to several advantages they offer over synthetic products. Such properties include higher biodegradability, lower toxicity, better environmental compatibility, high selectivity, and specific activity at extreme temperatures, pH, and salinity (Banat et al., 2000). Besides, these molecules can be produced in microorganisms using cheap renewable sources. The presence of both hydrophilic and hydrophobic functional groups may be responsible for its emulsifying property (Kavita et al., 2013). Several exopolymeric substances with emulsifying activity have been described from halophilic/haloalkaliphilic bacteria (Boujida et al., 2018; Gutierrez et al., 2007, 2008; Sran et al., 2019). Emulsification and surface-active properties of EPS make it suitable in bioremediation and other oil recovery and cosmetics-related applications (Gomes et al., 2018; Ibrahim et al., 2020; Radchenkova et al., 2018). The EPS emulsions from three moderately halophilic species, *Idiomarina fontislapidosi* F32^T, *Idiomarina ramblicola* R22^T, and *Alteromonas hispanica* F23^T were reported highly stable with small and uniform droplets. The suitability of such candidates has been proved as emulsifying agents in the food and oil industries (Mata et al., 2008).

10.5 Biofilm formation

Extracellular polysaccharides synthesized by the extremophilic microorganisms play important roles in adaptation, creating an environment that allows cell adhesion, retention of water, and concentration of nutrients (Rinaudi and Giordano, 2010). Biofilm formation is a complex and dynamic process, in which organized microbial communities are assembled in a matrix (Donlan, 2002; Renner and Weibel, 2011). It involves the initial attachment of cells to a solid support by EPS assistance followed by biofilm development (Norwood and Gilmour, 2001).

Halophilic/halotolerant bacteria forming biofilms are usually resistant to damage caused by high salt concentrations (Gagliano et al., 2017). A halophilic strain *Halomonas stenophila* HK30 from a saline wetland in Brikcha (Morocco) able to form biofilm in a medium with 5% (w/v) salt has been described (Amjres et al., 2015). Two halophiles; *Kocuria flava* AB402 and *Bacillus vietnamensis* AB403, effectively formed biofilms and produced a large amount of EPS under salt stress. These strains are also able to develop resistance and adsorb high concentrations of metal ions assisted by EPS (Mallick et al., 2018).

10.6 Salt tolerance

Studies have revealed that microbial polymers are involved in the tolerance to salt stress, not only for the producing microorganisms but also for the associated plants. The production of polymer by the

NaCl-tolerant isolates can decrease Na uptake by plants by trapping and decreasing the amount of ions available (Upadhyay et al., 2011). Therefore, the polymer prevents nutrient imbalance and osmotic stress, and thus assisting the survival of the microorganisms beneficial to the plant. *S. meliloti* strain EFBI severely reduces EPS production when grown with low salt concentrations. Since this strain was isolated from the nodules of a=plant growing in a salt marsh with a salinity level of 0.3 M, a lower amount of salt can be considered a stressful condition. However, the relevance of the EPS in survival and symbiosis is yet to be examined and investigated in detail (Lloret et al., 1998).

10.7 Drought protection

EPS production can confer advantages to microorganisms in drought stress. A high water-holding capacity was reported for an EPS produced by a *Pseudomonas* strain. The EPS added to a sandy soil enhances water holding capacity when compared to control (Roberson and Firestone, 1992; Sandhya et al., 2009). EPS protected the bacteria against desiccation, thereby providing the bacteria time for metabolic adjustments. The polymer acts as a reservoir of water and nutrients for bacterial survival (Costa et al., 2018). Cyanobacteria, *Nostoc calcicola* (Bhatnagar et al., 2014) and *Phormidium* 94a (Vicente-García et al., 2004) isolated from arid regions are capable to produce EPS, as a strategy for water/nutrient retention and survival.

10.8 Protection against antimicrobials

The matrix that surrounds microorganisms in biofilms decreases their susceptibility to antimicrobials (Costa et al., 2018). In general, the biofilm of halophilic bacteria provides a negative charge and therefore binds with the positively charged compounds, protecting the innermost cells. In addition, electrostatic repulsion can further prevent the entry of negatively charged antimicrobials through the biofilm (Everett and Rumbaugh, 2015). Many studies have established the inhibitory potential of bacterial EPS against antimicrobial compounds, particularly for clinically important bacterial strains. For instance, the slime produced by *Staphylococcus* sp. acts as an effective antagonist to antimicrobials vancomycin, perfloxacin, and teicoplanin by preventing diffusion of the compounds or even by interfering with their mode of action in the cell membrane (Farber et al., 1990; Souli and Giamarellou, 1998). EPS can also protect microorganisms against disinfection agents. Alginate produced by *P. aeruginosa* enhances bacterial survival in chlorinated water, and removal of the slime eliminates bacterial chlorine resistance (Grobe et al., 2001).

10.9 Nutrient trap

In addition to act as a carbon source, EPS can accumulate other nutrients and molecules. Further, EPS can also hold extracellular enzymes in the matrix forming an extracellular digestion system that captures compounds for their further use as nutrient and energy sources (Flemming and Wingender, 2010). The adsorption of metal ions by EPS for heavy-metal remediation and recovery of polluted environments has also been reported. In soils, microbial EPS can sorb, bind or entrap many soluble and insoluble metal species, as well as clay minerals, colloids, and oxides, which also have metal-binding properties (Gadd, 2009). In addition, EPS can establish networks with other EPS (Etemadi et al., 2003). The main factors influencing metal biosorption by EPS are related to the binding sites or their chemical nature, such as pH, metal content, ionic strength, surface properties, molecular weight, and branching (Costa et al., 2018; Fukushi, 2012; Guibaud et al., 2003).

10.10 Applications of EPS from halophiles

10.10.1 Food application

EPS producing microorganisms are widely used in dairy industries in providing stability to dairy products and displaying water binding ability to enhance the texture and taste of the products (Eddouaouda et al., 2012). Dextran is used in bakery products to improve moisture retention, viscosity and to prevent sugar crystallization (Rana and Upadhyay, 2020). Further, EPS stabilizing emulsions between water and hydrophobic compounds have potential applications as natural emulsifiers in the food industry (Freitas et al., 2009).

In industrial processes, emulsifiers are usually exposed to extremes of temperatures, pH, and salinity (Freitas et al., 2009). Bioemulsifiers from extremophiles have high emulsion stability over a wide range of temperature, pH, and salinity (Zheng et al., 2012). High pseudoplasticity is an attractive rheological characteristic in diverse food formulations, such as sauce, dairy, cake, salad dressing, syrup, and pudding (BahramParvar and Razavi, 2012; Han et al., 2014). The pseudoplastic property of EPS generates good sensory properties such as mouth feel and flavor release. It is also useful in food processes; mixing, pouring, and pumping at different operative shear rates (Han et al., 2014). Viscosity and pseudoplasticity of mauran (a halophilic EPS) solution were not affected by salts, sugar, surfactants, lactic acid, changes in pH, or freezing and thawing (Arias et al., 2003).

10.10.2 Cosmetics

Hyanify, a marine-derived EPS, boosts the synthesis of hyaluronic acid (HA) in the skin to reduce nasolabial folds. Hyaluronic acid content decreases with aging due to reduced synthesis and enhanced degradation, resulting in dehydration and volume loss in the skin (D'arrigo et al., 2012). Further, a newly synthesized exopolymer from a halophilic bacterium, *Chromohalobacter canadensis* 28, composed of carbohydrate (EPS) and a poly-gamma-glutamic acid (γ-PGA) has high hydrophilic and lipophilic properties (Radchenkova et al., 2018). These characteristics determined the formation of an effective stabilizing layer on the interfacial surface oil/water essential for its use in a dispersion system as an emulsifier, thickener, stabilizer, and water-binding agent. Due to its long-lasting hydrating effect, it can be used in skin cream preparation. Further, the presence of glucosamine in the unique EPS composition could contribute to reducing areas with increased pigmentation of the skin, making wrinkles and fine lines less visible (Radchenkova et al., 2018).

10.10.3 EPS-based nanoparticles

Nanotechnology is very important in the modern pharmaceutical field for a variety of applications (Sana et al., 2021). Inappropriate biocompatibility and biodegradability of the synthetic nanoparticles usually damage renal excretion and cause many side effects. Therefore, there is a continuous search for novel products with improved pharmaceutical functions. Halophilic EPSs have received increasing attention owing to their unique composition and properties (Raveendran et al., 2013).

Halophilic EPSs can be applied to nanoparticle technology in two ways: (1) to utilize EPS to directly form nanoparticles and (2) EPS encapsulation of the nanoparticles made from another material. Halophilic EPSs usually have high negatively charged groups to act as polyelectrolytes, which allow them to bind with positively charged biomolecules to form biodegradable nanoparticles. As a positively charged biopolymer, chitosan is often used for polyelectrolyte nanoparticle formation with negatively charged EPSs (Deepak et al., 2015; Karlapudi et al., 2016). Halophilic EPS mauran-chitosan hybrid nanoparticles, produced through an ionic-gelation method, manifested stable drug release and biocompatibility when used for antitumor drug encapsulation (Raveendran et al., 2013). Another halophilic EPS levan, produced by *Halomonas smyrnensis* AAD6T, was used for nanoparticle formation, and its suitability was validated as a nanocarrier for delivery of peptides and proteins (Sezer et al., 2011).

10.10.4 EPS-based materials using electrospinning

Electrospinning is a versatile and relatively cost-effective technique to fabricate a large variety of soluble or fusible synthetic and natural polymers into continuous fibers of the diameters in the submicron to nanometer range (Salem, 2007; Wang et al., 2019). The electrospinning method enables the production of novel biomaterials using naturally occurring biopolymers with complex molecular structures (Torres-Giner et al., 2008). The halophilic EPS mauran was blended with polyvinyl alcohol (PVA), and electrospun to generate a scaffold with continuous, uniform nanofibers (Raveendran et al., 2013). The mauran-based nanofiber was able to boost cellular adhesion, migration, proliferation, and differentiation of mammalian cells in vitro. The polyanionic nature of extremophilic EPSs increases the negative charge accumulation on the surface of the scaffold, which is helpful for protein adsorption and the ability to enhance cellular attachment (Raveendran et al., 2013).

10.10.5 Antitumor and immunoregulatory effect

The current state of the knowledge demonstrates that EPSs from extremophiles have a broad spectrum of biological activities, such as anti-cancer, anti-oxidant, and immunoregulatory properties, suggesting their promising biomedical applications (Abdelhamid et al., 2020). An oversulfated EPS produced by a halophilic bacterium *Halomonas stenophila* B100 could specifically induce apoptosis of leukemia cells from peripheral blood. The addition of sulfate moieties to the native EPS enhanced its antiproliferative efficacy (Ruiz-Ruiz et al., 2011). A halophilic EPS levan was modified through periodate oxidation to harbor aldehyde groups. The aldehyde-activated levan derivatives showed anti-cancer activity against several human tumor cell lines in vitro, being biocompatible to non-tumor cells. The antitumor efficacy was enhanced by increasing the oxidation degree of the EPS (Sarilmiser and Oner, 2014).

The immunoregulatory effect of halophilic EPSs leads to antiviral activity. The EPSs from halothermophilic strains *Bacillus licheniformis* B3–15 and *Bacillus licheniformis* T14 decreased the replication of the herpes simplex virus type 2 (HSV-2) in peripheral mononuclear blood cell (PMBC) by stimulating the expression of various proinflammatory cytokines. It suggests potential applications as therapy in herpes virus infection and immunocompromised hosts (Arena et al., 2006, 2009; Spanò and Arena, 2016).

10.10.6 Antioxidant effect

Antioxidant activities lead to scavenging reactive oxygen species (ROS), which generate oxidative stress to neuronal cells and are deeply associated with chronic and degenerative diseases such as neurodegenerative disorders (Xu et al., 2016). EPS plays a vital role as natural antioxidants for the prevention of oxidative damage in the human body (Wang et al., 2013). A Halophilic EPS isolated from *Halolactibacillus miurensis* demonstrated dose-dependent scavenging activity against DPPH (2,2-diphenyl-1-picrylhydrazyl), hydroxyl, and superoxide free radicals (Arun et al., 2017;

Wang et al., 2017a, b). Similarly, an EPS produced by halothermophilic bacterium *Halomonas nitroreducens* WB1 was reported to have antioxidant properties against hydroxyl and DPPH radicals (Chikkanna et al., 2018).

11 Conclusion

Halophilic bacteria possess unique metabolites and macromolecules to carry out various metabolic processes in saline habitats. Many of these molecules hold promising commercial values. Exopolysaccharides carry out a broad range of functions in the microorganisms including their adaptation capabilities in the ecosystem. The fundamental functions of the EPS include nutrient trapping, protection of microbial communities against the extremes of salt concentrations, pH, and temperatures. The EPSs are known to possess different biotechnological avenues in medical, cosmetics, pharmaceutical, food, and environmental applications. Overall, in view of the cellular implications and commercial potential of the EPS, there is a constant need to explore and investigate EPS from extremophiles including microorganisms from saline habitats.

Acknowledgments

The work cited in this review from the SPS laboratory at the Saurashtra University was supported by UGC-CAS program, DST-FIST, DBT-Multi-Intuitional Project, MoES (Government of India) Net Working Project, and the Saurashtra University. SPS acknowledges DST-SERB International Travel Fellowships to present his work in Hamburg (Germany), Cape Town (South Africa), and Kyoto (Japan). SPS also acknowledges the award of UGC BSR Faculty Fellowship. HBB acknowledges CSIR-Direct SRF and UGC BSR-Meritorious Fellowship during his doctoral research at Saurashtra University. HBB also acknowledges Infrastructure facilities at Atmiya University, Rajkot, India. VHR acknowledges DBT-JRF/SRF, SERB-Young Scientist Award, DST-SERB for the International Travel Fellowship, and UGC-Start up Research Grant. DB acknowledges SHODH (ScHeme Of Developing High Quality Research) fellowship, Education Department, Government of Gujarat, India. Further, VHR and DB acknowledge Infrastructure facilities at the Department of Microbiology and Biotechnology, Gujarat University, Ahmedabad, India.

References

Abdel-Fattah, A.M., Gamal-Eldeen, A.M., Helmy, W.A., Esawy, M.A., 2012. Antitumor and antioxidant activities of levan and its derivative from the isolate Bacillus subtilis NRC1aza. Carbohydr. Polym. 89, 314–322. https://doi.org/10.1016/j.carbpol.2012.02.041.

Abdelhamid, S.A., Mohamed, S.S., Selim, M.S., 2020. Medical application of exopolymers produced by marine bacteria. Bull. Natl. Res. Cent. 44 (1), 1–14. https://doi.org/10.1186/s42269-020-00323-x.

Agrawal, P.K., 1992. NMR spectroscopy in the structural elucidation of oligosaccharides and glycosides. Phytochemistry 31 (10), 3307–3330. https://doi.org/10.1016/0031-9422(92)83678-R.

Amjres, H., Béjar, V., Quesada, E., Abrini, J., Llamas, I., 2011. Halomonas rifensis sp. nov., an exopolysaccharide-producing, halophilic bacterium isolated from a solar saltern. Int. J. Syst. Evol. Microbiol. 61, 2600–2605. https://doi.org/10.1099/ijs.0.027268-0.

Amjres, H., Béjar, V., Quesada, E., Carranza, D., Abrini, J., Sinquin, C., Llamas, I., 2015. Characterization of haloglycan, an exopolysaccharide produced by Halomonas stenophila HK30. Int. J. Biol. Macromol. 72, 117–124. https://doi.org/10.1016/j.ijbiomac.2014.07.052.

Arena, A., Maugeri, T.L., Pavone, B., Iannello, D., Gugliandolo, C., Bisignano, G., 2006. Antiviral and immunoregulatory effect of a novel exopolysaccharide from a marine thermotolerant *Bacillus licheniformis*. Int. Immunopharmacol. 6 (1), 8–13. https://doi.org/10.1016/j.intimp.2005.07.004.

Arena, A., Gugliandolo, C., Stassi, G., Pavone, B., Iannello, D., Bisignano, G., Maugeri, T.L., 2009. An exopolysaccharide produced by Geobacillus thermodenitrificans strain B3-72: antiviral activity on immunocompetent cells. Immunol. Lett. 123 (2), 132–137. https://doi.org/10.1016/j.imlet.2009.03.001.

Arias, S., Del Moral, A., Ferrer, M.R., Tallon, R., Quesada, E., Bejar, V., 2003. Mauran, an exopolysaccharide produced by the halophilic bacterium Halomonas maura, with a novel composition and interesting properties for biotechnology. Extremophiles 7 (4), 319–326. https://doi.org/10.1007/s00792-003-0325-8.

Arun, J., Selvakumar, S., Sathishkumar, R., Moovendhan, M., Ananthan, G., Maruthiah, T., Palavesam, A., 2017. In vitro antioxidant activities of an exopolysaccharide from a salt pan bacterium Halolactibacillus miurensis. Carbohydr. Polym. 155, 400–406. https://doi.org/10.1016/j.carbpol.2016.08.085.

Ashtaputre, A.A., Shah, A.K., 1995. Emulsifying property of a viscous exopolysaccharide from Sphingomonas paucimobilis. World J. Microbiol. Biotechnol. 11 (2), 219–222. https://doi.org/10.1007/BF00704653.

Ates, Ö., Oner, E.T., Arga, K.Y., 2011. Genome-scale reconstruction of metabolic network for a halophilic extremophile, Chromohalobacter salexigens DSM 3043. BMC Syst. Biol. 5 (1), 1–13. https://doi.org/10.1186/1752-0509-5-12.

Ates, O., Arga, K.Y., Oner, E.T., 2013. The stimulatory effect of mannitol on levan biosynthesis: lessons from metabolic systems analysis of Halomonas smyrnensis AAD6T. Biotechnol. Prog. 29 (6), 1386–1397. https://doi.org/10.1002/btpr.1823.

Aydin, B., Ozer, T., Oner, E.T., Arga, K.Y., 2018. The genome-based metabolic systems engineering to boost levan production in a halophilic bacterial model. Omics 22 (3), 198–209. https://doi.org/10.1089/omi.2017.0216.

Azam, F., Malfatti, F., 2007. Microbial structuring of marine ecosystems. Nat. Rev. Microbiol. 5 (10), 782–791. https://doi.org/10.1038/nrmicro1747.

BahramParvar, M., Razavi, S.M., 2012. Rheological interactions of selected hydrocolloid–sugar–milk–emulsifier systems. Int. J. Food Sci. Technol. 47 (4), 854–860. https://doi.org/10.1111/j.1365-2621.2011.02918.x.

Bajaj, I.B., Survase, S.A., Saudagar, P.S., Singhal, R.S., 2007. Gellan gum: fermentative production, downstream processing and applications. Food Technol. Biotechnol. 45 (4), 341–354. https://hrcak.srce.hr/22370.

Banat, I.M., Makkar, R.S., Cameotra, S.S., 2000. Potential commercial applications of microbial surfactants. Appl. Microbiol. Biotechnol. 53 (5), 495–508. https://doi.org/10.1007/s002530051648.

Banik, R.M., Santhiagu, A., Upadhyay, S.N., et al., 2007. Optimization of nutrients for gellan gum production by Sphingomonas paucimobilis ATCC-31461 in molasses-based medium using response surface methodology. Bioresour. Technol. 98 (4), 792–797. https://doi.org/10.1016/j.biortech.2006.03.012.

BeMiller, J.N., 2007. Gum Arabic and other exudate gums. In: Carbohydrate Chemistry for Food Scientists. AACC International, St. Paul, MN, USA, pp. 313–320, https://doi.org/10.1016/B978-0-12-812069-9.00016-9.

Bhaskar, P.V., Bhosle, N.B., 2006. Bacterial extracellular polymeric substance (EPS): a carrier of heavy metals in the marine food-chain. Environ. Int. 32 (2), 191–198. https://doi.org/10.1016/j.envint.2005.08.010.

Bhatia, S.K., Gurav, R., Choi, Y.K., Choi, T.R., Kim, H.J., Song, H.S., Yang, Y.H., 2021. Bioprospecting of exopolysaccharide from marine Sphingobium yanoikuyae BBL01: production, characterization, and metal chelation activity. Bioresour. Technol. 324, 124674. https://doi.org/10.1016/j.biortech.2021.124674.

Bhatnagar, M., Pareek, S., Bhatnagar, A., Ganguly, J., 2014. Rheology and characterization of a low viscosity emulsifying exopolymer from desert borne *Nostoc calcicola*. Indian J. Biotechnol. 13 (2), 241–246. http://nopr.niscair.res.in/handle/123456789/29148.

Bhatt, H.B., Singh, S.P., 2020. Cloning, expression, and structural elucidation of a biotechnologically potential alkaline serine protease from a newly isolated Haloalkaliphilic *Bacillus lehensis* JO-26. Front. Microbiol. 11, 941. https://doi.org/10.3389/fmicb.2020.00941.

Bhatt, H.B., Begum, M.A., Chintalapati, S., Chintalapati, V.R., Singh, S.P., 2017. Desertibacillus haloalkaliphilus gen. nov., sp. nov., isolated from a saline desert. Int. J. Syst. Evol. Microbiol. 67 (11), 4435–4442. https://doi.org/10.1099/ijsem.0.002310.

Bhatt, H.B., Gohel, S.D., Singh, S.P., 2018. Phylogeny, novel bacterial lineage and enzymatic potential of haloalkaliphilic bacteria from the saline coastal desert of little Rann of Kutch, Gujarat, India. 3 Biotech 8 (1), 1–12. https://doi.org/10.1007/s13205-017-1075-0.

Biswas, J., Paul, A.K., 2017. Optimization of factors influencing exopolysaccharide production by Halomonas xianhensis SUR308 under batch culture. AIMS Microbiol. 3, 564. https://doi.org/10.3934/microbiol.2017.3.564.

Boujida, N., Palau, M., Charfi, S., El Moussaoui, N., Manresa, A., Miñana-Galbis, D., Abrini, J., 2018. Isolation and characterization of halophilic bacteria producing exopolymers with emulsifying and antioxidant activities. Biocatal. Agric. Biotechnol. 16, 631–637. https://doi.org/10.1016/j.bcab.2018.10.015.

Bragadeeswaran, S., Jeevapriya, R., Prabhu, K., Rani, S.S., Priyadharsini, S., Balasubramanian, T., 2011. Exopolysaccharide production by Bacillus cereus GU812900, a fouling marine bacterium. Afr. J. Microbiol. Res. 5 (24), 4124–4132. https://doi.org/10.5897/AJMR11.375.

Bramhachari, P.V., Dubey, S.K., 2006. Isolation and characterization of exopolysaccharide produced by Vibrio harveyi strain VB23. Lett. Appl. Microbiol. 43, 571–577. https://doi.org/10.1111/j.1472-765X.2006.01967.x.

Carrión, O., Delgado, L., Mercade, E., 2015. New emulsifying and cryoprotective exopolysaccharide from Antarctic *Pseudomonas* sp. ID1. Carbohydr. Polym. 117, 1028–1034. https://doi.org/10.1016/j.carbpol.2014.08.060.

Caruso, C., Rizzo, C., Mangano, S., Poli, A., Di Donato, P., Finore, I., Giudice, A.L., 2018. Production and biotechnological potential of extracellular polymeric substances from sponge-associated Antarctic bacteria. Appl. Environ. Microbiol. 84. https://doi.org/10.1128/AEM.01624-17.

Caruso, C., Rizzo, C., Mangano, S., Poli, A., Di Donato, P., Nicolaus, B., Giudice, A.L., 2019. Isolation, characterization and optimization of EPSs produced by a cold-adapted Marinobacter isolate from Antarctic seawater. Antarct. Sci. 31, 69–79. https://doi.org/10.1017/S0954102018000482.

Casillo, A., Lanzetta, R., Parrilli, M., Corsaro, M.M., 2018. Exopolysaccharides from marine and marine extremophilic bacteria: structures, properties, ecological roles and applications. Mar. Drugs 16 (2), 69. https://doi.org/10.3390/md16020069.

Chikkanna, A., Ghosh, D., Kishore, A., 2018. Expression and characterization of a potential exopolysaccharide from a newly isolated halophilic thermotolerant bacteria Halomonas nitroreducens strain WB1. PeerJ 6. https://doi.org/10.7717/peerj.4684, e4684.

Chin, W.C., Orellana, M.V., Verdugo, P., 1998. Spontaneous assembly of marine dissolved organic matter into polymer gels. Nature 391 (6667), 568–572. https://doi.org/10.1038/35345.

Cojoc, R., Merciu, S., Oancea, P., Pincu, E., Dumitru, L., Enache, M., 2009. Highly thermostable exopolysaccharide produced by the moderately halophilic bacterium isolated from a man-made young salt lake in Romania. Pol. J. Microbiol. 58 (4), 289–294.

Cosa, S., Mabinya, L.V., Olaniran, A.O., Okoh, O.O., Bernard, K., Deyzel, S., Okoh, A.I., 2011. Bioflocculant production by Virgibacillus sp. rob isolated from the bottom sediment of Algoa Bay in the eastern cape, South Africa. Molecules 16, 2431–2442. https://doi.org/10.3390/molecules16032431.

Costa, O.Y., Raaijmakers, J.M., Kuramae, E.E., 2018. Microbial extracellular polymeric substances: ecological function and impact on soil aggregation. Front. Microbiol. 9, 1636. https://doi.org/10.3389/fmicb.2018.01636.

Czaczyk, K., Wojciechowska, K., 2003. Formation of bacterial biofilms? The essence of the matter and mechanisms of interactions. Biotechnologia 3, 180–192.

D'arrigo, G., Di Meo, C., Geissler, E., Coviello, T., Alhaique, F., Matricardi, P., 2012. Hyaluronic acid methacrylate derivatives and calcium alginate interpenetrated hydrogel networks for biomedical applications: physico-chemical characterization and protein release. Colloid Polym. Sci. 290 (15), 1575–1582. https://doi.org/10.1007/s00396-012-2735-6.

Decho, A.W., 1990. Microbial exopolymer secretions in ocean environments: their role (s) in food webs and marine processes. Oceanogr. Mar. Biol. Annu. Rev. 28 (7), 73–153.

Decho, A.W., Herndl, G.J., 1995. Microbial activities and the transformation of organic matter within mucilaginous material. Sci. Total Environ. 165 (1–3), 33–42. https://doi.org/10.1016/0048-9697(95)04541-8.

Deepak, V., Pandian, S.R.K., Sivasubramaniam, S.D., Nellaiah, H., Sundar, K., 2015. Synthesis of polyelectrolyte nanoparticles from anticancer exopolysaccharide isolated from probiotic Lactobacillus acidophilus. Res. J. Microbiol. 10 (5), 193. https://doi.org/10.3923/jm.2015.193.204.

Delbarre-Ladrat, C., Sinquin, C., Lebellenger, L., Zykwinska, A., Colliec-Jouault, S., 2014. Exopolysaccharides produced by marine bacteria and their applications as glycosaminoglycan-like molecules. Front. Chem. 2, 85. https://doi.org/10.3389/fchem.2014.00085.

Di Donato, P., Poli, A., Taurisano, V., Abbamondi, G.R., Nicolaus, B., Tommonaro, G., 2016. Recent advances in the study of marine microbial biofilm: from the involvement of quorum sensing in its production up to biotechnological application of the polysaccharide fractions. J. Mar. Sci. Eng. 4, 34. https://doi.org/10.3390/jmse4020034.

Di Donato, P., Poli, A., Tommonaro, G., Abbamondi, G.R., Nicolaus, B., 2018. Exopolysaccharide productions from extremophiles: the chemical structures and their bioactivities. In: Extremophilic Microbial Processing of Lignocellulosic Feedstocks to Biofuels, Value-Added Products, and Usable Power, pp. 189–205, https://doi.org/10.1007/978-3-319-74459-9_10.

Dodia, M.S., Joshi, R.H., Patel, R.K., Singh, S.P., 2006. Characterization and stability of extracellular alkaline proteases from halophilic and alkaliphilic bacteria isolated from saline habitat of coastal Gujarat, India. Braz. J. Microbiol. 37, 276–282. https://doi.org/10.1590/S1517-83822006000300015.

Dodia, M.S., Bhimani, H.G., Rawal, C.M., Joshi, R.H., Singh, S.P., 2008a. Salt dependent resistance against chemical denaturation of alkaline protease from a newly isolated haloalkaliphilic Bacillus sp. Bioresour. Technol. 99, 6223–6227. https://doi.org/10.1016/j.biortech.2007.12.020.

Dodia, M.S., Rawal, C.M., Bhimani, H.G., Joshi, R.H., Khare, S.K., Singh, S.P., 2008b. Purification and stability characteristics of an alkaline serine protease from a newly isolated Haloalkaliphilic bacterium sp. AH-6. J. Ind. Microbiol. Biotechnol. 35, 121–131. https://doi.org/10.1007/s10295-007-0273-x.

Donlan, R.M., 2002. Biofilms: microbial life on surfaces. Emerg. Infect. Dis. 8 (9), 881. https://doi.org/10.3201/eid0809.020063.

Drouillard, S., Jeacomine, I., Buon, L., Boisset, C., Courtois, A., Thollas, B., Helbert, W., 2015. Structure of an amino acid-decorated exopolysaccharide secreted by a Vibrio alginolyticus strain. Mar. Drugs 13, 6723–6739. https://doi.org/10.3390/md13116723.

Dubois, M., Gilles, K.A., Hamilton, J.K., Rebers, P.T., Smith, F., 1956. Colorimetric method for determination of sugars and related substances. Anal. Chem. 28, 350–356.

Dudman, W.F., 1977. Role of surface polysaccharides in natural environments. In: Surface Carbohydrates of the Prokaryotic Cell. Academic Press, London.

Duus, J.Ø., Gotfredsen, C.H., Bock, K., 2000. Carbohydrate structural determination by NMR spectroscopy: modern methods and limitations. Chem. Rev. 100 (12), 4589–4614. https://doi.org/10.1021/cr990302n.

Eddouaouda, K., Mnif, S., Badis, A., Younes, S.B., Cherif, S., Ferhat, S., Sayadi, S., 2012. Characterization of a novel biosurfactant produced by Staphylococcus sp. strain 1E with potential application on hydrocarbon bioremediation. J. Basic Microbiol. 52 (4), 408–418. https://doi.org/10.1002/jobm.201100268.

Enikeev, R., 2012. Development of a new method for determination of exopolysaccharide quantity in fermented milk products and its application in technology of kefir production. Food Chem. 134, 2437–2441. https://doi.org/10.1016/j.foodchem.2012.04.050.

Etemadi, O., Petrisor, I.G., Kim, D., Wan, M.W., Yen, T.F., 2003. Stabilization of metals in subsurface by biopolymers: laboratory drainage flow studies. Soil Sediment Contam. 12 (5), 647–661. https://doi.org/10.1080/714037712.

Everett, J.A., Rumbaugh, K.P., 2015. Biofilms, quorum sensing and crosstalk in medically important microbes. In: Molecular Medical Microbiology. Academic Press, pp. 235–247, https://doi.org/10.1016/B978-0-12-397169-2.00012-3.

Farber, B.F., Kaplan, M.H., Clogston, A.G., 1990. Staphylococcus epidermidis extracted slime inhibits the antimicrobial action of glycopeptide antibiotics. J. Infect. Dis. 161 (1), 37–40. https://doi.org/10.1093/infdis/161.1.37.

Finore, I., Di Donato, P., Mastascusa, V., Nicolaus, B., Poli, A., 2014. Fermentation technologies for the optimization of marine microbial exopolysaccharide production. Mar. Drugs 12 (5), 3005–3024. https://doi.org/10.3390/md12053005.

Flemming, H.C., Wingender, J., 2010. The biofilm matrix. Nat. Rev. Microbiol. 8 (9), 623–633. https://doi.org/10.1038/nrmicro2415.

Freitas, F., Alves, V.D., Carvalheira, M., Costa, N., Oliveira, R., Reis, M.A., 2009. Emulsifying behaviour and rheological properties of the extracellular polysaccharide produced by Pseudomonas oleovorans grown on glycerol byproduct. Carbohydr. Polym. 78 (3), 549–556. https://doi.org/10.1016/j.carbpol.2009.05.016.

Fukushi, K., 2012. Phytoextraction of cadmium from contaminated soil assisted by microbial biopolymers. Agrotechnology 2 (110). https://doi.org/10.4172/2168-9881.1000110.

Gadd, G.M., 2009. Biosorption: critical review of scientific rationale, environmental importance and significance for pollution treatment. J. Chem. Technol. Biotechnol. 84 (1), 13–28. https://doi.org/10.1002/jctb.1999.

Gagliano, M.C., Ismail, S.B., Stams, A.J.M., Plugge, C.M., Temmink, H., Van Lier, J.B., 2017. Biofilm formation and granule properties in anaerobic digestion at high salinity. Water Res. 121, 61–71. https://doi.org/10.1016/j.watres.2017.05.016.

Gan, L., Li, X., Wang, H., Peng, B., Tian, Y., 2020a. Structural characterization and functional evaluation of a novel exopolysaccharide from the moderate halophile *Gracilibacillus* sp. SCU50. Int. J. Biol. Macromol. 154, 1140–1148. https://doi.org/10.1016/j.ijbiomac.2019.11.143.

Gomaa, M., Yousef, N., 2020. Optimization of production and intrinsic viscosity of an exopolysaccharide from a high yielding Virgibacillus salarius BM02: study of its potential antioxidant, emulsifying properties and application in the mixotrophic cultivation of Spirulina platensis. Int. J. Biol. Macromol. 149, 552–561. https://doi.org/10.1016/j.ijbiomac.2020.01.289.

Gomes, M.B., Gonzales-Limache, E.E., Sousa, S.T.P., Dellagnezze, B.M., Sartoratto, A., Silva, L.C.F., Oliveira, V.M., 2018. Exploring the potential of halophilic bacteria from oil terminal environments for biosurfactant production and hydrocarbon degradation under high-salinity conditions. Int. Biodeter. Biodegr. 126, 231–242. https://doi.org/10.1016/j.ibiod.2016.08.014.

Grivaud-Le Du, A., Zykwinska, A., Sinquin, C., Ratiskol, J., Weiss, P., Vinatier, C., Colliec-Jouault, S., 2017. Purification of the exopolysaccharide produced by Alteromonas infernus: identification of endotoxins and effective process to remove them. Appl. Microbiol. Biotechnol. 101, 6597–6606. https://doi.org/10.1007/s00253-017-8364-8.

Grobe, S., Wingender, J., Flemming, H.C., 2001. Capability of mucoid Pseudomonas aeruginosa to survive in chlorinated water. Int. J. Hyg. Environ. Health 204 (2–3), 139–142. https://doi.org/10.1078/1438-4639-00085.

Gu, D., Jiao, Y., Wu, J., Liu, Z., Chen, Q., 2017. Optimization of EPS production and characterization by a halophilic bacterium, Kocuria rosea ZJUQH from Chaka salt Lake with response surface methodology. Molecules 22, 814. https://doi.org/10.3390/molecules22050814.

Guibaud, G., Tixier, N., Bouju, A., Baudu, M., 2003. Relation between extracellular polymers' composition and its ability to complex Cd, Cu and Pb. Chemosphere 52 (10), 1701–1710. https://doi.org/10.1016/S0045-6535(03)00355-2.

Gutierrez, T., Mulloy, B., Black, K., Green, D.H., 2007. Glycoprotein emulsifiers from two marine Halomonas species: chemical and physical characterization. J. Appl. Microbiol. 103 (5), 1716–1727. https://doi.org/10.1111/j.1365-2672.2007.03407.x.

Gutierrez, T., Shimmield, T., Haidon, C., Black, K., Green, D.H., 2008. Emulsifying and metal ion binding activity of a glycoprotein exopolymer produced by Pseudoalteromonas sp. strain TG12. Appl. Environ. Microbiol. 74 (15), 4867–4876. https://doi.org/10.1128/AEM.00316-08.

Gutierrez, T., Morris, G., Green, D.H., 2009. Yield and physicochemical properties of EPS from Halomonas sp. strain TG39 identifies a role for protein and anionic residues (sulfate and phosphate) in emulsification of n-hexadecane. Biotechnol. Bioeng. 103 (1), 207–216. https://doi.org/10.1002/bit.22218.

Han, P.P., Sun, Y., Wu, X.Y., Yuan, Y.J., Dai, Y.J., Jia, S.R., 2014. Emulsifying, flocculating, and physicochemical properties of exopolysaccharide produced by cyanobacterium Nostoc flagelliforme. Appl. Biochem. Biotechnol. 172 (1), 36–49. https://doi.org/10.1007/s12010-013-0505-7.

Hassler, C.S., Schoemann, V., 2009. Bioavailability of organically bound Fe to model phytoplankton of the Southern Ocean. Biogeosciences 6 (10), 2281–2296. https://doi.org/10.5194/bg-6-2281-2009.

Hassler, C.S., Alasonati, E., Nichols, C.M., Slaveykova, V.I., 2011a. Exopolysaccharides produced by bacteria isolated from the pelagic Southern Ocean—role in Fe binding, chemical reactivity, and bioavailability. Mar. Chem. 123 (1–4), 88–98. https://doi.org/10.1016/j.marchem.2010.10.003.

Holmström, C., Kjelleberg, S., 1999. Marine Pseudoalteromonas species are associated with higher organisms and produce biologically active extracellular agents. FEMS Microbiol. Ecol. 30 (4), 285–293. https://doi.org/10.1111/j.1574-6941.1999.tb00656.x.

Hubbard, C., McNamara, J.T., Azumaya, C., Patel, M.S., Zimmer, J., 2012. The hyaluronan synthase catalyzes the synthesis and membrane translocation of hyaluronan. J. Mol. Biol. 418 (1–2), 21–31. https://doi.org/10.1016/j.jmb.2012.01.053.

Hussain, S.A., Sarker, M.I., Yosief, H.O., 2020. Efficacy of alkyltrimethylammonium bromide for decontaminating salt-cured hides from the red heat causing moderately halophilic bacteria. Lett. Appl. Microbiol. 70, 159–164. https://doi.org/10.1111/lam.13250.

Ibrahim, I.M., Konnova, S.A., Sigida, E.N., Lyubun, E.V., Muratova, A.Y., Fedonenko, Y.P., Elbanna, K., 2020. Bioremediation potential of a halophilic Halobacillus sp. strain, EG1HP4QL: exopolysaccharide production, crude oil degradation, and heavy metal tolerance. Extremophiles 24 (1), 157–166. https://doi.org/10.1007/s00792-019-01143-2.

Imran, M.Y.M., Reehana, N., Jayaraj, K.A., Ahamed, A.A.P., Dhanasekaran, D., Thajuddin, N., Muralitharan, G., 2016. Statistical optimization of exopolysaccharide production by Lactobacillus plantarum NTMI05 and NTMI20. Int. J. Biol. Macromol. 93, 731–745. https://doi.org/10.1016/j.ijbiomac.2016.09.007.

Insulkar, P., Kerkar, S., Lele, S.S., 2018. Purification and structural-functional characterization of an exopolysaccharide from Bacillus licheniformis PASS26 with in-vitro antitumor and wound healing activities. Int. J. Biol. Macromol. 120, 1441–1450. https://doi.org/10.1016/j.ijbiomac.2018.09.147.

Iyer, A., Mody, K., Jha, B., 2005. Biosorption of heavy metals by a marine bacterium. Mar. Pollut. Bull. 50 (3), 340–343. https://doi.org/10.1016/j.marpolbul.2004.11.012.

Jones, K.M., 2012. Increased production of the exopolysaccharide succinoglycan enhances Sinorhizobium meliloti 1021 symbiosis with the host plant Medicago truncatula. J. Bacteriol. 194 (16), 4322–4331. https://doi.org/10.1128/JB.00751-12.

Joshi, R.H., Dodia, M.S., Singh, S.P., 2008. Production and optimization of a commercially viable alkaline protease from a haloalkaliphilic bacterium. Biotechnol. Bioprocess Eng. 13, 552–559. https://doi.org/10.1007/s12257-007-0211-9.

Karlapudi, A.P., Kodali, V.P., Kota, K.P., Shaik, S.S., Kumar, N.S., Dirisala, V.R., 2016. Deciphering the effect of novel bacterial exopolysaccharide-based nanoparticle cream against Propionibacterium acnes. 3 Biotech 6 (1), 35. https://doi.org/10.1007/s13205-015-0359-5.

Kavita, K., Mishra, A., Jha, B., 2013. Extracellular polymeric substances from two biofilm forming Vibrio species: characterization and applications. Carbohydr. Polym. 94 (2), 882–888. https://doi.org/10.1016/j.carbpol.2013.02.010.

Kennedy, A.F., Sutherland, I.W., 1987. Analysis of bacterial exopolysaccharides. Biotechnol. Appl. Biochem. 9 (1), 12–19. https://doi.org/10.1111/j.1470-8744.1987.tb00458.x.

Küçükaşik, F., Kazak, H., Güney, D., Finore, I., Poli, A., Yenigün, O., Öner, E.T., 2011. Molasses as fermentation substrate for levan production by Halomonas sp. Appl. Microbiol. Biotechnol. 89, 1729–1740. https://doi.org/10.1007/s00253-010-3055-8.

Kumar, C.G., Joo, H.-S., Choi, J.-W., Koo, Y.-M., Chang, C.-S., 2004. Purification and characterization of an extracellular polysaccharide extremophiles 1 3 from haloalkalophilic Bacillus sp. I-450. Enzyme Microb. Technol. 34, 673–681.

Larsen, F.H., Engelsen, S.B., 2015. Insight into the functionality of microbial exopolysaccharides by NMR spectroscopy and molecular modeling. Front. Microbiol. 6, 1374. https://doi.org/10.3389/fmicb.2015.01374.

Le Costaouëc, T., Cérantola, S., Ropartz, D., Ratiskol, J., Sinquin, C., Colliec-Jouault, S., Boisset, C., 2012. Structural data on a bacterial exopolysaccharide produced by a deep-sea Alteromonas macleodii strain. Carbohydr. Polym. 90, 49–59. https://doi.org/10.1016/j.carbpol.2012.04.059.

Lelchat, F., Cérantola, S., Brandily, C., Colliec-Jouault, S., Baudoux, A.C., Ojima, T., Boisset, C., 2015. The marine bacteria Cobetia marina DSMZ 4741 synthesizes an unexpected K-antigen-like exopolysaccharide. Carbohydr. Polym. 124, 347–356. https://doi.org/10.1016/j.carbpol.2015.02.038.

Li, P., Liu, Z., Xu, R., 2001. Chemical characterisation of the released polysaccharide from the cyanobacterium Aphanothece halophytica GR02. J. Appl. Phycol. 13, 71–77. https://doi.org/10.1023/A:1008109501066.

Liu, S.B., Chen, X.L., He, H.L., Zhang, X.Y., Xie, B.B., Yu, Y., Zhang, Y.Z., 2013. Structure and ecological roles of a novel exopolysaccharide from the Arctic sea ice bacterium *Pseudoalteromonas* sp. strain SM20310. Appl. Environ. Microbiol. 79 (1), 224–230. https://doi.org/10.1128/AEM.01801-12.

Liu, Y., Bellich, B., Hug, S., Eberl, L., Cescutti, P., Pessi, G., 2020. The exopolysaccharide cepacian plays a role in the establishment of the Paraburkholderia phymatum–Phaseolus vulgaris symbiosis. Front. Microbiol. 11, 1600. https://doi.org/10.3389/fmicb.2020.01600.

Llamas, I., Mata, J.A., Tallon, R., Bressollier, P., Urdaci, M.C., Quesada, E., Béjar, V., 2010. Characterization of the exopolysaccharide produced by Salipiger mucosus A3T, a halophilic species belonging to the Alphaproteobacteria, isolated on the Spanish mediterranean seaboard. Mar. Drugs 8 (8), 2240–2251. https://doi.org/10.3390/md8082240.

Llamas, I., Amjres, H., Mata, J.A., Quesada, E., Béjar, V., 2012. The potential biotechnological applications of the exopolysaccharide produced by the halophilic bacterium Halomonas almeriensis. Molecules 17, 7103–7120. https://doi.org/10.3390/molecules17067103.

Lloret, J., Wulff, B.B., Rubio, J.M., Downie, J.A., Bonilla, I., Rivilla, R., 1998. Exopolysaccharide II production is regulated by salt in the halotolerant strain Rhizobium meliloti EFB1. Appl. Environ. Microbiol. 64 (3), 1024–1028. https://doi.org/10.1128/AEM.64.3.1024-1028.1998.

Logan, B.E., Hunt, J.R., 1987. Advantages to microbes of growth in permeable aggregates in marine systems 1. Limnol. Oceanogr. 32 (5), 1034–1048. https://doi.org/10.4319/lo.1987.32.5.1034.

Malick, A., Khodaei, N., Benkerroum, N., Karboune, S., 2017. Production of exopolysaccharides by selected Bacillus strains: optimization of media composition to maximize the yield and structural characterization. Int. J. Biol. Macromol. 102, 539–549. https://doi.org/10.1016/j.ijbiomac.2017.03.1514.

Mallick, I., Bhattacharyya, C., Mukherji, S., Dey, D., Sarkar, S.C., Mukhopadhyay, U.K., Ghosh, A., 2018. Effective rhizoinoculation and biofilm formation by arsenic immobilizing halophilic plant growth promoting bacteria (PGPB) isolated from mangrove rhizosphere: a step towards arsenic rhizoremediation. Sci. Total Environ. 610, 1239–1250. https://doi.org/10.1016/j.scitotenv.2017.07.234.

Mata, J.A., Béjar, V., Llamas, I., Arias, S., Bressollier, P., Tallon, R., Quesada, E., 2006. Exopolysaccharides produced by the recently described halophilic bacteria Halomonas ventosae and Halomonas anticariensis. Res. Microbiol. 157 (9), 827–835. https://doi.org/10.1016/j.resmic.2006.06.004.

Martínez-Checa, F., Bejar, V., Martínez-Cánovas, M.J., Llamas, I., Quesada, E., et al., 2005. *Halomonas almeriensis* sp. nov., a moderately halophilic, exopolysaccharide-producing bacterium from Cabo de Gata, Almería, south-east Spain. Int. J. Syst. Evol. Microbiol. 55 (5), 2007–2011. https://doi.org/10.1099/ijs.0.63676-0.

Mata, J.A., Béjar, V., Bressollier, P., Tallon, R., Urdaci, M.C., Quesada, E., Llamas, I., 2008. Characterization of exopolysaccharides produced by three moderately halophilic bacteria belonging to the family Alteromonadaceae. J. Appl. Microbiol. 105 (2), 521–528. https://doi.org/10.1111/j.1365-2672.2008.03789.x.

Matou, S., Colliec-Jouault, S., Galy-Fauroux, I., Ratiskol, J., Sinquin, C., Guezennec, J., Helley, D., 2005. Effect of an oversulfated exopolysaccharide on angiogenesis induced by fibroblast growth factor-2 or vascular endothelial growth factor in vitro. Biochem. Pharmacol. 69 (5), 751–759. https://doi.org/10.1016/j.bcp.2004.11.021.

Maugeri, T.L., Gugliandolo, C., Caccamo, D., Panico, A., Lama, L., Gambacorta, A., Nicolaus, B., 2002. A halophilic thermotolerant Bacillus isolated from a marine hot spring able to produce a new exopolysaccharide. Biotechnol. Lett. 24, 515–519. https://doi.org/10.1023/A:1014891431233.

Mende, S., Peter, M., Bartels, K., Rohm, H., Jaros, D., 2013. Addition of purified exopolysaccharide isolates from S. thermophilus to milk and their impact on the rheology of acid gels. Food Hydrocoll. 32, 178–185. https://doi.org/10.1016/j.foodhyd.2012.12.011.

More, T.T., Yadav, J.S.S., Yan, S., Tyagi, R.D., Surampalli, R.Y., 2014. Extracellular polymeric substances of bacteria and their potential environmental applications. J. Environ. Manage. 144, 1–25. https://doi.org/10.1016/j.jenvman.2014.05.010.

Nichols, C.M., Guezennec, J., Bowman, J.P., 2005a. Bacterial exopolysaccharides from extreme marine environments with special consideration of the Southern Ocean, sea ice, and deep-sea hydrothermal vents: a review. Marine Biotechnol. 7 (4), 253–271. https://doi.org/10.1007/s10126-004-5118-2.

Nicolaus, B., Kambourova, M., Oner, E.T., 2010. Exopolysaccharides from extremophiles: from fundamentals to biotechnology. Environ. Technol. 31 (10), 1145–1158. https://doi.org/10.1080/09593330903552094.

Norwood, D.E., Gilmour, A., 2001. The differential adherence capabilities of two listeria monocytogenes strains in monoculture and multispecies biofilms as a function of temperature. Lett. Appl. Microbiol. 33 (4), 320–324. https://doi.org/10.1046/j.1472-765X.2001.01004.x.

Notararigo, S., Nácher-Vázquez, M., Ibarburu, I., Werning, M.L., de Palencia, P.F., Dueñas, M.T., Prieto, A., 2013. Comparative analysis of production and purification of homo-and hetero-polysaccharides produced by lactic acid bacteria. Carbohydr. Polym. 93, 57–64. https://doi.org/10.1016/j.carbpol.2012.05.016.

Nowlan, B., Dodia, M.S., Singh, S.P., Patel, B.K., 2006. Bacillus okhensis sp. nov., a halotolerant and alkali tolerant bacterium from an Indian saltpan. Int. J. Syst. Evol. Microbiol. 56, 1073–1077. https://doi.org/10.1099/ijs.0.63861-0.

Nwodo, U.U., Green, E., Okoh, A.I., 2012. Bacterial exopolysaccharides: functionality and prospects. Int. J. Mol. Sci. 13 (11), 14002–14015. https://doi.org/10.3390/ijms131114002.

Onbasli, D., Aslim, B., 2008. Determination of antimicrobial activity and production of some metabolites by Pseudomonas aeruginosa B1 and B2 in sugar beet molasses. Afr. J. Biotechnol. 7, 4614–4619. https://doi.org/10.5897/AJB08.691.

Paerl, H.W., 1976. Specific associations of the bluegreen algae Anabaena and Aphanizomenon with bacteria in freshwater blooms 1. J. Phycol. 12 (4), 431–435. https://doi.org/10.1111/j.1529-8817.1976.tb02867.x.

Pandey, S., Singh, S.P., 2012. Organic solvent tolerance of an α-amylase from haloalkaliphilic bacteria as a function of pH, temperature, and salt concentrations. Appl. Biochem. Biotechnol. 166, 1747–1757. https://doi.org/10.1007/s12010-012-9580-4.

Pandey, S., Rakholiya, K.D., Raval, V.H., Singh, S.P., 2012. Catalysis and stability of an alkaline protease from a haloalkaliphilic bacterium under non-aqueous conditions as a function of pH, salt and temperature. J. Biosci. Bioeng. 114 (3), 251–256. https://doi.org/10.1016/j.jbiosc.2012.03.003.

Parikh, A., Madamwar, D., 2006. Partial characterization of extracellular polysaccharides from cyanobacteria. Bioresour. Technol. 97, 1822–1827. https://doi.org/10.1016/j.biortech.2005.09.008.

Patel, A., Prajapat, J.B., 2013. Food and health applications of exopolysaccharides produced by lactic acid bacteria. Adv. Dairy Res., 1–8. https://doi.org/10.4172/2329-888X.1000107.

Patel, R.K., Dodia, M.S., Joshi, R.H., Singh, S.P., 2006. Purification and characterization of alkaline protease from a newly isolated haloalkaliphilic Bacillus sp. Process Biochem. 41 (9), 2002–2009. https://doi.org/10.1016/j.procbio.2006.04.016.

Patel, A.K., Michaud, P., Singhania, R.R., Soccol, C.R., Pandey, A., 2010. Polysaccharides from probiotics: new developments as food additives. Food Technol. Biotechnol. 48, 451–463. https://doi.org/10.17113/ftb.

Poli, A., Kazak, H., Gürleyendağ, B., Tommonaro, G., Pieretti, G., Öner, E.T., Nicolaus, B., 2009. High level synthesis of levan by a novel Halomonas species growing on defined media. Carbohydr. Polym. 78, 651–657. https://doi.org/10.1016/j.carbpol.2009.05.031.

Poli, A., Anzelmo, G., Nicolaus, B., 2010. Bacterial exopolysaccharides from extreme marine habitats: production, characterization and biological activities. Mar. Drugs 8 (6), 1779–1802. https://doi.org/10.3390/md8061779.

Poli, A., Di Donato, P., Abbamondi, G.R., Nicolaus, B., 2011. Synthesis, production, and biotechnological applications of exopolysaccharides and polyhydroxyalkanoates by archaea. Archaea 2011. https://doi.org/10.1155/2011/693253.

Poli, A., Di Donato, P., Tommonaro, G., Abbamondi, G.R., Finore, I., Nicolaus, B., 2018. Exopolysaccharide-producing microorganisms from extreme areas: chemistry and application. In: Extremophiles in Eurasian Ecosystems: Ecology, Diversity, and Applications. Springer, Singapore, pp. 405–433, https://doi.org/10.1007/978-981-13-0329-6_15.

Pooja, K.P., Chandra, T.S., 2009. Production and partial characterization of a novel capsular polysaccharide KP-EPS produced by Paenibacillus pabuli strain ATSKP. World J. Microbiol. Biotechnol. 25 (5), 835–841. https://doi.org/10.1007/s11274-009-9954-0.

Purohit, M.K., Singh, S.P., 2013. A metagenomic alkaline protease from saline habitat: cloning, over-expression and functional attributes. Int. J. Biol. Macromol. 53, 138–143. https://doi.org/10.1016/j.ijbiomac.2012.10.032.

Qin, Q.L., Zhang, X.Y., Wang, X.M., Liu, G.M., Chen, X.L., Xie, B.B., Zhang, Y.Z., 2010. The complete genome of Zunongwangia profunda SM-A87 reveals its adaptation to the deep-sea environment and ecological role in sedimentary organic nitrogen degradation. BMC Genomics 11, 1–10. http://www.biomedcentral.com/1471-2164/11/247.

Quesada, E., Béjar, V., Ferrer, M.R., Calvo, C., Llamas, I., Martínez-Checa, F., del Moral, A., 2004. Moderately halophilic, exopolysaccharide-producing bacteria. In: Halophilic Microorganisms, pp. 297–314, https://doi.org/10.1007/978-3-662-07656-9.

Radchenkova, N., Boyadzhieva, I., Atanasova, N., Poli, A., Finore, I., Di Donato, P., Kambourova, M., 2018. Extracellular polymer substance synthesized by a halophilic bacterium Chromohalobacter canadensis 28. Appl. Microbiol. Biotechnol. 102 (11), 4937–4949. https://doi.org/10.1007/s00253-018-8901-0.

Ragavan, M.L., Das, N., 2019. Optimization of exopolysaccharide production by probiotic yeast Lipomyces starkeyi VIT-MN03 using response surface methodology and its applications. Ann. Microbiol. 69, 515–530. https://doi.org/10.1007/s13213-019-1440-9.

Raval, V.H., Purohit, M.K., Singh, S.P., 2013. Diversity, population dynamics and biocatalytic potential of cultivable and non-cultivable bacterial communities of the saline ecosystems. In: Marine Enzymes for Biocatalysis, pp. 165–189, https://doi.org/10.1533/9781908818355.2.165.

Rana, S., Upadhyay, L.S., 2020. Microbial exopolysaccharides: synthesis pathways, types and their commercial applications. Int. J. Biol. Macromol. 157, 577–583. https://doi.org/10.1016/j.ijbiomac.2020.04.084.

Raval, V.H., Bhatt, H.B., Singh, S.P., 2018. Adaptation strategies in halophilic bacteria. In: Extremophiles: From Biology to Biotechnology, p. 137, https://doi.org/10.1201/9781315154695-7.

Raveendran, S., Poulose, A.C., Yoshida, Y., Maekawa, T., Kumar, D.S., 2013. Bacterial exopolysaccharide based nanoparticles for sustained drug delivery, cancer chemotherapy and bioimaging. Carbohydr. Polym. 91 (1), 22–32. https://doi.org/10.1016/j.carbpol.2012.07.079.

Renner, L.D., Weibel, D.B., 2011. Physicochemical regulation of biofilm formation. MRS Bull. 36 (5), 347–355. https://doi.org/10.1557/mrs.2011.65.

Riedel, A., Michel, C., Gosselin, M., 2006. Seasonal study of sea-ice exopolymeric substances on the Mackenzie shelf: implications for transport of sea-ice bacteria and algae. Aquat. Microb. Ecol. 45 (2), 195–206. https://doi.org/10.3354/ame045195.

Rinaudi, L.V., Giordano, W., 2010. An integrated view of biofilm formation in rhizobia. FEMS Microbiol. Lett. 304 (1), 1–11. https://doi.org/10.1111/j.1574-6968.2009.01840.x.

Roberson, E.B., Firestone, M.K., 1992. Relationship between desiccation and exopolysaccharide production in a soil Pseudomonas sp. Appl. Environ. Microbiol. 58 (4), 1284–1291. https://doi.org/10.1128/AEM.58.4.1284-1291.1992.

Ruiz-Ruiz, C., Srivastava, G.K., Carranza, D., Mata, J.A., Llamas, I., Santamaría, M., Molina, I.J., 2011. An exopolysaccharide produced by the novel halophilic bacterium *Halomonas stenophila* strain B100 selectively induces apoptosis in human T leukaemia cells. Appl. Microbiol. Biotechnol. 89 (2), 345–355. https://doi.org/10.1007/s00253-010-2886-7.

Sahana, T.G., Rekha, P.D., 2020. A novel exopolysaccharide from marine bacterium Pantoea sp. YU16-S3 accelerates cutaneous wound healing through Wnt/β-catenin pathway. Carbohydr. Polym. 238. https://doi.org/10.1016/j.carbpol.2020.116191, 116191.

Salem, D.R., 2007. Electrospinning of nanofibers and the charge injection method. In: Nanofibers and Nanotechnology in Textiles. Woodhead Publishing, pp. 3–21, https://doi.org/10.1533/9781845693732.1.3.

Sam, S., Kucukasik, F., Yenigun, O., Nicolaus, B., Oner, E.T., Yukselen, M.A., 2011. Flocculating performances of exopolysaccharides produced by a halophilic bacterial strain cultivated on agro-industrial waste. Bioresour. Technol. 102, 1788–1794. https://doi.org/10.1016/j.biortech.2010.09.020.

Sana, S.S., Singh, R.P., Sharma, M., Srivastava, A.K., Manchanda, G., Rai, A.R., Zhang, Z.-J., 2021. Biogenesis and application of nickel nanoparticles: a review. Curr. Pharm. Biotechnol. 22 (6), 808–822. https://doi.org/10.2174/1389201022999210101235233.

Sandhya, V.Z.A.S., Grover, M., Reddy, G., Venkateswarlu, B., 2009. Alleviation of drought stress effects in sunflower seedlings by the exopolysaccharides producing *Pseudomonas putida* strain GAP-P45. Biol. Fertil. Soils 46 (1), 17–26. https://doi.org/10.1007/s00374-009-0401-z.

Sarilmiser, H.K., Oner, E.T., 2014. Investigation of anti-cancer activity of linear and aldehyde-activated levan from Halomonas smyrnensis AAD6T. Biochem. Eng. J. 92, 28–34. https://doi.org/10.1016/j.bej.2014.06.020.

Sarilmiser, H.K., Ates, O., Ozdemir, G., Arga, K.Y., Oner, E.T., 2015. Effective stimulating factors for microbial levan production by *Halomonas smyrnensis* AAD6T. J. Biosci. Bioeng. 119, 455–463. https://doi.org/10.1016/j.jbiosc.2014.09.019.

Satpute, S.K., Banat, I.M., Dhakephalkar, P.K., Banpurkar, A.G., Chopade, B.A., 2010. Biosurfactants, bioemulsifiers and exopolysaccharides from marine microorganisms. Biotechnol. Adv. 28 (4), 436–450. https://doi.org/10.1016/j.biotechadv.2010.02.006.

Schmid, J., Sieber, V., Rehm, B., 2015. Bacterial exopolysaccharides: biosynthesis pathways and engineering strategies. Front. Microbiol. 6, 496. https://doi.org/10.3389/fmicb.2015.00496.

Sezer, A.D., Kazak, H., Öner, E.T., Akbuğa, J., 2011. Levan-based nanocarrier system for peptide and protein drug delivery: optimization and influence of experimental parameters on the nanoparticle characteristics. Carbohydr. Polym. 84 (1), 358–363. https://doi.org/10.1016/j.carbpol.2010.11.046.

Singh, S.P., Raval, V., Purohit, M.K., 2013. Strategies for the salt tolerance in bacteria and archeae and its implications in developing crops for adverse conditions. In: Plant Acclimation to Environmental Stress. Springer, New York, NY, pp. 85–99, https://doi.org/10.1007/978-1-4614-5001-6.

Singha, T.K., 2012. Microbial extracellular polymeric substances: production, isolation and applications. IOSR J. Pharm. 2 (2), 271–281. https://doi.org/10.9790/3013-0220276281.

Souli, M., Giamarellou, H., 1998. Effects of slime produced by clinical isolates of coagulase-negative staphylococci on activities of various antimicrobial agents. Antimicrob. Agents Chemother. 42 (4), 939–941. https://doi.org/10.1128/AAC.42.4.939.

Spanò, A., Arena, A., 2016. Bacterial exopolysaccharide of shallow marine vent origin as agent in counteracting immune disorders induced by herpes virus. J. Immunoass. Immunochem. 37 (3), 251–260. https://doi.org/10.1080/15321819.2015.1126602.

Sran, K.S., Sundharam, S.S., Krishnamurthi, S., Choudhury, A.R., 2019. Production, characterization and bio-emulsifying activity of a novel thermostable exopolysaccharide produced by a marine strain of Rhodobacter johrii CDR-SL 7Cii. Int. J. Biol. Macromol. 127, 240–249. https://doi.org/10.1016/j.ijbiomac.2019.01.045.

Staley, J.T., Gosink, J.J., 1999. Poles apart: biodiversity and biogeography of sea ice bacteria. Annu. Rev. Microbiol. 53 (1), 189–215. https://doi.org/10.1146/annurev.micro.53.1.189.

Suja, L.D., Chen, X., Summers, S., Paterson, D.M., Gutierrez, T., 2019. Chemical dispersant enhances microbial exopolymer (EPS) production and formation of marine oil/dispersant snow in surface waters of the subarctic Northeast Atlantic. Front. Microbiol. 10, 553. https://doi.org/10.3389/fmicb.2019.00553.

Suresh Kumar, A., Mody, K., Jha, B., 2007. Bacterial exopolysaccharides—a perception. J. Basic Microbiol. 47, 103–117. https://doi.org/10.1002/jobm.200610203.

Survase, S.A., Saudagar, P.S., Singhal, R.S., 2007a. Use of complex media for the production of scleroglucan by Sclerotium rolfsii MTCC 2156. Bioresour. Technol. 98, 1509–1512. https://doi.org/10.1016/j.biortech.2006.05.022.

Survase, S.A., Saudagar, P.S., Singhal, R.S., 2007b. Enhanced production of scleroglucan by Sclerotium rolfsii MTCC 2156 by use of metabolic precursors. Bioresour. Technol. 98, 410–415. https://doi.org/10.1016/j.biortech.2005.12.013.

Suryawanshi, N., Naik, S., Eswari, J.S., 2019. Extraction and optimization of exopolysaccharide from Lactobacillus sp. using response surface methodology and artificial neural networks. Prep. Biochem. Biotechnol. 49, 987–996. https://doi.org/10.1080/10826068.2019.1645695.

Sutherland, I.W., 1982. Biosynthesis of microbial exopolysaccharides. Adv. Microb. Physiol. 23, 79–150. https://doi.org/10.1016/S0065-2911(08)60336-7.

Tansel, B., 2018. Morphology, composition and aggregation mechanisms of soft bioflocs in marine snow and activated sludge: a comparative review. J. Environ. Manage. 205, 231–243. https://doi.org/10.1016/j.jenvman.2017.09.082.

Tian, Y., 2008. Behaviour of bacterial extracellular polymeric substances from activated sludge: a review. Int. J. Environ. Pollut. 32, 78–89. https://doi.org/10.1504/IJEP.2008.0169.

Torres-Giner, S., Ocio, M.J., Lagaron, J.M., 2008. Development of active antimicrobial fiber-based chitosan polysaccharide nanostructures using electrospinning. Eng. Life Sci. 8 (3), 303–314. https://doi.org/10.1002/elsc.200700066.

Upadhyay, S.K., Singh, J.S., Singh, D.P., 2011. Exopolysaccharide-producing plant growth-promoting rhizobacteria under salinity condition. Pedosphere 21 (2), 214–222. https://doi.org/10.1016/S1002-0160(11)60120-3.

Van Colen, C., Underwood, G.J., Serôdio, J., Paterson, D.M., 2014. Ecology of intertidal microbial biofilms: mechanisms, patterns and future research needs. J. Sea Res. 92, 2–5. https://doi.org/10.1016/j.seares.2014.07.003.

Vancoppenolle, M., Meiners, K.M., Michel, C., Bopp, L., Brabant, F., Carnat, G., Van Der Merwe, P., 2013. Role of sea ice in global biogeochemical cycles: emerging views and challenges. Quat. Sci. Rev. 79, 207–230. https://doi.org/10.1016/j.quascirev.2013.04.011.

Vanhooren, P.T., Vandamme, E.J., 1999. L-Fucose: occurrence, physiological role, chemical, enzymatic and microbial synthesis. J. Chem. Technol. Biotechnol. 74 (6), 479–497. https://doi.org/10.1002/(SICI)1097-4660(199906)74:6<479::AIDJCTB76>3.0.CO;2-E.

Vicente-García, V., Ríos-Leal, E., Calderón-Domínguez, G., Cañizares-Villanueva, R.O., Olvera-Ramírez, R., 2004. Detection, isolation, and characterization of exopolysaccharide produced by a strain of *Phormidium* 94a isolated from an arid zone of Mexico. Biotechnol. Bioeng. 85 (3), 306–310. https://doi.org/10.1002/bit.10912.

Vu, B., Chen, M., Crawford, R.J., Ivanova, E.P., 2009. Bacterial extracellular polysaccharides involved in biofilm formation. Molecules 14 (7), 2535–2554. https://doi.org/10.3390/molecules14072535.

Wang, H., Liu, Y.M., Qi, Z.M., Wang, S.Y., Liu, S.X., Li, X., Xia, X.C., 2013. An overview on natural polysaccharides with antioxidant properties. Curr. Med. Chem. 20 (23), 2899–2913. https://doi.org/10.2174/0929867311320230006.

Wang, K., Li, W., Rui, X., Chen, X., Jiang, M., Dong, M., 2014. Characterization of a novel exopolysaccharide with antitumor activity from Lactobacillus plantarum 70810. Int. J. Biol. Macromol. 63, 133–139. https://doi.org/10.1016/j.ijbiomac.2013.10.036.

Wang, L., Zhang, H., Yang, L., Liang, X., Zhang, F., Linhardt, R.J., 2017a. Structural characterization and bioactivity of exopolysaccharide synthesized by Geobacillus sp. TS3–9 isolated from radioactive radon hot spring. Adv. Biotechnol. Microbiol. 4, 1–8. https://doi.org/10.19080/AIBM.2017.04.555635.

Wang, X., Shao, C., Liu, L., Guo, X., Xu, Y., Lü, X., 2017b. Optimization, partial characterization and antioxidant activity of an exopolysaccharide from Lactobacillus plantarum KX041. Int. J. Biol. Macromol. 103, 1173–1184. https://doi.org/10.1016/j.ijbiomac.2017.05.118.

Wang, J., Salem, D.R., Sani, R.K., 2019. Extremophilic exopolysaccharides: a review and new perspectives on engineering strategies and applications. Carbohydr. Polym. 205, 8–26. https://doi.org/10.1016/j.carbpol.2018.10.011.

Whitney, J.C., Howell, P.L., 2013. Synthase-dependent exopolysaccharide secretion in gram-negative bacteria. Trends Microbiol. 21 (2), 63–72. https://doi.org/10.1016/j.tim.2012.10.001.

Willis, L.M., Stupak, J., Richards, M.R., Lowary, T.L., Li, J., Whitfield, C., 2013. Conserved glycolipid termini in capsular polysaccharides synthesized by ATP-binding cassette transporter-dependent pathways in gram-negative pathogens. Proc. Natl. Acad. Sci. U. S. A. 110 (19), 7868–7873. https://doi.org/10.1073/pnas.1222317110.

Xu, X., Bi, D., Wan, M., 2016. Characterization and immunological evaluation of low-molecular-weight alginate derivatives. Curr. Top. Med. Chem. 16 (8), 874–887. https://doi.org/10.2174/1568026615666150827101239.

Yadav, V., Prappulla, S.G., Jha, A., Poonia, A., 2011. A novel exopolysaccharide from probiotic Lactobacillus fermentum CFR 2195: production, purification and characterization. Biotechnol. Bioinf. Bioeng. 1 (4), 415–421.

Yang, S.T. (Ed.), 2011. Bioprocessing for Value-Added Products From Renewable Resources: New Technologies and Applications. Elsevier, https://doi.org/10.1016/B978-0-444-52114-9.X5000-2.

Yim, J.H., Kim, S.J., Aan, S.H., Lee, H.K., 2004. Physicochemical and rheological properties of a novel emulsifier, EPS-R, produced by the marine bacterium Hahella chejuensis. Biotechnol. Bioprocess Eng. 9, 405.

Yim, J.H., Ahn, S.H., Kim, S.J., Lee, Y.K., Park, K.J., Lee, H.K., 2005. Production of novel exopolysaccharide with emulsifying ability from marine microorganism, Alteromonas sp. strain 00SS11568. In: Key Engineering Materials. vol. 277. Trans Tech Publications Ltd, pp. 155–161, https://doi.org/10.4028/www.scientific.net/KEM.277-279.155.

Zhang, Z., Chen, Y., Wang, R., Cai, R., Fu, Y., Jiao, N., 2015. The fate of marine bacterial exopolysaccharide in natural marine microbial communities. PLoS One 10 (11). https://doi.org/10.1371/journal.pone.0142690, e0142690.

Zheng, C., Li, Z., Su, J., Zhang, R., Liu, C., Zhao, M., 2012. Characterization and emulsifying property of a novel bioemulsifier by Aeribacillus pallidus YM-1. J. Appl. Microbiol. 113 (1), 44–51. https://doi.org/10.1111/j.1365-2672.2012.05313.x.

CHAPTER 4

Microorganisms in metal recovery—Tools or teachers?

Jana Sedlakova-Kadukova

Department of Ecochemistry and Radioecology, Faculty of Natural Sciences, University of Ss. Cyril and Methodius in Trnava, Trnava, Slovakia

Abstract

Environmental pollution, waste accumulation, or lack of critical resources has become a common part of our life in society. With metals as one of the main groups of pollutants—there is a problem when they are where they should not be, however, there is also a problem when they are not where we want them to be. Surprisingly, microorganisms offer us a helping hand to solve this problem. Many of them are already part of established biotechnological processes but many of them still surprise us with their abilities resulting from their roles in metal biogeochemical cycles. Their exploitation in metal recovery and waste treatment is widely accepted, nowadays, as a sustainable alternative to conventional processes. The aim of the chapter is to show the main mechanisms microorganisms use in metal recovery within the natural processes they use to obtain or detoxify metal ions, which of these processes are already used in metal recovery from the environment or waste processing within the environmental biotechnology process.

Keywords: Microorganisms, Metals, Biotechnology, Recycling, Biogeochemical cycles

1 Introduction

Growing environmental pollution is one of the most serious global challenges of the present time. A huge amount of toxic compounds, including toxic metals, pharmaceuticals, pesticides, dyes, detergents is released into the environment each year (Vardhan et al., 2019). In 2010 more than 37% of soil pollution was caused by metals, representing approximately 100,000 ha of seriously polluted soil in Europe and the USA (Vamerali et al., 2010; Kadukova and Kavulicova, 2010). Today, heavy metals are still considered the main pollutants of European soils and groundwater (Vareda et al., 2019). It was estimated that metal processing and mining activities are responsible for 48% of the total release of contaminants by the European industrial sector (Panagos et al., 2013). Except for metal processing, fossil fuel burning, vehicle traffic, chemical manufacturing or sewage irrigation significantly contribute to an increasing level of pollution (Wei et al., 2021). Consequently, increased anthropogenic industrial activities have led to an increase in the landfilling of various kinds of wastes ranging from sewage sludge, mining waste, electronic waste to nuclear waste (Fornalczyk et al., 2013) spreading pollution indirectly. Fast growth in waste amount can be documented by the increase in electrical and electronic waste that is produced in the amount of 40–50 million tonnes annually (Zhang et al., 2017). Finally, heavy metals create not only environmental pollution, but they affect food generation and quality and well-being, as well (Vardhan et al., 2019).

Microbial Syntrophy-mediated Eco-enterprising. https://doi.org/10.1016/B978-0-323-99900-7.00002-X

A huge amount of produced pollution or artificial composition unknown for natural systems are just a few examples of problems related to pollution. Elements, which were originally built in stable, little soluble compounds stored in the subsoil, reach Earth surface and become parts of the food chain where they can circulate for a long time (Han and Gu, 2010). On the other hand, microorganisms depend on these elements that represent a nutrient source and a very important energy source for microorganisms. Both these parts (microorganisms and elements) co-exist for a very long time as inseparable parts of the ecosystem. The ecosystem as a complex system forming a complicated net of organisms on Earth is mutually connected (Madsen, 2008; Crowther et al., 2019). With increasing knowledge, we are more and more surprised how much is the net connected with the abiotic environment.

Microorganisms had to learn to live with metals—they require them for their metabolism, but at higher concentrations, many of them are toxic for cells. So, the complex net of interactions exists among microorganisms and metals (Fomina and Skorochod, 2020). This ability relies on the huge genetic and metabolic diversity of microorganisms. An in-depth understanding of these interactions can become a tool in fighting against increasing pollution. In this way, microorganisms become our very good teachers as well as our very efficient tools in the war against waste.

Learning from the natural ecosystem, since the success of each ecosystem depends on its recycling ability, is crucial in the present time as many natural metal resources are already exhausted but on the other hand huge amount of metal-bearing waste is produced each year. Adaptation of circular economy principles based on the natural biogeochemical systems seems to be the most sustainable way toward zero-waste and less energy-dependent technologies.

2 Microorganisms as key-players in biogeochemical cycling

The movement of matter between the major reservoirs (the geosphere—the biosphere—the atmosphere) termed biogeochemical cycles is crucial for the existence of life on the Earth. The main aim of this cycling is to transform matter into usable forms necessary for the ecosystems function (Fisher, 2018; Crowther et al., 2019). All elements necessary for the existence of life are parts of a variety of chemical forms both organic and inorganic. Some of the biogenic elements do not undergo chemical changes within the cycling, e.g., phosphorus is cycling always in the form of phosphate ions. On the other hand, other elements change their oxidation state, type of compound, even organic and inorganic forms, e.g., iron, sulfur etc. The biochemical activity of living organisms is the main process responsible for these changes (Han and Gu, 2010; Billah et al., 2019).

Inorganic nutrient availability in the biosphere is limited so the existence of cycling is important for the survival of the whole ecosystem. The system without nutrient recycling can exist only while limiting factor (electron source or acceptor) is available (Amils et al., 2007). Autotrophic organisms synthesize organic compounds from simple inorganic/mineral compounds. They gain energy from the sun in the process of photosynthesis or energy released by chemical processes (chemosynthesis). Autotrophic organisms (plants) provide nutrients for animals (including humans) and all heterotrophic organisms. Living organisms provide nutrients and energy for parasitic organisms. Dead biomass is mineralized by microorganisms which result in the formation of an inorganic form of nutrients (Šimonovičová et al., 2013).

Key players in biochemical cycles are microorganisms. They exist in every ecosystem on our planet and contribute to its geochemistry, cycling of elements and waste degradation. Their degradation activities are many times higher than in other organisms what relates to their growth rate and amount in all parts of ecosystems (Konhausser, 2007; Yamanaka, 2008). Biochemical cycles run all together and bacteria are important in one of them can be equally important in others. Up to existing knowledge bacteria from S and Fe cycles are the most important in metal recovery, however, bacterial involvement was observed in the cycles of all metals even in the cycle of gold (Kaduková et al., 2012; Rea et al., 2016).

The simplified *sulfur* cycle is shown in Fig. 1. Microorganisms play important role in the transformation of both inorganic and organic sulfur. Inorganic sulfur is typically assimilated by organisms in the form of sulfate ions and then incorporated into organic sulfur compounds. Some of the transformations (e.g., aerobic H_2S and S^0 oxidation) can take place abiotically without microbial action just very slowly. On the other hand, at least two of the transformations anaerobic H_2S and S^0 oxidation to H_2SO_4 and sulfate reduction to H_2S do not run abiotically under typical conditions on the Earth easily. Sulfate reduction is considered one of the most important mechanism of anaerobic organic carbon mineralization (Ehrlich and Newman, 2009). Sulfur-oxidating and sulfur-reducing bacteria are the most important groups active in metal recovery, as well.

In sulfur oxidation, only prokaryotic microorganisms from domain *Archaea* or *Bacteria* can oxidize sulfur compounds in geochemically important amounts. From all these groups just bacteria from genera *Acidithiobacillus* and *Thiobacillus* produce sulfate directly without elemental sulfate accumulation under normal partial pressure of oxygen in the environment. Other microorganisms accumulate sulfur and just in the lack of hydrogen sulfate oxidize stored elemental sulfur further to sulfates (Kaduková et al., 2012).

Among bacteria reducing sulfur compounds, sulfate-reducing bacteria represent a very important group. The majority of known members belong to the domain *Bacteria*, but at least two species are from the domain *Archaea* (*Archeoglobus fulgidus* and *A. profundus*). These organisms can use various

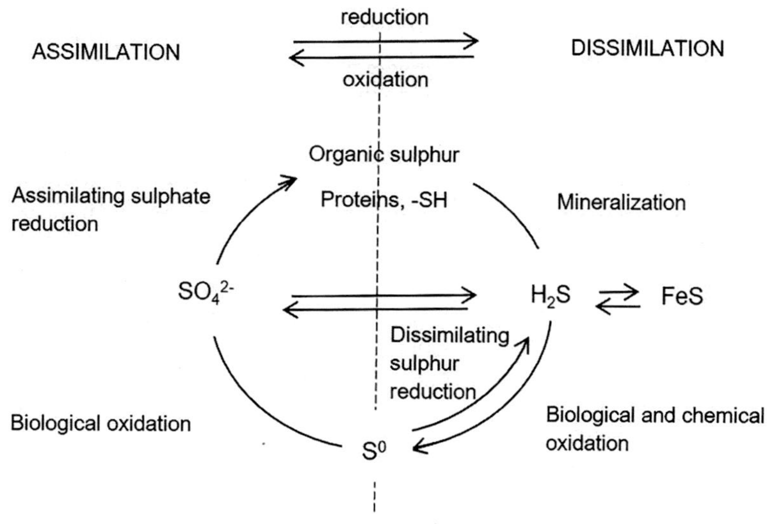

FIG. 1

Simplified sulfur cycle.

carbohydrates as well as hydrogen as electron sources and sulfur as a terminal electron acceptor in their respiratory chain producing hydrogen sulfate (Luptakova et al., 2016).

Microbial *iron* transformations via oxidation and reduction play important role in natural iron cycling. They affect iron mobility as well as accumulation in the environment. The iron cycle starts with the weathering of iron-bearing minerals. Microbial activity is the most important in mineral weathering, chemical/abiotic processes play just a minor role. The main cycle represents oxidation of Fe(II) into Fe(III) followed by the reduction back to Fe(II) (Yamanaka, 2008; Sedlakova-Kadukova et al., 2017). Iron oxidation from Fe(II) to Fe(III) at pH above 5 can take place abiotically or microbially, however, below pH 4 bacterial oxidation is dominant. Several enzymatic or non-enzymatic processes were found responsible for iron oxidation by bacteria. The Fe-containing minerals serve as electron donors under aerobic conditions (Ehrlich and Newman, 2009). However, under anaerobic conditions, trivalent iron serves as an electron acceptor and is reduced back to divalent ferrous iron. The reduction of Fe(III) under anaerobic conditions is mainly a microbially mediated process (Li and Zhou, 2020). These bacteria are very important in the carbon cycle as well.

3 Metabolism in metal recovery

Close microorganisms—metals connection is important for good function of ecosystem but in deeper understanding, it serves as a basis for biotechnological processes exploited in metal processing, waste recycling or environmental cleaning up. In a simple way, main requirement of microorganisms can be described as the need for nutrients, electrons (energy) and carbon. Several metabolic processes used by microorganisms to fulfil these needs already form a basis of biotechnological routes in metal recovery (Table 1).

For example, biosorption and bioaccumulation—processes basically used in nutrient accumulation—are on a large-scale study in wastewater or groundwater treatment technologies. In some cases, working technologies were already developed based on them. Production of secondary metabolites (e.g., organic acids, siderophores, cyanides etc.) necessary to obtain nutrients from the environment can be used in bioleaching of non-ferrous and precious metals (Fig. 2).

Table 1 Summary of cell requirements and possibilities of their exploitation in biotechnological processes.

Cell requirement	Metabolic process	Biotechnological process
Nutrient	Nutrient acquisition	Biosorption
		Bioaccumulation
		Bioleaching with organic compounds
Energy	Electron source	Bioleaching with inorganic acid and Fe^{3+} ions
	Electron acceptor	Bioprecipitation
		Biological nanoparticle formation
Detoxification	Ionic pumps	none
	Biomethylation	Volatilization or accumulation
	Bioreduction	Biological nanoparticle formation
	Bioprecipitation	Bioprecipitation

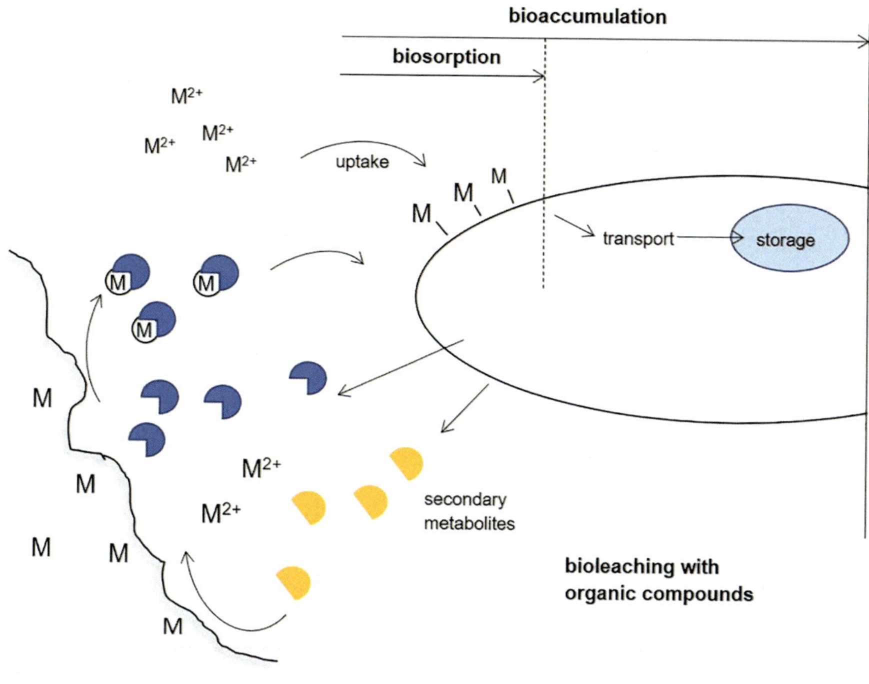

FIG. 2

Scheme of nutrient acquisition and its relation to biotechnological application (M—metal).

Processes where metal-bearing compounds serve as electron sources for an aerobic respiratory chain, mainly sulfur and iron oxidation, are widely used in non-ferrous metals bioleaching from low-grade ores or waste material. And metabolic pathways where metal ions serve as terminal electron acceptors belong among bioprecipitation or biological nanoparticles formation processes (Kaduková and Štofko, 2006; Kaduková et al., 2012).

Since not all metals (e.g., As) have any known useful exploitation for microorganisms, several existing processes are used in metal detoxification and are not part of cell metabolisms (Hussain et al., 2021).

The application of microorganisms in waste treatment is even wider not limited just to metal-bearing compounds processing. Aerobic composting and anaerobic fermentation are the main processes in organic compounds biodegradation. Without them, Earth surface would be covered with several kilometres thick layers of various wastes (Ehrlich and Newman, 2009). Both these processes are part of the carbon cycle. Microorganisms from the carbon cycle can be used in processes of CO_2 capturing resulting in the decrease of greenhouse emissions and production of various biofuels. The exploitation of waste as a microbial energy source leads to other important processes in pollution reduction, e.g., ammonia oxidation results in nitrification, one of the main processes in water treatment, and/or methane oxidation can contribute to greenhouse emissions reduction, as well (Di Maria and Sisani, 2017). However, the aim of the chapter is to focus on microorganism-metal interactions with the potential to be applied in metal recycling so just this part of interactions will be discussed further.

4 Microbial applications in metal recovery

4.1 Processes connected with nutrient metabolism

All organisms, from unicellular microorganisms to complex multicellular organisms, such as humans, every day accumulate huge amounts of various compounds. Nutrient accumulation is a fundamental process allowing cells to obtain necessary inorganic and organic compounds essential for their survival (Han and Gu, 2010). Nutrient accumulation is the net result of several interactions that can be classified as uptake, storage and elimination of chemicals. Within the uptake mechanisms nutrients are basically taken up by cell surface and then by active or passive transport systems transferred inside the cells (Tottey et al., 2007). Concentrations of such elements in the environment are usually lower than are the needs of organisms. Some of the elements necessary for cell survival such as phosphorous are accumulated by purpose; however, others such as arsenic are accumulated by chance just because of their similar physico-chemical properties with the key elements (arsenic or stibium with phosphorus, cadmium with zinc etc.). Mechanisms of these processes do not distinguish between essential and toxic chemicals (Volesky, 2003; Billah et al., 2019). It means that very often toxic metal accumulation is a non-selective process using a normal cell transport system for nutrients (Elbaz-Poulichet et al., 2000; Nanda et al., 2019).

The majority of those mechanisms represent processes responsible for metal/element uptake, its transport into the cell and inside the cell and storage, as well. These mechanisms together can be classified as processes called bioaccumulation and biosorption. While biosorption represents the first part of the sequestration process localized on the cell surfaces, bioaccumulation is the second step combining transport inside the cell and storage in organelles (vacuoles, magnetosomes) or compounds (metallothioneins, oxalates) (Nanda et al., 2019; Amor et al., 2020). Both may involve active and passive steps, however, biosorption is mostly considered a passive metal immobilization process independent of cell metabolisms. On the other hand, bioaccumulation is always an active process consuming energy. Differences between biosorption and bioaccumulation were found in their kinetics, activation energy necessary for the process or in their capacity (Kaduková and Virčíková, 2005). Biosorption as metabolisms independent process cannot be inhibited by metabolic inhibitors while bioaccumulation is sensitive to several groups of compounds—toxic metal ions, metabolic inhibitors, low nutrient concentrations, pH, temperature etc. (Michalak et al., 2013; Kadukova, 2016). Biosorption is a process based on the physical-chemical interactions between metal ions and various functional groups of the cell wall. The process is fast and reversible (Zabochnicka-Świątek and Krzywonos, 2014). There are three basic mechanisms of biosorption (Volesky, 2003):

1. physical adsorption—the process is based on intermolecular van der Waals forces. It takes place on the solid surface; however, molecules of adsorbate are not bound to the specific surface place. Multilayer and multimolecular adsorption can be present. Equilibrium is reached fast. The process is reversible, so desorption is possible.
2. chemisorption—the process is based on chemical interaction between adsorbate and adsorbent and it can take place just on specific sites on the cell surface and thus one layer of molecules can be present.
3. ionic exchange—electron exchange between adsorbate and adsorbent on the basis of higher chemical affinity of ions dissolved in the surrounding solution (adsorbate). The main equation of the process for bivalent ions can be recorded as follows:

$$2\,H-OOC-R+M^{2+}\rightarrow M(OOC-R)_2+2\,H^+ \tag{1}$$

where HOOC-R represent carboxylic functional group on the cell surface and M represent the metal ion.

Due to the complexity of cell surfaces, the biosorption process is probably more intricate, however, it is still not understood in detail. Very often the formation of precipitates can take place after the metal sequestration on the cell wall in the process called microprecipitation (Jin et al., 2016).

Intracellular bioaccumulation is a very complex and not well-understood process consisting of various mechanisms. Among already known bioaccumulation mechanisms belong:

1. binding of metal ions with intracellular compounds—compounds such as metallothioneins are biosynthesized as transport or storage molecules (Blindauer, 2011; Garcia-Garcia et al., 2016; Etesami, 2018).
2. intracellular precipitation—metal ions are reduced to metallic form (granules of Cu) (Kimber et al., 2018) or precipitated in the form of oxalates (Pueschel, 2019), citrates (Ye et al., 2018), carbonates (Cam et al., 2017) etc. and stored in the cell cytoplasm. In some cases, specific organelles are formed, such as magnetosomes in magnetotactic bacteria (Muñoz et al., 2020).

Despite the crucial role of bioaccumulation for the life of all organisms, the ability of organisms to sequester and store various elements, including toxic ions, represent a threat for them, as well. Chemicals which enter the cell in this way enter the food chain and consequently circulate within the individual trophic levels and directly threaten all food chain parts including human beings (Volesky, 2003).

However, in the environment, microelements are often present in the insoluble not bioavailable state and thus organisms need mechanisms enhancing the availability of many elements in the favourable form prior bioaccumulation. Microorganisms solubilize minerals, acidify the surrounding environment, or release various compounds enhancing nutrient bioavailability (Etesami, 2018). For example, iron is an essential nutrient for almost all organisms, however, in the environment it is present in an insoluble form (Nadell et al., 2016). Since iron can be sequestered by the cell just in the form of dissolved ion, many microorganisms evolved the ability to synthesize chelating agents able to dissolve sufficient amount to fulfil their nutrient needs. The common term of all the groups of ferric iron-chelating agents is siderophores. The group consist of various organic compounds such as organic acids, catecholates, carboxylates or hydroxamates (Khan et al., 2018; Manoj et al., 2020). Similarly, phosphate-solubilising microbes, widely distributed in the plant rhizosphere, enable changes in insoluble cation phosphate complexes unavailable for microorganisms or plants and change them into soluble phosphate ions (Mitra et al., 2020).

Many of these compounds such as citric, gluconic and/or oxalic acids have been proved very good solvents dissolving insoluble metal forms into soluble ones that are used in metal recovery from ores or waste materials. Several mechanisms were found to be responsible for metal dissolution by these compounds among them acidolysis (by proton action), complexolysis (formation of complexes) and redoxolysis (reduction of metals by organic acids) are the most important (Fomina et al., 2005; Arshadi et al., 2019). If microscopic fungi are responsible for the process, additional biomechanical deterioration of solid material by hyphae penetration or mucus production enhance the metal recovery (Gadd, 2010; Sedlakova-Kadukova et al., 2020).

All the above-mentioned processes are nowadays widely studied or even applied for metal recovery from wastewater or solid waste. Biosorption and bioaccumulation are studied for metal recovery from various industrial effluents, wastewater or drinking water with the aim to recover target metal ions or clean up the particular environmental component (Sharma and Shukla, 2021). To facilitate industrial application the clear distinction between biosorption and bioaccumulation is made based on the living state of used cells. Thus, from the technological point of view, biosorption is considered a process using non-living or dead biomass while bioaccumulation is a process where living cells are applied (Volesky, 2003). Several very good review studies on the practical application of these processes can be found in the literature (Chojnacka, 2010; Fomina and Gadd, 2014; de Freitas et al., 2019; Qin et al., 2020).

Application of organic acid or other secondary metabolites in metal recovery from various solid materials including ores and waste material is widely studied on a laboratory scale with some emerging applications on a larger scale (Garcia-Garcia et al., 2016; Srichandana et al., 2019).

4.2 Processes connected with energetic metabolism

Energy represents a key term in metabolisms. It is a basis of all metabolic processes. Humans and animals obtain energy chemotrophically degrading organic molecules and obligatory depend on oxygen as the terminal electron acceptor in their respiratory processes. However, microorganisms are the only group of organisms on the Earth that utilize a much broader range of compounds as terminal electron acceptor. Besides, they can utilize many various compounds as electron sources (substrates) which donates electrons at the beginning of their energy gaining processes. These processes belong to cell respiration processes. Thus, in energy metabolisms two basic groups of compounds can be distinguished: substrates which donate electrons and acceptors which accept electrons within the respiratory chain including the terminal electron acceptor at the end of the whole process. For aerobic organisms the terminal electron acceptor is always oxygen, however, for anaerobic organisms a whole group of molecules or elements can serve at this point. According to this classification the processes used in metal recovery can be divided into:

1. processes where electrons are obtained from inorganic substrates,
2. processes where electrons are donated to the specific electron acceptors.

In the first group in the majority of cases the electron source is an inorganic compound and the terminal electron acceptor is oxygen. The whole process is termed autotropic aerobic respiration. According to the nature of oxidized substrate, the autotrophic aerobic respiration can be divided into several groups:

1. ammonia oxidation (nitrification),
2. sulfur compounds oxidation,
3. iron compounds oxidation,
4. hydrogen oxidation,
5. methane oxidation.

One microorganism or consortium of microorganisms can be responsible for the process.

The second group of processes used in metal recovery belongs among anaerobic respiration processes where NO_3^-, Fe(III), Mn(IV), SO_4^{2-}, CO_2 or other reducible compounds serve as terminal electron acceptors instead of oxygen. Again, one or more microorganisms can be responsible for the process.

From the first group, the most important in metal recovery are processes of sulfur and iron oxidation referred to as bioleaching. The process is often called bioleaching with inorganic acids although inorganic acids have just a minor role in this process in comparison with oxidation/reduction processes. Direct and indirect mechanisms are considered responsible for this type of bioleaching. Besides, there are other processes that contribute to metal dissolution from metal-bearing materials such as acid leaching and galvanic leaching with a minor role (Sand et al., 2001; Kaduková et al., 2012). The direct mechanism is carried out by direct contact of bacteria and sulphides and it does not depend on the content of Fe^{3+} ions in the environment (Crundwell, 2003; Kaduková and Štofko, 2006). Mineral degradation takes place by enzymatic bacterial oxidation (Borovec et al., 1990; Zhang and Fang, 2005) and thus directly gain the energy from sulphidic mineral (Sharma et al., 2003). The nature of enzymatic oxidants (products of bacterial metabolisms) released into solution is not known yet, however, several compounds were suggested such as phospholipids or cysteine (Crundwell, 2003). Enzymatic oxidation of insoluble sulphides by chemoautotrophic sulfur-oxidating bacteria into soluble sulfates may be summarized as follows:

$$MS + 2O_2 \rightarrow MSO_4 \tag{2}$$

Oxygen is required for the process because of sulfuric acid formation (Lizama et al., 2003). The mechanism is probably responsible for the oxidation of pyrite (FeS_2), coveline (CuS), sphalerite (ZnS), galena (PbS) and some others (Mačingová, 2008). As an example, oxidation of pyrite by a direct mechanism can be summarized as follows:

$$2\,FeS_2 + 7\,O_2 + 2\,H_2O \rightarrow 2\,FeSO_4 + H_2SO_4 \tag{3}$$

Within indirect mechanisms bacteria do not oxidize minerals directly, they oxidize ferrous iron in an aqueous solution into ferric iron. Consequently, ferric iron (Fe^{3+}) serves as a strong oxidation agent. Bacterial oxidation of Fe^{2+} ions is much faster in comparison with any chemical oxidation (Barrett et al., 1993). The substrate is oxidized by Fe^{3+} ions forming metal sulfate, ferrous sulfate and sulfuric acid. The process can be summarized as follows:

$$Fe_2(SO_4)_3 + 2\,MS + 4\,H_2O + 2\,O_2 \rightarrow 2\,MSO_4 + 8\,FeSO_4 + 4\,H_2SO_4 \tag{4}$$

$$2\,FeSO_4 + 0.5\,O_2 + H_2SO_4 \rightarrow Fe_2(SO_4)_3 + H_2O \tag{5}$$

In indirect mechanisms just reaction 5 is carried out by bacteria. The first one represents the process of chemical leaching by a strong oxidation agent. However, without bacteria the process would stop after all Fe^{3+} ions are exhausted in oxidation. Bacteria regenerate leaching agents and thus oxidize Fe^{2+} ions into Fe^{3+} ions. In the pH range of 1.5–2.5 bacterial oxidation is approximately 10^5–10^6 faster than chemical oxidation (Munoz et al., 1995). Produced acid increase the bioleaching efficiency and help maintain the pH where Fe^{3+} ions can exist in a dissolved state. As metals in indirect mechanisms are dissolved by-product of bacterial metabolism (Fe^{3+} ions and H_2SO_4) direct contact of bacteria and mineral or waste is not necessary. Thus, this process can take place:

- without contact bacteria—metal-bearing material (two-step or spent medium bioleaching),
- with contact when bacteria attach to the material surface (one step bioleaching).

Microorganisms studied in this type of bioleaching belong to several genera such as *Acidithiobacillus* or *Leptospirillum*. Among them, *A. ferrooxidans* and *A. thiooxidans* as well as *A. caldus* are the most often applied. From other genera members of the genus *Leptospirillum*, such as *L. ferriphilum* and *L. ferrooxidans* are widely exploited (Shah et al., 2015; Auerbach et al., 2019; Zhang et al., 2019).

On the other side of energy metabolism are processes of accepting electrons at the end of the respiratory chain. Dissimilatory metal-reducing microorganisms use metals as terminal electron acceptors what often result in metal precipitation, nanoparticle formation or production of metal ions with different oxidation states with lower toxicity.

Reduction of Cr(VI) into Cr(III) by bacteria leads to the formation of less toxic and less soluble chromium compounds with a strong impact on Cr bioremediation (Sathishkumar et al., 2017). Dissimilating Fe^{3+}- reducing bacteria, e.g., *Geobacter metallireducens* or *Shewanella putrefaciens*, are able to use other metals as terminal electron acceptors, as well. They were found to reduce Tc, Pt, Pd, Mn, Co and many others. The process often results in the production of nanoparticles of particular metals (Wright et al., 2016; Belli and Tailefert, 2018; Valdiva-Gonzalez et al., 2018). Bacteria *Thauera selenatis* reduce soluble Se(VI) into insoluble elementary selenium (Butler et al., 2012).

Sulfate reducing bacteria do not use metal as the terminal electron acceptor, however, they have wide application in metal recovery. The most often sulfates (occasionally sulphites, thiosulfates or tetrathionates) serve as an electron acceptor in their respiratory chain. The product of sulfate respiration is hydrogen sulfate—with a very strong ability to precipitate metal ions present in its environment (Luptáková et al., 2002; Rudzanova et al., 2019). The main mechanisms of sulfate reduction can be described as follows:

$$4\,H_2 + SO_4^{\,2-} \rightarrow H_2S + 2\,H_2O + 2\,OH^- \,(\text{autotrophic sulfate reduction}) \tag{6}$$

$$2\,C_3H_5O_3Na + MgSO_4 \rightarrow 2\,CH_3COONa + CO_2 + MgCO_3 + H_2S + H_2O\,(\text{heterotrophic sulfate reduction}) \tag{7}$$

Produced hydrogen sulphide easily react with metal ions resulting in the formation of insoluble metal sulphides:

$$Me^{2+} + H_2S \rightarrow MeS + 2\,H^+ \tag{8}$$

Application of processes connected with energy metabolisms in metal recovery is summarized in many review articles such as Macingova and Luptakova (2012), Isosaari and Sillanpaa (2017), Qian et al. (2019), Baniasadi et al. (2019), and/or Arya and Kumar (2020).

4.3 Detoxication processes

The survival of organisms depends on their ability to obtain all necessary nutrients and energy as well as on their ability to cope with toxic characteristics of xenobiotics present in their environment. Organisms developed various mechanisms that enable them to maintain low concentrations of toxic elements inside their cells. Parts of long-term resistance developed over the organisms phylogeny is incorporated in their nucleic DNA, however, recent resistance often developed along with intensifying human industrial activities is mostly coded in plasmids or other mobile genetic elements (Jasmine et al., 2012). Since nutrient accumulating systems are mainly focused on essential elements accumulation (many of them are metals) they cannot distinguish some of the similar toxic metals and thus nonselectively accumulate them, as well. However, the ability of organisms to detoxify these elements is very specific for a particular organism and metals. Metal ions excretion via *ionic pumps* is a very selective and energy-dependent process. For example, in the case of arsenic, pentavalent arsenic in form of arsenate is sequestered by organisms based on its similarity with phosphate ion. However, inside the cell its difference and toxicity are distinguished, arsenic is reduced into more toxic trivalent ion and excreted via an ionic pump into the surrounding environment (Malik, 2004; Nanda et al., 2019). The application of this process in metal recovery is unknown.

Other basic mechanism belonging to detoxification processes is *biomethylation*. It is the ability of organisms to produce alkyl and aryl-alkyl compounds. Depending on the organism volatile compounds released into the environment or molecules stored inside the organisms can be formed (Kaduková and Štofko, 2006). These processes are often part of biogeochemical cycles of several metals such as Bi, Hg, Sn and metalloids (As, Sb, Se, Te). Volatile compounds such as arsane, mono-, di- and trimethylarsane, trimethylbismuth, stibane, mono-, di-trimethylstibane, tetramethylstannane may be produced (Xu et al., 2016). Volatile compounds are very often produced by anaerobic bacteria present in sewage sludge or other anaerobic environments. Among the main producers of volatile compounds belong methanogenic bacteria (*Methanobacterium formicicum*, *Methanosarcina barkeri*, *Methanobacterium thermoautotrophicum*), sulfate-reducing bacteria (*Desulfovibrio vulgaris*, *D. gigas*, *Desulfosporosinus auripigmentum*) and/or peptolytic bacteria (*Clostridium collagenovorans*) (Michalke et al., 2000; Andres and Bertin, 2016; Qiao et al., 2018). The majority of volatile metal compounds is more toxic than their inorganic forms. The reason is that organic derivates are lipophilic and thus more bioactive (Hussain et al., 2021). Methylated mercury compounds are also more toxic since HgS as the most common Hg compound in nature is insoluble but methylated mercury is soluble in the lipids, thus, available for organisms (Tang et al., 2020; Durand et al., 2020). On the other hand, methylation of arsenic or selenium by eucaryotic organisms produces less toxic compounds. Metal (As, Se) ions are built-in sugars or phospholipids replacing nitrogen resulting in compounds with significantly lower toxicity (Cullen and Reiner, 1989; Hussain et al., 2021).

Several aerobic organisms were found to reduce metal ions in solutions, however, the mechanisms cannot be related to respiratory processes as oxygen serves as a terminal electron acceptor for these organisms (Singh et al., 2020). The process is probably connected with the detoxification of metal ions in the aqueous environment and it may result in *nanoparticle formation*. Production of silver nanoparticles by living algal cells of *Parachlorella kessleri* was suggested as the main detoxification mechanism of silver ions present in the algal environment. A combination of two main mechanisms—rapid biosorption onto the cell surface and extracellular bioreduction resulting in nanoparticles formation—was used by the algal cell for reduction of silver toxicity. Formed nanoparticles exhibited lower antialgal activity in comparison with silver ions. The only minor part of silver (less than 2%) was accumulated inside the cell, however, again reduction and nanoparticle production were confirmed by TEM microscopy (Kadukova, 2016). Silver precipitates/nanoparticles were observed in media containing *Candida albicans*, *Saccharomyces cerevisiae*, *Streptomyces* sp., *Aspergillus niger* and many others (Korbekandi et al., 2016; Zomorodian et al., 2016; Adiguzel et al., 2018; Halbandge et al., 2019).

Overproduction of enzyme phosphatase resulting in excretion of HPO_4^{2-} ions into solution represent again probably the detoxification mechanisms used by several bacteria such as *Citrobacter* sp. to decrease the toxicity of metal ions in their environment. *Bioprecipitation* of Cd^{2+} ions in the form of $CdHPO_4$ or UO^{2+} ions in the form of $NH_4UO_2PO_4$ or $[CaU(PO_4)_2{\cdot}H_2O]$ (Macaskie et al., 2000; Voda and Danihelka, 2001; Rui et al., 2013) was observed. Sharma et al. (2018) have recorded extracellular production of pyromorphite $[Pb_5(PO_4)_3Cl]$ within Pb detoxification by *Achromobacter xylosoxidans* SJ11 in the presence of elevated phosphatase activity.

5 Conclusion

The application of environmental biotechnology is growing within last years. We understand their great advantages and try to overcome their limitations. However, our understanding is often separated from the reality of microbial life as a common part of the environment. One of the reasons is an ongoing separation of research divided into technological studies and pure biological or microbiological studies. Also, with increasing interest in these technologies a pool of information is very fast growing just recently. It is very important to realize that microorganisms exist on the Earth for a very long time, longer than we are and that their co-existence with the abiotic environment is both, long-term and very close. Therefore, metal presence is nothing new for them and they had a long time to learn how to survive together. They developed several, often shocking, ways how to cope with metal existence so they can be our teachers in the way how to change our technologies and modify our approach toward metal processing or waste production. We can learn from microorganisms to apply to recycling everywhere possible, change waste into a valuable source for new technologies as well as cooperation in various successive processes preferably in close proximity. And the same microorganisms can be our tools in developing new technologies in accordance with natural processes.

References

Adiguzel, A.O., Adiguzel, S.K., Mazmanci, B., et al., 2018. Silver nanoparticle biosynthesis from newly isolated streptomyces genus from soil. Mater. Res. Express 5 (4), 045402.

Amils, R., González-Toril, E., Fernández-Remolar, D., et al., 2007. Extreme environments as Mars terrestrial analogs: the Rio Tinto case. Planet. Space Sci. 55, 370–381.

Amor, M., Ceballos, A., Wan, J., et al., 2020. Magnetotactic bacteria accumulate a large pool of iron distinct from their magnetite. Appl. Environ. Microbiol. 86. https://doi.org/10.1128/AEM.01278-20, e01278-20.

Andres, J., Bertin, P.N., 2016. The microbial genomics of arsenic. FEMS Microbiol. Rev. 40, 299–322.

Arshadi, M., Nili, S., Yaghmaei, S., 2019. Ni and Cu recovery by bioleaching from the printed circuit boards of mobile phones in non-conventional medium. J. Environ. Manage. 250, 109502.

Arya, S., Kumar, S., 2020. Bioleaching: urban mining option to curb the menace of E-waste challenge. Bioengineered 11 (1), 640–660.

Auerbach, R., Ratering, S., Bokelmann, K., 2019. Bioleaching of valuable and hazardous metals from dry discharged incineration slag. An approach for metal recycling and pollutant elimination. J. Environ. Manage. 232, 428–437.

Baniasadi, M., Vakilchap, F., Bahaloo-Horeh, N., et al., 2019. Advances in bioleaching as a sustainable method for metal recovery from e-waste: a review. J. Ind. Eng. Chem. 76, 75–90.

Barrett, J., Hughes, M.N., Karavaiko, G.I., Spencer, P.A., 1993. Metal Extraction by Bacterial Oxidation of Minerals. Ellis Horwood Lmt, Chichester, England.

Belli, K.M., Tailefert, M., 2018. Geochemical controls of the microbially mediated redox cycling of uranium and iron. Geochim. Cosmochim. Acta 235, 431–449.

Billah, M., Khan, M., Bano, A., et al., 2019. Phosphorus and phosphate solubilizing bacteria: keys for sustainable agriculture. Geomicrobiol J. 36, 10.

Blindauer, C.A., 2011. Bacterial metallothioneins: past, present, and questions for the future. J. Biol. Inorg. Chem. 16 (7), 1011–1024. https://doi.org/10.1007/s00775-011-0790-y.

Borovec, Z., Doležal, J., Fediuk, F., et al., 1990. Úvod do biotechnologie nerostných hmot (Introduction to mineral biotechnology). SPN, Praha.

Butler, C.S., Debieux, C.M., Dridge, E.J., et al., 2012. Biomineralization of selenium by the selenate-respiring bacterium *Thauera selenatis*. Biochem. Soc. Trans. 40 (6), 1239–1243.

Cam, N., Benzerara, K., Georgelin, T., et al., 2017. Cyanobacterial formation of intracellular Ca-carbonates in undersaturated solutions. Geobiology 16 (1), 49–61.

Chojnacka, K., 2010. Biosorption and bioaccumulation—the prospects for practical application (review). Environ. Int. 36 (3), 299–307.

Crowther, T.W., van den Hoogen, J., Wan, J., et al., 2019. The global soil community and its influence on biogeochemistry. Science 365 (6455), eaav0550.

Crundwell, F.K., 2003. How do bacteria interact with minerals? Hydrometallurgy 71, 75–81.

Cullen, R.W., Reiner, K.J., 1989. Arsenic speciation in the environment. Chem. Rev. 89, 713–764.

de Freitas, G.R., da Silva, M.G.C., Vieira, M.G.A., 2019. Biosorption technology for removal of toxic metals: a review of commercial biosorbents and patents. Environ. Sci. Pollut. Res. 26 (19), 19097–19118.

Di Maria, F., Sisani, F., 2017. A life cycle assessment of conventional technologies for landfill leachate treatment. Environ. Technol. Innov. 8, 411–422.

Durand, A., Maillard, F., Foulon, J., et al., 2020. Interaction between Hg and soil microbes: microbial diversity and mechanisms, with an emphasis on fungal processes. Appl. Microbiol. Biotechnol. 104 (23), 9855–9876.

Ehrlich, H.L., Newman, D.K., 2009. Geomicrobiology, fifth ed. CRC Press, Taylor and Francis, London.

Elbaz-Poulichet, F., Dupuy, C., Cruzado, A., et al., 2000. Influence of sorption processes by iron oxides and algae fixation on arsenic and phosphate cycle in an acidic estuary (Tinto River, Spain). Water Res. 34 (12), 3222–3230.

Etesami, H., 2018. Bacterial mediated alleviation of heavy metal stress and decreased accumulation of metals in plant tissues: mechanisms and future prospects. Ecotoxicol. Environ. Saf. 147, 175–191.

Fisher, M.R. (Ed.), 2018. Environmental Biology. Pressbooks, Oregon.

Fomina, M., Gadd, G.M., 2014. Biosorption: current perspectives on concept, definition and application. Bioresour. Technol. 160, 3–14.

Fomina, M., Skorochod, I., 2020. Microbial interaction with clay minerals and its environmental and biotechnological implications. Fortschr. Mineral. 10, 861.

Fomina, M.A., Alexander, I.J., Colpaert, J.V., et al., 2005. Solubilization of toxic metal minerals and metal tolerance of mycorrhizal fungi. Soil Biol. Biochem. 37, 851–866.

Fornalczyk, A., Willner, J., Francuz, K., et al., 2013. E-waste as a source of valuable metals. Arch Mater Sci Eng 63 (2), 87–92.

Gadd, G.M., 2010. Metals, minerals and microbes: geomicrobiology and bioremediation. Microbiology 156, 609–643.

Garcia-Garcia, J.D., Sanchez-Tomas, R., Moreno-Sanchez, R., 2016. Bio-recovery of non-essential heavy metals by intra- and extracellular mechanisms in living microorganisms. Biotechnol. Adv. 34 (5), 859–873.

Halbandge, S.D., Jadhav, A.K., Jangid, P.M., et al., 2019. Molecular targets of biofabricated silver nanoparticles in *Candida albicans*. J. Antibiot. 72, 640–644.

Han, X., Gu, J.-D., 2010. Sorption and transformation of toxic metals by microorganisms. In: Mitchel, R., Gu, J.-D. (Eds.), Environmental Microbiology, second ed. Wiley-Blackwell, New York, pp. 153–176.

Hussain, M.M., Wang, J., Bibi, et al., 2021. Arsenic speciation and biotransformation pathways in the aquatic ecosystem: the significance of algae. J. Hazard. Mater. 403, 124027.

Isosaari, P., Sillanpaa, M., 2017. Use of sulfate-reducing and bioelectrochemical reactors for metal recovery from mine water. Sep. Purif. Rev. 46 (1), 1–20.

Jasmine, R., Venkadesan, B., Ragul, K., 2012. Identification and characterization of heavy metal-resistant *Pseudomonas aeruginosa* and its potential for bioremediation. Am. J. Pharm. Tech. Res. 2 (4), 783–787.

Jin, Y., Wang, X., Zang, T., et al., 2016. Biosorption of Lead (II) by arthrobacter sp. 25: process optimization and mechanism. J. Microbiol. Biotechnol. 26 (8), 1428–1438.

Kadukova, J., 2016. Surface sorption and nanoparticle production as a silver detoxification mechanism of the freshwater alga *Parachlorella kessleri*. Bioresour. Technol. 216, 406–413.

Kadukova, J., Kavulicova, J., 2010. Phytoremediation and Stress, Evaluation of Heavy Metal-Induced Stress in Plants. Nova Science Publishers, Inc., New York.

Kaduková, J., Štofko, M., 2006. Environmentálne biotechnológie pre hutníkov (Environmental Biotechnology for Metallurgists). TU Košice.

Kaduková, J., Štofko, M., Mražíková, A., 2012. Základy biometalurgie (Biolúhovanie kovov autotrofnými mikroorganizmami) (Fundamentals of Biometallurgy – Bioleaching by Autotrophic Microorgnaisms), first ed. Equilibria, s.r.o, Košice.

Kaduková, J., Virčíková, E., 2005. Comparison of differences between copper bioaccumulation and biosorption. Environ. Int. 31 (2), 227–232.

Khan, A., Singh, P., Srivastav, A., 2018. Synthesis, nature and utility of universal iron-chelator—siderophore: a review. Microbiol. Res. 212–213, 103–111.

Kimber, R.L., Lewis, E.A., Parmeggiani, F., et al., 2018. Biosynthesis and characterization of copper nanoparticles using *Shewanella oneidensis*: application for click chemistry. Small 14 (10), 1703145.

Konhausser, K., 2007. In: Majzlan, et al. (Eds.), Introduction to Geomicrobiology, 2015ed. Edmonton University, Alberta, Canada, Slovak, Bratislava.

Korbekandi, H., Mohseni, S., Jouneghani, R.M., et al., 2016. Biosynthesis of silver nanoparticles using *Sascharomyces cerevisiae*. Artif. Cell Nanomed. Biotechnol. 44 (1), 235–239.

Li, T., Zhou, Q., 2020. The key role of *Geobacter* in regulating emissions and biogeochemical cycling of soil-derived greenhouse gases. Environ. Pollut. 266, 115135.

Lizama, H.M., Fairweather, M.J., Dai, Z., et al., 2003. How does bioleaching start? Hydrometallurgy 69, 109–116.

Luptáková, A., Kušnierová, M., Fečko, P., 2002. Minerálne biotechnológie II (Mineral Biotechnology). VŠB-TU Ostrava.

Luptakova, A., Macingova, E., Kotulicova, et al., 2016. Sulphates removal from acid mine drainage. IOP Conf. Ser. Earth Environ. Sci. 44 (5), 124957.

Macaskie, L.E., Bonthrone, K.M., Yong, P., et al., 2000. Enzymically mediated bioprecipitation of uranium by a *Citrobacter* sp.: a concerted role for exocellular lipopolysaccharide and associated phosphatase in biomineral formation. Microbiology 146, 1855–1867.

Mačingová, E., 2008. Možnosti aplikácie bioremediačných metód pri eliminácii environmentálnych a priemyselných záťaží. Disertaion Thesis, SAV, Ústav geotechniky, Košice.

Macingova, E., Luptakova, A., 2012. Recovery of metals from acid mine drainage. Chem. Eng. Trans. 28, 109–114.

Madsen, E.L., 2008. Environmental Microbiology (from Genomes to Biogeochemistry), first ed. Blackwell Publishing, London.

Malik, A., 2004. Metal bioremediation trough growing cells. Environ. Int. 30, 261–278.

Manoj, S.R., Karthik, C., Kadirvelu, K., et al., 2020. Understanding the molecular mechanisms for the enhanced phytoremediation of heavy metals through plant growth promoting rhizobacteria: a review. J. Environ. Manage. 254, 109779. https://doi.org/10.1016/j.jenvman.2019.109779.

Michalak, I., Chojnacka, K., Witek-Krowiak, A., 2013. State of the art for the biosorption process—a review. Appl. Biochem. Biotechnol. 170, 1389–1416. https://doi.org/10.1007/s12010-013-0269-0.

Michalke, K., Wickenheiser, E.B., Mehring, M., et al., 2000. Production of volatile derivatives of metal(loid)s by microflora involved in anaerobic digestion of sewage sludge. Appl. Environ. Microbiol. 66 (7), 2791–2796.

Mitra, D., Anđelković, S., Panneerselvam, P., et al., 2020. Phosphate-solubilizing microbes and biocontrol agent for plant nutrition and protection: current perspective. Commun. Soil Sci. Plant Anal. 51 (5), 645–657.

Muñoz, D., Marcano, L., Martin-Rodriguez, R., et al., 2020. Magnetosomes could be protective shields against metal stress in magnetotactic bacteria. Sci. Rep. 10, 11430.

Munoz, J.A., Gonzalez, F., Blazquez, M.L., et al., 1995. A study of the bioleaching of a Spanish uranium ore. Part I: a review of the bacterial leaching in the treatment of uranium ores. Hydrometallurgy 38, 39–57.

Nadell, C.D., Drescher, K., Foster, K.R., 2016. Spatial structure, cooperation and competition in biofilms. Nat. Rev. Microbiol. 14, 589–600.

Nanda, M., Kumar, V., Sharma, D.K., 2019. Multimetal tolerance mechanisms in bacteria: the resistance strategies acquired by bacteria that can be exploited to 'clean-up' heavy metals contaminants from water. Aquat. Toxicol. 212, 1–10.

Panagos, P., Van Liedekerke, M., Yigini, Y., et al., 2013. Contaminated sites in Europe: review of the current situation based on data collected through a European network. J. Environ. Public Health 7309, 158764. https://doi.org/10.1155/2013/158764.

Pueschel, C.M., 2019. Calcium oxalate mineralisation in the algae. Phycologia 58 (4), 331–350.

Qian, Z., Tianwei, H., Mackey, H.R., van Loosdrecht, M.C.M., et al., 2019. Recent advances in dissimilatory sulfate reduction: from metabolic study to application. Water Res. 150, 162–181.

Qiao, J.-T., Li, X.-M., Li, F.-B., 2018. Roles of different active metal-reducing bacteria in arsenic release from arsenic-contaminated paddy soil amended with biochar. J. Hazard. Mater. 344, 958–967.

Qin, H., Hu, T., Zhai, Y., et al., 2020. The improved methods of heavy metals removal by biosorbents: a review. Environ. Pollut. 258, 113777.

Rea, M.A., Zammit, C.M., Reith, F., 2016. Bacterial biofilms on gold grains—implications for geomicrobial transformation of gold. FEMS Microbiol. Ecol. 92, fiw082.

Rudzanova, D., Luptakova, A., Macingova, E., 2019. The possibilities of using sulphate-reducing bacteria for phenol degradation. Physicochem. Probl. Miner. Process. 5 (55), 1148–1155.

Rui, X., Kwon, M.J., EJ, O.´.L., et al., 2013. Bioreduction of hydrogen uranyl phosphate: mechanisms and U(IV) products. Environ. Sci. Technol. 47 (11), 5668–5678.

Sand, W., Gehrke, T., Jozsa, P.-G., et al., 2001. (Bio)chemistry of bacterial leaching-direct vs. indirect bioleaching. Hydrometallurgy 59 (2–3), 159–175.

Sathishkumar, K., Murugan, M., Benelli, G., et al., 2017. Bioreduction of hexavalent chromium by *Pseudomonas stutzeri* L1 and *Acinetobacter baumannii* L2. Ann. Microbiol. 67, 91–98.

Sedlakova-Kadukova, J., Marcincakova, R., Luptakova, A., Vojtko, M., Fujda, M., Pristas, P., 2020. Comparison of three different bioleaching systems for Li recovery from Lepidolite. Sci. Rep. 10 (1), 14594.

Sedlakova-Kadukova, J., Marcincakova, R., Mrazikova, A., et al., 2017. Closing the loop: key role of iron in metal-bearing waste recycling. Arch. Metall. Mater. 62 (3), 1459–1466.

Shah, M.B., Tipre, D.R., Purohit, M.S., Dave, S.R., 2015. Development of two-step process for enhanced biorecovery of Cu-Zn-Ni from computer printed circuit boards. J. Biosci. Bioeng. 120 (2), 167–173.

Sharma, B., Shukla, P., 2021. Lead bioaccumulaiton by *Bacillus cereus* BPS-9 from an industrial waste contaminated site encoding heavy metal resistance genes and their transporter. J. Hazard. Mater. 401, 123285.

Sharma, J., Shamim, K., Dubey, S.K., 2018. Phosphatase mediated bioprecipitation of lead as pyromorphite by *Achromobacter xylosoxidans*. J. Environ. Manage. 217, 754–761.

Sharma, P.K., Das, A., Hanumantha Rao, K., et al., 2003. Surface characterization of *Acidithiobacillus ferrooxidans* cells grown under different conditions. Hydrometallurgy 71, 285–292.

Šimonovičová, A., Piecková, E., Ferianc, P., et al., 2013. Environmentálna mikrobiológia (Enviromental Microbiology), first ed. Commenius University, Bratislava.

Singh, R.P., Anwar, M.N., Singh, D., Bahuguna, V., Manchanda, G., Yang, Y., 2020. Deciphering the key factors for heavy metal resistance in gram-negative bacteria. In: Singh, R.P., et al. (Eds.), Microbial Versatility in Varied Environments. Springer, Singapore, pp. 101–116, https://doi.org/10.1007/978-981-15-3028-9_7.

Srichandana, H., Mohapatrab, R.K., Parhib, P.K., et al., 2019. Bioleaching approach for extraction of metal values from secondary solid wastes: a critical review. Hydrometallurgy 189, 105122.

Tang, W.-L., Liu, Y.-R., Guan, W.-Y., et al., 2020. Understanding mercury metathylation in the changing environment: recent advances in assessing microbial methylators and mercury bioavailability. Sci. Total Environ. 714, 136827.

Tottey, S., Harvie, D.R., Robinson, N.J., 2007. Understanding how cell allocate metals. In: Nies, D.H., Silver, S. (Eds.), Molecular Microbiology of Heavy Metals. Springer-Verlag, Berlin, Heidelberg, pp. 4–35.

Valdiva-Gonzalez, M.A., Diaz-Vasquez, W.A., Ruiz-León, D., et al., 2018. A comparative study of tellurite detoxification by member of the genus *Shewanella*. Arch. Microbiol. 200, 267–273.

Vamerali, T., Bandiera, M., Mosca, G., 2010. Field crops for phytoremediation of metal-contaminated land. A review. Environ. Chem. Lett. 8, 1–17.

Vardhan, K.H., Kumar, P.S., Panda, R.C., 2019. A review on heavy metal pollution toxicity and remedial measures: current trends and future perspectives. J. Mol. Liq. 290, 111197.

Vareda, J.P., Valente, A.J.M., Duraes, L., 2019. Assessment of heavy metal pollution from anthropogenic activities and remediation strategies: a review. J. Environ. Manage. 246, 101–118.

Voda, V., Danihelka, P., 2001. Bioeliminace toxických kovů z odpadních vod, Recyklace odpadů V. In: Conference Proceedings, Ostrava, pp. 43–49.

Volesky, B., 2003. Sorption and Biosorption. BV Sorbex, Inc, Montreal, Canada.

Wei, Z., Le, Q.V., Peng, W., et al., 2021. A review on phytoremediation of contaminants in air, water and soil. J. Hazard. Mater. 403, 123658.

Wright, M.H., Farooqui, S.M., White, A.R., et al., 2016. Production of manganese oxide nanoparticles by *Shewanella* species, production of manganese oxide nanoparticles by *Shewanella* species. Appl. Environ. Microbiol. 82, 5402–5409.

Xu, L., Wu, X., Wang, S., et al., 2016. Speciation change and redistribution of arsenic in soil under anaerobic microbial activities. J. Hazard. Mater. 301, 538–546.

Yamanaka, T., 2008. Chemolithoautotrophic Bacteria (Biochemistry and Environmental Biology), first ed. Springer.

Ye, B., Luo, Y., He, J., et al., 2018. Investigation of lead bioimmobilization and transformation by *Penicillium oxalicum* SL2. Bioresour. Technol. 264, 206–210.

Zabochnicka-Świątek, M., Krzywonos, M., 2014. Potentials of biosorption and bioaccumulation processes for heavy metal removal. Pol. J. Environ. Stud. 23 (2), 551–561.

Zhang, G., Fang, Z., 2005. The contribution of direct and indirect actions in bioleaching of pentlandite. Hydrometallurgy 80, 59–66.

Zhang, M., Guo, X., Tian, B., et al., 2019. Improved bioleaching of copper and zinc from brake pad waste by low-temperature thermal pretreatment and its mechanisms. Waste Manag. 87, 629–635.

Zhang, S., Ding, Y., Liu, B., et al., 2017. Supply and demand of some critical metals and present status of their recycling in WEEE. Waste Manag. 65, 113–127.

Zomorodian, K., Pourshahid, S., Sadatsharifi, A., et al., 2016. Biosynthesis and characterisation of silver nanoparticles by *Aspergillus* species. BioMed Res. Int. 2016. https://doi.org/10.1155/2016/5435397, 5435397.

CHAPTER

5 Endophytic microbes mitigate biotic-abiotic stresses and modulate secondary metabolite pathway in plants

Sucheta Singh[a], Suman Singh[b], Akanksha Singh[c], and Alok Kalra[c]

[a]*Molecular Biology and Biotechnology Division, CSIR-National Botanical Research Institute, Lucknow, Uttar Pradesh, India,*

[b]*Department of Botany, University of Lucknow, Lucknow, Uttar Pradesh, India,*

[c]*Division of Crop Protection and Production, CSIR-Central Institute of Medicinal and Aromatic Plants, Lucknow, Uttar Pradesh, India*

Abstract

Microorganisms in nature are found as parasites, symbionts, and mutualists or as pathogens. It is the plant immunity which decides whether the interaction with microbes will be friendly or hostile. Friendly interactions may have a unique way of mutual interrelations for a resource contribution. One of them is called plant-endophytes symbiotic or mutualistic relation. Endophytes are known to positively impact the development, growth, fitness, uptake of nutrients, and plant secondary metabolite pathway by a complex set of directly or indirectly influencing mechanisms. Secondary metabolites of the plant (having medicinal properties or aromatic compounds of perfumery interest) are of tremendous use for human welfare and commercial importance. Many factors such as plant habitat, plant genotype, biotic and abiotic factors, seasonal changes, and ecological conditions may affect plant growth, development, and productivity. Therefore, keeping in mind the above constraints it is necessary to develop alternative methods for improving plant productivity. In this context, a suitable alternative source like endophytes could be helpful in improving the plant products and meeting their demands for human welfare.

Keywords: Endophytes, Biotic and abiotic factors, Plant-endophytes symbiotic relation, Secondary metabolites

1 Introduction

Microorganisms in nature are found as parasites, symbionts, and mutualists or as pathogens. It is the plant immunity which decides whether the interaction with microbes will be friendly or hostile (Zeilinger et al., 2016). Friendly interactions may have a unique way of mutual interrelations for a resource contribution. One of them is called plant-endophytes symbiotic or mutualistic relation (Kusari et al., 2012a, b). De Bary (1866) was the first to coin the term endophyte (Gr. *endon*, within; *phyton*, plant). An endophyte is a microorganism, which spends the whole or a part of its life cycle inside the healthy tissues, inter- and/or intra-cellularly, of the host plant, without causing any apparent symptoms of disease (Wilson, 1995). Microorganisms requiring living plant tissues to complete their life

Microbial Syntrophy-mediated Eco-enterprising. https://doi.org/10.1016/B978-0-323-99900-7.00001-8

cycle are termed as "obligate." On the other hand, endophytes that mainly thrive outside plant tissues (epiphytes) and sometimes enter the plant endosphere are "opportunistic" (Hardoim et al., 2012). These endophytes are often rhizospheric in nature and originate from microbial division in the soil environment. Only specialized community adapted or merely opportunistic microbes survive (Hardoim et al., 2012) and enter through germinating radicles (Gagne et al., 1987), plant secondary roots (Agarwal and Shende, 1987), stomata (Roos and Hattingh, 1983), wounds, hydathodes and may be vertically transmitted from parent to offspring via seeds. Root has higher colonization, mainly through the epidermis of lateral root emergence or below the root hair zone, than through phyllosphere or through stomata. The fact is that the below-ground root system in contrast to the above-ground organ system of the plants is more frequently and systemically colonized by the microbes. It may be due to the fact that roots closely interact with the soil environment that harbors plant growth-promoting microbes which actively provide water and nutrients to the plants (Schulz and Boyle, 2005). Many studies have suggested that mutualistic interactions between microorganisms and roots developed because plant roots act as natural carbon sink which provides dual and multi-organism symbioses with nutrients (Yang et al., 2020). In return, the microorganisms provide water and minerals to the plants. Additionally, in contrast to the shoot; plant roots have fewer xeromorphic tissue structures (epidermal wall, wax, etc.). Since an endophyte cannot directly improve the nutrient level in photosynthetic organs, therefore, mutualistic systemic interaction with the roots system is found to have more potential than shoot system interaction. On the molecular front, mutualistic interaction of roots with microorganisms has been proved with various techniques (Imaizumi-Anraku et al., 2005; Nakagawa and Imaizumi-Anraku, 2015). Colonization of the endophytes depends on their efficiency and ability to enter and establish inside the different organs of the host plant. Endophytic microbes can be classified under the categories into different recognizable classes which are generally dependent upon their plant organ source, with the major groups as follows: (1) endophytic Clavicipitaceae; (2) fungal endophytes of dicots; (3) fungal endophytes of lichens; (4) endophytic fungi of bryophytes and ferns; (5) fungal endophytes of root; (6) endophytic fungi of tree bark; (7) fungal endophytes of xylem; (8) fungal endophytes of gall and cysts; (9) other systemic fungal endophytes; and (10) prokaryotic endophytes of plants (includes endophytic bacteria and actinomycetes) (Zhang et al., 2006a, b). Fossilized tissues of plant stems and leaves have also shown the microbial-plant association which suggests that endophytic association with host plants was generated in times when higher plants initially emerged on the earth (Strobel and Daisy, 2003). Successful colonization and adaptation of endophytes within host plants depend on many variables, including microbial taxon, strain type, plant tissue type, plant genotype, biotic and abiotic environmental conditions. Colonization and population of endophytes in the plant and the application of their beneficial effects involved from direct impact to indirect influential mechanisms under different conditions (Glick, 2012).

Endophyte-host interaction involves asymptomatic colonization with "balanced antagonism," viz. endophytes lives within plant system without activating host defense and improve their self-resistance through self-production of plant-like metabolites, and/or by avoiding the effect of plant toxic metabolites (Schulz et al., 1999). It is irrespective of the host organ specificity but the balance between plant and endophytes is influenced by colonization pattern, infection modes, physiological status, life-history strategy, developmental and evolutionary stages (Schulz and Boyle, 2005). Often, it is difficult to know about life-history strategies or the importance of endophytic phase of microbes. Different approaches are used to recognize bacterial cell and fungal mycelium in plant tissues. Autofluorescent proteins (AFP) are the key methods and tools for the visualization and study of plant-microbe interaction and biofilm formation. Techniques for visualization and gene expression studies include in vivo expression technology (IVET), GFP (green fluorescent protein), glucuronidase (GUS) reporter system, and recombination of in vivo expression technology (Leveau and Lindow, 2001; Preston et al., 2001; Zhang et al., 2006a, b). *GFP* gene integrated into the chromosome of bacteria or plasmid containing *GFP* cloned in a bacterial cell can be visualized either by confocal laser scanning microscopy or epifluorescence microscopy (Villacieros et al., 2003; Germaine et al., 2004). *GFP* biomarker is a better technique in terms that it does not require a substrate or cofactor for their fluoresce expression. In GUS reporter system, staining occurs where bacteria move.

2 Plant endophytes symbiosis and their impact on plant phytochemical changes during biotic and abiotic stresses

2.1 Stress in plants

Any change in physiological condition alternating the equilibrium state generates a stressful situation. This is also known as strain because the thermodynamic optimal point shifts forcefully during chemical and physical changes. The development of plant parts and other related activities are influenced by a variety of stresses and these stresses have been categorized into biotic and abiotic stress. The growth of the plant is decreased due to the effect of these stresses due to alternation in plants morphology, physiology, biochemistry, and molecular biology. The detailed mechanisms operating in plant system under the influence of biotic and abiotic stresses have been discussed below.

2.1.1 Abiotic and biotic stresses

Abiotic stresses are the environmental circumstances which reduce the growth, productivity, and fertility of plants rather than the action of other organisms. Abiotic stresses have been classified into salinity, drought, floods, high or low temperatures, poor soil quality (due to nutrient status, acidic or alkaline soil, heavy metal contamination in soil), long-term exposure to UV radiation, or low light intensity, etc. A number of agricultural crops having economic values are vulnerable to abiotic stress and losses recorded up to 82.1% for *Triticum aestivum*, 65.8% for *Zea mays*, 69.3% for *Glycine max*, and 54.1% for *Solanum tuberosum* have been recorded by a group of workers (Wang et al., 2013).

Biotic stresses are caused by living organisms, for example, microorganisms including bacteria fungi, viruses, insect pests, and parasites, etc., which are responsible for the plant diseases damaging the morphological, biochemical, and molecular mechanisms in plants. Losses reported from various biotic stresses are also huge as loss up to 28.2% *in Triticum aestivum*, 37.4% *Oryza sativa*, 31.2% *Zea mays*, 40.3% *Solanum tuberosum*, 26.3% *Glycine max*, and 28.8% *Gossypium* spp. have been reported by scientists (Wang et al., 2013). Among these microorganism fungi are the most harmful biotic stress factor that causes disease in plants and severely affected crop yields.

2.2 Tactics adapted to mitigate environmental stress in plants

Overpopulation and depletion of natural resources have increased global food demand which is anticipated to almost double by 2050. Although the Green Revolution has cooperated in enhancing agricultural productivity worldwide in the past, however, limited resources such as arable land, eco-friendly fertilizers, climate, and water create a serious challenge for global food security in the future. To fulfill the demands, several methods such as breeding practices, crop rotation, and use of genetically engineered stress-tolerant plant varieties have been evaluated for the enhancement of crop productivity but have largely proved to be less effective in totality (Ceccarelli, 2015). Physical and chemical methods including excavation, in situ fixation or stabilization (addition of chemicals to modify the metal structure so that it does not get absorbed by the plants) employed for the clean-up of contaminated sites and soil cleansing, have also been tested (Bradl and Xenidis, 2005). However, these methods have proved to be slow, expensive and especially, caused genetic erosion of native species. Therefore cost-effective, reproducible, and environmentally friendly approaches are needed with confirmed tolerance to abiotic and biotic stresses, improvising quality of nutrition, and the ability to acclimatize under varying environment for constant agricultural sustainability. Applications of plant growth-promoting microbes (PGPM) are most promising in recent years to increase the productivity of crops and reduce the use of chemical fertilizers (Zhang et al., 2020). Among these PGPM, endophytes are well-known efficient inoculants for agriculture, as these microorganisms make beneficial associations with plants and promote plant growth through a number of PGP traits fulfilling the global food demand (Fadiji and Babalola, 2020).

2.3 Role of endophytes in reducing abiotic stresses in plants

Rise and fall in environmental conditions as well as global warming distress the plant growth. Various literature has surfaced around the globe which has reported that abiotic stresses such as water-logging, drought, heat, cold, salinity, and metal toxicities have severely reduced the agricultural crop yields. The responses of plants to abiotic stresses are vibrant and complicated; it can be permanent (irreversible) or temporary (reversible) (Cramer et al., 2011). Endurance against abiotic effect depends on the plant tissues or parts affected by the stress. Moreover, the duration and intensity of stress like acute or chronic also play a crucial role in the survival of the affected plants (Pinheiro and Chaves, 2011).

2.3.1 Drought stress

Drought is generally defined as scarcity of water which takes place in warm and dry areas where less rainfall occurs. For the survival of plants, water is a vital ingredient for transporting essential nutrients for supporting the growth and development of plants. Therefore, scarcity of water leads to drought stress reducing water levels in plant cells (Skirycz and Inze, 2010). In this process, the meristematic cells cover the xylem and hinder the water flow to the growing cells in the epidermal region thereby accentuating plants to develop higher water potential (Yadav et al., 2021). Continuous stress limits the stomatal activity for CO_2 uptake and increases photoinhibition making it difficult to disperse light energy, thus declining the photosynthesis rate (Pinheiro and Chaves, 2011). Energy metabolism in the plant is affected by severe drought stress which consequently results in complex changes in several physiological, chemical, biological, and molecular processes taking place in response to the stress (Lei et al., 2015; Min et al., 2016). The effect of inoculation of various endophytes on a variety of plant growth attributes has been investigated under drought stress. Under drought stress, bacterial endophytes *Burkholderia phytofirmans* (PsJN) and *Enterobacter* sp. (FD17) were applied to two maize cultivars which showed an increase in biomass, photosynthetic efficiency, and higher leaf relative water content thereby minimizing the effect of drought stress in bacterized seedlings over the non-bacterized control plants (Naveed et al., 2014). Likewise, Ullah et al. (2017) have reported that endophytic bacteria improved plant yield through the synthesis of plant hormones such as IAA, ABA, and gibberellins which in turn helped in uptake of nutrients and also reduced osmotic stress. In general plants are able to tolerate drought stress through their increased root growth by which water and nutrient content reaches the plants (Timmusk et al., 2014). Impact of an endophytic fungus *Paraconiothyrium variabile* on *Cephalotaxus harringtonia* leaf secondary metabolome was studied and it was found that fungus metabolized the glycosylated flavonoids of the plants into aglycone moieties which play a significant role in hyphal growth of endophyte's germinated spores (Tian et al., 2014). Another study stated that *Neotyphodium coenophialum* infected *Festuca arundinacea* (Schreb.) plants showed an increased level of phenols in roots and shoots over non-inoculated plants (Malinowski et al., 1998). Therefore, endophytic microbes associated with plants are capable of reducing drought effects through various mechanisms such as phytohormone synthesis, compatible solute formation, ACC deaminase, and indirectly by improving the relative water content, reduction of oxidative stress and also by inducing synthesis of secondary metabolites as discussed later in this chapter.

2.3.2 Salt stress

In dried areas, when solutes from the irrigation water accumulate in soil and ultimately reach a level that badly affects the soil fertility and plant growth is called salt stress. It has been estimated that 800 million hectares of agricultural land are affected by salt stress worldwide; highly found in central and north Asia, South America, and Australia (Yasin et al., 2018). Based on the duration and severity of salt stress; various physiological and metabolic mechanisms are affected in plants. Salt stress creates an ionic imbalance in plants, induces nutrient deficiency thereby reducing the level of water in the root system due to osmotic stress. Total photosynthetic capacity of the plant is decreased by the salinity stress which reduces leaf growth and limits their ability to grow (Hashem et al., 2016; Pan et al., 2019). Tolerance approaches in plants against salt stress are dynamic and complex which can affect any biosynthetic mechanism in the life cycle of plants. For instance, under salt stress, photosynthesis was found to be reduced by stomatal closure in soybean and wheat (He et al., 2014; Szopko et al., 2017); decreased biomass and RUBISCO activity recorded in soybean sensitive to salt (Carrell and Frank, 2014); decreased content of chlorophyll in maize leaves (Chen et al., 2018); decreased enzyme activity, protein content and downregulated genes for RUBISCO in *Elaeagnus aungustifolia* L. (Lin et al., 2018). Various salt-tolerant endophytic species have been studied to increase plant growth under salt stress (Gangwar and Singh, 2018). In report colonization by *Glomus* sp. mixture (AMF) improved nutrient uptake and growth of tomato (*Solanum lycopersicum* L.) plants under salinity over the untreated salinized plants (Balliu et al., 2015). Another similar study exhibited that AMF colonization ameliorated the salt stress effect in *Cucumis sativus* by increasing their root growth and photosynthetic efficiency over the untreated salinized plants (Sallaku et al., 2019). In many other cases, endophyte colonized plants have been shown to hold greater osmolytes content which has the ability to reduce the harmful effects of ROS (reactive oxygen species) under salt stress (Ahmad et al., 2015; Song et al., 2015). Besides, some secondary metabolites such as carotenoids, tocopherols, polyamines, phenols and flavonoids have been reported to be produced in plants as a protective agent against oxidative stress by the inoculation of endophytes (Khan et al., 2014). For example, *Penicillium minioluteum* associated soybean plants increased growth attributes, nitrogen assimilation under salt stress over the untreated non-salinized plants. The endophyte-soybean symbiosis reduced the ABA content and increased salicylic acid secretion in plants. Moreover, they also enhanced the daidzein and genistein contents in the soybean over the untreated salinized plants (Khan et al., 2011). Another study revealed that inoculation of *Trichoderma harzianum* improved barley plant growth by modulating the content of metabolites under salt stress for sustainable agriculture.

2.3.3 Temperature

Plants living in varying temperature ranges (high and low) experience sudden changes in metabolic routes, and the higher temperature may trigger early leaves senescence, changes in phytochemicals and gene expression (Thomsen et al., 2012). Plant growth, photosynthesis and other developmental processes are impaired due to overproduction of ROS (reactive oxygen species) under high temperature. Denaturation of heat shock proteins, alteration in enzymatic activities and nucleic acid damages also occur by heat stress which eventually causes cell death (Ihsan et al., 2019). As normal proteins get reduced due to heat stress, plants release heat shock proteins (Schulz and Boyle, 2005) for the maintenance and conformation of other proteins in response to resistance against stress (Wahid et al., 2007). Cold stress also badly affects the plant growth in terms of stunted seedlings, chlorosis, leaf curling, wilting which ultimately leads to necrosis. Cold stress majorly damages the cell membrane due to acute dehydration and incurs alteration in biochemical and molecular mechanisms in plants (Yadav, 2010). To protect against temperature (heat or cold) stress, plants release various enzymatic antioxidants like superoxide dismutase (SOD), peroxidase (POD-APx/GPx), catalase (CAT), glutathione reductase (GR) and non-enzymatic antioxidants including ascorbic acid, tocopherols, glutathione, β-carotene and polyphenols (Strimbeck et al., 2015; Askari-Khorasgani et al., 2019). Evidences for plant-endophyte mediated extreme temperature stress tolerance and related mechanism are also available. It has been reported that endophyte and plants cannot tolerate individually against high temperature stress up to 40 °C. However, when both are grown symbiotically they are able to tolerate up to 65 °C temperature (Márquez et al., 2007). Likewise another study showed that in the case of *F. culmorum* which colonizes coastal dunegrass (*L. mollis*), both host plant and endophyte survived under salt stress (Rodriguez et al., 2009). Another case also revealed that fungal endophyte *Curvularia protuberata* increased the efficacy of monocot *Dichanthelium lanuginosum* (panic grass) by resisting the geothermal effect and both symbiotic partners were also able to tolerate salinity stress (300–500 mM NaCl) (Rodriguez et al., 2009). These observations pointed toward the fact that endophyte and plants both mutually benefit each other for escaping from the harsh effects of abiotic stress (Rodriguez et al., 2009). Likewise, in another study endophytic fungi residing in plants are reported to secrete specific molecules in response to stress as seen in case of *Paraphaeosphaeri aquadriseptata* which induced the synthesis of monocillin I (heat shock protein 90 (HSP90) inhibitor) enhancing heat stress tolerance in *Arabidopsis* (Mclellan et al., 2007).

Under low temperature conditions, plants-endophytic partnership undergoes acclimatization processes that stimulate metabolic pathways and regulates the expression of genes changing the osmolytes, proline, proteins, sugars, and lipid composition in membrane and peroxidation in plants (Acuna-Rodriguez et al., 2020). For example, *Glomus etunicatum* in association with maize plants showed positive effects in the form of an increase in ROS-scavenging enzymes like catalase and peroxidise in maize plants at 5 °C and over the uninoculated plants (Zhu et al., 2010). Likewise another study showed *Rhizophagus irregularis* reduced the MDA content and enhanced the antioxidant enzyme activities in *Digitaria erianthea* under cold stress (4 °C) (Pedranzani et al., 2016). Another endophyte *Burkholderia phytofirmans* enhanced the acclimation of *Vitis vinifera* to cold conditions (4 °C) by inducing the level of carbohydrates such as trehalose, raffinose or starch (Fernandez et al., 2012a, b). Accumulation of trehalose or soluble sugar acts as osmoprotectants maintaining the water balance in plant cells under osmotic stress (Thalmann and Santelia, 2017). Ait-Barka et al. illustrated that elevated levels of proline in endophyte-associated plants provided osmotic benefits under cold stress (Ait-Barka et al., 2006).

2.3.4 Light

Light is also an important abiotic factor responsible for the growth and several metabolic processes in plants. Photoperiod or short duration also influences the plant growth associated with changes in photosynthetic rate, secondary metabolite production and gene expression (Ghosh et al., 2018). Sunlight includes UV-A, UV-B, and infrared radiation but plants can utilize only little fraction of UV-B, a component of daylight spectrum, for the growth and development, depending on their exposure time (Caldwell et al., 2003). The light is also observed as a limiting factor that affects plants growth and metabolic processes (Isah et al., 2018). Earlier researches illustrated that high light intensity can induce parasitism in the endophyte associated with a plant (Álvarez-Loayza et al., 2011; Davitt et al., 2010). It has also been reported previously that light has an essential role in which low light intensity appears to be beneficial in plant endophyte association (Gangwar and Singh, 2018; Gao et al., 2010) while strong light intensity could trigger some physiological alteration in the endophytic fungus inducing pathogenicity such as the formation of reactive oxygen species in plants (Álvarez-Loayza et al., 2011; Egan et al., 2007).

2.3.5 Heavy metal stress

Heavy metals (HMs) such as Pb, Cd, Cr, As, Ni, Cu, Hg, etc., are dangerous pollutants released into the atmosphere from mostly anthropogenic activities (Lajayer et al., 2017). HMs in high quantities are observed to be highly injurious for plants (Kumar et al., 2016). The noticeable signs of HMs toxicity in plants are stunted growth, root browning, leaf yellowing, and ultimately death of the plants (Shah et al., 2015). The direct effect of HMs on plants causes cell toxicity through increased synthesis of ROS (reactive oxygen species), which disturb photosynthesis, respiration, and homeostasis (Nanda and Agrawal, 2016; Rui et al., 2016). Many physiological and biochemical processes including growth and development of plants are negatively influenced by HMs (Munzuroglu and Geckil, 2002). This condition lowers the genetic potential of plants affecting yield and quality of food (Bücker-Neto et al., 2017). Although plants become accustomed to HMs stress, but resistance depends on the potential to impede oxidative stress and firm regulation of antioxidant activities (Lajayer et al., 2017; Choudhury et al., 2017). To overcome HMs stress, endophyte-associated plants develop a variety of physiochemical and molecular mechanisms. Wan et al. have reported that endophytes have the ability to reduce oxidative damage due to HMs by lowering lipid peroxidation and enhancing antioxidant enzyme activities (Wan et al., 2012). In a study, *Methylobacterium oryzae* and *Burkholderia* sp. inoculation in tomato plants showed lower accumulation of Cd and Ni reducing the toxicity in plants (Madhaiyan et al., 2007). Studies also suggested that endophytes also protected plants from the drastic effects of herbicides (Ngigi et al., 2012). For example, *Pseudomonas* sp. associated plants showed no accumulation of herbicides in plant tissues upon exposure to 2,4-dichlorophenoxyacetic acid (Germaine et al., 2006). There may be some metabolic changes occur during stress tolerance in plant due to association of endophytes (Lafi et al., 2017). Endophytes also alter the host secondary metabolite production, but the mechanisms adopted by the plants are not clear. An endophytic fungus *Eupenicillium parvum* was found to induce Azadirachtin syntheisis which is a natural insecticide from *Azadirachta indica* plant (Kusari et al., 2012a, b). In a case study, *B. amyloliquefaciens* inoculation significantly increased growth attributes of rice seedlings over non-endophytic seedlings in Cu stress. It reduced Cu accumulation by lowering the abscisic acid (ABA) and jasmonic acid level and also increased antioxidant enzyme activities, amino acid synthesis as well as carbohydrate metabolism (Raheem et al., 2019).

2.4 Mode of action of endophytes in plants under abiotic stress

Plant growth-promoting microbes are known to enhance crop productivity and protect plants from the harmful effect of abiotic stress (Vurukonda et al., 2016). As endophytes reside inside the plants, they are likely to collaborate more strongly with their host plant as compared to rhizospheric bacteria (Miliute et al., 2015). It has been illustrated that plants act very fast in response to stress when inoculated with the endophytes (Compant et al., 2005). For example, plant growth-promoting endophytes *B. phytofirmans* and *Enterobacter* sp. were found to enhance growth and reduce the effect of abiotic and biotic stress in some crops and vegetables (Mitter et al., 2013). The reasonable mechanism induced by endophytes in plants is to produce a number of antistress metabolites in response to tolerance against environmental stresses, such as: (1) phytohormones like indole-3-acetic acid (IAA), gibberellic acid, cytokinins, and abscisic acid (ABA); (2) ACC deaminase for decreasing the ethylene content; (3) IST (induced systemic tolerance) by microbial metabolites (induce host stress response); (4) formation of osmoprotectants; and (5) antioxidant defense mechanisms in plants for alleviating the stress effects (Singh et al., 2011).

Hormones produced by endophytes also play a critical role in tolerances to abiotic stress in plants (Hashem et al., 2016). Many literatures have shown that phytohormones (IAA, gibberellins, ethylene, abscisic acid, and cytokinins) are important for the survival of plants under stressful conditions (Skirycz and Inze, 2010; Egamberdieva, 2013; Fahad et al., 2015). For example, inoculation of *Azospirillum lipoferum* in maize plants showed amelioration of drought stress effects due to the formation of abscisic acid (ABA) and gibberellins (Cohen et al., 2008). ABA is a stress hormone released by the plants during drought stress which plays an important role in controlling water loss with the help of managing stomatal closure and regulating stress-induced signal transduction mechanisms (Kaushal and Wani, 2015). Another study illustrated that *A. terreus* associated *Pennisetum glaucum* seedlings increased IAA production and improved plant growth under salt stress (Khushdil et al., 2019). Moreover, several osmolytes like sugars, proline, polyamines, polyhydric alcohols, quaternary ammonium compounds, betaines and dehydrins (water stress protein) are also reported to be accumulated by plants to combat the effects of abiotic stress (Vurukonda et al., 2016). Endophytes are also able to secrete these osmotic products in response to stress and perform synergistically with plant-produced osmolytes and improve the growth of plants (Paul et al., 2008). A study supporting the above mode of action showed that when microbial consortia involving *Arthrobacter nitroguajacolicus*, *Pseudomonas jessenii*, and *Pseudomonas synxantha* were inoculated in two rice cultivars (Sahbhagi and IR-64) enhanced growth under drought stress was observed. Consortia inoculated plants increased proline content in plants under osmotic stress, thus ameliorating the stress effects (Gusain et al., 2015). Another study showed that inoculation of endophyte *B. phytofirmans* in grapevine plantlets enhanced starch, proline, and phenolics content inducing cold tolerance mechanism in plants (Fernandez et al., 2012a, b). Similarly, trehalose which is a nonreducing disaccharide has the potential to maintain the dehydrating enzymes and cell membranes acting as an osmoprotectant (Yang et al., 2010). Similar effects were examined when *A. brasilense* was inoculated in maize plants which showed upregulation of trehalose biosynthesis gene and induced signal stress tolerance pathways in the plants (Rodriguez et al., 2009). Choline, another osmolyte, plays an important role in enhancing glycine betaine (GB) accumulation reducing the stress effects (Zhang et al., 2010). A study supported the role of choline together with GB in *Arabidopsis* plants inoculated with *Bacillus subtilis* which showed that choline and GB accumulation was enhanced by overexpression of PEAMT gene in osmotically stressed plants as compared to uninoculated plants (Zhang et al., 2010).

Ethylene levels play an important part in various plants' activities and their synthesis is regulated by biotic and abiotic stresses (Hardoim et al., 2008). Environemntal stress increased the synthesis of ethylene in plants causing severe damage to plant growth and health (Reinhold-Hurek and Hurek, 2011). During ethylene biosynthesis, SAdoMet (*S*-adenosylmethionine) is converted into ACC (1-aminocyclopropane-1-carboxylate) by ACS (1-aminocyclopropane-1-carboxylate synthase) which is a precursor of ethylene. Plant ACC is accumulated and degraded by ACC deaminase producing endophytic bacteria thereby removing excess ACC content and reducing the damaging effect of ethylene (Glick, 2005). For example, ACC deaminase producing bacteria *Achromobacter piechaudii* helped in enhancing biomass of tomato and pepper plants and also lowered the accumulation of ethylene under water deficit conditions (Mayak et al., 2004).

ROS are the signaling molecules employed in various activities in plants such as cell division, development, stress resistance and apoptosis and also promote the activities of subsequent genes. During abiotic stress, ROS consisting of H_2O_2 (hydrogen peroxide), O^{2-} (superoxide anion radicals), OH (hydroxyl radicals), O^{12} (singlet oxygen), and RO (alkoxy radicals) are synthesized which modify the amino acid sequence and degrade nucleotides and fatty acids, etc., eventually being linked up with the oxidative stress (Gill and Tuteja, 2010). Association of endopytes with host plant increased ROS scavenging property majorly inside plants under stress situations. A study supported that endophytic colonization in tomato plants induced low-level production of ROS in plants over the control under drought stress (Azad and Kaminskyj, 2016). Another study reported that association of *P. indica* with barley showed a higher accumulation of ascorbate protecting plants from the adverse effect of salinity (Khan et al., 2013). Rodriguez et al. (2009) had illustrated that endophyte treated plants showed an increase in antioxidant enzymes such as CAT (catalase), APX (ascorbate peroxidise), GR (glutathione reductase), DHAR (dehydroascorbate reductase), and MDHAR (mono-dehydroascorbate reductase) under salt stress. Under cadmium (Cd) stress, endophyte *Exophiala pisciphila* colonizing maize increased antioxidant enzymes which converted accumulated Cd into inactive form in cell wall by inducing root and shoot gene regulation mechanisms resulting in increased maize growth and Cd tolerance (Wang et al., 2016).

It has been reported that seeds inoculated with *Trichoderma harzianum* T22 increased seedling vigor and alleviated stress by decreased lipid peroxides (Mastouri et al., 2010). In comparison to non-endophytic plants, plants inoculated with endophytic fungi had the capability to retrieve a higher quantity of water from sources under stress conditions (Khan et al., 2013). Aquaporin, dehydrin, and malondialdehyde-related genes were upregulated during abiotic stress when rice plants were inoculated with endophytic *T. harzianum* (Pandey et al., 2016c). In general fungal endophytes have characteristic features of good colonization capability, improved plant growth, producing secondary metabolites against environmental stress imparting resistance to the host plant (Khan et al., 2011). In a report, the effects of an endophyte *B. pumilus* improved *Glycyrrhiza uralensis* growth, increased glycyrrhizic acid, total flavonoids, polysaccharide and antioxidants while reduced lipid peroxidation level under drought stress (Xie et al., 2019). A list of plant endophytes ameliorating the detrimental effect of abiotic stress in various plants has been given in Table 1.

2.5 Role of endophytes in reducing biotic stress in plants

Diverse living organisms, such as fungi, bacteria, insects, virus, nematodes, and weeds infect plants. Due to infection, growth and development of plants are stunted and cause biotic stress (Lattanzio et al., 2006). Fungi are considered as the most destructive among other biotic organisms (Waller et al., 2005).

Table 1 Impact of endophytes on plants under abiotic stress.

Endophytes	Host plant	Abiotic stress	Effect on plant	References
Bacillus amyloliquefaciens RWL-1	*Oryza sativa*	Various heavy metals (Cu, Cr, Pb, and Cd)	Increased growth of seedlings; reduced Cu uptake; improved carbohydrate levels; enhanced antioxidant activity (POD, PPO, and GHS), amino acids regulation, reduction of phytohormones (ABA and JA)	Raheem et al. (2019)
Serratia PRE01 or *Arthrobacter* PRE05	*Brassica juncea*	Vanadium-contaminated soil	Improved rhizosphere soil quality and mobilized heavy metals, enhanced biomarkers related to PGP in endosphere and enhanced phtyoremediation of V-contaminated soil	Wang et al. (2020)
Serratia sp. IU01 and *Enterobacter* sp.	*B. juncea*	Cadmium (Cd) stress	Promoted plant growth; enhanced antioxidant enzymes (PPO, POD, CAT, SOD, ADH, GSH, MDA) and metabolites including flavonoid and polyphenolic contents	Ullah et al. (2019)
Pseudomonas koreensis AGB-1	*Miscanthus sinensis*	Zn, Cd, As, and Pb	Showed effective tolerance against As, Cd, Pb, and Zn by extracellular sequestration, increased CAT and SOD activities	Babu et al. (2015)
Yarrowia lipolytica	*Zea mays*; *Hordeum brevisubulatum*	Salt stress	Increased plant growth by inducing controlled metabolism and hormonal secretions (ABA and IAA)	Gul Jan et al. (2019)
Epichloë bromicola	*Hordeum brevisubulatum*	Salt stress	Enhanced dry weight and increased content of spermidine and spermine, but decreased putrescine content	Chen et al. (2019)
Penicillium brevicompactum and *P. chrysogenum*	*Lycopersicum esculentum* and *Lactuca sativa*	Salt stress	Provoked a higher efficiency in photosynthetic energy production and diminished the effects of salt stress; induced higher up-/downregulation of ion homeostasis by enhanced expression of the *NHX1* gene	Molina-Montenegro et al. (2020)
Pseudomonas pseudoalcaligenes along with *Bacillus pumilus*	*Oryza sativa* GJ	Salt stress	Increased shoot biomass, accumulation of glycine betaine by reduction in proline content by 5%	Jha and Subramanian (2014)
Pantoea agglomerans	*Zea mays mexicana*	Salt stress	Increased dry biomass upregulated expression of the aquaporin gene family specifically plasma membrane integral protein genes (*ZmPIP*)	Gond et al. (2015)

Table 1 Impact of endophytes on plants under abiotic stress—cont'd

Endophytes	Host plant	Abiotic stress	Effect on plant	References
P. indica	*O. sativa*	Salt stress	Increased glycerol concentration, *PiHOG1* gene upregulated	Jogawat et al. (2016)
P. pseudoalcaligenes	*Arabidopsis thaliana*	Salt stress	Increased plant biomass, lateral roots density, and chlorophyll content, lowered Na^+/K^+ ratio and less pronounced accumulation of anthocyanin, and showed overexpression of *HKT1*, *KAT1*, *KAT2* transcript levels	Abdelaziz et al. (2017)
Alternaria sp., *Penicillium brevicompactum*, *Penicillium chrysogenum*, *Phaeosphaeria* sp., and *Eupenicillium osmophilum*	*Colobanthus quitensis*	Drought stress	Modulated the biochemical (TSS, proline, and TBARS) and morphological (stomatal opening), as well as molecular (*CqNCED1*, *CqABCG25*, and *CqRD22*) levels in presence of endophytes in plant	Hereme et al. (2020)
Phialophora sp. and *Leptosphaeria* sp.	*Hedysarum scoparium*	Drought stress	Improved biomass, nutrient content, and antioxidant enzymes	Li et al. (2019)
Gluconacetobacter diazotrophicus	*Saccharum officinarum* cv. SP701143	Drought stress	Increased biomass, IAA and proline production, overexpression of *ERD15 DREB1A/CBF3* and *DREB1B/CBF*	Vargas et al. (2014)
Trichoderma harzianum TH56	*Oryza sativa*	Drought stress	Modulated proline, SOD, lipid peroxidation, drought related *DHN/AQU* transcript level and the growth attributes	Pandey et al. (2016c)
Burkholderia phytofirmans PsJN	*Arabidopsis thaliana*	Cold stress	Promoted plant growth, accumulation of pigments and induced cold response pathway, downregulation of *Rbc L* and *COR78*	Su et al. (2015)
Pseudomonas vancouverensis OB155-gfp, *Pseudomonas frederiksbergensis OS261-gfp*	*Solanum lycopersicum*	Cold stress	Reduced membrane damage and reactive oxygen species level, improved antioxidant activity in leaf tissues, and upregulated expression of cold acclimation genes *LeCBF1* and *LeCBF3*	Subramanian et al. (2015)
Fusarium culmorum, *Curvularia protuberata*	*Oryza sativa*	Drought, salt and cold stress	Reduced water consumption by 20%–30% and increased growth rate, reproductive yield, and biomass of greenhouse grown plants	Redman et al. (2011)
Epichloe sp.	*Poa leptocoma*	Flooding	Showed overall benefits of fungal symbiosis with plants with improved plant growth	Adams et al. (2017)

Continued

Table 1 Impact of endophytes on plants under abiotic stress—cont'd

Endophytes	Host plant	Abiotic stress	Effect on plant	References
Epichloe sp.	*H. brevisubulatum*	Flooding	Increased root biomass, tiller production, and chlorophyll content, along with lower leaf wilt rates, enhanced proline and MDA content, and electrolyte leakage affecting plant osmotic potential	Song et al. (2015b)
Funneliformis mosseae	*Prunus persica*	Waterlogging stress	Modulated chlorophyll and proline content	Tuo et al. (2015)

PPO, polyphenol oxidase; POD, peroxidase; CAT, catalase; SOD, superoxide dismutase; ADH, alcohol dehydrogenase; GSH, glutathione; MDA, malondialdehyde; TSS, total soluble sugar; TBARS, thiobarbituric acid-reactive substances; IAA, indole acetic acid; ABA, abscisic acid; Zn, zinc; Cd, cadmium; As, arsenic; Cr, chromium; Cu, copper; Pb-lead; PGP, plant growth promtion.

Cell membrane of plants has some extracellular receptor which has an ability to recognize the infection sites caused by microbial attack (Chisholm et al., 2006). On the other hand, pathogen also has an ability to develop the mechanism for hindering the signaling of specific proteins into cytosol and to decrease the resistance or expression of defense response by the plants (Chisholm et al., 2006). As pathogen develops the ability to suppress the main resistances of plants, plants activate their potential metabolic processes to sense microorganisms, called as ETI (effector-triggered immunity) (Bilgin et al., 2010). In the ETI condition, the receptors (host) identify the effectors (pathogen) which brings about resistance capability in plants (Miedes et al., 2015). Various researchers have observed that endophytes act as biocontrol agents to protect plants from pathogenic attack and control the deterioration caused by pathogenic microbes (Pavithra et al., 2020). The endophytic microbes have the ability to fortify their host plant against biotic stress by inducing secondary metabolites synthesis or by activating plant endogenous signaling pathways. Based on these activities of endophytes, plant growth is improved which paves the way for them being used as biocontrol agents against agricultural pathogens (Zivanovic and Rodgers, 2018; Wang et al., 2013). Another case reported that fungal endophyte from *Artemisia annua* induced synthesis of n-butanol and ethylacetate which have the potential role in suppressing the activities of plant pathogens (Liu et al., 2001). Likewise a report also showed that *Phomopis cassia* isolated from *Cassia spectabilis* synthesized important antifungal compounds against *Cladosporium cladsporioides* and *Cladosporium sphaerospermum* (Silva et al., 2006). In the case it was found that *Epichloë festucae* induced the synthesis of anti-fungal protein for restricting the function of *Sclerotinia homoeocarpa* in *Festuca rubra* (Tian et al., 2017). Endophytes also produce some phytohormones like gibberellins, IAA which play major role in enhancing resistance mechanism (salicylic acid and jasmonic acid) in plants against various plant pathogens (Khare et al., 2016; Waqas et al., 2015). In a study, it was reported that black pepper-associated endophytic *Pseudomonas putida* (BP25) synthesized organic volatile substances inhibiting various phytopathogens such as *Athelia rolfsii*, *Colletotrichum gloeosporioides*, *Pythium myriotylum*, *Phytophthora capsici*, *Gibberella moniliformis*, *Rhizoctonia solani*, and plant parasitic nematode, *Radopholus similis* (Sheoran et al., 2015). Similarly, antagonistic strains of endophytic *Bacillus*, *Pseudomonas*, *Pantoea*, *Stenotrophomonas*, and *Serratia* isolated from wild pistachio showed inhibitory activity against *Pseudomonas syringae* and *Pseudomonas tolaasii* (Etminani and Harighi, 2018).

2.6 Mode of action of endophytes in plants under biotic stress

Endophytes including fungi and bacteria can survive in plant tissue without causing any pathogenic infection to the plants (Sandhya et al., 2017). Fungal endophytes due to possessing antimicrobial activity have been utilized as biocontrol agents to resist biotic stresses (Rodriguez et al., 2009; Jaber, 2015). A study has reported that *Penicillium citrinum* can act against the pathogen *Fusarium oxysporum* by synthesizing higher content of related enzymes (Ting et al., 2012). In the same way, entamopathogenic fungi *Metarhizium anisopliae* and *Beauveria bassiana* colonized the inner olive tissues and acted as biocontrol agents for *Verticillium* wilt in olive (Schneider and Ait, 2012). Another study showed that *Burkholderia phytofirmans* induced ROS signaling and defense-related gene expression signaling pathways in grapevine which could protect the plants from the necrotrophic pathogen *Botrytis cinerea* (Miottovilanova et al., 2016). Similarly, other studies have illustrated that fungal endophytes induce phytoalexins which can save plants from biotic stresses by the overexpression of secondary metabolites including phenolic compounds, phytohormones, and defense enzymes and improve gene expression (Gao et al., 2010; Kumar et al., 2014). A study supported the role of secondary metabolites synthesis in disease control which stated that endophytic isolates *Cryptosporiopsis* sp. and *Fusarium oxysporum* protected larch (*Larix decidua*) and barley (*Hordeum vulgare*) respectively against virulent pathogens due to increased levels of phenolic compounds (Rodriguez et al., 2009). However, more researches are required to understand the exact mechanisms involved in the alleviation of biotic stress by endophytic fungi. Similarly, bacterial endophytes have also biocontrol properties to reduce diseases caused by fungus, bacteria, and nematodes (Sandhya et al., 2017; Adam et al., 2014; Upreti and Thomas, 2015). It has been found that the bacterial endophytes secrete structural or toxic compounds to suppress the pathogenic infection by blocking the susceptible tissues of pathogen (Pageni et al., 2014). Another research showed that endophytic bacteria stimulate ISR (induced systemic resistance) arbitrated by ethylene and/or jasmonic acid (JA), which is useful against pathogens (Ryan et al., 2008). Likewise one study illustrated that *Bacillus subtilis* can inhibit root-knot nematode growth by the activation of ISR in tomatoes including the oligosaccharides of the core-region of the bacterial lipopolysaccharides (Adam et al., 2014). Bacterial endophytes *Bacillus* and *Paenibacillus* in symbiosis with *Lonicera japonica* showed inhibitory activity against phytopathogenic fungi (Zhao et al., 2015).

A theory has been proposed by Schulz and Boyle (2005) which explains how endophytes trigger host defenses. There may be pathogenesis, competition, parasitism, ISR (induced systemic resistance), and repellence mechanisms of endophytes which involve suppression of plant-pathogenic interaction (Ryan et al., 2008). Another possible mechanism involved by endophytes against biotic stresses is by production of secondary metabolites, microbial toxins, antibiotics, enzymes, phytohormone, and siderophores reducing ROS (reactive oxygen species) damage and improving gene expression (Sandhya et al., 2017; Zhang et al., 2019). A list of plant endophytes ameliorating the detrimental effect of biotic stress in various plants has been given in Table 2.

3 Plant-endophytes interactions under changing climate

Plant harbors a number of endophytes whose colonization in the endosphere has been found to be dependent on crop cultivars and changing season (VanOverbeek and VanElsas, 2008; Manter et al., 2010; Baldan et al., 2014; Bulgari et al., 2014). Climate change is a key factor in altering endophyte colonization and distribution in plants. Distribution of endophytes is related to their functional capability which

Table 2 Impact of endophytes on plants under biotic stress.

Fungal endophyte	Host	Target pathogens	Effect	References
Penicillium citrinum LWL4 and *Aspergillus terreus* LWL5	*Helianthus annuus* L.	Stem rot caused by *Sclerotium rolfsii*	Increased plant biomass, chrolophyll, photosynthesis, transpiration; reduced the effect of fungal infection as shown by low level of salicylic acid and jasmonic acid	Waqas et al. (2015)
Epichloe typhina	*Puccinella distans*	*Aspergillus flavus*	Reduced the effect of fungal infection in endophyte associated plants in comparison to endophyte free plants	Gorzynska et al. (2017)
Epichloe sp.	*Leymus chinensis*	*Bipolaris sorokiniana*	Endophyte associated plants had smaller lesion area on the leaves in comparison to endophyte free plants	Wang et al. (2016)
Epichloe gansuensis	*Achnatherum inebrians*	*Blumeria graminis*	Endophyte associated plants had minor disease incidence, higher tiller numbers biomass and greater photosynthetic ability	Liu et al. (2015)
Epichloe occultans	*Lolium muliflorum*	*Claviceps purpurea*	Reduced the effect of fungal infection in endophyte associated plants in comparison to endophyte free plants	Perez et al. (2017)
Epichloe coenophialum	*Festuca arundinacea*	*Curvularia lunata*	Endophyte associated plants had smaller lesion number and spore concentration of pathogen in comparison to endophyte free plants	Chen et al. (2017)
Pseudomonas putida BP25	*Piper nigrum*	*Phytophthora capsici* and *Radopholus similis*	Airborne volatiles such as released by bacterium which inhibited pathogens	Sheoran et al. (2015)
Alternaria alternata (MF693801) and *Fusarium fujikuroi* (MF693802)	*Lycopersicon esculentum*	*Fusarium oxysporum f.* sp. *radicis lycopersici (FORL)* Fusarium Crown and Root Rot	Increased plant growth attributes thereby reducing the fungal disease	Ahlem et al. (2018)
Bacillus, *Pseudomonas*, *Klebsiella* and *Citrobacter*	*Lycopersicon esculentum*	*Fusarium* wilt disease in tomato	50% of the isolates showed antagonistic activity against *Fusarium* pathogen	Nandhini et al. (2012)
Pseudomonas spp.	*Cucumis sativus*	*F. oxysporum f.* sp. *cucumerinum*	Enhanced plant growth and acted as a biocontrol agent for protecting the plants from Fusarial wilt of cucumber	Ozaktan et al. (2015)

Table 2 Impact of endophytes on plants under biotic stress—cont'd

Fungal endophyte	Host	Target pathogens	Effect	References
B. subtilis, *B. cereus*, and *Arthrobotrys cladodes*	*Abelmoschus esculentus*	*M. incognita*	Increased germination %, vigor index and biomass, reduced soil infestation of *M. incognita*	Vetrivelkalai et al. (2009)
Bacillus sp. *8CE*	*Cucumis sativus*	*Pseudoperonospora cubensis* causing downy mildew disease	Had higher efficacy to control pathogen	Sun et al. (2013)
Streptomyces sp.	*Solanum tuberosum*	*Streptomyces scabies*	Efficacy to control pathogen	Flatley et al. (2015)
Bacillus subtilis (UFLA28), Paenibacillus lentimorbus (MEN2)	*Gossypium* sp.	Bacterial blight and damping-off caused by *Xanhomonas axonopodis* pv. *malvacearum* and *Colletotrichum gossypii* var. *cephalosporioides*	Recorded the lowest bacteria blight incidence by 26%	De Medeiros et al. (2015)
Bacillus subtilis E1R-j	*Triticum aestivum*	*Puccinia striiformis* f. sp. *tritici* (*Pst*) causing wheat stripe rust	Strong inhibitory effect on wheat stripe rust	Li et al. (2013)
Pseudomonas putida (C4r4), *Achromobacter* spp. (Gcr1), *Rhizobium* sp. (Klr4), *Ochromobactrum* sp. (Klc2), *Rhizobium* sp. (Lpr2) and *Bacillus flexus* (Tvpr1)	*Musa* spp.	*Fusarium oxysporum* causing wilt of banana	Increased plant growth and had an efficacy to control the pathogen	Thangavelu and Muthukathan (2015)

is defined as the key development of a predictive framework for endophyte functional characteristics in symbiosis. Therefore, Different selection of different microbes from plants and their distribution sorted via environmental filtering and local adaptation may play a role in endophyte functioning in some stressful habitats (Classen et al., 2015). It is very difficult to know about (1) how endophytic community distribution is related to environmental factors and spatial processes and (2) how can we identify the functional role of endophytes in symbiosis which is reflected under local environmental conditions. Under climate changes, a number of factors such as altered CO_2, moisture, temperature, precipitation, warming, season, day length, frost are involved to affect plant endophytes interactions (Brosi et al., 2011; Fuchs et al., 2017).

A microbiome study reveals that seasonal variation plays a significant role in determining optimal variations in species abundance and diversity within different microbial communities (Mocali et al., 2003). Bacterial community composition and dominant species among different mulberry (*Morus* L.)

cultivars over seasonal shifting reveal that spring samples have a higher number of bacterial operational taxonomic units (OTUs), α-diversity, and bacterial community complexity in comparison with the autumn samples. In a study the taxonomic composition analysis demonstrated that the majority of endophytic communities have been classified into Proteobacteria (*Methylobacterium*) and *Actinobacteria* in spring while Proteobacteria (*Pantoea* and *Pseudomonas*) were abundant in autumn season (Ou et al., 2019).

Different authors have reported contradicting results about the endophytic community analysis from both culture-dependent and independent methods which indicate that endophytic diversity in similar habitats had the same species-specific communities which were more sensitive to seasonal differences (Carrell and Frank, 2014). The study suggests the notion of dominant species-specific endophyte relationship with a specific plant which might be the reason for conferring special adaptive characteristics to the host plant. In another report, the diversity of bacterial endophytes in grape plants was found to be dependent on temperature as it is the main factor governing the seasonal changes (Baldan et al., 2014; Bulgari et al., 2014). In another study, temperature changed the community of bacteria in plant and physiology of plants which was reflected as their respective effect on plant growth (Jansson and Douglas, 2007). It has been suggested that changes in physiology of plants directly affect the concentration of primary metabolites like soluble sugar, amino acids, organic acids, and other plant nutrients as well as secondary metabolites in plants which is indirectly enhance or diminishes community of microbes in plant endosphere (Cox and Stushnoff, 2001). Root symbionts such as mycorrhizal fungi are found to be associated with all the land plants which affect plant productivity by changing carbon and nutrient-cycling processes. Common mycorrhizal networks by mycorrhizal fungi interconnect plants uptake nutrient and carbon at different rates by a variety of mechanisms, including changes to mycorrhizosphere or hyposphere chemistry and increased physical access to the soil (Walder and van der Heijden, 2015). The impacts of mycorrhizal fungi on plant are not always symbiotic and the functional characteristics of mycorrhizal fungi can be changed with varying environmental factors or even plant stress (Johnson et al., 1997; Treseder, 2004). Mycorrhizal community composition can substantially differ with climatic changes such as temperature (Deslippe and Simard, 2011). In rising temperature, mycrorrhizal fungi foraging increases which leads to increased carbon allocation which in turn may alter symbiotic partnership to a parasitic one (Nuccio et al., 2013; Clemmensen et al., 2013; Leifheit et al., 2015; Moore et al., 2015; Hawkes et al., 2008). Shifts in these mycorrhizal-plant interactions indirectly cascade to alteration in soil microbial communities in mycorrhizosphere through releasing organic acids and production of enzymes (de Boer et al., 2005; Nazir et al., 2010) and activity (Leifheit et al., 2015).

Climate change directly affects the function, physiology, temperature sensitivity, growth rates and abundance of soil microbial communities in soils which is in turn further alters the colonization properties of microbes within the plants (Giauque and Hawkes, 2013). In colonization, endophytes increased the plasticity of plants compared to the plants without colonization. Increased plasticity facilitates plant flexibility to grow in new environmental conditions (Ghalambor et al., 2007). It has been shown that from dried to wetter region environment, moisture plays an important role in endophytic colonization, distribution, and adaptation inside the plants. Some endophytes are well adapted to local moisture conditions and their effects on different plants traits have been optimized for such environment from which they were isolated (Conover et al., 2009). In a report, endophytes isolated from the dried growing plant showed more drought resistance to the plants than the wetter growing plants (Conover et al., 2009; Kawecki and Ebert, 2004).

A well-known endophyte, *Epichloe festucae* var. *lolii*, a vertically transmitted fungus, belonging to Clavicipitaceae family (Cheplick and Faeth, 2009), associated their cool-season grass, *Lolium perenne* (perennial ryegrass). *E. festucae* var. *lolii* produces anti-insect alkaloids peramine and lolines (Schardl et al., 2004). Alkaloids of endophyte vary in quantity which is due to grass and endophytic genotype combination (Schardl et al., 2013; Ryan et al., 2015), as well as affected differently by seasonal variation (temperature) (Salminen et al., 2005; McCulley et al., 2014) and drought (Bush et al., 1993; Bush et al., 1997). According to a study plant age and seasonal rhythm determined *E. festucae* colonization, growth and also affected the alkaloid biosynthesis (Fuchs et al., 2017). Fuchs et al. (2017) examined that the seasonal dynamics of fungal and alkaloids concentration in corresponding to different climatic conditions of the four seasons in a temperate region which showed strong changes in both fungal and alkaloids concentration with prominent peaks during the summer season (July–September) (Fuchs et al., 2017).

4 Endophytes modulates plant secondary metabolites biosynthesis

Endophytes are considered for the rich source of secondary metabolites which act as anticancer, antioxidant, antidiabetic, immunosuppressive, antifungal, antioomycete, antibacterial, insecticidal, nematicidal, and antiviral agents for human welfare (Zhang et al., 2006a; Gunatilaka, 2006; Aly et al., 2011; Verma et al., 2009). Majority of endophytic secondary metabolites are similar to their host plants. Besides the plant like similar secondary metabolite production, endophytes are actively involved in mechanisms of signaling, defense, genetic regulation of plants and are able to affect the secondary metabolism of their host plant (Zhang et al., 2006b) (Table 3). Some compounds are not synthesized by the host plant, but in presence of endophytes the plant is reported to synthesize the secondary metabolites. It was examined in one of the studies that the strawberry plants inoculated with a *Methylobacterium* species strain, greatly influenced the biosynthesis of furanones compounds in the host plants resulting in the enhancement in the flavor of strawberry (Zabetakis and Holden, 1997; Koutsompogeras et al., 2007; Verginer et al., 2010). Nasopoulou et al. (2014) reinforced that bacterial endophytes contained bacterial alcohol dehydrogenase enzyme that was expressed for furanone biosynthesis and localized in achenes cell and vascular tissues of strawberry receptacles (Nasopoulou et al., 2014). Similarly, a fungal endophyte *Paraphaeosphaeria* sp. enhanced the flavonoid biosynthesis and increased the richness of phenolic acids, flavan-3-ols, and oligomeric proanthocyanidins in wild berries of bilberry (*Vaccinium myrtillus* L.) plants (Koskimäki et al., 2009). Endophyte has characteristics which affect the expression of secondary metabolites biosynthetic pathway genes for enhancing the content of the end product. Endophytic actinobacterium, *Pseudonocardia* sp. strain YIM 63111 enhanced the artemisinin in *Artemisia annua* by the up-regulation of cytochrome P450 monooxygenase (*CYP71AV1*) and cytochrome P450 oxidoreductase (*CPR*) gene. In a similar study, *Colletotrichum gloesporioides* could also elevate artemisinin production in in vitro grown hairy-root cultures of *A. annua* (Wang et al., 2006). The best known example of a high valuable anticancer compound, taxol was found in endophytic fungi *Taxomyces andreanae* that was isolated from anticancer synthesizing *Taxus brevifolia* plant (Stierle et al., 1995). In another report analysis revealed that endophytes influence plant tissue-specific biosynthetic pathway genes and its secondary metabolite content. For example, foliar endophytes enhanced primary metabolites and capsule endophytes upregulated key genes of BIA biosynthesis in *Papaver somniferum*. Capsule endophytes *Marmoricola sp.* (SM3B) and *Acinetobacter* (SM1B) were isolated from capsule tissue

Table 3 Impact of endophytes on enhancement of host plant secondary metabolite content.

Endophytes	Host plant	Enhanced secondary metabolites	References
Curvularia sp., *Choanephora infundibulifera*	*Catharanthus roseus*	Vindoline	Pandey et al. (2016b)
Curvularia sp., *Choanephora infundibulifera*, *Aspergillus japonicus*, *Pseudomonas* sp.	*C. roseus*	Ajmalicine and serpentine	Singh et al. (2020)
Micrococcus sp.	*C. roseus*	Vindoline, serpentine, ajmalicine	Tiwari et al. (2010)
Pseudonocardia sp.	*Artemisia annua*	Artemisinin	Li et al. (2012)
Bacillus amyloliquefaciens, *Pseudomonas fluorescens*	*Withania somnifera*	Withaferin A, withanolide A, withanolide B	Mishra et al. (2018)
Trichoderma atroviride	*Salvia miltiorrhiza*	Tanshinones	Ming et al. (2013)
Aspergillus terreus	*W. somnifera*	Withanolide A	Kushwaha et al. (2019)
Aspergillus terreus, *Penicillium oxalicum*, *Sarocladium kiliense*	*W. somnifera*	Withanolide A	Kushwaha et al. (2019)
Bacillus muralis, *Bacillus megaterium*, *Pseudomonas* sp., *Streptomyces* sp., *Pantoea* sp.	*W. somnifera*	Withanolide-A, withaferin-A, 12-deoxy withstramonolide	Pandey et al. (2018)
Marmoricola sp., *Janibacter*, *Acinetobacter* sp., *Kocuria* sp.	*Papaver somniferum*	Morphine, thebaine, papaverine, noscapine	Pandey et al. (2016a)
Bacillus subtilis	*Ocimum sanctum*	Ocimum oil	Tiwari et al. (2010)
Gilmaniella sp.	*Atractylodes lancea*	Volatile oil accumulation	Ren and Dai (2012), Yuan et al. (2016a, b), and Wang et al. (2012)
Piriformospora indica	*Curcuma longa*	Curcumin	Bajaj et al. (2014)
P. indica	*Bacopa monnieri*	Bacoside	Prasad et al. (2013)
P. indica	*Centella asiatica*	Asiaticoside	Satheesan et al. (2012)
Marmoricola sp., *Acinetobacter* sp.	*P. somniferum*	Morphine and thebaine	Ray et al. (2019)
Fusarium redolens, *Phialemoniopsis cornearis*, *Macrophomina pseudophaseolina*	*Coleus forskohlii*	Forskolin content	Mastan et al. (2019)

of *P. somniferum.* SM1B inoculated plant showed upregulated expression of all key genes for BIA pathway except thebaine 6-*O*-demethylase (*T6ODM*) and codeine Odemethylase (*CODM*) while SM3B inoculated plant showed upregulation of *T6ODM* and *CODM* genes (Pandey et al., 2016a). It is indicative that endophytes act as upregulators of specific gene expression for tissue-specific roles. Further study in same plant reported consortium of endophytes that modulated multiple BIA pathway genes for increasing the desired end product of the plant (Ray et al., 2019). Furthermore, many experiments showed that endophytic function varied only on plant tissue basis but it also depends on plant genotype and family to modulate biosynthetic pathway gene expression. In a recent report a study interpreted that endophytes also modulate the expression of pathway-specific transcription factors. In case of low alkaloid yielding *Catharanthus roseus* cultivar Prabal, *Curvularia* sp. and *Choanephora infundibulifera* fungal endophytes enhanced *in planta* content of vindoline by modulation of terpenoid indole alkaloids (TIAs) biosynthetic pathway structural key genes geraniol 10-hydroxylase (*G10H*), tryptophan decarboxylase (*TDC*), strictosidine synthase (*STR*), 16-hydoxytabersonine-*O*methyltransferase (*16OMT*), desacetoxyvindoline-4-hydroxylase (*D4H*), deacetylvindoline-4-*O*acetyltransferase (*DAT*). Not only TIAs biosynthetic pathway genes, endophyte also upregulated the expression of transcriptional activator known as octadecanoid-responsive Catharanthus AP2-domain protein (*ORCA3*) and downregulated the expression of Cys2/His2-type zinc finger protein family transcriptional repressors (*ZCTs*) (Pandey et al., 2016b). Similar result were obtained by Singh et al. (2020) who reported that *Pseudomonas* sp. bacterial and *Curvularia* sp., *Choanephora infundibulifera* and *Aspergillus japonicus* fungal endophytes modulated the transcript abundance of TIAs biosynthetic pathway genes in *C. roseus* root for enhancing the content of ajmalicine and serpentine (Singh et al., 2020). Novel exopolysaccharide mannan was discovered in a *Gilmaniella* sp. AL12 endophytic fungus that enhanced the content of volatile oil in *Atractylodes lancea* medicinal plant (Chen et al., 2016). Further study revealed that *Gilmaniella* sp. AL12 fungal treatment improved the photosynthetic rate as well as elevated the carbohydrate levels and chlorophyll content in host leaves while its elicitors enhanced a total content of volatile oil in in vitro grown *A. lancea* plantlets (Wang et al., 2012). In tissue culture system, endophytes cannot be applied directly but extract of endophytes act as biotic elicitors for effective production of secondary metabolites. Likewise, *Trichoderma atroviride* D16, an endophytic fungus was isolated from *Salvia miltiorrhiza* root which enhanced tanshinone I (T-I) and tanshinone IIA (T-IIA) content. Extract of fungal endophytes upregulated expression of 1-deoxyd-xylulose 5-phosphate reductoisomerase (*DXR*), geranylgeranyl diphosphate synthase (*GGPPS*), hydroxymethylglutaryl-CoA reductase (*HMGR*), copalyl diphosphate synthase (*CPS*), and kaurene synthase-like (*KSL*) gene in tanshinone biosynthetic pathway. Additional report for stimulating the production of tropane alkaloids (TA) biosynthesis, bacterial elicitors inhibited *H6H* (hyoscyamine 6β-hydoxylase) expression leading to enhance the productivity of scopolamine in in vitro grown adventitious hairy root cultures of *Scopolia parviflora* (Jung et al., 2003). However, which types of potent elicitors are responsible for the activation of pathway gene in host plants is still unclear?

5 Commercialization of endophytes

The outcome of plant-microbe interaction studies has generated some commercial products which have positively been incorporated into food production, adaptation to environmental stresses and also for creating income and employment. Although plants have been extensively investigated for therapeutically important chemical compounds, endophytic microbes also form a stockroom of new secondary metabolites that can provide a good source of drugs for antimicrobial and insecticidal activities (Godstime et al., 2014). It is well reported that in presence of beneficial endophytes, plants showed a significant increase in biomass and also facilitate cost-effective agricultural production (Shen et al.,

2019; Santoyo et al., 2016). The detection of new secondary metabolites with higher drug potential from endophytic microorganisms is the most important substitute to conquer various major pathogenic infections (Godstime et al., 2014). Some endophytic products have been commercialized. For instance, *Epichlo¨e* endophytes have been effectively commercialized in perennial ryegrass and tall fescue with different qualities and pastoral benefits for Australia, New Zealand, the United States of America, and South America (Young et al., 2013; Johnson et al., 2013). They have proved their potential in improving the plant survival by protecting them from both abiotic and biotic stresses and were thus consequently exploited in agricultural production. Currently, new endophytic efficient strains have been quickly accepted by the farmers (Caradus et al., 2013) as in New Zealand alone they are contributing to the economy by about US$ 130 million per annum (Johnson et al., 2013).

As endophytes are known for increasing agricultural productivity, nutrient uptake efficiency, reducing oxidative damages in the host and protecting plants from pathogenic infections, herbivores, etc., they have been suggested to be used in place of agrochemicals involving chemical fertilizers, fungicides, herbicides, and insecticides in the crop cultivation. However, due to domestication and extensive cultivation, the endophytic microbes are removed from the plants which could be restored by inoculation of endophytes isolated from wild relatives of crop species. For example, due to seed cleansing and continuous cultivation of *Nicotiana attenuate*, loss of microbes occurred from plants leading toward pathogenic infections (Santhanam et al., 2015). Regain of beneficial microbes from wild tobacco and inoculation in seedlings showed an increase in tolerance to the disease. Similar results were also seen in the case of endophytic remedial of cotton (Irizarry and White, 2017) and maize plants (Johnston-Monje et al., 2016). Therefore, the development of endophytic formulations should have the characteristic of improved agricultural yield, compatible with other rhizobacteria, ecofriendly, easy mass multiplication capability, and increased shelf life as well as the property of alleviating biotic and abiotic stresses, which pave the way for commercialization of the technology rapidly (White et al., 2019).

5.1 Carrier-based formulations

In many literature, direct application of cell suspensions of endophytes to phyllosphere, spermosphere or rhizosphere have been reported but on large scale (field conditions), it could not be used as a cell suspension. It should be bound in some suitable carriers such as talc, charcoal, clay, vermiculite, sand or peat, alginate beads, vermicompost, etc., for long-term storage, easy inoculation, and commercialization purpose. Albeit each carrier-based bioformulation has its own benefits and restrictions. A good carrier material should have more surface area, rich in organic matter, high water holding capacity, neutral pH, economical and easily available that supports the stability of endophytic microbes for a longer duration (Arangarasan et al., 1998). Similar to plant growth-promoting rhizobacterial formulations, endophytic formulations should be standardized for the management of pest and diseases. A study has reported chitosan-based *Bacillus pumilus* endophytic formulation in tomato plants for inducing resistence against *F. oxysporum* pathogen (Benhamou et al., 1998). In another case, talcum powder-based *P. indica* formulation was found to increase growth parameters of *Phaseolus vulgaris* under pot study (Tripathi et al., 2015). Similarly, talcum and vermiculite-based formulations of endophytic fungus (*P. indica*) and bacterial (Pseudomonads strains) consortia were found to be effective in improving biomass of *Vigna mungo* under glasshouse as well as field conditions (Kumar et al., 2012). Globally peat-based bioformulation is generally used (Gopalakrishnan et al., 2016) and in India, talc-based bioformulations are popular (Basheer et al., 2018; Selim et al., 2017). Another study reported that wheat bran-based endophytic formulations (fungal endophytes isolated from *Coleus forskohlii*) significantly increased biomass, chlorophyll contents as compared to talc-based formulation in *C. forskohlii* (Mastan et al., 2019). Among various carrier materials, vermicompost is identified as a good delivery source for bioinoculants due to iys cheaper, ease in production, high nutrient content, and superior water holding capacity (Puttanna et al., 2010). Waste enrichment has now become significant for the development of sustainable farming. The use of earthworms (*Eisenia foetida*) for the deterioration of organic wastes and production of vermicompost is close to commercialization because the minimum loss of nitrogen occurs during vermicomposting of organic matter from agricultural wastes and dung. Earlier studies have reported that vermicompost produced from organic waste material supports the survival of beneficial microbes (Hussain and Abbasi, 2018; Kalra et al., 2010; Singh et al., 2012). Higher yields in case of crops like patchouli (Singh et al., 2012), tomato (Al-Karaki, 2006), and *Coleus forskohlii* (Singh et al., 2009) have been reported when plants were inoculated with vermicompost-based microbial consortia in comparison to chemical fertilizers, vermicompost, and compost individually (Singh et al., 2009). Another literature showed that residual waste material after sugar production has been used for vermicomposting and found as a good preservative material for diazotrophic bacteria (Martinez-Balmori et al., 2013). Solid and liquid both bioformulations are available in the market. However, producers mostly prefer liquid bioformulation due to ease in diluting and spraying on crops.

5.2 Stability and shelf life study

Stability assessment of carrier-based bioinoculants at different temperature is very important and desirable for commercialization because the viability and biological efficacy of bioinoculants in a prescribed formulation must be checked during 6 months of storage period (Bazilah et al., 2011). The most favourable moisture level of 35%–50% and a temperature of 30 °C are needed for the survival of microbial cells in bioformulation for longer duration (Bajpai et al., 1978). For example, a talcum powder-based formulation (5%) was studied to be the most constant at 30 °C with $10^8\,CFU\,g^{-1}$ and successful for a storage period of 6 months (Tripathi et al., 2015). However, loss in moisture content and purity of bioinoculants also influence the stability of microbial cells in bioformulation during long-term storage. After inoculation in carrier material bioinoculants are kept at different temperature (25 °C, 30 °C, 35 °C, and 40 °C) and surviving populations (colony forming unit) were checked at different time interval for 6 months (Dalia et al., 2013). The shelf life of bioformulations can be increased by using suitable carrier material, retaining moisture content and use of appropriate additives such as calliterpenone, cacilum chloride, glycerol, etc., which can sustain higher population of bioinoculants while transferring in the agricultural field (Tripathi et al., 2015). For commercial utilization of bioformulation, rapid decrease in shelf life should be avoided and at least 6-month survivality of microbial cells is necessary for application in agriculture. To retain the vitality of cells for longer duration storage of bioformulations in freezers (4 °C or − 20 °C) are suggested (Berger et al., 2018). However, some studies have reported a shelf life of 1 or 2 years at 25 °C (Sallam Nashwa et al., 2013; Arora et al., 2008).

5.3 Single vs multiple strains formulation

Many researchers have suggested that microbial consortia can improve plant growth more efficiently in comparison to single inoculated microbes (Singh et al., 2021). For example, Bharti et al. (2016) reported that vermicompost-based bioinoculants (*Dietzia natronolimnaea* and *Glomus intraradices*) improved the growth and microbial community of *Ocimum basilicum* plants in comparison to single applications of either of the bio-inoculants. Another microbial consortium involving *Bacillus subtilis* OTPB1 and *Trichoderma harzianum* OTPB3 when inoculated in tomato showed increased growth parameters, antioxidant enzymes and induced systemic resistence in comparison to their individual application (Kumar et al., 2015). Similar results were observed in case of *Pogostemon cablin* (patchouli) when treated with a vermicompost-based consortium of microbes, viz. *Trichoderma harzianum*, *Pseudomonas monteilii*, *Bacillus megaterium*, and *Azotobacter chroococcum* reduced the disease index % and improved essential oil yield (Singh et al., 2013). The commercial use of microbial consortia in crop production is the current need of the world and thus preferred over single inoculants due to their multiple capabilities from improving plant growth, disease resistance capacity to maintaining soil health (Sekar et al., 2016; Bradáˇcová et al., 2019; Singh et al., 2019).

5.4 Patent and registration

The success of microbial formulation to be used in agriculture is dependent on its authenticity and awareness among the people. For authenticity and awareness patent approval and registration of the developed formulation is necessary from the legal agency (EPA). Furthermore, approved efficiency of bioformulation as a product is also required which makes it possible for the transfer of technology from laboratory to field level. Therefore, patent protection and registration of bioformulation should be essential to the industrialist for the commercialization of the product. Although, the full-scale registration process and production of formulation take 3–6 years and local production process takes less than 1 year (White et al., 2019).

5.5 Policies to support commercialization

Selection of appropriate bioinoculants, carrier material, large-scale production process involving mixing, curing, preservation of inocula viability, and superior quality control is the imperative steps required for the development of commercial bioformulations (Malik et al., 2005). Commercial utilization of developed formulation could be supported by spreading awareness among the producers, farmers and also build connection with the industries. Communication gap among the farmers or producers about the development and usage of biofertilizers is a considerable restriction in the popularization of these bioformulations. Community programmes, field demonstration, biovillage adoption projects and organizing training programmes are remarkable in motivating the growers about proper use of bioformulation (Pindi and Satyanarayana, 2012). Besides, industries should also focus on the quality control and registration of these products. Industry should establish research and development unit to standardize the dosage, storage, and delivery systems. Government should also have to make positive policies to support entrepreneurs so that demand and supply chain of products increases. Recently, U.S. Government Agriculture Improvement Act (2018 Farm Bill) in the United States and regulatory documents in the European Union have clearly addressed bioinoculants usage guideline to facilitate the application of microbes in agriculture (Ricci et al., 2019). The legal, scientific and doctrinaire structure in applications of endophytic microbes in agriculture prepares major opportunity in the form of efficient products development.

For instance, in the United States commercial product of *B. subtilis* (Kodiak) when inoculated in cotton crops was found to be effective against soilborne pathogens such as *Fusarium* and *Rhizoctonia* (Nakkeeran et al., 2005). This product is also found suitable for growing small grains, corns, peanut, vegetables, and soybeans (Backman et al., 1997). China has also developed commercial products known as yield increasing bacteria (YIB) and applied more than 20 million hectares land of different crops (Chen et al., 1996; Kilian et al., 2000). In India, over 40 stakeholders from different regions have registered for large production of bioformulations with Central Insecticide Board (CIB), Faridabad, and Haryana through cooperation with Tamil Nadu Agricultural University, Coimbatore, India for the practical support and information (Ramakrishnan et al., 2001). Some other endophytic products such as *BioEnsure* R-*Corn* and *BioEnsure* R-*Rice* have been developed to improve agricultural production concerning environmental changes like drought, salinity, waterlogging, cold and high temperature stress. (http://www.adaptivesymbiotictechnologies.com/executiveteam.html). *BioEnsure* R-*Rice has been found to* support the plant yield under drought and salt stresses with reduced water usage approximately by 25%–40%. These products are traded as liquid formulations that are sprayed on seeds and the strains remain latent until germination, thereafter establish a symbiotic relationship with seedlings (http://www.adaptivesymbiotictechnologies.com/press–publications.html).

6 Conclusion and perspectives

This chapter gives an insight into the fact that all species of plant harbor a single endophyte or endophytic community as endosymbiont. Each endophyte has its own functional characteristics which affect the plant functions. The use of endophytes is a promising tool for plant protection, yield of crops under biotic and abiotic stresses along with improvising the physiological performance of the plant. Functional endophytes could be a sustainable bioinoculant for the enhancement of therapeutically important secondary metabolites biosynthesis and their pathway-specific regulation in plant. Apart from that, we need to understand the variation in endophytic community composition and functions in response to climate effect for their proper exploitation. For this, we need to determine the best approaches to observe quantifiy and assess the degree of plant endophyte mutualism interactions in symbiosis over altered climate change. In future, plant age, genotype, cultivation and endophyte species, and its community as well as climate effect must be considered in plant tissue culture system and field-grown plants studies to determine the function of plant endophyte association. In sum, the actual benefits and viability of endophytes are usually determined in terms of the yield potential and its economics value. A sustainable step of recruiting native endophytes could be non-chemical agri-input alternatives for green agriculture. Additional benefits are that the metabolic potential of the endophytic microorganisms could be harnessed as microbial cell factory for obtaining novel and non-plant-based bioactive metabolic compounds having broader applications against the suppression of pathogens, insects, and physiological disorders of plants. The potential of endophytic organisms, having desirable functional traits will be a valuable key to farmers for the establishment of modern agriculture in time to come.

Acknowledgments

Sucheta Singh acknowledges the SERB-National Post Doctoral Fellowship (SERB-NPDF) for financial support (PDF/2019/001174). Suman Singh is grateful to the Department of Science and Technology, New Delhi, India for financial support during her Ph.D. work (Grant no. DST/DISHA/SoRf/PM-016/2013). Akanksha Singh is thankful to CSIR for funding the Focused Basic Research (FBR) Endophyte Network Project (Plg/MLP-0048/NCP-FBR 2020 CSIR-NBRI).

References

Abdelaziz, M.E., Kim, D., Ali, S., Fedoroff, N.V., Al-Babili, S., 2017. The endophytic fungus *Piriformospora indica* enhances *Arabidopsis thaliana* growth and modulates Na^+/K^+ homeostasis under salt stress conditions. Plant Sci. 263, 107–115.

Acuna-Rodriguez, I.S., Newsham, K.K., Gundel, P.E., Torres-Diaz, C., Molina –Montenegro, M.A., 2020. Functional roles of microbial symbionts in plant cold tolerance. Ecol. Lett. 23, 1034–1048.

Adam, M., Heuer, H., Hallmann, J., 2014. Bacterial antagonists of fungal pathogens also control root-knot nematodes by induced systemic resistance of tomato plants. PLoS One 9 (2). https://doi.org/10.1371/journal.pone.0090402, e90402.

Adams, A.E., Kazenel, M.R., Rudgers, J.A., 2017. Does a foliar endophyte improve plant fitness under flooding? Plant Ecol. 218, 711–723.

Agarwal, S., Shende, S.T., 1987. Tetrazolium reducing microorganisms inside the root of *Brassica species*. Curr. Sci. 56, 187–188.

Ahlem, N., Rania, A.B.A., Hayfa, J.K., Nawaim, A., Lamia, S., Walid, H., Rabiaa, H., Mejda, D.R., 2018. Biostimulation of tomato growth and suppression of *Fusarium* crown and root rot disease using fungi naturally associated to *Lycium arabicum*. J. Agri. Sci. Food Res. 9, 1.

Ahmad, P., Hashem, A., Abd-Allah, E.F., Alqarawi, A.A., John, R., Egamberdieva, D., Gucel, S., 2015. Role of *Trichoderma harzianum* in mitigating NaCl stress in Indian mustard (*Brassica juncea* L.) through antioxidative defense system. Front. Plant Sci. 6, 868.

Ait-Barka, E.A., Nowak, J., Clement, C., 2006. Enhancement of chilling resistance of inoculated grapevine plantlets with a plant growth-promoting Rhizobacterium, *Burkholderia phytofirmans* strain PsJN. Appl. Environ. Microbiol. 72, 7246–7252.

Al-Karaki, G.N., 2006. Nursery inoculation of tomato with arbuscular mycorrhizal fungi and subsequent performance under irrigation with saline water. Sci. Hortic. 109, 1–7.

Álvarez-Loayza, P., Jr White, J.F., Torres, M.S., et al., 2011. Light converts endosymbiotic fungus to pathogen, influencing seedling survival and niche-space filling of a common tropical tree, *Iriartea deltoidea*. PLoS One 6, e16386.

Aly, A.H., Debbab, A., Proksch, P., 2011. Fungal endophytes: unique plant inhabitants with great promises. Appl. Microbiol. Biotechnol. 90, 1829–1845.

Arangarasan, V., Palaniappan, S.P., Chelliah, S., 1998. Inoculation effects of diazotrophs and phosphobacteria on rice. Ind. J. Microbiol. 38, 111–112.

Arora, N.K., Khare, E., Naraian, R., Maheshwari, D.K., 2008. Sawdust as a superior carrier for production of multipurpose bioinoculant using plant growth promoting rhizobial and pseudomonad strains and their impact on productivity of *Trifolium repense*. Curr. Sci. 95, 90.

Askari-Khorasgani, O., Hatterman-Valenti, H., Flores, F.B., Pessarakli, M., 2019. Managing plant-environment-symbiont interactions to promote plant performance under low temperature stress. J. Plant Nutr., 1–18.

Azad, K., Kaminskyj, S., 2016. A fungal endophyte strategy for mitigating the effect of salt and drought stress on plant growth. Symbiosis 68, 73–78.

Babu, A.G., Shea, P.J., Sudhakar, D., Jung, I.B., Oh, B.T., 2015. Potential use of Pseudomonas koreensis AGB-1 in association with *Miscanthus sinensis* to remediate heavy metal (loid)-contaminated mining site soil. J. Environ. Manage. 151, 160–166.

Backman, P.A., Wilson, M., Murphy, J.F., 1997. Bacteria for biological control of plant diseases. In: Rechcigl, N.A., Rechecigl, J.E. (Eds.), Environmentally Safe Approaches to Crop Disease Control. Lewis Publishers, Boca Raton, FL, pp. 95–109.

Bajaj, R., Agarwal, A., Rajpal, K., Asthana, S., Prasad, R., Kharkwal, A.C., Kumar, R., Sherameti, I., Oelmüller, R., Varma, A., 2014. Co-cultivation of *Curcuma longa* with *Piriformospora indica* enhances the yield and active ingredients. Am. J. Curr. Microbiol. 2, 6–17.

Bajpai, P.D., Gupta, B.R., Ram, B., 1978. Studies on survival of *Rhizobium leguminosarum* in two carriers as affected by moisture and temperature conditions. Ind. J. Agric. Res. 112, 39–43.

Baldan, E., Nigris, S., Populin, F., Zottini, M., Squartini, A., Baldan, B., 2014. Identification of culturable bacterial endophyte community isolated from tissues of *Vitis vinifera"Glera"*. Plant Biosyst. 148, 508–516.

Balliu, A., Sallaku, G., Rewald, B., 2015. AMF inoculation enhances growth and improves the nutrient uptake rates of transplanted, salt-stressed tomato seedlings. Sustainability 7, 15967–15981.

Basheer, J., Ravi, A., Mathew, J., Krishnankutty, R.E., 2018. Assessment of plant-probiotic performance of novel endophytic *Bacillus* sp. in talc-based formulation. Probiotics Antimicrob. Proteins. https://doi.org/10.1007/s12602-018-9386-y.

Bazilah, A.B.I., Sariah, M., Zainal, M.A., Abidin, Yasmeen, S., 2011. Effect of carrier and temperature on the viability of *Burkholderia sp.* (UPMB3) and *Pseudomonas sp.* (UPMP3) during storage. Int. J. Agric. Biol. 13, 198–202.

Benhamou, N., Kloepper, J.W., Tuzun, S., 1998. Induction of resistance against *Fusarium* wilt of tomato by combination of chitosan with an endophytic bacterial strain: ultrastructural and cytochemistry of the host response. Planta 204, 153–168.

Berger, B., Patz, S., Ruppel, S., Dietel, K., Faetke, S., Junge, H., Becker, M., 2018. Successful formulation and application of plant growth-promoting *Kosakonia radicincitans* in maize cultivation. Biomed. Res. Int. https://doi.org/10.1155/2018/6439481, 6439481.

Bharti, N., Barnawal, D., Wasnik, K., Tewari, S.K., Kalra, A., 2016. Co-inoculation of *Dietzia natronolimnaea* and *Glomus intraradices* with vermicompost positively influences *Ocimum basilicum* growth and resident microbial community structure in salt affected low fertility soils. Appl. Soil Ecol. 100, 211–225. ISSN 0929-1393 https://doi.org/10.1016/j.apsoil.2016.01.003.

Bilgin, D.D., Zavala, J.A., Zhu, J.I.N., Clough, S.J., Ort, D.R., Delucia, E., 2010. Biotic stress globally downregulates photosynthesis genes. Plant Cell Environ. 33 (10), 1597–1613.

Bradáˇcová, K., Florea, A.S., Bar-Tal, A., Minz, D., Yermiyahu, U., Shawahna, R., Kraut-Cohen, J., Zolti, A., Erel, R., Dietel, K., Weinmann, M., Zimmermann, B., Berger, N., Ludewig, U., Neumann, G., Posta, G., 2019. Microbial consortia versus single strain inoculants: an advantage in PGPM-assisted tomato production? Agronomy 9, 105. https://doi.org/10.3390/agronomy9020105.

Bradl, H., Xenidis, A., 2005. Remediation techniques. Interface Sci. Technol. 6, 165–261.

Brosi, G.B., McCulley, R.L., Bush, L.P., Nelson, J.A., Classen, A.T., Norby, R.J., 2011. Effects of multiple climate change factors on the tall fescue-fungal endophyte symbiosis: infection frequency and tissue chemistry. New Phytol. 189, 797e805.

Bücker-Neto, L., Paiva, A.L.S., Machado, R.D., Arenhart, R.A., Margis-Pinheiro, M., 2017. Interactions between plant hormones and HMs responses. Genet. Mol. Biol. 40, 373–386.

Bulgari, D., Casati, P., Quaglino, F., Bianco, P.A., 2014. Endophytic bacterial community of grape vine leaves influenced by sampling date and *phytoplasma* infection process. BMC Microbiol. 14, 198.

Bush, L.P., Fannin, F.F., Siegel, M.R., Dahlman, D.L., Burton, H.R., 1993. Chemistry, occurrence and biological effects of saturated pyrrolizidine alkaloids associated with endophyte-grass interactions. Agric. Ecosyst. Environ. 44, 81e102.

Bush, L.P., Wilkinson, H.H., Schardl, C.L., 1997. Bioprotective alkaloids of grass-fungal endophyte symbioses. Plant Physiol. 114, 1–7.

Caldwell, M.M., Ballaré, C.L., Bornman, J.F., et al., 2003. Terrestrial ecosystems, increased solar ultraviolet radiation and interactions with other climatic change factors. Photochem. Photobiol. Sci. 2, 29–38.

Caradus, J., Lovatt, S., Belgrave, B., 2013. Adoption of forage technologies by New Zealand farmers–case studies. In: Proceedings of the International Grasslands Congress. vol. 22. NSW Department of Primary Industries, Sydney, pp. 1843–1845.

Carrell, A.A., Frank, A.C., 2014. *Pinus flexilis* and *Piceaeen gelmannii* share a simple and consistent needle endophyte microbiota with a potential role in nitrogen fixation. Front. Microbiol. 5, 333. https://doi.org/10.3389/fmicb.2014.00333.

Ceccarelli, S., 2015. Efficiency of plant breeding. Crop. Sci. 55, 87–97. https://doi.org/10.2135/cropsci2014.02.0158.

Chen, Y., Mei, R., Lu, S., Liu, L., Kloepper, J.W., 1996. The use of yield increasing bacteria (YIB) as plant growth-promoting rhizobacteria in Chinese agriculture. In: Utkhede, R.S., Gupta, V.K. (Eds.), Management of Soil Borne Diseases. Kalyani Publishers, New Delhi, India, pp. 165–176.

Chen, L., Zhang, Q.Y., Jia, M., Ming, Q.L., Yue, W., Rahman, K., 2016. Endophytic fungi with antitumor activities: their occurrence and anticancer compounds. Crit. Rev. Microbiol. 42, 454–473.

Chen, W., Liu, H., Han, W., Gao, Y., Card, S.D., Ren, A., 2017. The advantages of endophyte infected over unidentified tall fescue in the growth and pathogen resistence are counteracted by elevated CO_2. Sci. Rep. 7, 6952.

Chen, Y., et al., 2018. Wheat microbiome bacteria can reduce virulence of a plant pathogenic fungus by altering histone acetylation. Nat. Commun. 9, 3429.

Chen, T., Li, C., White, J.F., et al., 2019. Effect of the fungal endophyte *Epichloë bromicola* on polyamines in wild barley (*Hordeum brevisubulatum*) under salt stress. Plant Soil 436, 29–48. https://doi.org/10.1007/s11104-018-03913-x.

Cheplick, G.P., Faeth, S.H., 2009. Ecology and Evolution of the Grass-Endophyte. Symbiosis. Oxford University Press, Oxford, UK.

Chisholm, S.T., Coaker, G., Day, B., Staskawicz, B.J., 2006. Host-microbe interactions: shaping the evolution of the plant immune response. Cell 124 (4), 803–814.

Choudhury, F.K., Rivero, R.M., Blumwald, E., Mittler, R., 2017. Reactive oxygen species, abiotic stress and stress combination. Plant J. 90, 856–867.

Classen, A., Sundqvist, M., Henning, J., Newman, G., Moore, J., et al., 2015. Direct and indirect effects of climate change on soil microbial and soil microbial-plant interactions: what lies ahead? Ecosphere 6 (8). https://doi.org/10.1890/ES15-00217.1.

Clemmensen, K.E., Bahr, A., Ovaskainen, O., Dahlberg, A., Ekblad, A., Wallander, H., Stenlid, J., Finlay, R.D., Wardle, D.A., Lindahl, B.D., 2013. Roots and associated fungi drive long-term carbon sequestration in boreal forest. Science 339, 1615–1618.

Cohen, A.C., Bottini, R., Piccoli, P.N., 2008. *Azosprillium brasilense* Sp 245 produces ABA in chemically defined culture medium and increases ABA content in *Arabidopsis* plants. Plant Growth Regul. 54, 97–103.

Compant, S., Reiter, B., Sessitsch, A., et al., 2005. Endophytic colonization of *Vitis vinifera* L. by a plant growth-promoting bacterium, *Burkholderia* sp. strain PsJN. Appl. Environ. Microbiol. 71, 1685–1689.

Conover, D.O., Duffy, T.A., Hice, L.A., 2009. The covariance between genetic and environmental influences across ecological gradients. Ann. N. Y. Acad. Sci. 1168, 100–129.

Cox, S.E., Stushnoff, C., 2001. Temperature-related shifts in soluble carbohydrate content during dormancy and cold acclimation in *Populus tremuloides*. Can. J. For. Res. 31, 730–737.

Cramer, G.R., Urano, K., Delrot, S., et al., 2011. Effects of abiotic stress on plants: a systems biology perspective. BMC Plant Biol. 11, 163. https://doi.org/10.1186/1471-2229-11-163.

Dalia, A., El-Fattah, A., Eweda, W.E., Zayed, M.S., Mosaad, K., 2013. Hassanein Effect of carrier materials, sterilization method, and storage temperature on survival and biological activities of *Azotobacter chroococcum* inoculants. Ann. Agric. Sci. 58 (2), 111–118.

Davitt, A.J., Stansberry, M., Rudgers, J.A., 2010. Do the costs and benefits of fungal endophyte symbiosis vary with light availability? New Phytol. 188, 824–834.

De Bary, A., 1866. Morphologie und Physiologie der Pilze, Flechten und Myxomyceten. vol. 2 Hofmeister's Handbook of Physiological Botany, Leipzig.

de Boer, W., Folman, L., Summerbell, R., Boddy, L., 2005. Living in a fungal world: impact of fungi on soil bacterial niche development. FEMS Microbiol. Rev. 29, 795–811.

De Medeiros, F.H.V.D., Souza, R.M.D., Ferro, H.M., Zanotto, E., Machado, J.D.C., 2015. Screening of endospore-forming bacteria for cotton seed treatment against bacterial blight and damping-off. Adv. Plants Agricult. Res. 2 (4), 56–61.

Deslippe, J.R., Simard, S.W., 2011. Below-ground carbon transfer among Betula nana may increase with warming in Arctic tundra. New Phytol. 192, 689–698.

Egamberdieva, D., 2013. The role of phytohormone producing bacteria in alleviating salt stress in crop plants. In: Miransari, M. (Ed.), Biotechnological Techniques of Stress Tolerance in Plants. Stadium Press LLC, USA, pp. 21–39.

Egan, M.J., Wang, Z.Y., Jones, M.A., et al., 2007. Generation of reactive oxygen species by fungal NADPH oxidases is required for rice blast disease. Proc. Natl. Acad. Sci. U. S. A. 104, 11772–11777.

Etminani, F., Harighi, B., 2018. Isolation and identification of endophytic bacteria with plant growth promoting activity and biocontrol potential from wild pistachio trees. Plant Pathol. J. 34, 208–217. https://doi.org/10.5423/PPJ.OA.07.2017.0158.

Fadiji, A.E., Babalola, O.O., 2020. Elucidating mechanisms of endophytes used in plant protection and other bioactivities with multifunctional prospects. Front. Bioeng. Biotechnol. 8, 467. https://doi.org/10.3389/fbioe.2020.00467.

Fahad, S., Hussain, S., Bano, A., et al., 2015. Potential role of phytohormones and plant growth-promoting rhizobacteria in abiotic stresses: consequences for changing environment. Environ. Sci. Pollut. Res. 22, 4907–4921.

Fernandez, O., Theocharis, A., Bordiec, S., Feil, R., Jacquens, L., Clement, C., et al., 2012a. *Burkholderia phytofirmans*-PsJN acclimates grapevine to cold by modulating carbohydrate metabolism. Mol. Plant Microbe Interact. 25, 496–504.

Fernandez, O., Vandesteene, L., Feil, R., Baillieul, F., Lunn, J.E., Clement, C., 2012b. Trehalose metabolism is activated upon chilling in grapevine and might participate in *Burkholderia phytofirmans* induced chilling tolerance. Planta 236 (2), 355–369.

Flatley, A., Ogle, L., Noel, A., Fraley, E., Goodman, A., Donna, B., 2015. Isolation of Possible Biocontrol Endophytic Bacteria From *Solanum tuberosum* Effective Against *Streptomyces scabies*. Poster Sessions 7.

Fuchs, B., Krischke, M., Martin, J., Mueller, B., Krauss, J., 2017. Plant age and seasonal timing determine endophyte growth and alkaloid biosynthesis. Fungal Ecol. 29, 52e58.

Gagne, S., Richard, C., Roussean, H., Antoun, H., 1987. Xylem-residing bacteria in alfalfa roots. Can. J. Microbiol. 33, 996–1000.

Gangwar, O., Singh, A.P., 2018. *Trichoderma* as an efficacious bioagent for combating biotic and abiotic stresses of wheat—a review. Agric. Rev. 39, 49–54.

Gao, F.K., Dai, C.C., Liu, X.Z., 2010. Mechanisms of fungal endophytes in plant protection against pathogens. Afr. J. Microbiol. Res. 4 (13), 1346–1351.

Germaine, K., Keogh, E., Borremans, B., et al., 2004. Colonization of poplar trees by gfp expressing bacterial endophytes. FEMS Microbiol. Ecol. 48, 109–118.

Germaine, K., Liu, X., Cabellos, G., Hogan, J., Ryan, D., Dowling, D.N., 2006. Bacterial endophyte-enhanced phyto-remediation of the organochlorine herbicide 2,4-dichlorophenoxyacetic acid. FEMS Microbiol. Ecol. 57, 302–310. https://doi.org/10.1111/j.1574-6941.2006.00121.x.

Ghalambor, C.K., Mckay, J.K., Carroll, S.P., Reznick, D.N., 2007. Adaptive versus non-adaptive phenotypic plasticity and the potential for contemporary adaptation in new environments. Funct. Ecol. 21, 394–407.

Ghosh, S., Watson, A., Gonzalez-Navarro, O.E., et al., 2018. Speed breeding in growth chambers and glasshouses for crop breeding and model plant research. Nat. Protoc. 13 (12), 2944–2963.

Giauque, H., Hawkes, C.V., 2013. Climate affects symbiotic fungal endophyte diversity and performance. Am. J. Bot. 100 (7), 1435–1444.

Gill, S.S., Tuteja, N., 2010. Reactive oxygen species and antioxidant machinery in abiotic stress tolerance in crop plants. Plant Physiol. Biochem. 48, 909–930.

Glick, B.R., 2005. Modulation of plant ethylene levels by the bacterial enzyme ACC deaminase. FEMS Microbiol. Lett. 251, 1–7.

Glick, B.R., 2012. Plant growth-promoting bacteria: mechanisms and applications. Scientifica 15, 963401. https://doi.org/10.6064/2012/963401.

Godstime, O.C., Enwa, F.O., Augustina, J.O., Christopher, E.O., 2014. Mechanisms of antimicrobial actions of phytochemicals against enteric pathogens—a review. J. Pharm. Chem. Biol. Sci. 2, 77–85.

Gond, S.K., Torres, M.S., Bergen, M.S., Helsel, Z., White, J.F., 2015. Induction of salt tolerance and up-regulation of aquaporin genes in tropical corn by rhizobacterium *Pantoea agglomerans*. Lett. Appl. Microbiol. 60, 392–399.

Gopalakrishnan, S., Sathya, A., Vijayabharathi, R., Sriniva, V., 2016. Formulations of plant growth-promoting microbes for field applications. In: Microbial Inoculants in Sustainable Agricultural Productivity, pp. 239–251, https://doi.org/10.1007/978-81-322-2644-4_15.

Gorzynska, K., Ryszka, P., Anielska, T., Turnau, K., Lembicz, M., 2017. Effect of *Epichlo¨e typhina* fungal endophyte on the diversity and incidence of other fungi in *Puccinellia distans* wild grass seeds. Flora 228, 60–64.

Gul Jan, F., Hamayun, M., Hussain, A., et al., 2019. An endophytic isolate of the fungus *Yarrowia lipolytica* produces metabolites that ameliorate the negative impact of salt stress on the physiology of maize. BMC Microbiol. 19, 3. https://doi.org/10.1186/s12866-018-1374-6.

Gunatilaka, A.A.L., 2006. Natural products from plant-associated microorganisms: distribution, structural diversity, bioactivity, and implications of their occurrence. J. Nat. Prod. 69, 509–526.

Gusain, Y.S., Singh, U.S., Sharma, A.K., 2015. Bacterial mediated amelioration of drought stress in drought tolerant and susceptible cultivars of rice (*Oryza sativa* L.). Afr. J. Biotechnol. 14, 764–773.

Hardoim, P.R., Van Overbeek, L.S., Van Elsas, J.D., 2008. Properties of bacterial endophytes and their proposed role in plant growth. Trends Microbiol. 16, 463–471.

Hardoim, P.R., Hardoim, C.C.P., van Overbeek, L.S., van Elsas, J.D., 2012. Dynamics of seed-borne rice endophytes on early plant growth stages. PLoS ONE 7 (2), e30438. https://doi.org/10.1371/journal.pone.0030438.

Hashem, A., Abd-Allah, E., Alqarawi, A., Al-Huqail, A., Shah, M., 2016. Induction of osmoregulation and modulation of salt stress in *Acacia gerrardii* Benth. By arbuscular mycorrhizal fungi and Bacillus subtilis (BERA 71). Biomed. Res. Int., 6294098.

Hawkes, C.V., Hartley, I.P., Ineson, P., Fitter, A.H., 2008. Soil temperature affects carbon allocation within arbuscular mycorrhizal networks and carbon transport from plant to fungus. Glob. Chang. Biol. 14, 1181–1190.

He, Y., Yu, C., Zhou, L., Chen, Y., Liu, A., Jin, J., Hong, J., Qi, Y., Jiang, D., 2014. Rubisco decrease is involved in chloroplast protrusion and Rubisco-containing body formation in soybean (*Glycine max*) under salt stress. Plant Physiol. Biochem. 4, 118–124.

Hereme, R., Morales-Navarro, S., Ballesteros, G., Barrera, A., Ramos, P., Gundel, P.E., Molina-Montenegro, M.A., 2020. Fungal endophytes exert positive effects on *Colobanthus quitensis* under water stress but neutral under a projected climate change scenario in Antarctica. Front. Microbiol. https://doi.org/10.3389/fmicb.2020.00264.

Hussain, N., Abbasi, S.A., 2018. Efficacy of the vermicomposts of different organic wastes as "clean" ferlilizers: state of the art. Sustainability 10, 1205. https://doi.org/10.3390/su10041205.

Ihsan, M.Z., Daur, I., Alghabari, F., Alzamanan, S., Rizwan, S., Ahmad, M., Waqas, M., Waqas, S., 2019. Heat stress and plant development: role of sulphur metabolites and management strategies. Acta Agric. Scand. Sect. B Soil Plant Sci. 69 (4), 332–342. https://doi.org/10.1080/09064710.2019.1569715.

Imaizumi-Anraku, H., Takeda, N., Charpentier, M., Perry, J., Miwa, H., Umehara, Y., Kouchi, H., Murakami, Y., et al., 2005. Plastid proteins crucial for symbiotic fungal and bacterial entry into plant roots. Nature 433, 527–531.

Irizarry, I., White, J.F., 2017. Application of bacteria from non-cultivated plants to promote growth alters root architecture and alleviates salt stress of cotton. J. Appl. Microbiol. 122, 1110–1120.

Isah, T., Umar, S., Mujib, A., et al., 2018. Secondary metabolism of pharmaceuticals in the plant in vitro cultures: strategies, approaches, and limitations to achieving higher yield. Plant Cell Tiss. Org. Cult. 132 (2), 239–265.

Jaber, L.R., 2015. Grapevine leaf tissue colonization by the fungal entomopathogen *Beauveria bassiana* and its effect against downy mildew. BioControl 60, 103–112.

Jansson, S., Douglas, C.J., 2007. *Populus*: a model system for plant biology. Annu. Rev. Plant Biol. 58, 435–458.

Jha, Y., Subramanian, R.B., 2014. PGPR regulate caspaselike activity, programmed cell death, and antioxidant enzyme activity in paddy under salinity. Physiol. Mol. Biol. Plants 20, 201–207.

Jogawat, A., Vadassery, J., Verma, N., Oelmuller, R., Dua, M., Nevo, E., Johri, A.K., 2016. PiHOG1, a stress regulator MAP kinase from the root endophyte fungus *Piriformospora indica*, confers salinity stress tolerance in rice plants. Sci. Rep. 6, 36765.

Johnson, N.C., Graham, J.H., Smith, F.A., 1997. Functioning of mycorrhizal associations along the mutualism–parasitism continuum. New Phytol. 135, 575–585.

Johnson, L.J., De Bonth, A.C.M., Briggs, L.R., et al., 2013. The exploitation of epichloae endophytes for agricultural benefit. Fungal Divers. 60, 171–188.

Johnston-Monje, D., Lundberg, D.S., Lazarovits, G., Reis, V.M., Raizada, M.N., 2016. Bacterial populations in juvenile maize rhizospheres originate from both seed and soil. Plant Soil 405, 337–355.

Jung, H.K., Kanga, S.M., Kanga, Y.M., Kanga, M.J., Yun, D., Bahk, J.D., Yang, J.K., Choi, M.S., 2003. Enhanced production of scopolamine by bacterial elicitors in adventitious hairy root cultues of *Scopalia parviflora*. Enzyme Microb. Technol. 33, 987–990.

Kalra, A., Chandra, M., Awasthi, A., Singh, A.K., Khanuja, S.P.S., 2010. Natural compound enhancing growth and survival of rhizobial inoculants in vermicompost based formulation. Biol. Fertil. Soils 46, 521–524.

Kaushal, M., Wani, S.P., 2015. Plant-growth-promoting rhizobacteria: drought stress alleviators to ameliorate crop production in drylands. Ann. Microbiol., 1–8.

Kawecki, T.J., Ebert, D., 2004. Conceptual issues in local adaptation. Ecol. Lett. 7, 1225–1241.

Khan, A.L., Hamayun, M., Kim, Y.H., Kang, S.M., Lee, I.J., 2011. Ameliorative symbiosis of endophyte *Penicillium funiculosum* LH 06 under salt stress elevated plant growth of *Glycine max* L. Plant Physiol. Biochem. 49 (8), 852–861.

Khan, A.L., Javid, H., Ahmed, A.H., Ahmed, A.R., In-Jung, L., 2013. Endophytic fungi: resource for gibberellins and crop abiotic stress resistance. Crit. Rev. Biotechnol. 35, 62–74.

Khan, T.A., Mazid, M., Quddusi, S., 2014. Role of organic and inorganic chemicals in plant-stress mitigation. In: Gaur, R.K., Sharma, P. (Eds.), Approaches to Plant Stress and Their Management. Springer, Germany, pp. 39–52.

Khare, E., Kim, K.M., Lee, K.J., 2016. Rice OsPBL1 (*Oryza sativa* Arabidopsis PBS1-LIKE 1) enhanced defense of *Arabidopsis* against *Pseudomonas syringae* DC3000. Eur. J. Plant Pathol. 146, 901–910.

Khushdil, F., Jan, F.G., Hamayun, M., Iqbal, A., Hussain, A., Bibi, N., 2019. Salt stress alleviation in *Pennisetum galucam* through secondary metabolites modulation by *Aspergillus terreus*. Plant Physiol. Biochem. 144, 127–134.

Kilian, M., Steiner, U., Krebs, B., Junge, H., Schmiedeknecht, G., Hain, R., 2000. FZB24 *Bacillus subtilis*-mode of action of a microbial agent enhancing plant vitality. Pflanzenschutz-Nachrichten, Bayer 100 (1), 72–93.

Koskimäki, J.J., Hokkanen, J., Jaakola, L., Suorsa, M., Tolonen, A., Mattila, S., Pirttilä, A.M., Hohtola, A., 2009. Flavonoid biosynthesis and degradation play a role in early defence responses of bilberry (*Vaccinium myrtillus*) against biotic stress. Eur. J. Plant Pathol. 125, 629–640.

Koutsompogeras, P., Kyriacou, A., Zabetakis, I., 2007. The formation of 2,5-dimethyl-4-hydroxy-2H-furan-3-one by cell-free extracts of *Methylobacterium extorquens* and strawberry (*Fragaria ananassa* cv. Elsanta). Food Chem. 104, 1654–1661.

Kumar, V., Sarma, M.V.R.K., Saharan, K., et al., 2012. Effect of formulated root endophytic fungus *Piriformospora indica* and plant growth promoting rhizobacteria fluorescent pseudomonads R62 and R81 on *Vigna mungo*. World J. Microbiol. Biotechnol. 28, 595–603. https://doi.org/10.1007/s11274-011-0852-x.

Kumar, S., Aharwal, R.P., Shukla, H., Rajak, R.C., Sandhu, S.S., 2014. Endophytic fungi: as a source of antimicrobials bioactive compounds. World J. Pharm. Pharm. Sci. 3, 1179–1197.

Kumar, S.P.M., Chowdappa, P., Krishna, V., 2015. Development of seed coating formulation using consortium of *Bacillus subtilis* (OTPB1) and *Trichoderma harzianum* (OTPB3) for plant growth promotion and induction of systemic resistance in field and horticultural crops. Indian Phytopath. 68 (1), 25–31.

Kumar, R., Mishra, R.K., Mishra, V., Qidwai, A., Pandey, A., Shukla, S.K., Pandey, M., Pathak, A., Dikshit, A., 2016. Detoxification and tolerance of HMs in plants. In: Ahmad, P. (Ed.), Plant Metal Interaction. Elsevier, Amsterdam, pp. 335–359.

Kusari, S., Hertweck, C., Spiteller, M., 2012a. Chemical ecology of endophytic fungi: origins of secondary metabolites. Chem. Biol. 19, 792–798.

Kusari, S., Verma, V.C., Lamshoeft, M., Spiteller, M., 2012b. An endophytic fungus from *Azadirachta indica A. Juss.* that produces Azadirachtin. World J. Microbiol. Biotechnol. 28, 1287–1294. https://doi.org/10.1007/s11274-011-0876-2.

Kushwaha, R.K., Singh, S., Pandey, S.S., et al., 2019. Compatibility of inherent fungal endophytes of *Withania somnifera* with *Trichoderma virdie* and its impact on plant growth and withanolide content. J. Plant Growth Regul. 38, 1228–1242.

Lafi, F.F., AlBladi, M.L., Salem, N.M., AlBanna, L., Alam, I., Bajic, V.B., 2017. Draft genome sequence of the plant growth-promoting *Pseudomonas punonensis* strain D1–6 isolated from the desert plant *Erodium hirtum* in Jordan. Genome Announc. 5. https://doi.org/10.1128/genomeA.01437-16. e1437–e1416.

Lajayer, A.B., Ghorbanpour, M., Nikabadi, S., 2017. HMs in contaminated environment: Destiny of secondary metabolite biosynthesis, oxidative status and phytoextraction in medicinal plants. Ecotoxicol. Environ. Saf. 145, 377–390.

Lattanzio, V., Lattanzio, V.M., Cardinali, A., 2006. Role of phenolics in the resistance mechanisms of plants against fungal pathogens and insects. Phytochem. Adv. Res. 661, 23–67.

Lei, L., Shi, J., Chen, J., Zhang, M., Sun, S., Xie, S., Li, X., Zeng, B., Peng, L., Hauck, A., et al., 2015. Ribosome profiling reveals dynamic translational landscape in maize seedlings under drought stress. Plant J. 84 (6), 1206–1218.

Leifheit, E.F., Verbruggen, E., Rillig, M.C., 2015. Arbuscular mycorrhizal fungi reduce decomposition of woody plant litter while increasing soil aggregation. Soil Biol. Biochem. 81, 323–328.

Leveau, J.H.J., Lindow, S.E., 2001. Appetite of an epiphyte: quantitative monitoring of bacterial sugar consumption in the phyllosphere. Proc. Natl. Acad. Sci. U. S. A. 98 (6), 3446–3453.

Li, J., Zhao, G.Z., Varma, A., Qin, S., Xiong, Z., Huang, H.Y., et al., 2012. An endophytic *Pseudonocardia* species induces the production of artemisinin in *Artemisia annua*. PLoS One 7. https://doi.org/10.1371/journal.pone.0051410, e51410.

Li, H., Zhao, J., Feng, H., Huang, L.L., Kang, Z.S., 2013. Biological control of wheat stripe rust by an endophytic *Bacillus subtilis* strain E1R-j in greenhouse and field trials. Crop Prot. 43, 201–206.

Li, X., He, X.L., Zhou, Y., Hou, Y.T., Zuo, Y.L., 2019. Effects of dark septate endophytes on the performance of *Hedysarum scoparium* under water deficit stress. Front. Plant Sci. 10, 903. https://doi.org/10.3389/fpls.2019.00903.

Lin, J., Li, J.P., Yuan, F., Yang, Z., Wang, B.S., Chen, M., 2018. Transcriptome profiling of genes involved in photosynthesis in *Elaeagnus angustifolia* under salt stress. Photosynthetica 56, 998–1009.

Liu, C.H., Zou, W.X., Lu, H., Tan, R.X., 2001. Antifungal activity of *Artemisia annua* endophyte cultures against phytopathogenic fungi. J. Biotechnol. 88 (3), 277–282.

Liu, J., Xie, B., Shi, X., Ma, J., Guo, C., 2015. Effects of two plant growth-promoting rhizobacteria containing 1-aminocyclopropane-1-carboxylate deaminase on oat growth in petroleum-contaminated soil. Int. J. Environ. Sci. Technol. 12, 3887–3894. https://doi.org/10.1007/s13762-015-0798-x.

Madhaiyan, M., Poonguzhali, S., Sa, T., 2007. Metal tolerating methylotrophic bacteria reduces nickel and cadmium toxicity and promotes plant growth of tomato (*Lycopersicon esculentum* L.). Chemosphere 69 (2), 220–228.

Malik, K.A., Hafeez, F.Y., Mirza, M.S., Hameed, S., Rasul, G., Bilal, R., 2005. Rhizospheric plant—microbe interactions for sustainable agriculture. In: Wang, Y.P., Lin, M., Tian, Z.X., Elmerich, C., Newton, W.E. (Eds.), Biological Nitrogen Fixation. Sustainable Agriculture and the Environment. Springer, pp. 257–260.

Malinowski, D.P., Alloush, G.A., Belesky, D.P., 1998. Evidence for chemical changes on the root surface of tall fescue in response to infection with the fungal endophyte *Neotyphodium coenophialum*. Plant Soil 205, 1–12. https://doi.org/10.1023/A:1004331932018.

Manter, D.K., Delgado, J.A., Holm, D.G., Stong, R.A., 2010. Pyrosequencing reveals a highly diverse and cultivar-specific bacterial endophyte community in potato roots. Microb. Ecol. 60, 157–166.

Márquez, L.M., Redman, R.S., Rodriguez, R.J., Roossinck, M.J., 2007. A virus in a fungus in a plant—three way symbiosis required for thermal tolerance. Science 315, 513–515.

Martinez-Balmori, D., Olivares, F.L., Spaccini, R., Aguiar, K.P., Araújo, M.F., Aguiar, N.O., Guridi, F., Canellas, L.P., 2013. Molecular characteristics of vermicompost and their relationship to preservation of inoculated nitrogen-fixing bacteria. J. Anal. Appl. Pyrolysis 104, 540–550. https://doi.org/10.1016/j.jaap.2013.05.015.

Mastan, A., Bharadwaj, R., Kushwaha, R.K., Vivek Babu, C.S., 2019. Functional fungal endophytes in *Coleus forskohlii* regulate labdane diterpene biosynthesis for elevated forskolin accumulation in roots. Microb. Ecol. 78 (4), 914–926.

Mastouri, F., Bjorkman, T., Harman, G.E., 2010. Seed treatment with *Trichoderma harzianum* alleviates biotic, abiotic and physiological stresses in germinating seeds and seedlings. Phytopathology 100, 1213–1221.

Mayak, S., Tirosh, T., Glick, B.R., 2004. Plant growth promoting bacteria that confer resistence to water stress in tomatos and peppers. Plant Sci. 166 (2), 525–530.

McCulley, R.L., Bush, L.P., Carlisle, A.E., Ji, H., Nelson, J.A., 2014. Warming reduces tall fescue abundance but stimulates toxic alkaloid concentrations in transition zone pastures of the U.S. Front. Chem. 2, 88.

Mclellan, C.A., Turbyville, T.J., Wijeratne, E.M.K., Kerschen, A., Vierling, E., Queitsch, C., Whitesell, L., Gunatilaka, A.A.L., 2007. A rhizosphere fungus enhances *Arabidopsis* thermotolerance through production of an HSP90 inhibitor. Plant Physiol. 145, 174–182.

Miedes, E., Vanholme, R., Boerjan, W., Molina, A., 2015. The role of the secondary cell wall in plant resistance to pathogens. Plant Cell Wall Pathog. Parasit. Symb. 78, 213–221.

Miliute, I., Buzaite, O., Baniulis, D., Vidmantas, S., 2015. Bacterial endophytes in agricultural crops and their role in stress tolerance: a review. Zemdirbyste 102, 465–478.

Min, H., Chen, C., Wei, S., Shang, X., Sun, M., Xia, R., Liu, X., Hao, D., Chen, H., Xie, Q., 2016. Identification of drought tolerant mechanisms in maize seedlings based on transcriptome analysis of recombination inbred lines. Front. Plant Sci. 7, 1080.

Ming, Q., Su, C., Zheng, C., Jia, M., Zhang, Q., Zhang, H., Rahman, K., Han, T., Qin, L., 2013. Elicitors from the endophytic fungus *Trichoderma atroviride* promote *Salvia miltiorrhiza* hairy root growth and tanshinone biosynthesis. J. Exp. Bot. 64, 5687–5694.

Miottovilanova, L., Jacquard, C., Courteaux, B., Wortham, L., Michel, J., Clement, C., 2016. *Burkholderia phytofirmans* PsJN confers grapevine resistance against *Botrytis cinerea* via a direct antimicrobial effect combined with a better resource mobilization. Front. Plant Sci. 7, 1236.

Mishra, A., Singh, S.P., Mahfooz, S., Singh, S.P., Bhattacharya, A., Mishra, N., Nautiyal, C.S., 2018. Endophyte-mediated modulation of defenserelated genes and systemic resistance in *Withania somnifera* (L.) Dunal under *Alternaria alternate* stress. Appl. Environ. Microbiol. 84, e02845-17.

Mitter, B., Petric, A., Shin, M.W., et al., 2013. Comparative genome analysis of *Burkholderia phytofirmans* PsJN reveals a wide spectrum of endophytic lifestyles based on interaction strategies with host plants. Front. Plant Sci. 4, 120.

Mocali, S., Bertelli, E., DiCello, F., Mengoni, A., Sfalanga, A., Viliani, F., et al., 2003. Fluctuation of bacteria isolated from elm tissues during different seasons and from different plant organs. Res. Microbiol. 154, 105–114.

Molina-Montenegro, M.A., Acuña-Rodríguez, I.S., Torres-Díaz, C., et al., 2020. Antarctic root endophytes improve physiological performance and yield in crops under salt stress by enhanced energy production and Na^+ sequestration. Sci. Rep. 10, 5819. https://doi.org/10.1038/s41598-020-62544-4.

Moore, J.A., Jiang, M.J., Post, W.M., Classen, A.T., 2015. Decomposition by ectomycorrhizal fungi alters soil carbon storage in a simulation model. Ecosphere 6 (3), 29.

Munzuroglu, O., Geckil, H., 2002. Effects of metals on seed germination, root elongation, and coleoptile and hypocotyl growth in *Triticum aestivum* and *Cucumis sativus*. Arch. Environ. Contam. Toxicol. 43, 203.

Nakagawa, T., Imaizumi-Anraku, H., 2015. Rice arbuscular mycorrhiza as a tool to study the molecular mechanisms of fungal symbiosis and a potential target to increase productivity. Rice 8, 32.

Nakkeeran, S., Fernando, W.G.D., Siddiqui, Z.A., 2005. Plant growth promoting rhizobacteria formulations and its scope in commercialization for the management of pests and diseases. In: Siddiqui, Z.A. (Ed.), PGPR: Biocontrol and Biofertilization. Springer, Dordrecht, https://doi.org/10.1007/1-4020-4152-7_10.

Nanda, R., Agrawal, V., 2016. Elucidation of zinc and copper induced oxidative stress, DNA damage and activation of defence system during seed germination in *Cassia angustifolia* Vahl. Environ. Exp. Bot. 125, 31–41.

Nandhini, S., Sendhilvel, V., Babu, S., 2012. Endophytic bacteria from tomato and their efficacy against *Fusarium oxysporum f.* sp. *lycopersici*, the wilt pathogen. J. Biopest. 5, 178–185.

Nasopoulou, C., Pohjanen, J., Koskimaki, J.J., Zabetakis, I., Pirttilä, A.M., 2014. Localization of strawberry (*Fragaria ananassa*) and *Methylobacterium extorquens* genes of strawberry flavor biosynthesis in strawberry tissue by *in situ* hybridization. J. Plant Physiol. 171, 1099–1105.

Naveed, M., Hussain, M.B., Zahir, Z.A., Mitter, B., Sessitsch, A., 2014. Drought stress amelioration in wheat through inoculation with *Burkholderia phytofirmans* strain PsJN. Plant Growth Regul. 73, 121–131.

Nazir, R., Warmink, J.A., Boersma, H., van Elsas, J.D., 2010. Mechanisms that promote bacterial fitness in fungal-affected soil microhabitats. FEMS Microbial. Ecol. 71, 169–185.

Ngigi, A.N., Getenga, Z.M., Boga, H.I., Ndalut, P.K., 2012. Biodegradation of s-triazine herbicide atrazine by *Enterobacter cloacae* and *Burkholderia cepacia* from long-term treated sugarcane-cultivated soils in Kenya. J. Environ. Sci. Health B 47, 769–778. https://doi.org/10.1080/03601234.2012.676364.

Nuccio, E.E., Hodge, A., Pett-Ridge, J., Herman, D.J., Weber, P.K., Firestone, M.K., 2013. An arbuscular mycorrhizal fungus significantly modifies the soil bacterial community and nitrogen cycling during litter decomposition. Environ. Microbiol. 15, 1870–1881.

Ou, T., Xua, W.F., Wanga, F., Strobel, G., Zhou, Z., Xiang, Z., Liu, J., Xie, J., 2019. A microbiome study reveals seasonal variation in endophytic bacteria among different Mulberry cultivars. Comput. Struct. Biotechnol. J. 17, 1091–1100.

Ozaktan, H., Çakır, B., Gül, A., Yolageldi, L., Akköprü, A., 2015. Isolation and evaluation of endophytic bacteria against *Fusarium oxysporum f.* sp. *cucumerinum* infecting cucumber plants. Austin J. Plant Biol. 1 (1), 1003.

Pageni, B.B., Lupwayi, N.Z., Akter, Z., et al., 2014. Plant growth-promoting and phytopathogen-antagonistic properties of bacterial endophytes from potato (*Solanum tuberosum* L.) cropping systems. Can. J. Plant Sci. 94, 835–844.

Pan, J., Peng, F., Xue, X., You, Q., Zhang, W., Wang, T., et al., 2019. The growth promotion of two salt-tolerant plant groups with PGPR inoculation: a meta-analysis. Sustainability 11 (2), 378.

Pandey, S.S., Singh, S., Babu, C.S., Shanker, K., Srivastava, N.K., Kalra, A., 2016a. Endophytes of opium poppy differentially modulate host plant productivity and genes for the biosynthetic pathway of benzylisoquinoline alkaloids. Planta 243, 1097–1114.

Pandey, S.S., Singh, S., Babu, C.S., Shanker, K., Srivastava, N.K., Shukla, A.K., et al., 2016b. Fungal endophytes of *Catharanthus roseus* enhance vindoline content by modulating structural and regulatory genes related to terpenoid indole alkaloid abiosynthesis. Sci. Rep. 6, 26583. https://doi.org/10.1038/srep26583.

Pandey, V., Ansari, M.W., Tula, S., Yadav, S., Sahoo, R.K., Shukla, N., Kumar, A., 2016c. Dose dependent response of *Trichoderma harzianum* in improving drought tolerance in rice genotypes. Planta 243 (5), 1251–1264.

Pandey, S.S., Singh, S., Pandey, H., Srivastava, M., Ray, T., Soni, S., Pandey, A., et al., 2018. Endophytes of *Withania somnifera* modulate *in planta* content and the site of withanolide biosynthesis. Sci. Rep. 8, 5450.

Paul, M.J., Primavesi, L.F., Jhurreea, D., Zhang, Y., 2008. Trehalose metabolism and signaling. Annu. Rev. Plant Biol. 59, 417–441.

Pavithra, G., Bindal, S., Rana, M., Srivastava, S., 2020. Role of endophytic microbes against plant pathogens: a review. Asian J. Plant Sci. 19, 54–62.

Pedranzani, H., Rodrıguez-Rivera, M., Gutierrez, M., Porcel, R., Hause, B., Ruiz-Lozano, J.M., 2016. Arbuscular mycorrhizal symbiosis regulates physiology and performance of *Digitaria eriantha* plants subjected to abiotic stresses by modulating antioxidant and jasmonate levels. Mycorrhiza 26, 141–152.

Perez, L.I., Gundel, P.E., Marrero, H.J., et al., 2017. Symbiosis with systemic fungal endophytes promotes host escape from vector-borne disease. Oecologia 184, 237–245. https://doi.org/10.1007/s00442-017-3850-3.

Pindi, P.k., Satyanarayana, S.D.V., 2012. Liquid microbial consortium—a potential tool for sustainable soil health. J. Biofertil. Biopestici. 3, 4.

Pinheiro, C., Chaves, M.M., 2011. Photosynthesis and drought: can we make metabolic connections from available data? J. Exp. Bot. 62 (3), 869–882. https://doi.org/10.1093/jxb/erq340.

Prasad, R., Kamal, S., Sharma, P.K., Oelmüller, R., Varma, A., 2013. Root endophyte *Piriformospora indica* DSM 11827 alters plant morphology, enhances biomass and antioxidant activity of medicinal plant *Bacopa monniera*. J. Basic Microbiol. 53, 1016–1024.

Preston, G.M., Bertrand, N., Rainey, P.B., 2001. Type III secretion in plant growth-promoting *Pseudomonas fluorescens* SBW25. Mol. Microbiol. 41, 999–1014.

Puttanna, K., Prakasa Rao, E.V.S., Parameswaran, T.N., Singh, R., Kalra, A., 2010. Effect of organic and inorganic fertilizers and *Trichoderma harzianum* on patchouli (*Pogostemon cablin*) herb yield. J. Med. Arom Plant Sci. 32, 50–52.

Raheem, S., Saqib, B., Muhammad, I., Khan, A.L., Alosaimi, A.A., Al-Shwyeh, H.A., Almahasheer, H., Rehman, S., Lee, J., 2019. Amelioration of heavy metal stress by endophytic *Bacillus amyloliquefaciens* RWL-1 in rice by regulating metabolic changes: potential for bacterial bioremediation. Biochem. J. 476 (21), 3385–3400. https://doi.org/10.1042/BCJ20190606.

Ramakrishnan, G., Nakkeeran, S., Chandrasekar, G., Doraiswamy, S., 2001. Biocontrol agents-novel tool to combat plant diseases. In: The III Asia Pacific Crop Protection Conference, New Delhi, India, pp. 20–39.

Ray, T., Pandey, S.S., Pandey, A., Srivastava, M., Shanker, K., Kalra, A., 2019. Endophytic consortium with diverse gene-regulating capabilities of benzylisoquinoline alkaloids biosynthetic pathway can enhance endogenous morphine biosynthesis in *Papaver somniferum*. Front. Microbiol. 10, 925.

Redman, R.S., Kim, Y.O., Woodward, C.J., Greer, C., Espino, L., Doty, S.L., et al., 2011. Increased fitness of rice plants to abiotic stress via habitat adapted symbiosis: a strategy for mitigating impacts of climate change. PLoS One 6 (7), e14823.

Reinhold-Hurek, B., Hurek, T., 2011. Living inside plants: bacterial endophytes. Curr. Opin. Plant Biol. 14, 435–443.

Ren, C.G., Dai, C.C., 2012. Jasmonic acid is involved in the signaling pathway for fungal endophyte-induced volatile oil accumulation of *Atractylodes lancea* plantlets. BMC Plant Biol. 12, 128.

Ricci, M., Tilbury, L., Daridon, B., Sukalac, K., 2019. General principles to justify plant biostimulant claims. Front. Plant Sci. 10, 494. https://doi.org/10.3389/fpls.2019.00494.

Rodriguez, R.J., White, J.F., Arnold, A.E., Redman, R.S., 2009. Fungal endophytes: diversity and functional roles. New Phytol. 182, 314–330. https://doi.org/10.1111/j.1469-8137.2009.02773.x.

Roos, I.M.M., Hattingh, M.J., 1983. Scanning electron microscopy of *Pseudomonas syringae* pv. *morsprunorum* on sweet cherry leaves. Phytopathology 108, 18–25.

Rui, H., Chen, C., Zhang, X., Shen, Z., Zhang, F., 2016. Cd-induced oxidative stress and lignification in the roots of two *Vicia sativa* L. varieties with different Cd tolerances. J. Hazard. Mater. 301, 304–313.

Ryan, R.P., Germaine, K., Franks, A., Ryan, D.J., Dowling, D.N., 2008. Bacterial endophytes: recent developments and applications. FEMS Microbiol. Lett. 278, 1–9.

Ryan, G.D., Rasmussen, S., Parsons, A.J., Newman, J.A., 2015. The effects of carbohydrate supply and host genetic background on Epichlo€e endophyte and alkaloid concentrations in perennial ryegrass. Fungal Ecol. 18, 115–125.

Sallaku, G., Sandén, H., Babaj, I., Kaciu, S., Balliu, A., Rewald, B.J., 2019. Specific nutrient absorption rates of transplanted cucumber seedlings are highly related to RGR and influenced by grafting method, AMF inoculation and salinity. Sci. Hortic. 243, 177–188.

Sallam Nashwa, M., Riad Shaimaa, N., Mohamed, M.S., Seef Eleslam, A., 2013. Formulations of *Bacillus* spp and *Pseudomonas fluorescens* for biocontrol of cantaloupe root rot caused by *Fusarium solani*. J. Plant Prot. Res. 53, 275–300.

Salminen, S.O., Richmond, D.S., Grewal, S.K., Grewal, P.S., 2005. Influence of temperature on alkaloid levels and fall armyworm performance in endophytic tall fescue and perennial ryegrass. Entomol. Exp. Appl. 115, 417–426.

Sandhya, V., Ali, S., Vurukonda, S.S.K.P., Shrivastava, M., 2017. Plant growth promoting endophytes and their interaction with plants to alleviate abiotic stress. Curr. Biotechnol. 6, 252–263.

Santhanam, R., Luu, V.T., Weinhold, A., Goldberg, J., Oh, Y., Baldwin, I.T., 2015. Native root-associated bacteria rescue a plant from a sudden-wilt disease that emerged during continuous cropping. Proc. Natl. Acad. Sci. U. S. A. 112, 5013–5020.

Santoyo, G., Moreno-Hagelsieb, G., Del, C., Orozco-Mosqueda, M., Glick, B.R., 2016. Plant growth-promoting bacterial endophytes. Microbiol. Res. 183, 92–99. https://doi.org/10.1016/j.micres.2015.11.008.

Satheesan, J., Narayanan, A.K., Sakunthala, M., 2012. Induction of root colonization by *Piriformospora indica* leads to enhanced asiaticoside production in *Centella asiatica*. Mycorrhiza 22, 195–202.

Schardl, C.L., Leuchtmann, A., Spiering, M.J., 2004. Symbioses of grasses with seed borne fungal endophytes. Annu. Rev. Plant Biol. 55, 315–340.

Schardl, C., Young, C., Pan, J., Florea, S., Takach, J., Panaccione, D., Farman, M., Webb, J., Jaromczyk, J., Charlton, N., Nagabhyru, P., Chen, L., Shi, C., Leuchtmann, A., 2013. Currencies of mutualisms: sources of alkaloid genes in vertically transmitted epichloae. Toxins 5, 1064–1088.

Schneider, C., Ait, B.E., 2012. Endophytes in biotechnology and agriculture. In: Current Aspects of European Endophyte Research. University of Reims, France, pp. 28–30.

Schulz, B., Boyle, C., 2005. The endophytic continuum. Mycol. Res. 109, 661–686. https://doi.org/10.1017/S095375620500273X.

Schulz, B., Roemmert, A.K., Dammann, U., Aust, H.J., Strack, D., 1999. The endophyte-host interaction: a balanced antagonism? Mycol. Res. 103 (10), 1275–1383.

Sekar, J., Raj, R., Prabavathy, V.R., 2016. Microbial consortial products for sustainable agriculture: commercialization and regulatory issues in India. In: Singh, H.B., Sarma, B.K., Keswani, C. (Eds.), Agriculturally Important Microorganisms. Springer Science+Business Media, Singapore, pp. 107–131.

Selim, H.M., Gomaa, N.M., Essa, A.M., 2017. Application of endophytic bacteria for the biocontrol of *Rhizoctonia solani* (Cantharellales: ceratobasidiaceae) damping-off disease in cotton seedlings. Biocontrol Sci. Technol. 27 (1), 81–95.

Shah, F.U.R., Ahmad, N., et al., 2015. Heavy metal toxicity in plants. In: Ashraf, M., Ozturk, M., Ahmad, M.S.A. (Eds.), Plants Adaptation and Phytoremediation. Springer, New York.

Shen, F.T., Yen, J.H., Liao, C.S., Chen, W.C., Chao, Y.T., 2019. Screening of rice endophytic biofertilizers with fungicide tolerance and plant growth-promoting characteristics. Sustainability 11, 1133. https://doi.org/10.3390/su11041133.

Sheoran, N., Nadakkakath, A.V., Munjal, V., Kundu, A., Subaharan, K., Venugopal, V., Rajamma, S., Eapen, S.J., Kumar, A., 2015. Genetic analysis of plant endophytic *Pseudomonas putida* BP25 and chemo-profiling of its antimicrobial volatile organic compounds. Microbiol. Res. 173, 66–78 (ISSN 0944-5013).

Silva, G.H., Teles, H.L., Zanardi, L.M., Young, M.C.M., Eberlin, M.N., Hadad, R., et al., 2006. Cadinane sesquiterpenoids of *Phomopsis cassiae*, an endophytic fungus associated with *Cassia spectabilis* (Leguminosae). Phytochemistry 67, 1964–1969. https://doi.org/10.1016/j.phytochem.2006.06.004.

Singh, R., Parameswaran, T.N., Kalra, A., Puttanna, K., Prakasa rao E.V.S., Divya, S., 2009. Colonization efficacy of *Trichoderma harzianum* and selected bio-inoculants on patchouli (distilled and vermicomposted). In: 5th International Conference on Plant Pathology in Globalized Era, New Delhi, India, p. 343.

Singh, L.P., Gill, S.S., Tuteja, N., 2011. Unravelling the role of fungal symbionts in plant abiotic stress tolerance. Plant Signal. Behav. 6, 175–191.

Singh, R., Divya, S., Awasthi, A., Kalra, A., 2012. Technology for efficient and successful delivery of vermicompost colonized bioionculants in *Pogostemon cablin* (patchouli) Benth. World J. Microbiol. Biotechnol. 28, 323–333.

Singh, R., Singh, R., Soni, S.K., Singh, S.P., Chauhan, U.K., Kalra, A., 2013. Vermicompost from biodegraded distillation waste improves soil properties and essential oil yield of *Pogostemon cablin* (patchouli) Benth. Appl. Soil Ecol. 70, 48–56.

Singh, S., Tripathi, A., Maji, D., Awasthi, A., Vajpayee, P., Kalra, A., 2019. Evaluating the potential of combined inoculation of *Trichoderma harzianum* and *Brevibacterium halotolerans* for increased growth and oil yield in *Mentha arvensis* under greenhouse and field conditions. Ind. Crop Prod. 131, 173–181.

Singh, S., Pandey, S.S., Shanker, K., Kalra, A., 2020. Endophytes enhance the production of root alkaloids ajmalicine and serpentine by modulating the terpenoid indole alkaloid pathway in *Catharanthus roseus* roots. J. Appl. Microbiol. 128 (4), 1128–1142.

Singh, S., Pandey, S.S., Tiwari, R., Pandey, A., Shanker, K., Kalra, A., 2021. Endophytic consortium with plant growth promoting and alkaloid enhancing capabalities enhance key terpenoid indole alkaloids of *Catharanthus roseus* in winter and summer seasons. Ind. Crop Prod. 166, 113437.

Skirycz, A., Inze, D., 2010. More from less: plant growth under limited water. Curr. Opin. Biotechnol. 1 (2), 197–203. https://doi.org/10.1016/j.copbio.2010.03.002.

Song, M., Chai, Q., Li, X., Yao, X., Li, C., Christensen, M.J., Nan, Z., 2015. An asexual Epichloë endophyte modifies the nutrient stoichiometry of wild barley (*Hordeum brevisubulatum*) under salt stress. Plant Soil 387, 153–165.

Song, M., Li, X., Saikkonen, K., Li, C., Nan, Z., 2015b. An asexual Epichloe¨ endophyte enhances waterlogging tolerance of *Hordeum brevisubulatum*. Fungal Ecol. 13, 44–52.

Stierle, A., Strobel, G., Stierle, D., Grothaus, P., Bignami, G., 1995. The search for a taxol-producing microorganism among the endophytic fungi of the Pacific yew, *Taxus brevifolia*. J. Nat. Prod. 58, 1315–1324.

Strimbeck, G.R., Schaberg, P.G., Fossdal, C.G., Schroder, W.P., Kjellsen, T.D., 2015. Extreme low temperature tolerance in woody plants. Front. Plant Sci. 884 (6), 1–15. https://doi.org/10.3389/fpls.2015.00884.

Strobel, G., Daisy, B., 2003. Bioprospecting for microbial endophytes and their natural products microbial. Microbiol. Mol. Biol. Rev. 67 (4), 491–502.

Su, F., Jacquard, C., Villaume, S., Michel, J., Rabenoelina, F., Clement, C., Barka, E.A., Dhondt-Cordelier, S., et al., 2015. *Burkholderia phytofirmans* PsJN reduces impact of freezing temperatures on photosynthesis in *Arabidopsis thaliana*. Front. Plant Sci. 6, 810.

Subramanian, P., Mageswari, A., Kim, K., Lee, Y., Sa, T., 2015. Psychrotolerant endophytic *Pseudomonas* sp. strains OB155 and OS261 induced chilling resistance in tomato plants (*Solanum Lycopersicum* Mill.) by activation of their antioxidant capacity. Mol. Plant Microbe Interact. 28, 1073–1081.

Sun, Z., Yuan, X., Zhang, H., Wu, L., Liang, C., Feng, Y., 2013. Isolation, screening and identification of antagonistic downy mildew endophytic bacteria from cucumber. Eur. J. Plant Pathol. 137, 847–857.

Szopko, D., Molnar, I., Kruppa, K., Halo, B., Vojtko, A., Molnar-Lang, M., 2017. Photosynthetic responses of a wheat (Asakaze)–barley (Manas) 7h addition line to salt stress. Photosynthetica 55, 317–328.

Thalmann, M., Santelia, D., 2017. Starch as a determinant of plant fitness under abiotic stress. New Phytol. 214, 943–951.

Thangavelu, R., Muthukathan, G., 2015. Field suppression of Fusarium wilt disease in banana by the combined application of native endophytic and rhizospheric bacterial isolates possessing multiple functions. Phytopathol. Mediterr. 54, 241–252.

Thomsen, M.G., Galambosi, B., Galambosi, Z., Uusitalo, M., Mordal, R., Heinonen, A., 2012. Harvest time and drying temperature effect on secondary metabolites in *Rhodiola rosea*. Acta Hortic. 2012 (955), 243–252.

Tian, Y., Amand, S., Buisson, D., Kunz, C., Hachette, F., Dupont, J., Nay, B., Prado, S., 2014. The fungal leaf endophyte *Paraconiothyrium* variabile specifically metabolizes the host-plant metabolome for its own benefit. Phytochemistry 108, 95–101.

Tian, B., Zhang, C., Ye, Y., Wen, J., Wu, Y., Wang, H., et al., 2017. Beneficial traits of bacterial endophytes belonging to the core communities of the tomato root microbiome. Agric. Ecosyst. Environ. 247, 149–156.

Timmusk, S., Abd El-Daim, I.A., Copolovici, L., Tanilas, T., Kannaste, A., Behers, L., Niinemets, U., 2014. Drought-tolerance of wheat improved by rhizosphere bacteria from harsh environments: enhanced biomass production and reduced emissions of stress volatiles. PLoS One 9, e96086.

Ting, A.S.Y., Mah, S.W., Tee, C.S., 2012. Evaluating the feasibility of induced host resistance by endophytic isolate *Penicillium citrinum* BTF08 as a control mechanism for Fusarium wilt in banana plantlets. Biol. Control 61, 155–159.

Tiwari, R., Kalra, A., Darokar, M.P., Chandra, M., Aggarwal, N., Singh, A.K., Khanuja, S.P., 2010. Endophytic bacteria from *Ocimum sanctum* and their yield enhancing capabilities. Curr. Microbiol. 60, 167–171.

Treseder, K.K., 2004. A meta-analysis of mycorrhizal responses to nitrogen, phosphorus, and atmospheric CO_2 in field studies. New Phytol. 164, 347–355.

Tripathi, S., Das, A., Chandra, A., et al., 2015. Development of carrier-based formulation of root endophyte *Piriformospora indica* and its evaluation on *Phaseolus vulgaris* L. World J. Microbiol. Biotechnol. 31, 337–344. https://doi.org/10.1007/s11274-014-1785-y.

Tuo, X.Q., Li, S., Wu, Q.S., Zou, Y.N., 2015. Alleviation of waterlogged stress in peach seedlings inoculated with Funneliformis mosseae: changes in chlorophyll and proline metabolism. Sci. Hortic. 197, 130–134.

Ullah, A., Mushtaq, H., Fahad, S., Shah, A., Chaudhary, H.J., 2017. Plant growth promoting potential of bacterial endophytes in novel association with *Olea ferruginea* and *Withania coagulans*. Microbiology 86, 119–127.

Ullah, I., Al-Johny, B.O., Al Ghamdi, K., Al-Zahrani, H.A.A., Anwar, Y., Firoz, A., Al-kenani, N., Almatry, M.A.A., 2019. Endophytic bacteria isolated from *Solanum nigrum* L., alleviate cadmium (Cd) stress response by their antioxidant potentials, including SOD synthesis by sodA gene. Ecotoxicol. Environ. Saf. 174, 197–207.

Upreti, R., Thomas, P., 2015. Root-associated bacterial endophytes from *Ralstonia solanacearum* resistant and susceptible tomato cultivars and their pathogen antagonistic effects. Front. Microbiol. https://doi.org/10.3389/fmicb.2015.00255.

VanOverbeek, L., VanElsas, J.D., 2008. Effects of plant genotype and growth stage on the structure of bacterial communities associated with potato (*Solanum tuberosum* L.). FEMS Microbiol. Ecol. 64, 283–296.

Vargas, L., Santa Brıgida, A.B., Mota Filho, J.P., de Carvalho, T.G., Rojas, C.A., Vaneechoutte, D., Van Bel, M., Farrinelli, L., et al., 2014. Drought tolerance conferred to sugarcane by association with *Gluconacetobacter diazotrophicus*: a transcriptomic view of hormone pathways. PLoS One 9, e114744.

Verginer, M., Siegmund, B., Cardinale, M., Muelle, H., Choi, Y., Miguez, C.B., Leitner, E., Berg, G., 2010. Monitoring the plant epiphyte *Methylobacterium extorquens* DSM 21961 by real-time PCR and its influence on the strawberry flavor. FEMS Microbiol. Ecol. 74, 136–145.

Verma, V.C., Kharwar, R.N., Strobel, G.A., 2009. Chemical and functional diversity of natural products from plant associated endophytic fungi. Nat. Prod. Commun. 4, 1511–1532.

Vetrivelkalai, P., Sivakumar, M., Jonathan, E.I., 2009. Biocontrol potential of endophytic bacteria on *Meloidogyne incognita* and its effect on plant growth in bhendi. J. Biopest. 3, 452–457.

Villacieros, M., Power, B., Sanchez-Contreras, B., et al., 2003. Colonization behaviour of *Pseudomonas fluorescens* and *Sinorhizobium meliloti* in the alfalfa (*Medicago sativa*) rhizosphere. Plant Soil 251, 47–54.

Vurukonda, S.S.K.P., Vardharajula, S., Shrivastava, M., Ali, S.Z., 2016. Enhancement of drought stress tolerance in crops by plant growth promoting rhizobacteria. Microbiol. Res. 184, 13–24.

Wahid, A., Gelani, S., Ashraf, M., Foolad, M.R., 2007. Heat tolerance in plants: an overview. Environ. Exp. Bot. 61, 199–223.

Walder, F., van der Heijden, M., 2015. Regulation of resource exchange in the arbuscular mycorrhizal symbiosis. Nat. Plants 1, 15159.

Waller, F., Achatz, B., Baltruschat, H., Fodor, J., Becker, K., Fischer, M., Franken, P., 2005. The endophytic fungus *Piriformospora indica* reprograms barley to salt-stress tolerance, disease resistance, and higher yield. Proc. Natl. Acad. Sci. U. S. A. 102 (38), 13386–13391.

Wan, Y., et al., 2012. Effect of endophyte-infection on growth parameters and Cd-induced phytotoxicity of Cd-hyperaccumulator *Solanum nigrum* L. Chemosphere 89 (6), 743–750.

Wang, J.W., Zheng, L.P., Xiang, T.R., 2006. The preparation of an elicitor from a fungal endophyte to enhance artemisinin production in hairy root cultures of *Artemisia annua* L. Chin. J. Biotechnol. 22, 829–834.

Wang, Y., Dai, C.C., Cao, J.L., Xu, D.S., 2012. Comparison of the effects of fungal endophyte *Gilmaniella* sp. and its elicitor on *Atractylodes lancea* plantlets. World J. Microbiol. Biotechnol. 28, 575–584.

Wang, X., Radwan, M.M., Tarawneh, A.H., Gao, J., Wedge, D.E., 2013. Antifungal activity against plant pathogens of metabolites from the endophytic fungus *Cladosporium cladosporioides*. J. Agric. Food Chem. 61, 4551–4555.

Wang, J.L., Li, T., Liu, G.Y., Smith, J.M., Zhao, Z.W., 2016. Unraveling the role of dark septate endophyte (DSE) colonizing maize (*Zea mays*) under cadmium stress: physiological, cytological and genic aspects. Sci. Rep. 6 (1), 22028.

Wang, L., Lin, H., Dong, Y., Li, B., He, Y., 2020. Effects of endophytes inoculation on rhizosphere and endosphere microecology of Indian mustard (*Brassica juncea*) grown in vanadium-contaminated soil and its enhancement on phytoremediation. Chemosphere 240, 124891.

Waqas, M., Khan, A.L., Hamayun, M., et al., 2015. Endophytic fungi promote plant growth and mitigate the adverse effects of stem rot: an example of *Penicillium citrnum* and *Aspergillus terreus*. J. Plant Interact. 10, 280–287.

White, J.F., Kingsley, K.L., Zhang, Q., et al., 2019. Review: endophytic microbes and their potential applications in crop management. Pest Manag. Sci. 75 (10), 2558–2565. https://doi.org/10.1002/ps.5527.

Wilson, D., 1995. Endophyte-the evolution of a term, and clarification of its use and definition. Oikos 73, 274–276.

Xie, Z., Chu, Y., Zhang, W., Lang, D., Zhang, X., 2019. *Bacillus pumilus* alleviates drought stress and increases metabolite accumulation in *Glycyrrhiza uralensis Fisch*. Environ. Exp. Bot. 158, 99–106.

Yadav, S.K., 2010. Cold stress tolerance mechanisms in plants. A review. Agron. Sustain. Dev. 30, 515–527. https://doi.org/10.1051/agro/2009050.

Yadav, A., Singh, R.P., Singh, A.L., Singh, M., 2021. Identification of genes involved in phosphate solubilization and drought stress tolerance in chickpea symbiont *Mesorhizobium ciceri* Ca181. Arch. Microbiol. 203 (3), 1167–1174. https://doi.org/10.1007/s00203-020-02109-1.

Yang, S., Vanderbeld, B., Wan, J., Huang, Y., 2010. Narrowing down the targets, towards successful genetic engineering of drought-tolerant crops. Mol. Plant 3, 469–490.

Yang, Y., Liu, L., Singh, R.P., Meng, C., Ma, S., Jing, C., Li, Y., Zhang, C., 2020. Nodule and root zone microbiota of salt-tolerant wild soybean in coastal sand and saline-alkali soil. Front. Microbiol. 11, 2178. https://doi.org/10.3389/fmicb.2020.523142.

Yasin, N.A., Akram, W., Khan, W.U., Ahmad, S.R., Ahmad, A., Ali, A., 2018. Halotolerant plantgrowth promoting rhizobacteria modulate gene expression and osmolyte production to improve salinity tolerance and growth in *Capsicum annum* L. Environ. Sci. Pollut. Res. 25 (23), 23236–23250. https://doi.org/10.1007/s11356-018-2381-8.

Young, C.A., Hume, D.E., McCulley, R.L., 2013. Forages and pastures symposium: fungal endophytes of tall fescue and perennial ryegrass: pasture friend or foe? J. Anim. Sci. 91, 2379–2394.

Yuan, J., Sun, K., Deng-Wang, M.Y., Dai, C.C., 2016a. The mechanism of ethylene signaling induced by endophytic fungus *Gilmaniella* sp. AL12 mediating sesquiterpenoids biosynthesis in *Atractylodes lancea*. Front. Plant Sci. 7, 361.

Yuan, J., Zhou, J.Y., Li, X., Dai, C.C., 2016b. The primary mechanism of endophytic fungus *Gilmaniella* sp AL12 promotion of plant growth and sesquiterpenoid accumulation in *Atractylodes lancea*. Plant Cell Tiss. Org. Cult. 125, 571–584.

Zabetakis, I., Holden, M.A., 1997. Strawberry flavour: analysis and biosynthesis. J. Sci. Food Agric. 74, 421–434.

Zeilinger, S., Gupta, V.K., Dahms, T.E., Silva, R.N., Singh, H.B., Upadhyay, R.S., Gomes, E.V., Tsui, C.K., Nayak, S.C., 2016. Friends or foes? Emerging insights from fungal interactions with plants. FEMS Microbiol. Rev. 40 (2), 182–207.

Zhang, H.W., Song, Y.C., Tan, R.X., 2006a. Biology and chemistry of endophytes. Nat. Prod. Rep. 23, 753–771.

Zhang, N., Castlebury, L.A., Miller, A.N., Huhndorf, S.M., Schoch, C.L., Seifert, K.A., Rossman, A.Y., Rogers, J.D., Kohlmeyer, J., Volkmann-Kohlmeyer, B., Sung, G.H., 2006b. An overview of the systematics of the *Sordariomycetes* based on a four-gene phylogeny. Mycologia 98, 1076–1087.

Zhang, H., Murzello, C., Sun, Y., Kim, X., Mi-S, R., Jeter, R.M., Zak, J.C., Scot, E., Dowd, P.W., 2010. Choline and osmotic-stress tolerance induced in *Arabidopsis* by the soil microbe *bacillus subtilis* (GB03). Mol. Plant Microbe Interact. 23, 1097–1104.

Zhang, Y., Yu, X., Zhang, W., Lang, D., Zhang, X., Cui, G., Zhang, X., 2019. Interactions between endophytes and plants: beneficial effect of endophytes to ameliorate biotic and abiotic stresses in plants. J. Plant Biol. 62, 1–13. https://doi.org/10.1007/s12374-018-0274-5.

Zhang, J., Peng, S., Shang, Y., Brunel, B., Li, S., Zhao, Y., Liu, Y., Chen, W., Wang, E., Singh, R.P., James, E.K., 2020. Genomic diversity of chickpea-nodulating rhizobia in Ningxia (north central China) and gene flow within symbiotic *Mesorhizobium muleiense* populations. Syst. Appl. Microbiol. 43 (4), 126089. https://doi.org/10.1016/j.syapm.2020.126089.

Zhao, L., Xu, Y., Lai, X.H., Shan, C., Deng, Z., Ji, Y., 2015. Screening and characterization of endophyte *Bacillus* and *Paenibacillus* strains from medicinal plant *Lonicera japonica* for use as potential plant growth promoters. Braz. J. Microbiol. 46, 977–989.

Zhu, X.C., Song, F.B., Xu, H.W., 2010. Influence of arbuscular mycorrhiza on lipid peroxidation and antioxidant enzyme activity of maize plants under temperature stress. Mycorrhiza 20, 325–332.

Zivanovic, A., Rodgers, L., 2018. The role of fungal endophytes in plant pathogen resistance. Bios 89, 192–197.

CHAPTER

Biotechnological approaches for upgrading of unconventional crude oil

6

Wael A. Ismail[a], Abdul Salam Abdul Raheem[a], and Dawoud Bahzad[b]

[a]*Environmental Biotechnology Program, Life Sciences Department, College of Graduate Studies, Arabian Gulf University, Manama, Bahrain,*

[b]*Petroleum Research Center, Kuwait Institute for Scientific Research, Kuwait City, Kuwait*

Abstract

Most of the current crude oil reserves worldwide are of the low-quality unconventional type including heavy and extra-heavy crudes. Compared to the light (conventional) crudes, unconventional oils are characterized by higher viscosity and density due to the higher content of the heavy components asphaltenes and resins, besides higher concentration of heavy metals and heteroatoms (sulfur and nitrogen). Biotechnological upgrading has the potential to overcome the environmental, economic and technical shortcomings associated with the recovery, transportation and refining of heavy crude oils. Biocatalytic upgrading is based on the catabolic capabilities of dedicated hydrocarbonoclastic microorganisms which are equipped with enzymes that can catalyze diverse biodegradation and biotransformation reactions with the different hydrocarbon components of the heavy crude. Cleavage or degradation of asphaltenes into lighter fragments and removal of sulfur via biodesulfurization are among the bioupgrading-relevant microbial activities that can lead to viscosity reduction, promoting recovery and processing.

Keywords: Asphaltenes, Biotransformation, Heavy crude oil, Biodesulfurization, Viscosity, Bioupgrading, Aromatic compounds, 4S pathway, Biosurfactants

1 Introduction

It has become evident that crude oil will continue to be the major energy source worldwide through the 21st century. The International Energy Agency has predicted a 70% increase in global energy consumption by 2030. About 90% of this increase is expected to be covered by fossil fuels such as oil and natural gas (Dehghani et al., 2009). Accordingly, the oil industry has expanded oil production and processing operations to meet the accelerating growth in global energy demands. Unfortunately, the reserves of the conventional light crude oil are depleting, leaving behind huge resources of unconventional low-quality heavy and extra-heavy crudes, bitumen and oil sands (Vazquez-Duhalt and Quintero-Ramirez, 2004). Estimates indicate that the global heavy oil reserves are seven times higher than the remaining conventional light crude reserves. Global deposits of heavy hydrocarbons are estimated to be almost 6 trillion barrels, of which only 500–1000 billion barrels are recoverable with conventional technologies (Lavania et al., 2012).

It is, therefore, obvious that the refinery feed will gradually switch from the sweet light crude to the sour low-quality heavy crude as the production of heave oil increases (Vazquez-Duhalt and Quintero-Ramirez, 2004; León and Kumar, 2005). The production and processing of unconventional

Microbial Syntrophy-mediated Eco-enterprising. https://doi.org/10.1016/B978-0-323-99900-7.00009-2

heavy crudes generate many operational, economic, and environmental problems due to their challenging physicochemical properties which are attributed to the presence of higher content of asphaltenes, resins, heteroatoms (S, N), and metals (V and Ni), compared to the light crudes (Gudiña and Teixeira, 2017). Therefore, the quality of unconventional crudes must be improved via several upgrading treatments to facilitate further processing. Conventional thermochemical technologies for production, transportation, and refining are not suitable for heavy oils.

The physicochemical technologies that are adopted upstream for the recovery of heavy crude oils include non-thermal (flooding with water, gases or polymers, and solvents injection) and thermal methods (injecting hot fluids or steam into the reservoir). These techniques are costly, troublesome, and environmentally harmful (Sen, 2008; Al-Bahry et al., 2016). On the other hand, technologies that are implemented in the refinery (downstream) for handling the heavy crude oil or residue such as carbon rejection processes (non-catalytic, thermal, and low-pressure methods) produce a low yield of light products, albeit at low capital and operating costs. The other catalytic technologies that are also implemented for upgrading heavy crudes in refineries like hydrogen addition technologies (delayed coking, solvent de-asphalting, gasification, residue hydrocracking, and visbreaking) produce high yields of the light products, albeit at high capital and operating costs (Dehghani et al., 2009; Shi et al., 2019).

There has been increasing interest in the development of novel heavy oil upgrading technologies that could enable the oil industry to (i) meet the growing market demand for cleaner fuels; (ii) comply with the strict environmental regulations; and (iii) overcome the rising costs of coking and hydrotreatment processes. Biological treatment of heavy crude oil has emerged as a green and economic alternative or adjunct oil biocatalytic upgrading approach. As compared to the conventional thermochemical and chemical treatments, biological processes are selective, specific, environmentally friendly, and proceed under mild conditions (Vazquez-Duhalt and Quintero-Ramirez, 2004; León and Kumar, 2005; Kilbane, 2006; Montiel et al., 2009). These desirable features can lead to huge savings in capital and operational costs. Biocatalytic upgrading has the potential to improve the physicochemical properties of the treated heavy crudes, thus facilitating or promoting production and transportation. This can be achieved with dedicated microorganisms that have the capabilities to degrade or transform high-boiling fractions like asphaltenes into smaller fragments and reduce the oil viscosity (Vazquez-Duhalt and Quintero-Ramirez, 2004; León and Kumar, 2005). Furthermore, biodesulfurization of heavy crude oils can help remove sulfur that is mostly concentrated in the heavy fractions of the crude (León and Kumar, 2005). In this chapter, we highlight the latest advances in the field, identify the obstacles or challenges and discuss the prospects.

2 Unconventional crude oil resources

The continuous increase in the global population and the economic developments have led to higher consumption of energy because of the increase in worldwide energy demand (Mirchi et al., 2012; Tang et al., 2019). Crude oil (petroleum) remains the main energy source worldwide despite the efforts invested in developing various renewable energies (Shibulal et al., 2014; Clemente, 2015; Doman, 2016; Alaei et al., 2017). It was predicted that between 2000 and 2030 the annual demand for crude oil would increase by an average of 1.7% in terms of the number of oil barrels (Elraies and Tan, 2012; Silva et al., 2014). The growth of global oil demand in 2017 increased by 1.53 million barrels per year (OPEC, 2017). Furthermore, according to Höök et al. (2009), the light and medium crude oil reserves are declining annually by an average of 4.5%. The discovery of conventional (light and medium) crude

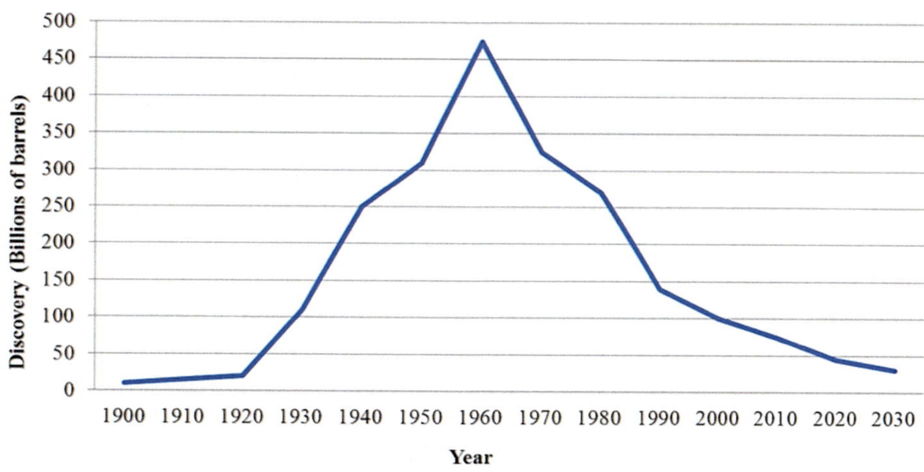

FIG. 1

History and forecast of world discoveries of conventional crude oil (Tverberg, 2007).

oil reserves has declined in different parts of the world, and 1960 witnessed the peak of production (Fig. 1) (Tverberg, 2007). This decline occurred because almost all oil fields have been exploited for a long time (Brown et al., 2000; Sen, 2008) and 90% of global conventional crude oil resources have been already discovered (Heinberg, 2005).

It has been anticipated that the remaining conventional crude oil reserves can maintain the global energy supply for around 40 years only (Elraies and Tan, 2012; Silva et al., 2014), and the API (American Petroleum Institute) gravity of the crude oil is continuously decreasing (Ancheyta et al., 2009). This situation has provoked the need for alternative energy sources, among which the unconventional crude oil reserves (heavy, extra heavy, bitumen, shale oil, and oil sand) have attracted interest (Tang et al., 2019).

The current conventional crude oil reserves worldwide are estimated to be two trillion barrels, of which 63% exists in the Arab Countries (Yernazarova et al., 2016) (Table 1). The unconventional crude oil reserves are estimated to be more than six trillion barrels, thus constituting 70% of all resources of fossil fuels (Santos et al., 2014; Al-Bahry et al., 2016; Tang et al., 2019). This vast energy resource can satisfy the global demand throughout the 21st century. The unconventional crude oil reserves are located and distributed in more than 70 countries (Lavania et al., 2012; Shibulal et al., 2014; He et al., 2015; Tang et al., 2019). Canada, Venezuela, and Russia together hold more than 80% of the heavy and extra heavy crude oil reserves (Al-Sulaimani et al., 2011; De Sena et al., 2013; Omajali et al., 2017), and the Middle East harbors 970 billion barrels of heavy and extra heavy crude oil (Schlumberger, 2010) (Table 1).

Obviously, the production and processing of unconventional crude oil will increase to meet the future energy demand and maintain world economic growth (Lavania et al., 2012; Kang et al., 2018). Moreover, in the near future, refineries in countries such as Canada, Venezuela, and Mexico will replace light crude oil with heavy or extra-heavy crude oils (Ancheyta et al., 2009). However, this cannot be achieved with the currently applied technologies for the production, transportation, and refining of light crude oil. Unconventional oil resources need unconventional (novel, efficient and economic) technologies due to their challenging physicochemical characteristics (high density, high viscosity, and compositional complexity) (Kang et al., 2018; Joshi et al., 2019; Tang et al., 2019).

Table 1 Estimated reserves of unconventional crude oils in some countries.

Country	Total reserves (billion barrels)	Unconventional crude oil type	Estimated unconventional reserves (billion barrels)	API of unconventional crude oil	Reference
Canada	167.8	Oil sands	162.3	< 10°	BPE Outlook (2019) Natural Gas Intelligence (2019)
Kuwait	101.5	Heavy	13	11–20°	The Economist Intelligence Unit (2019) Oxford Business Groups (2017)
Mexico	7.7	Heavy and extra heavy	4.7	< 20°	Castorena-Cortés et al. (2012)
Oman	5.4	Heavy	7	10–19°	Al-Bahry et al. (2016)
Saudi Arabia	297.7	Heavy	37	< 27°	Offshore Technology (2019)
Venezuela	303.3	Heavy and extra heavy	261.4	4–16°	BPE Outlook (2019) Schenk et al. (2009)

3 Characteristics of unconventional crude oils

Crude oil is composed of a complex and heterogeneous mixture of compounds with different molecular weights (Fig. 2). Heavy crude oil was broadly defined as any liquid petroleum having API gravity $<20°$, specific gravity >0.933, and sulfur content $>2\%$ (w/w) (Speight, 2006; Santos et al., 2014). Nonetheless, there is no specific definition for unconventional oils. Still, their high viscosity, high density, and low API gravity are the main properties that can be used to distinguish them when compared

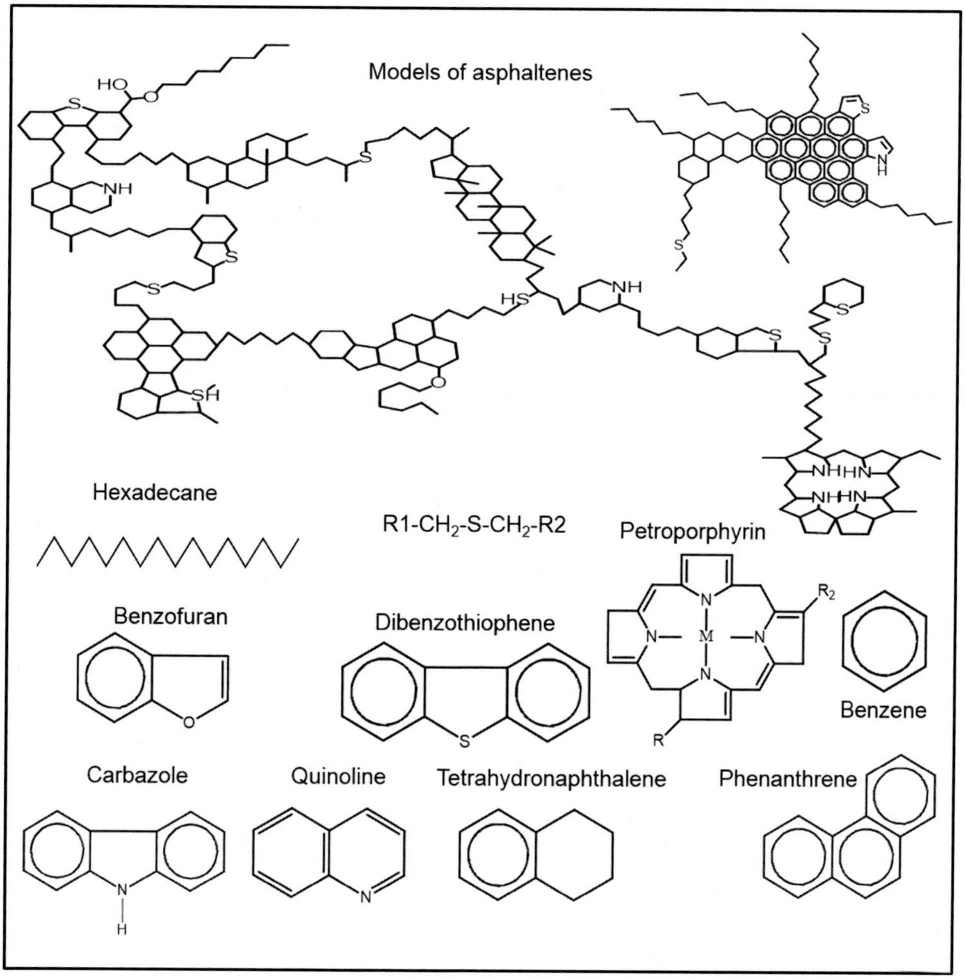

FIG. 2

Examples of crude oil chemical components.

Reprinted from Ismail, W., Mohamed, M. El-S., Awadh, M. N., Obuekwe, C., & El Nayal, A.M. 2017. Simultaneous valorization and biocatalytic upgrading of heavy vacuum gas oil by the biosurfactant-producing Pseudomonas aeruginosa AK6U. J. Microbial. Biotechnol. 10 (6), 1628–1639. Copyright (2017). The authors.

with the conventional crudes (Joshi et al., 2019; Tang et al., 2019). The API gravity is used to classify oils according to their density compared to water. Light crude oil has an API gravity $>31.1°$, whereas extra-heavy crude oil has an API gravity $<10°$ which means that it is in the solid or semisolid state (Castro and Vázquez, 2009; Joshi et al., 2019).

Unconventional crude oils are characterized by a higher content of high-molecular weight components and smaller amounts of low-molecular weight structures that make their recovery from subsurface reservoirs more challenging. Unconventional crude oil's viscosity may range from <20 to >1.000,000 cP, whereas the conventional oil viscosity may range from one to ten cP (Joshi et al., 2019). In addition, unconventional crude oils contain a high concentration of heteroatoms such as nitrogen, oxygen, and sulfur, in addition to high levels of heavy metals such as vanadium and nickel when compared with the light and medium crude oils. These heteroatoms and heavy metals are commonly found associated with the heavier fractions of the crude (Madden and Morawski, 2011; He et al., 2015; Alaei et al., 2017; Gudiña and Teixeira, 2017; Kang et al., 2018).

The molecular characterization of unconventional crude oils is a difficult and costly task due to their structural complexity. Several analytical methods for crude oil characterization have been developed based on the main hydrocarbon classes. However, SARA analysis (Saturates, Aromatics, Resins, Asphaltenes) is the most commonly applied method (Riazi and Eser, 2013; Al-Sayegh et al., 2016). SARA analysis separates all kinds of crude oils into four main fractions, namely; Saturates, Aromatics, Resins, and Asphaltenes, based on their solubility and polarity (Muhammad et al., 2013; Joshi et al., 2019). The SARA content of conventional and unconventional crude oils differs remarkably from one reservoir to another. The light component that is found in different crude oils is the saturates fraction which comprises mainly non-polar alkanes with linear or branched chains and alicyclic paraffins (Muhammad et al., 2013; Gudiña and Teixeira, 2017). The aromatic components are made of one or more aromatic rings and usually contain embedded heteroatoms and heavy metals. Resins are not soluble in liquid propane but soluble in heptane and pentane (Muhammad et al., 2013; Gudiña and Teixeira, 2017). They are composed of aromatic rings and aliphatic side chains and are rich in heavy metals and heteroatoms. The molecular weights of resins range from 700 to 950 g/mol (Castro and Vázquez, 2009; He et al., 2015; Gudiña and Teixeira, 2017).

The fourth fraction of crude oils is the asphaltenes which is the heaviest and most polar fraction of crude oil (Muhammad et al., 2013; Gudiña and Teixeira, 2017). Asphaltenes are soluble in aromatic solvents such as xylene, benzene and toluene, but insoluble in *n*-alkanes such as heptane and hexane (Muhammad et al., 2013; Gudiña and Teixeira, 2017). The asphaltene fraction contains a higher concentration of heavy metals and heteroatoms than the resins fraction, though their chemical constituents are similar to each other (Akbarzadeh et al., 2007; Mullins et al., 2007; Muhammad et al., 2013; Gudiña and Teixeira, 2017). The asphaltene structure has been a mystery for many decades because it is made up of complex components that have a tendency to combine and form aggregates and large clusters, thus rendering their characterization difficult (Mullins et al., 2007; Gudiña and Teixeira, 2017). In general, unconventional crude oils contain low levels of saturates and a high concentration of aromatics, resins, and asphaltenes. In addition, unconventional crude oils contain high levels of heteroatoms (S, N) and metals (Ni, V). Therefore, the higher the level of resins and asphaltenes, the heavier and more viscous the crude oil will be. Table 2 compares the API gravity and composition of different oil types (Castorena-Cortés et al., 2012; Kok and Gul, 2013; Sanchez-Minero et al., 2013; Strubinger et al., 2015).

Table 2 API gravity and composition of conventional, heavy and extra heavy crude oils (Castorena-Cortés et al., 2012; Kok and Gul, 2013; Sanchez-Minero et al., 2013; Strubinger et al., 2015).

	Crude oil type			
	Light	**Medium**	**Heavy**	**Extra heavy**
Source	Turkey	Mexico	Mexico	Venezuela
API gravity (°)	31.5	27.1	11.6	8.1
Saturates (%)	41.6	43.1	10.8	17.0
Aromatics (%)	46.0	29.9	42.3	33.1
Resins (%)	8.3	18.2	23.1	35.6
Asphaltenes (%)	4.0	8.7	23.7	14.2
Sulfur (%)	0.4	2.5	4.8	2.6

4 Problems associated with the production and processing of unconventional crudes

Extraction, transportation, and processing of heavy and extra heavy crude oils are more complicated and challenging when compared with the corresponding processes for the conventional crude oils, and need the use of specialized and expensive technologies (Gudiña and Teixeira, 2017). The problems generated by the production and processing of unconventional crude oils are attributed to the presence of high content of resins, asphaltenes, waxes, heteroatoms (S, N, O) and metals (Ni, V). Unconventional oils are mostly found trapped in tight places and inside small porous spaces of the reservoir matrix (Gudiña and Teixeira, 2017). Moreover, the presence of high-molecular weight components in a high concentration, such as asphaltenes and resins, increases the density and viscosity of the unconventional crude. Therefore, it is difficult, challenging and costly to recover the unconventional crude oils because these heavy components impede the flow of the oil and block the equipment and pipelines that are used for crude oil extraction (Harner et al., 2011; Silva et al., 2014; He et al., 2015; Gudiña and Teixeira, 2017; Tang et al., 2019).

The presence of asphaltenes and their strong inter/intramolecular interactions in unconventional oils are the leading causes of undesirable oil features such as high viscosity and tendency to form polymers, coke and emulsions (García-Arellano et al., 2004). Asphaltenes can cause problems in extraction, transport and processing (Pineda-Flores and Mesta-Howard, 2001; Lavania et al., 2012; Kang et al., 2018). They can block the porous spaces of the oil deposits which reduces the permeability and oil mobility. Precipitation of asphaltenes causes the formation of "asphaltenic mud" which deposits in pipelines of oil transport, leading to obstruction and blocking of the free flow of crude oils (Pineda-Flores and Mesta-Howard, 2001; Kang et al., 2018). Solvents such as xylene and toluene are usually used to dissolve the mud and enable free flow of crude oils. In addition to the high production costs, this process also generates residues with high toxicity (Kirkwood et al., 2004; León and Kumar, 2005).

Deposition problems in the reservoirs, pipelines, as well as storage and processing equipment are frequently encountered due to asphaltenes and waxes (Kirkwood et al., 2004; Choi et al., 2018). Moreover, the high content of heteroatoms, metals and asphaltenes causes environmental pollution, corrosion problems and poisoning of the refinery catalysts (Pineda-Flores and Mesta-Howard, 2001;

Pineda-Flores et al., 2004; León and Kumar, 2005). Resins are interfacially active and cause the stabilization of water/oil emulsions (Kirkwood et al., 2004). Environmental pollution with spills and leakages of heavy crude oils can be hazardous due to the resistance of asphaltenes to microbial biodegradation (Pineda-Flores and Mesta-Howard, 2001).

5 Recovery and processing of unconventional crude oils

Two main approaches are adopted upstream for the recovery of unconventional crude oils, namely, non-thermal and thermal methods. In the non-thermal methods, flooding with water, gases or polymers, solvents injection, hydraulic fracturing and surface mining are used to release oil from the substrate. Thermal production techniques are applied by injecting hot fluids or steam into the reservoir and/or in-situ combustion to increase the temperature of the oil, thus promoting the flow of oil to production wells using mechanical pumps or gas lifts (Sen, 2008; Al-Bahry et al., 2016). Unfortunately, these physicochemical technologies are costly, troublesome and environmentally hazardous (Sen, 2008; Al-Bahry et al., 2016). On the other hand, the technologies that are used in the refinery (downstream) for handling the heavy crude oil or residue are of main two types. The first comprises carbon rejection processes (non-catalytic, thermal, and low-pressure methods) that break the high-molecular weight components into smaller fragments. These methods produce a low yield of light products at low capital and operating costs. The other technologies that are used for upgrading heavy crude oil by refineries are catalytic and known as hydrogen addition technologies, such as delayed coking, solvent de-asphalting, gasification, residue hydrocracking and visbreaking, which need high capital and operating costs and produce high yields of the light products (Dehghani et al., 2009; Shi et al., 2019).

The presence of large quantities of non-distillable hydrocarbons, heteroatoms and heavy metals in heavy and extra heavy crudes reduces the yields of light fractions (distillates) and increases the refining costs, thus impacting the economic benefit (Ancheyta et al., 2009; Alaei et al., 2017). To gain more distillates and refined fuels and to reduce the volume of the refining residues, it is important to break up the large and heavy (high-boiling) components of the heavy and extra heavy crude oil into smaller (low-boiling) fractions (Gudiña and Teixeira, 2017). This treatment is designated as upgrading and it consumes a high amount of hydrogen and catalysts and requires high pressure (> 100 psi) and temperature (> 538°C), which makes it a costly process (Ancheyta et al., 2009; Alaei et al., 2017). Accordingly, unconventional oils should be subjected to rigorous upgrading processes (hydrocracking) before they can be refined to produce commercial fuels and valuable products (Alaei et al., 2017).

The high asphaltene and heavy metal concentrations reduce the refining catalysts' life during hydrotreating which increases the cost of hydroprocessing. Siddiqui (2003) reported that the high level of V and Ni in asphaltenes increases when the crude oil gets heavier. Therefore, the first step in hydroprocessing is to remove the excess of asphaltene and metals from the heavy crude oil, followed by upgrading via hydrocracking to improve the quantity and quality of distillates and reduce the corresponding vacuum residues (Le Borgne and Quintero, 2003; García-Arellano et al., 2004; Pineda-Flores et al., 2004; León et al., 2007; Rana et al., 2007). The operational, economic and environmental drawbacks associated with the production and processing of unconventional heavy crudes have provoked the need for environmentally friendly, economic and efficient technologies to cope with the challenging features of heavy crudes (Rana et al., 2007; Harner et al., 2011). In particular, biotechnology-based biocatalytic upgrading has gained increasing interest.

6 Bioupgrading of heavy crude oils

Biocatalytic upgrading of heavy crudes has emerged as a novel approach to circumvent the shortcomings of the conventional upgrading and oil recovery processes. As compared to traditional physicochemical technologies, biotechnological processes are environmentally friendly and cost-effective (Le Borgne and Quintero, 2003; Kilbane, 2006; Shibulal et al., 2018). The concept was coined based on the known metabolic capabilities of some dedicated microorganisms, some of which are natural inhabitants of the oil formations. Some bacteria can produce organic acids, polymers, gases and biosurfactants which can be used to reduce the viscosity and density of the heavy crude and, hence, promote mobility and recovery. Moreover, some microbes have the ability to degrade/transform the various hydrocarbon and heterocyclic components of heavy crudes, including asphaltenes (Biria et al., 2013; Elshafie et al., 2015; Ni'matuzahroh et al., 2015; Patel et al., 2015; Shibulal et al., 2017). Hydrocarbon biodegradation (hydrocarbonoclastic) bacteria include members of the genera *Achromobacter, Rhodococcus, Arthrobacter, Bacillus, Acinetobacter, Actinomyces, Nocardia, Corynebacterium, Pseudomonas, Brevibacterium, Alcaligenes, Flavobacterium, Moraxella, Vibrio, Micromonospora*. These bacteria differ in their substrate range and preferences (Britton, 1984; Joshi et al., 2019).

The literature has many reports on the isolation of heavy crude oil-degrading bacteria from oil-field samples. For instance, Gao et al. (2017) isolated two *Pseudomonas aeruginosa* strains from oil-contaminated soil using crude oil as a carbon source. In addition, Shahebrahimi et al. (2020) isolated microbial consortia from crude oil, oil-contaminated soil and oil sludge using asphaltenes as a carbon source. Hao and Lu (2009) isolated a heavy crude oil-degrading halophilic bacterial strain from an oil reservoir.

A successful upgrading process should deal with the problematic components of the heavy crude. Broadly speaking, bioupgrading encompasses all biological activities and products that facilitate the extraction, transport and processing operations in addition to improving the quality and, consequently, the economic profit of the treated feedstock (Kirkwood et al., 2004; León and Kumar, 2005; Morales et al., 2010). Bioupgrading treatment exploits the astonishing biocatalytic machinery and metabolic diversity of oil-adapted microorganisms to perform desirable biochemical transformations on various petroleum hydrocarbons. Several areas have been proposed where biocatalysis can lead to favorable changes in the physicochemical properties of heavy crudes (Fig. 3).

As mentioned previously, the heavy crude oil's high viscosity is mainly due to the highest molecular weight component, asphaltenes. A promising biotechnology-based approach toward viscosity reduction is the application of biosurfactants and bioemulsifiers (Kirkwood et al., 2004; Assadi and Tabatabaee, 2010; Perfumo et al., 2010). These microbial products can enhance oil mobility and recovery by reducing oil-rock and oil-brine interfacial tension, modifying the wettability of porous media and emulsifying crude oil. Moreover, biosurfactants can promote the accessibility of hydrocarbons to the microbial cell, thus facilitating metabolism which releases lighter hydrocarbons and improves oil mobility. From the application perspective, biosurfactants or biosurfactants-producing bacteria can be directly injected into the oil reservoirs. Alternatively, relevant nutrients can be supplied to stimulate biosurfactants production by indigenous (reservoir-inhabiting) biosurfactants producers (Perfumo et al., 2010). It is known that oil/water emulsions occur either incidentally or deliberately in the oil industry throughout. Emulsifiers could be oil-borne amphipathic components including resins, naphthenic acids, asphaltenes, fine solids, clay, scale, wax crystals, or microorganisms. Regardless of the origin of

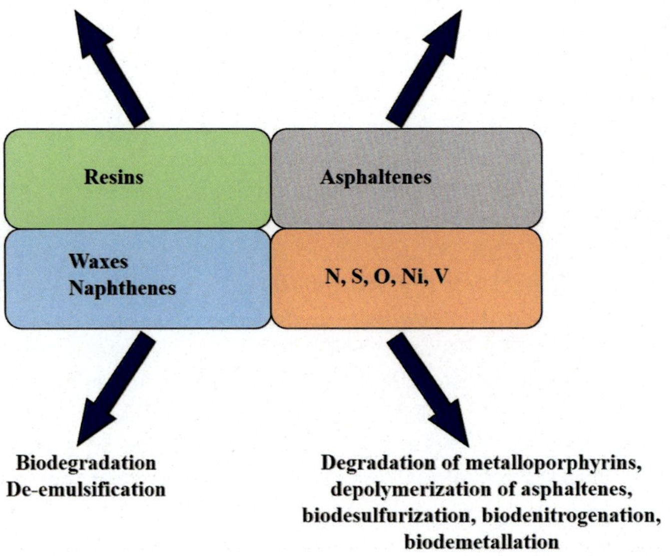

FIG. 3

Problematic components of unconventional heavy crudes and potential bioupgrading opportunities (León and Kumar, 2005).

these emulsions, they are very complex and must be resolved to avoid corrosion, scale formation and sludge accumulation in storage tanks. In addition, emulsions can negatively affect the viscosity and flow properties of the oil and alter the efficiency of distillation. Biologically assisted de-emulsification can be performed with some types of microbes which have variable cell surface hydrophobic properties. Moreover, some microbial products such as glycolipids, polysaccharides, and glycoproteins can destabilize crude oil emulsions (Banat et al., 2000; Kirkwood et al., 2004; Khire, 2010).

Biosurfactants are surface-active microbial products having superior properties as compared to synthetic surfactants (less or no toxicity, biodegradability, higher selectivity and specific activity under extreme conditions, greater structural diversity and biological activity). In addition, biosurfactants can be produced from renewable resources (Hausmann and Syldatk, 2015). All these features make biosurfactants environmentally compatible and increase the interest in these microbial products (Banat et al., 2000; Santos et al., 2016). They can lower surface and interfacial tension and emulsify various hydrocarbons. A powerful biosurfactant can reduce the surface tension of water from 72 mN/m to ~27 mN/m (Christofi and Ivshina, 2002). Based on their chemical structure, biosurfactants can be classified as glycolipids, lipoproteins, lipopeptides, lipoheteropolysaccharides (polymeric surfactants) and particulate biosurfactants (Mulligan, 2005; Santos et al., 2016). They are produced by many microorganisms such as various species of *Pseudomonas*, *Bacillus*, *Acinetobacter*, *Corynebacterium*, *Lactobacillus*, *Arthrobacter*, *Candida*, etc. (Banat et al., 2014; Fracchia et al., 2015).

Interest in biosurfactants has increased due to their superior physicochemical properties, compared to synthetic surfactants, and diverse environmental, industrial, agricultural and biomedical applications. In the petrochemical industry, they can be applied in bioremediation of oil spills, soil washing, cleaning of storage tanks, upgrading of heavy oils and refinery residues, petrochemical formulations and heavy metal removal (Perfumo et al., 2010; Banat et al., 2014; Fracchia et al., 2015).

Viscosity reduction can also be achieved by breaking the high-molecular weight asphaltenes into smaller fragments, thus decreasing the average molecular weight. Microbial splitting of the aliphatic chains in asphaltenes can affect the reduction in molecular weight (Kirkwood et al., 2004). Furthermore, microbial hydrogenation and oxidation of the aromatic rings break the compact structures and inhibit the stacking interactions, thus reducing viscosity.

Beneficial chemical changes in heavy crude oils can be performed via enzyme-catalyzed biotransformations. These include removal of metals and heteroatoms, increasing the H/C ratio through hydrogenation of the aromatic rings and dearomatization through aromatic ring cleavage. All these biochemical reactions can improve the quality of the treated feedstock in terms of distillate yield and fuel combustion properties (Kirkwood et al., 2004; León and Kumar, 2005).

Biocatalysis can also contribute as a solution to the deposition problems caused by waxes and asphaltenes using microbial metabolites that can render waxes or asphaltenes more soluble. In addition, biotransformation and biodegradation of waxes and asphaltenes can lead to products that are more soluble and eliminate or reduce the content of these compounds in the biotreated oil or deposits (Kirkwood et al., 2004).

6.1 Biodegradation and biotransformation of asphaltenes

Asphaltenes constitute the oil fraction that is most resistant to microbial attack. This is due to the challenging and complex chemical structure, physical properties and high molecular weight that can be in the range of 600 to 2,000,000 g/mol (Pineda-Flores et al., 2004; Tavassoli et al., 2012). Moreover, asphaltenes consist mainly of high-molecular weight polycyclic aromatic hydrocarbons with boiling points higher than 500°C, and the asphaltene average molecule contains between 40 and 70 aromatic rings. This heavy fraction of crude oil is rich with heteroatoms (N, O, S) heavy metals (V, Ni) and some naphthenic structures bearing alkyl chains (Pineda-Flores et al., 2004; Akbarzadeh et al., 2007; Mullins et al., 2007; Ancheyta et al., 2009; Tavassoli et al., 2012).

Asphaltenes can be divided into two types based on their structure: the continental (island) type, consisting of a large core of condensed polycyclic aromatic compounds bearing peripheral aliphatic chains, and the archipelago-type which is characterized by the presence of several polycyclic aromatic structures connected by alkane bridges of different sizes (Ancheyta et al., 2009). Recently, Schuler et al. (2015) studied coal-derived and petroleum asphaltenes using Atomic Force and Scanning Tunneling Microscopy. The main finding is that asphaltenes consist mainly of a central aromatic core with peripheral alkane chains (Figs. 4 and 5). In some cases, this central component is broken up into smaller polycyclic aromatic cores linked by a single bridge, confirming the archipelago-type model. Both petroleum and coal-derived asphaltenes exhibit similar architecture in terms of the overall shape of the polycyclic aromatic cores and the type and number of the aromatic rings. However, petroleum asphaltenes bear longer side groups and more substituted aromatic rings.

A strong evidence showing unequivocally that microorganisms could degrade asphaltenes has been lacking for many years, and asphaltenes were considered refractory to microbial attack due to high

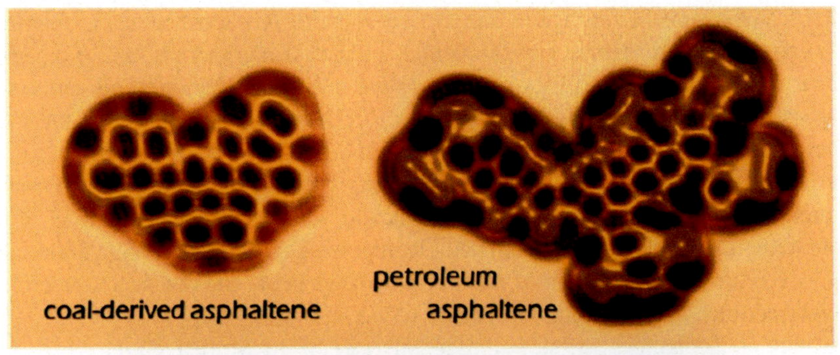

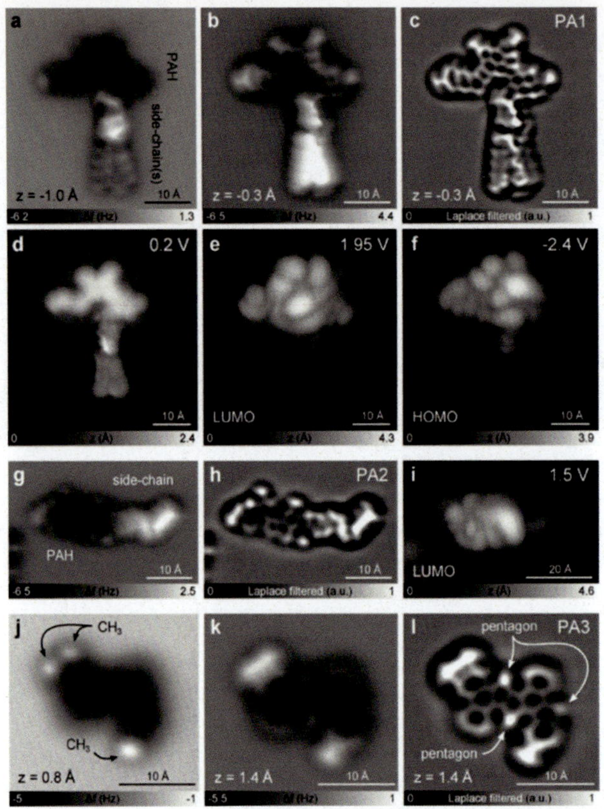

FIG. 4

Atomic force and scanning tunneling microscopy images of petroleum asphaltenes.

Reprinted with permission from Schuler, B., Meyer, G., Peña, D., Mullins, O. C., Gross, L. 2015. Unraveling the molecular structures of asphaltenes by atomic force microscopy. J. Am. Chem. Soc. 137 (31), 9870–9876. Copyright (2015) American Chemical Society.

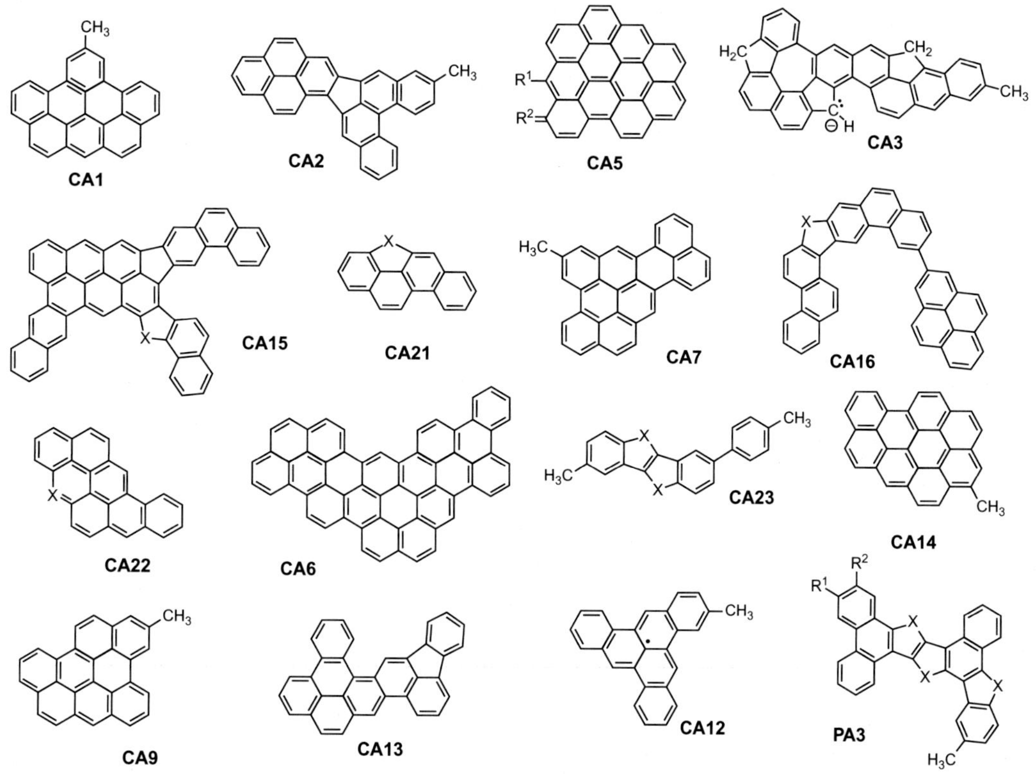

FIG. 5

Chemical structure of proposed polycyclic aromatic molecules found in asphaltenes. The structures are based on atomic force microscopy measurements and scanning tunneling microscopy orbital images. *X*, unknown moiety, *R*, unknown side group.

Reprinted with permission from Schuler, B., Meyer, G., Peña, D., Mullins, O. C., Gross, L. 2015. Unraveling the molecular structures of asphaltenes by atomic force microscopy. J. Am. Chem. Soc. 137 (31), 9870–9876. Copyright (2015) American Chemical Society.

hydrophobicity and molecular weight, in addition to a complex structure that limits their mass transfer in aqueous media, rendering them less preferred nutrients for microorganisms (Hernández-López et al., 2015a; Gudiña and Teixeira, 2017). There is now an accumulating evidence supporting biodegradation or biotransformation of asphaltenes by some bacteria and fungi, a phenotype that could be beneficial for bioupgrading of unconventional crude oils (Lavania et al., 2012; Tavassoli et al., 2012; Jahromi et al., 2014; Hernández-López et al., 2015a; Ayala et al., 2017). It is, however, worth noting that the majority of the studies used asphaltenes fraction precipitated from heavy crudes as a source of carbon for microbial metabolism. Precipitation may lead to a fraction that is structurally different from the original asphaltenes fraction, as it exists in the crude oil. Consequently, the results might not be applicable to real heavy oils (Hernández-López et al., 2015a). Therefore, to obtain conclusive results, bioupgrading studies should be performed on real heavy crude.

For most studies, only a few analyses were made, which is not relevant to a highly complex structure like that of asphaltenes, and the extent of asphaltenes biodegradation or biotransformation was usually measured by gravimetry, overlooking the potential effect of oxidized molecules or biosurfactants (produced by the degrading microbes) on the precipitation behavior of asphaltenes (Hernández-López et al., 2015a). Moreover, in some studies, asphaltenes were added to the culture medium with other co-substrates such as yeast extract and molasses which could also be utilized as a carbon source (Lavania et al., 2012; Jahromi et al., 2014).

Theoretically, asphaltenes can serve as a substrate for microbial metabolism since they contain elements (C, H, S, N) that are essential for microbial growth and metabolism. A few laboratory studies have addressed asphaltenes biodegradation from an upgrading perspective, and most of the available data are based on the use of model compounds or asphaltenes fraction precipitated from crude oil as substrates. Earlier studies on oil biodegradation claimed that mixed cultures could degrade the asphaltene fraction (Rontani et al., 1985). Nevertheless, none of these studies presented the results from proper abiotic controls (Morales et al., 2010) and knowledge on degradation products and mechanisms is lacking.

Claims that asphaltenes can support bacterial growth, according to the initial studies on asphaltenes biodegradation by different microbial consortia, were not sufficiently backed by experimental evidence. In those experiments, the extent of asphaltenes biodegradation was usually measured by gravimetric methods after precipitation with *n*-alkanes. However, the changes in asphaltenes content reported by those studies could be due to production of biosurfactants which disrupted the asphaltenic matrix, thus releasing entrapped hydrocarbons. Accordingly, most of the decrease that has been observed in the asphaltene content during bacterial growth could be due to abiotic losses (Lacotte et al., 1996; Premuzic et al., 1999; García-Arellano et al., 2004; Hernández-López et al., 2015a). The application of more precise analytical tools, such as Fourier Transform-Infrared Spectroscopy (FT-IR) or Nuclear Magnetic Resonance (NMR), allowed more in-depth analysis and gave more conclusive data regarding asphaltenes degradation (Desando and Ripmeester, 2002; Tavassoli et al., 2012; Jahromi et al., 2014).

Pineda-Flores et al. (2004) applied respirometry to assess asphaltene biodegradation with a microbial consortium in a mineral medium. The authors reported that the asphaltenes cultures produced 800 μmol CO_2 in 13 days which was higher than the amount produced by the control cultures (200–300 μmol CO_2). However, the inoculum used in these experiments originated from crude oil cultures. This indicates that part of the produced CO_2 could be due to utilization of hydrocarbon carry-over from the precultures.

A biosurfactant-producing *Pseudomonas* strain capable of degrading S and N heterocyclic compounds was used for the degradation of heavy oil under anaerobic conditions. Growing cultures of this strain degraded asphaltenes and heavy crude oil by 22.5% and 38.27%, respectively, after 50 days. In a field trial, the strain reduced the viscosity of heavy oil by 50%–65% and promoted oil recovery from 48 to 566 tons (Xia et al., 2021). Pan et al. (2015) used pyrolysis gas chromatography to study the composition of alkyl moieties in asphaltene fractions precipitated from biodegraded bitumen. The authors reported that asphaltenes-bound alkyl substituents, *n*-fatty acids and aliphatic alcohols were biodegraded.

Hao et al. (2004) studied the bioconversion of different crude oils using the thermophilic strain *Thermus* sp. TH-2 which was isolated from an oil reservoir. The authors reported decrease in the viscosity of the biotreated oils, in addition to reduction in the content of aromatic hydrocarbons, resins and asphaltenes accompanied by an increase in the light fractions. Infrared spectroscopy and gas chromatography showed that cultures of *Serratia liquefasciens* and *Bacillus* sp. strains, isolated from petroleum hydrocarbon-polluted soils, removed 50% of asphaltenes after 168 h of treatment (Rojas-Avelizapa et al., 2002). Tavassoli et al. (2012) reported up to 48% biodegradation of asphaltenes after two months of treatment with pure and mixed cultures of *Bacillus firmus*, *Pseudomonas* sp., *Bacillus lentus*, *Bacillus licheniformis* and *Bacillus cereus*, isolated from oil-polluted soil samples from an Iranian oil reservoir.

The bacterial strain *Garciaella petrolearia* TERIG02, isolated from a sea-buried oil pipeline, preferentially degraded asphaltenes and reduced the viscosity of asphalt by 42% when it was mixed with molasses. After 30 days of incubation at 50°C under anaerobic conditions, the strain TERIG02 degraded 45% of the aromatic fraction, 55% of the asphaltenic fraction and 25% of the aliphatic fraction (Lavania et al., 2012). In addition to studies with pure cultures, there have been some reports on the biodegradation of asphaltenes and bioupgrading of heavy crude oil using bacterial consortia. In an earlier study, Venkateswaran et al. (1995) isolated different bacteria from sediments and used them to degrade fractions from Arabian light crude oil using thin layer chromatography coupled to flame ionization detector (TLC-FID) for analysis. The authors reported that a mixed culture degraded 35% of resins after 15 days in seawater medium containing resins as the sole carbon source, and 50% of the saturates and aromatic fractions of crude oil after 7 days. A *Pseudomonas* strain isolated form the mixed culture degraded only 30% of these fractions. A microbial consortium obtained from formation fluid reduced crude oil viscosity from 665 to 10 cP at 70°C after 30 days of incubation under anaerobic conditions and degraded 56% of the aromatic fraction (Lavania et al., 2015). A laboratory study showed that a bacterial consortium could degrade 48% of an asphaltenic fraction after 2 months of incubation at ambient temperature (Tavassoli et al., 2012).

Jahromi et al. (2014) used different consortia of bacterial strains isolated from oil-contaminated soils and sludge, including *P. aeruginosa*, *P. fluorescens*, *Citrobacter amalonaticus*, *Enterobacter cloacae*, *Staphylococcus hominis*, *B. cereus* and *Lysinibacillus fusiformis*, to study asphaltene biodegradation. The authors reported 21%–51% of degradation and FT-IR analysis of the biotreated asphaltenic fraction showed the presence of aldehyde groups as compared to alkynes, aldehydes and alkenes which were present in the untreated asphaltenes. Shahebrahimi et al. (2020) isolated different bacterial consortia form an oil field and used them for biodegradation of asphaltenes precipitated from crude oil under different conditions of temperature, salinity, pH, and initial asphaltene concentration. The maximum asphaltene biodegradation of 46.4% was attained after 60 days with a consortium consisting of *Staphylococcus saprophyticus* and *Bacillus cereus* at 45°C, salinity 160 g L^{-1}, pH 6.5, and 25 g L^{-1} initial asphaltene concentration. The authors also noted reduction in C, H, N, and S content in the biotreated asphaltene. Moreover, Zargar et al. (2021) isolated a microbial consortium from oil-contaminated soil and used it for biotransformation of asphaltenes fraction and upgrading of Maya crude oil in shake flasks and bioreactor experiments. The microbial consortium consisted of bacteria belonging to the genera *Arthrobacter*, *Rhodococcus*, *Bacillus*, *Lysinibacillus*, *Sporosarcina*, *Micrococcus*, *Paenibacillus*, and *Barrientosiimonas*. The authors reported 72% and 75% biotransformation of asphaltenes by growing cells in shake flasks and a bioreactor, respectively. This was accompanied by 80% reduction in S and N content. In oil upgrading experiments, the viscosity of biotreated Maya crude oil was reduced by 91% after 2 weeks. Enzymes like catechol 2,3-dioxygenase, catechol 1,2-dioxygenase, lignin peroxidase and manganese peroxidase were detected in the culture medium. This study represents one of the few rigorous investigations reported on heavy crude bioupgrading and asphaltenes biotransformation. The study was very well controlled by eight different experiments to identify the effects of abiotic factors and additional substrates that were added to the bioassays. Furthermore, several analytical tools were applied to get more conclusive data such as LC-MS, ^{1}H-NMR, ^{13}C-NMR, FT-IR, ICPMS, and elemental analysis.

Recently, Ismail et al. (2017) studied the bioupgrading of heavy vacuum gas oil (HVGO) with a strain of *P. aeruginosa*. After growth in a minimal medium containing HVGO as the sole carbon and sulfur source, the authors extracted the maltene fraction from HVGO for further analysis. Fractional distillation (SimDis) analysis of the biotreated HVGO revealed a relative increase in the light distillate fraction and decrease in the heavy fuel fraction compared to the abiotic control. They also reported compositional changes in the maltene fraction of the biotreated HVGO which was manifested in the higher number-average (Mn) and weight-average (Mw) molecular weights, as well as the absolute number of hydrocarbons and sulfur heterocycles. Furthermore, HVGO bioconversion was associated with the production of rhamnolipid biosurfactants. However, the authors did not present evidence supporting that these changes are due to direct biocatalytic activity on the asphaltene fraction.

Although most of the bioupgrading studies focused on bacteria, there are some reports on the upgrading of heavy and extra heavy crude oils using yeasts, filamentous fungi and fungal enzymes. For instance, a cell-free laccase produced by *Pestalotiopsis palmarum* BM-04 oxidized carbon and sulfur atoms in maltenes and asphaltenes from extra heavy crude oil (Naranjo-Briceño et al., 2013). Moreover, Oudot et al. (1993) reported degradation of resins (28%) and asphaltenes (40%) from crude oil by *Emericella nidulans*, *Graphium putredinis*, *Eupenicillium javanicum* and *Aspergillus flavipes*. In another study, *Paecilomyces variotii*, *Fusarium decemcellulare*, *Candida palmioleophila* and *Pichia guilliermondii* degraded between 10 and 15% of resins and asphaltenes (Chaillan et al., 2004). Naranjo et al. (2007) reported the ability of some non-white-rot fungi to utilize extra heavy crude oil via the production of extracellular lignin-degrading oxidative enzymes. However, the authors did not report any specific changes or improvements in the chemical or physical properties of the treated oil. Moreover, Uribe-Álvarez et al. (2011) isolated a fungal strain of *Neosartorya fischeri* that grew on purified asphaltenes as a sole source of carbon and energy. After 11 weeks of incubation, the fungus metabolized 15.5% of the asphaltenic carbon. Recently, Hernández-López et al. (2015b) investigated the differential gene expression in *Neosartorya fischeri* grown on either petroleum asphaltenes or glucose-peptone using microarrays. Genes encoding aromatic hydrocarbon monooxygenases were among those upregulated in asphaltenes-grown cells. Zhang et al. (2015) reported that extracellular enzymes from *Aspergillus* strains degraded 13%–35% of resins and 24%–34% of asphaltenes from crude oil after four days of treatment at 40°C, and the oil viscosity was reduced by 40%–90%.

Crude oil biodegradation studies commonly depend on SARA analysis to reveal compositional changes in the biotreated oil (Evdokimov, 2005). It is generally known that the SARA fractions of crude oil differ in their susceptibility to biodegradation or bioconversion due to differences in their polarity and structural complexity (Fuentes et al., 2014). Petroleum hydrocarbon components can be generally ranked as follows according to their biodegradability: linear alkanes > branched alkanes > low-molecularweight aromatics > cycloalkanes > polycyclic aromatics > resins > asphaltenes (Das and Chandran, 2011; Hernández-López et al., 2015a). However, the typical degradation pattern depends on the type of the involved microorganisms as well as the type and composition of the crude oil (Lavania et al., 2012; Hernández-López et al., 2015a).

Hydrocarbon-degrading microorganisms differ in their preferences for the substrates to be degraded (Premuzic et al., 1999; Fuentes et al., 2014). Some hydrocarbon degraders prefer light hydrocarbons, while others prefer heavy substrates (Das and Chandran, 2011; Lavania et al., 2012; Chronopoulou et al., 2014; Hazen et al., 2016; Zhou et al., 2018). From the perspective of heavy oil bioupgrading, bioconversion or biodegradation should lead to considerable increase in the concentration of saturates fraction in parallel with the reduction of the asphaltenes, resins and aromatics contents (Premuzic et al.,

1999; Hao and Lu, 2009; Pourfakhraei et al., 2018). This kind of hydrocarbon redistribution leads to bioconversion of the heavy hydrocarbons into lighter ones, thus improving the physicochemical properties of the biotreated heavy crude oil (Premuzic et al., 1999; Hernández-López et al., 2015a; Gudiña and Teixeira, 2017). Premuzic et al. (1999) noted that SARA analysis may provide an evidence for oil bioconversion if a large increase in the saturates fraction is coupled to any changes in the asphaltenes. The authors also reported that biocatalysis is expected to transform the polar macromolecular resins and asphaltenes into smaller, less polar, or nonpolar fragments.

In a previous study on bioconversion of heavy crude oils, Premuzic et al. (1999) performed SARA analysis to gain insight into the biochemically induced changes in SARA distribution and noted invariable increase in the saturates and decrease in the aromatic fractions. Recently, Pourfakhraei et al. (2018) studied heavy oil biodegradation by the fungus *Daedaleopsis* sp. The authors reported 44.8% increase in the saturates fraction, in addition to 88.7% and 38% decrease in the asphaltenes and aromatics, respectively. Viscosity reduction of heavy crude oil after biological treatment is usually correlated with the decrease in asphaltenes and increase in lighter hydrocarbons in addition to microbial products such as biosurfactants, acids and gases (Lavania et al., 2012, 2015).

Premuzic et al. (1999) explained the biochemically induced compositional shifts and hydrocarbon redistribution in biotreated crudes by proposing a structural model that depicts the SARA fractions as a molecular solution. The latter originates from inter-and intramolecular interactions within the SARA fractions to form inclusion complexes, charge transfer complexes, clathrates, coordination compounds and multiple-bridged polymeric structures stacked and connected via weak (complexes) and strong (heteroatom bridges) bonds. Biocatalytic activity on these molecular solutions may induce structural rearrangements which lead to unfolding and de-stacking of the initial three-dimensional structures, thus reversing the process that produced the SARA molecular solutions. These structural rearrangements produce lighter moieties and the embedded compounds are released, accompanied by hydrocarbon redistribution and fragmentation at the heteroatom sites (Premuzic et al., 1999).

It is noteworthy that biological treatments do not always improve the physicochemical properties of the treated oil. Strubinger et al. (2015) studied the bioconversion of two extra heavy crude oils from the Orinoco Oil Belt (Venezuela) using a microbial consortium consisting of *Ochrobactrum intermedium*, *Comamones testosterone* and *Pseudomonas putida*. After biological treatment, the authors noted that the initial boiling point of the oils increased and the light fractions were consumed. The API gravity, concentration of heteroatoms and asphaltenes remained unchanged. In addition, no significant degradation of heavy oil components was observed. A prerequisite for biocatalytic upgrading to be effective is the ability of the biocatalyst to degrade and/or transform the heavy ends of the crude oil into lighter and low-boiling fractions (Premuzic et al., 1999; Hernández-López et al., 2015a; Gudiña and Teixeira, 2017; Pourfakhraei et al., 2018).

From the application perspective, biocatalytic upgrading of unconventional oils can be implemented either in-situ (during recovery, within the reservoir) or ex-situ (after recovery). The in-situ approach has the advantage that it improves recovery and transportation (Gudiña and Teixeira, 2017). Either microorganisms or their enzymes can be beneficial for in-situ bioupgrading technology provided that they withstand the harsh environment of the reservoirs (lack of oxygen, high pressure, high temperature, high salinity, pH, etc.). Extremophilic microorganisms (including those isolated directly from the reservoir) can be beneficial for these applications (Lavania et al., 2012, 2015). The same principle applies to microbial enzymes. Those recovered from extremophilic microbes would be more suitable for in-situ bioupgrading applications. Zhang et al. (2015) reported that enzymatic preparations can be directly injected into oil reservoirs, leading to degradation of heavy oil fractions and viscosity reduction. Nevertheless, in-situ bioupgrading treatments suffer from variability between different reservoirs, and difficulty in controlling microbial growth. In contrast, ex-situ (surface) upgrading offers better control on the process parameters and conditions which could be more suitable to microbial activity (Gudiña and Teixeira, 2017).

It is known that oil reservoirs are rich in a diversity of microorganisms and many of these are unique to this deep subsurface environment. These microbes are acclimatized to the harsh conditions prevailing in oil reservoirs (high temperature, pressure, salinity and hydrocarbon loads) and are described to be polyextremophilic because their cellular components are inherently stable against high temperature, high pressure, solvents, detergents, extreme pH, and salinity. Therefore, oil reservoirs constitute an invaluable microbial resource that can be exploited to isolate microorganisms better suited for different petroleum biotechnology applications including microbial enhanced oil recovery and heavy crude bioconversion (Kotlar, 2012).

6.2 Sulfur removal and bioupgrading of unconventional crude oils

An important aspect of the heavy crude upgrading process deals with the removal of heteroatoms such as sulfur and nitrogen because they contribute to environmental pollution and interfere with the refining processes by causing catalyst poisoning and corrosion problems. Furthermore, the high content of sulfur increases the viscosity of the oil (Kirkwood et al., 2004; León and Kumar, 2005). As mentioned earlier, heavy crude oils have a higher content of sulfur as compared to the conventional light crudes, and sulfur is found more concentrated in the heavy fractions such as asphaltenes (León and Kumar, 2005; Gudiña and Teixeira, 2017). The sulfur-containing components in asphaltenes are structurally diverse and can be in the form of aliphatic sulfide bridges, cyclic sulfides and thiophenic sulfur compounds (aromatic sulfur heterocycles). Aliphatic sulfides act as bridges between aromatic cores of asphaltenes and linkages to smaller structures like alkanes. It was reported that molecular weight and viscosity can be reduced only when the sulfide bridges are broken. In contrast, the carbon backbone remains intact if the cyclic sulfides and thiophenes are degraded (Kirkwood et al., 2004; León and Kumar, 2005). Actually, oil prices and processing costs are largely determined by the sulfur content, which makes desulfurization a key and critical conversion process in most refineries (Javadli and de Klerk, 2012). This is particularly prominent for unconventional heavy crudes.

6.2.1 Sulfur in crude oil

Sulfur is the third most abundant element in crude oil after carbon and hydrogen and is found at a high level mostly associated with the heavy hydrocarbon fractions (Abbasian et al., 2016). Sulfur content in crude oil varies between 0.03 and 10% (weight %) based on the type and the source of the crude oil, and can be found as inorganic or organic forms (Mohebali and Ball, 2016). Inorganic sulfur species like elemental sulfur, sulfate, thiosulfate, hydrogen sulfide and pyrite are found in dissolved or suspended forms. Organosulfur compounds, more than 200, are classified into two categories: non-heterocyclic, which are found generally as aromatic or saturated forms such as thiols, sulfides and disulfides; whereas the heterocyclic ones are found as condensed polycyclic forms and constitute about 50–95% of the organosulfur compounds in crude oil (Abbasian et al., 2016; Mohebali and Ball, 2016) (Fig. 6). The most abundant sulfur heterocyclic compounds are the thiophenes, such as benzothiophene, dibenzothiophene and their alkylated derivatives (Mohebali

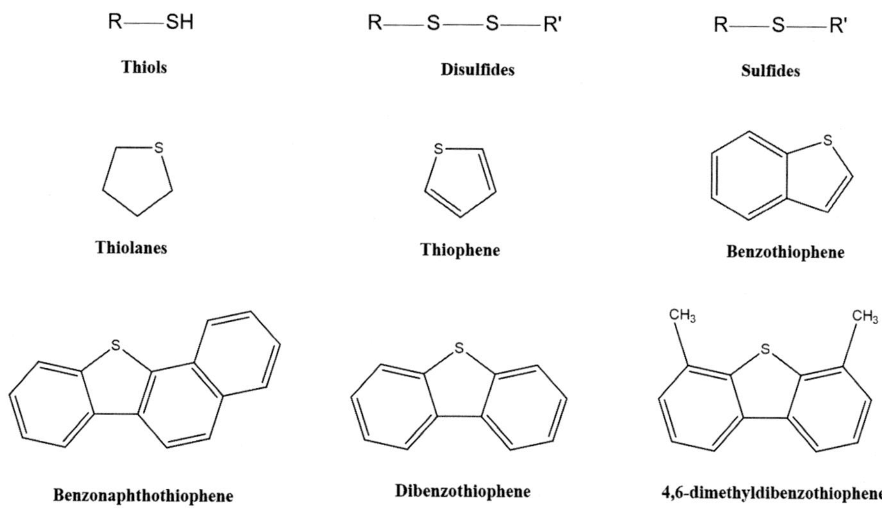

FIG. 6

Examples of sulfur compound classes found in crude oil. *R*, alkyl (Javadli and de Klerk, 2012).

and Ball, 2016). Although aliphatic and cyclic sulfides can be readily removed in the refinery via hydrodesulfurization, heterocyclic sulfur compounds (e.g., benzothiophene, dibenzothiophene, benzonaphthothiophene) are more resistant to conventional thermochemical treatments (Kilbane, 2006; Javadli and de Klerk, 2012).

Combustion of fossil fuels is the main source of sulfur oxide emissions which cause severe harm to the environment, economy and human health (Kilbane, 2006; Monticello and Finnerty, 1985; Kropp and Fedorak, 1998). Sulfur-containing heterocyclic compounds are themselves environmental pollutants as they can accumulate in living tissues and provoke mutagenic, carcinogenic and toxic effects (US Environmental Protection Agency, 2017). In addition to their well-documented environmental impact, thiophenic compounds in crude oil interfere with the refining process by causing corrosion, catalyst poisoning and deterioration of refining pumps (Singh et al., 2012). Moreover, as mentioned earlier, a high content of sulfur compounds increases the viscosity of heavy crudes (Kirkwood et al., 2004; León and Kumar, 2005).

Strict environmental regulations were imposed to limit SO_2 emission by forcing the refineries to drastically reduce the sulfur level in fuels (Mohebali and Ball, 2016; Porto et al., 2018). Hydrodesulfurization is the most common technology applied by refineries to remove sulfur from fuels during the refining process. This technology operates under high temperature (200–425°C) and hydrogen pressure (150–250 psi) depending on the desulfurization level that is needed and on the type of the hydrocarbon feedstock (Boniek et al., 2014; Mohebali and Ball, 2016). Unfortunately, hydrodesulfurization is associated with many operational, economic and environmental shortcomings. It is costly, energy-demanding and affects the quality of the treated fuel. It also generates hydrogen sulfide, carbon dioxide and huge amounts of hazardous spent catalysts that need proper disposal (Bhatia and Sharma, 2010). In addition, it is not sufficiently efficient since the bulk of sulfur in the treated fuels (thiophenic compounds) escapes hydrodesulfurization (Ohshiro and Izumi, 1999; Boniek et al., 2014; Mohebali and Ball, 2016).

The refineries' feedstock is getting heavier and contains higher sulfur concentrations. This, with the fact that the currently applied desulfurization technology is not suitable for refining the high-sulfur (sour) heavy crudes, in addition to the strict environmental legislations, impose economic and technical challenges on refineries (Kirkwood et al., 2004; León and Kumar, 2005; Kilbane, 2006). Therefore, it is important to develop green, efficient and cost-effective desulfurization technologies. To overcome the shortcomings of commonly applied oil desulfurization technologies, microbial biodesulfurization has been proposed and studied as a green and economic approach.

6.2.2 Biodesulfurization

Microorganisms need sulfur for growth because it is an essential element for the physiological activity and survival of microorganisms. Sulfur constitutes 0.5%–1% of the dry weight of a bacterial cell, and it is a structural component of some vitamins and enzyme cofactors such as coenzyme A and thiamin in addition to the sulfur-containing amino acids methionine and cysteine (Kertesz, 1999). To satisfy their sulfur requirements, microorganisms are endowed with diverse metabolic strategies to utilize a broad range of sulfur-containing compounds, either organic or inorganic. Accordingly, there has been an increasing interest in the application of some dedicated bacteria to remove sulfur from fossil fuels in a process known as biodesulfurization (Kilbane, 2006; Mohebali and Ball, 2016).

More than three decades ago, microbial biodesulfurization emerged as a promising alternative or complementary process for removing sulfur from fossil fuels (Kilbane, 1990; Kilbane and Jackowski, 1992; Denome et al., 1993). As compared to the thermochemical hydrodesulfurization, biodesulfurization is environmentally friendly, cost-effective, sulfur-specific and efficient toward thiophenic compounds such as dibenzothiophene and its alkyl derivatives which constitute the bulk of sulfur in crude oil and diesel (~70%) and escape the hydrodesulfurization treatment (Martinez et al., 2017; Sadare et al., 2017). The capital and operating costs of biodesulfurization were reported to be two times and 15% lower, respectively, than the conventional hydrodesulfurization (Kaufman et al., 1998; Linguist and Pacheco, 1999; Pacheco et al., 1999).

Biodesulfurization exploits the unique capability of some bacteria to utilize organosulfur compounds as a sole sulfur source without consuming their carbon skeleton. The 4S pathway is the most commonly investigated biodesulfurization mechanism for dibenzothiophene and related thiophenes (Fig. 7) (Kilbane, 2006; Martínez et al., 2017) and was originally discovered in the biodesulfurization prototype strain *Rhodococcus qingshengii* IGTS8 (formerly *R. erythropolis* IGTS8) (Thompson et al., 2020). The 4S pathway proceeds via four enzyme-catalyzed reactions which are initiated by two monooxygenases DszC and DszA. These two enzymes, supported by a flavin reductase DszD, transform dibenzothiophene to dibenzothiophene sulfone and hydroxyphenyl benzenesulfinate, respectively. Eventually, a desulfinase (DszB) removes the sulfur atom as sulfite (to be assimilated by the biodesulfurizing cells) and produces 2-hydroxybiphenyl as the dead-end product which can't be further metabolized by the 4S pathway-harboring biodesulfurizing bacteria, thus preserving the calorific value of the treated oil (Kilbane, 2006; Mohebali and Ball, 2016; Martínez et al., 2017; Sadare et al., 2017). Some biodesulfurization-competent strains can further transform 2-hydroxybiphenyl to either biphenyl or 2-methoxybiphenyl (Martínez et al., 2017; Xu et al., 2006). In the IGTS8 strain, the enzymes of the 4S pathway are encoded by three co-transcribed genes (*dszA*, *dszB*, *dszC*), which constitute the biodesulfurization *dsz* operon. These three genes are carried on a 120 kb mega linear plasmid found in the IGTS8 strain, while the fourth gene, *dszD,* is located on the chromosome and encodes the flavin reductase (Denome et al., 1994; Gray et al., 1996).

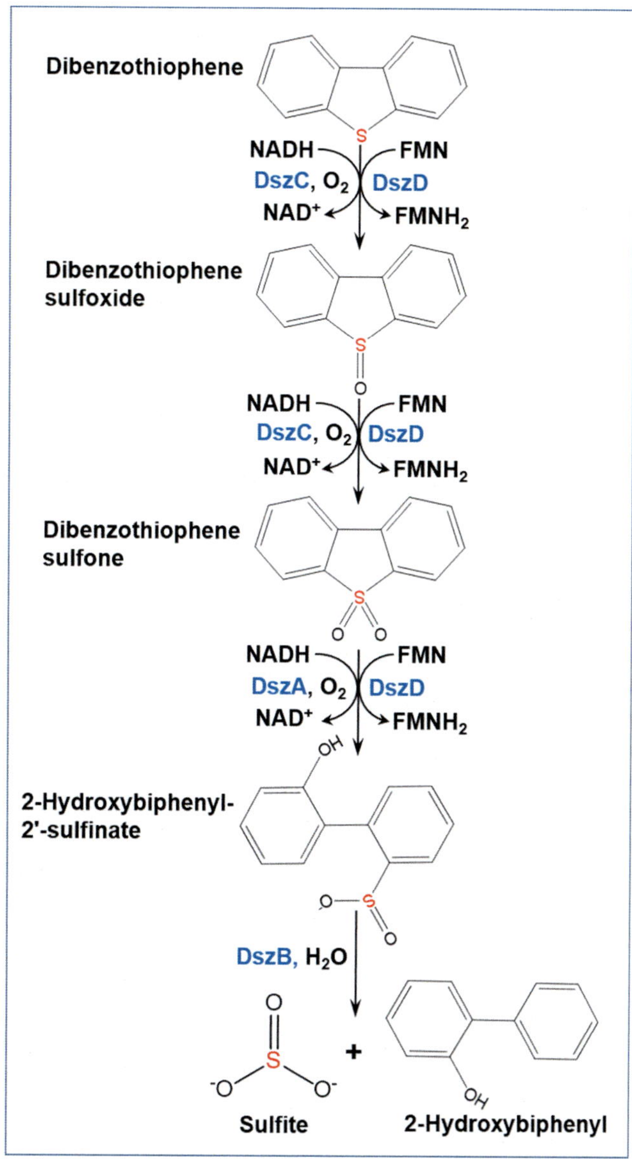

FIG. 7

The 4S biodesulfurization pathway.

Since the discovery of the 4S pathway in the IGTS8 strain (Kilbane and Jackowski, 1992; Gallagher et al., 1993), many biodesulfurizing bacteria were isolated and studied such as *Agrobacterium* sp. (Constanti et al., 1996; Gunam et al., 2017), *Achromobacter* sp. (Bordoloi et al., 2014), *Mycobacterium* spp. (Chen et al., 2008), *Gordonia* sp. (Alves et al., 2007) and thermophilic aerobic bacteria, e.g., *Bacillus subtilis* (Kirimura et al., 2001). These different kinds of desulfurizing strains harbor the 4S pathway and share a conserved nature of the *dsz* genotype (Kilbane, 2006; Mohebali and Ball, 2016). It was reported that *dsz* genes collected from different soil samples and different locations have a few variations in the DNA sequences. The majority of these variations were found in the *dszA* gene sequences which showed 95% or more homology to the *dszA* gene sequence of the IGTS8 strain (Duarte et al., 2001). Furthermore, different biodesulfurizing bacteria having identical *dsz* gene sequences have different biodesulfurization phenotypes in terms of substrate spectrum, desulfurization activity and yield of metabolites (Abbad-Andaloussi et al., 2003; Kilbane, 2006; Mohebali and Ball, 2016).

Biodesulfurization has been mostly investigated for light distillates like diesel and organosulfur model compounds like dibenzothiophene and its alkylated derivatives. On the contrary, there are a few reports on the biodesulfurization of heavy oils. Jiang et al. (2014) studied biodesulfurization of bunker heavy oil before and after de-asphalting with a microbial consortium enriched from oil sludge. The authors noted that the biodesulfurization rate before de-asphalting was 2.8% after 7 days, which increased to 36.5% for the de-asphalted oil. Desulfurization of crude oil via the 4S pathway was reported recently using a *Klebsiella oxytoca* SOB strain and was concomitant with sulfate formation (Mawad et al., 2021). Mohamed et al. (2015) reported that resting cells of a *Rhodococcus* strain could desulfurize heavy crude oil and hexane-soluble fraction of heavy crude oil by reducing the total sulfur content by 10% and 18%, respectively, after one week. Fig. 8 depicts a conceptual diagram for the process of crude oil biodesulfurization (Monticello, 2000).

Recently, Porto et al. (2018) studied the biodesulfurization of heavy gas oil using a commercial bacterial consortium consisting of *Alcaligenes faecalis*, *Acinetobacter* sp., *Bacillus* sp., *Komagataeibacter hansenii*, *Oceanobacillus iheyensis*, *Ochrobactrum anthropic*, *Paenibacillus lautus* and *Providencia rettgeri*. This consortium achieved 71.8% biodesulfurization rate after 2 h of treatment at a biomass load of 5% (w/w).

A potential intervention that might lead to viscosity reduction is the cleavage of the sulfide bridges in the asphaltene fraction. This would lead to depolymerization of the high-molecular weight asphaltenes into smaller assemblies or condensed structures. Van Hamme et al. (2004) used a low-molecular weight model compound to show that some microorganisms, under sulfur-limiting conditions, can cleave sulfide linkages in linear chains between larger aromatic structures. The authors identified bacteria capable of removing sulfur from *bis*-(3-pentafluorophenylpropyl)-sulfide (PFPS), which is recalcitrant toward terminal attack due to the presence of pentafluorinated aromatic rings. One bacterium, *Rhodococcus* sp. strain JVH1, was isolated from an oil-polluted environment and could remove sulfur from the sulfidic linkages in PFPS and other aryl-aryl/aryl-alkyl moieties such as dioctyl sulfide, didodecylsulfide, dibenzylsulfide and 1,4-dithiane. This biodesulfurization process involved oxidation of the sulfur atom to sulfoxide, sulfone, followed by C–S bond cleavage.

6.2.3 Obstacles facing biodesulfurization commercialization

Over the past three decades, extensive research and development efforts have been invested with the ultimate goal of establishing a commercially viable biodesulfurization technology for the oil industry (Chen et al., 2018). Many biodesulfurization-competent bacteria have been isolated and their biodesulfurization potential has been studied. In addition, extensive genetic engineering and physicochemical treatments have been implemented to enhance the biodesulfurization activity of different bacteria (Kilbane and Stark, 2016; Parveen et al., 2020; Sar et al., 2021). Despite these tremendous efforts, a bioprocess for biodesulfurization of fossil fuels has not been commercialized yet, mainly due to the very low catalytic activity and insufficient robustness of the applied biocatalysts/microbial hosts,

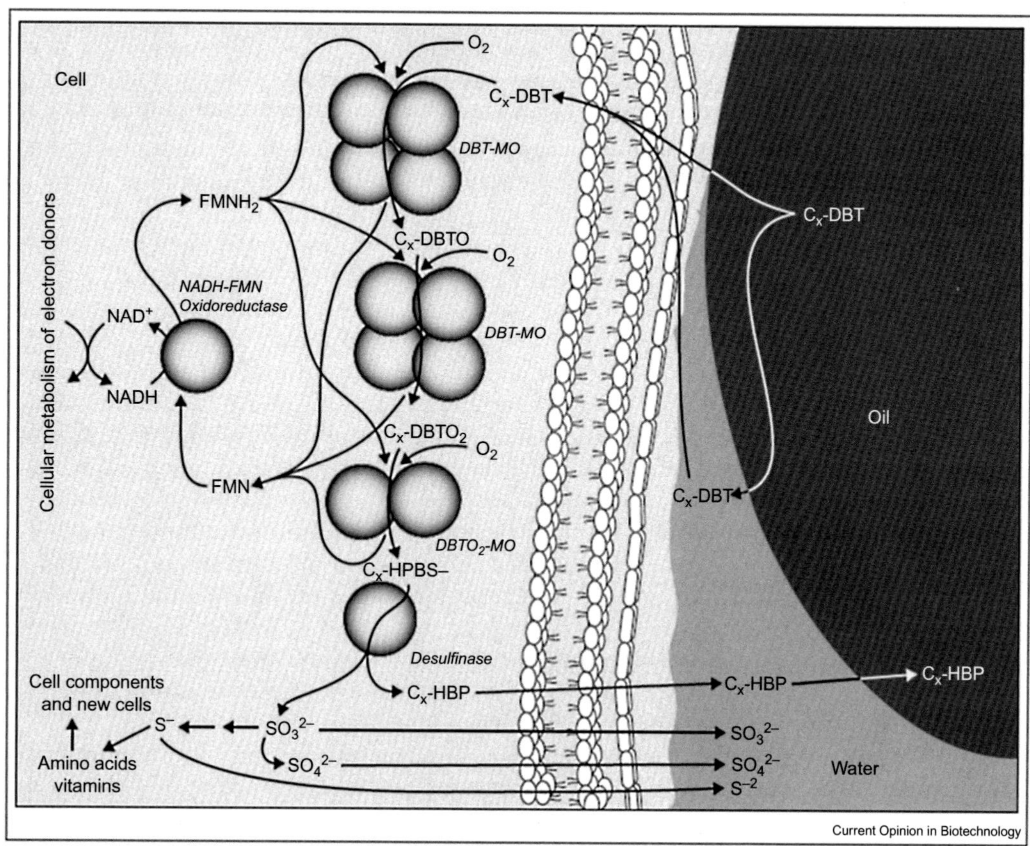

FIG. 8

A conceptual diagram of the crude oil biodesulfurization mechanism. *DBT*, dibenzothiophene; *DBTO*, dibenzothiophene sulfoxide; *$DBTO_2$*, dibenzothiophene sulfone; *DBT-MO*, dibenzothiophene monooxygenase; *C_x-DBT*, alkyl derivatives of DBT.

Reprinted with permission from Monticello, D.J. 2000. Biodesulfurization and the upgrading of petroleum distillates. Curr. Opin. Biotechnol. 11 (6), 540–546. Copyright (2000) Elsevier.

among other hurdles (Aggarwal et al., 2012, 2013; Ferreira et al., 2017). Even important issues like the uptake of the organosulfur substrates by biodesulfurizing cells have not been resolved.

The low biodesulfurization activities of the known biodesulfurizing bacteria can be attributed to different reasons. First, the *dsz* genes are strongly repressed by inorganic sulfate and other easily bioavailable sulfur sources such as cysteine, methionine and casamino acids (Lei and Tu, 1996; Denis-Larose et al., 1997; Caro et al., 2007). Therefore, as sulfite is one of the 4S pathway products, it is difficult to maintain a high level of biodesulfurization activity (Xu et al., 2002; Lee et al., 2006). Second, the accumulation of 2-hydroxybiphenyl as a product of the 4S pathway causes inhibition of bacterial growth and the biodesulfurization enzymes (Xu et al., 2002). In addition, hydroxyphenyl benzenesulfinate, an intermediate of the 4S pathway, is known to inhibit the activity of dibenzothiophene monooxygenase, the first enzyme of the 4S pathway (Abin-Fuentes et al., 2013). Besides, the DszB-catalyzed reaction is a rate-limiting step because DszB is produced in a minimal amount and is the slowest of the three 4S enzymes (Gupta et al., 2005; Jiang et al., 2014). Another obstacle limiting the applicability of the currently known biodesulfurizing bacteria is the narrow substrate spectrum (Kilbane, 2006). Additionally, mass transfer limitations due to the hydrophobicity of the thiophenic organosulfur substrates have been reported to impede the biocatalytic desulfurization process (Mohebali and Ball, 2016).

In addition to the above-mentioned shortcomings, there are huge knowledge gaps in our understanding of the physiology and metabolism of fuel-biodesulfurizing bacteria. To date, biodesulfurization research has focused almost exclusively on the central biodesulfurization mechanism namely, the 4S pathway, despite earlier indications that other cellular functions might have an impact on the biodesulfurization phenotype (Kilbane, 2006). For instance, Tanaka et al. (2002) reported that disruption of the cystathionine-β-synthase gene by transposon mutagenesis in *Rhodococcus erythropolis* KA2-5-1 led to a higher biodesulfurization activity in the presence of sulfate compared to the wild type, and suggested that this phenotype is due to reduced cysteine biosynthesis in the mutant. Moreover, Aggarwal et al. (2012) showed, by *in silico* metabolic modeling, that the biodesulfurization activity can be improved by manipulating sulfur metabolism enzymes such as sulfite oxidoreductase and sulfite reductase. Results from very recent studies backed those earlier reports. Dorado-Morales et al. (2020) showed significant improvement in the biodesulfurization activity of *R. erythropolis* by inducing biofilm formation via heterologous expression of a diguanylate cyclase which produced cyclic di-GMP, the secondary messenger and trigger of biofilm formation. In addition, systems biology investigations showed that *R. qingshengii* IGTS8 remodeled the sulfur assimilation pathways and oxidative stress proteins as well as exhibited sulfur sparing under biodesulfurization conditions (Hirschler et al., 2021). The authors concluded that the sulfur assimilation pathways are key determinants of the biodesulfurization phenotype and identified potential targets for metabolic engineering to construct more efficient recombinant biodesulfurizers. These studies underscore the importance of thorough understanding of the physiology and metabolism of fuel-biodesulfurizing bacteria, particularly under the challenge of using fuels as the sole sulfur source.

Obviously, the 4S pathway and the known biodesulfurizing microbes have many drawbacks that have impeded the commercial application of fuels biodesulfurization. Accordingly, there is a need to look for more efficient biodesulfurizing microbes and novel biodesulfurization pathways. The great majority of the currently known biodesulfurizing microbes have been isolated from environmental samples using traditional culture-dependent methods (Martín-Cabello et al., 2020). It is known that culturing recovers only a minor fraction (less than 10%) of the microbial content in a given sample due to the selectivity of the laboratory media and incubation conditions (Al-Mailem et al., 2014; Mason et al., 2014; Martín-Cabello et al., 2020). This means we can probably find novel biodesulfurization biocatalysts in environmental samples if we implement culture-independent techniques such as metagenomics to access the wealth of microbial diversity that is not accessible with the culture-dependent techniques.

6.2.4 Metagenomics and biodesulfurization

Metagenomics is a molecular approach that enables researchers to isolate and functionally screen and characterize the entire genomic contents of microbial communities inhabiting a specific environment (Abbasian et al., 2016). The environmental gene mining process helps to screen a high number of metagenomes of uncultured bacteria or other microorganisms. Therefore, metagenomics can detect both known genes in new microorganisms and new genes in known or unknown microorganisms.

The first step in gene mining is DNA isolation from a particular environment, followed by cloning the isolated DNA fragments into a suitable vector to construct a metagenomic library. Next, these vectors are propagated, usually in *Escherichia coli*, followed by functional screening for the phenotype of interest. In an earlier in silico study, Bhatia and Sharma (2010) screened the available genomic databases for new biodesulfurizing microorganisms using the protein sequences of the IGTS8 strain's biodesulfurization enzymes DszABCD as a query for BLAST search. The authors identified 13 putative new dibenzothiophene degraders belonging to 12 genera, which have homologs of the DszA, DszC and DszD proteins. Seven of the identified microbes did not have homologs of the desulfinase DszB. Two of the identified bacteria, *Burkholderia fungorum* and *Thermobifida fusca,* were tested experimentally for growth on dibenzothiophene as a sole sulfur source. Although the two cultures consumed dibenzothiophene, they did not produce 2-hydroxybiphenyl, the typical product of the 4S pathway.

A few publications reported metagenomics investigations on biodesulfurization. Abbasian et al. (2016) identified three bacterial genes (*dszA, dszB* and *dszC*) which constituted an operon for biodesulfurization of dibenzothiophene in a metagenome from crude oil. The sequences of the detected genes showed the following similarity to those of known *dsz* genes: *dszA* (78%), *dszB* (73%), *and dszC* (77%). The genes' sizes were 1167 bp (*dszA*) 1116 bp (*dszB*) and 1242 bp (*dszC*), including initiation and stop codons, and were designated as ADRO1, ADRO2 and ADRO3 in the GenBank. The function of these genes was investigated by heterologous expression in *E. coli* DH5α, and the authors showed that they encode Dsz enzymes that catalyze the desulfurization of dibenzothiophene via the 4S pathway. Nonetheless, based on the relatively low sequence similarity and different gene sizes as compared to known *dsz* genes, the authors proposed that the detected genes are new *dsz* genes. More recently, Martín-Cabello et al. (2020) detected a complete operon consisting of the 4S pathway genes *dszEABC* in the metagenome of oil-contaminated soil, and they also studied its transcriptional regulation.

These studies show the potential of culture-independent approaches such as metagenomics and in silico analyses as enabling tools for the detection of new biodesulfurizing microbes and pathways. To exploit the full potential, there should be systematic culture-independent studies on metagenomes from different ecosystems including oilfields and refineries.

6.3 Hydrocarbon biodegradation pathways

To date, specific biochemical mechanisms of heavy crude upgrading have not been revealed or investigated, which could be due to the complexity of the process and the multitude of intertwined biodegradation and biotransformation reactions that occur during crude bioconversion. Furthermore, crude oil in general and heavy crudes in particular are very heterogeneous and complex substrates in terms of physicochemical properties, which might hinder in-depth characterization of the underlying biochemical and molecular mechanisms. Nonetheless, it is worth noting that there is a wealth of knowledge on microbial biodegradation and biotransformation of organic substrates starting from simple alkanes and monocyclic aromatic compounds up to more complex substrates such as polycyclic aromatic hydrocarbons, steroids, lignin, polymers, etc. The biochemical principles underlying the microbial attack on these substrates can be largely applicable to more challenging ones like unconventional crude oils. Therefore, within the context of hydrocarbon metabolism, we present here a brief account of the main pathways of hydrocarbon biodegradation. For more details on the microbiology, enzymology and molecular biology of those pathways, we refer the readers to recent publications (Fuentes et al., 2014; Pérez-Pantoja et al., 2019; Widdel and Musat, 2019; Boll et al., 2020a, 2020b; Boll and Estelmann, 2020; Wilkes and Rabus, 2020).

The wide distribution of hydrocarbons in different ecosystems makes them a good carbon and energy source for many dedicated microorganisms. A large body of references has reported the ability of diverse microbes to utilize several aliphatic and aromatic hydrocarbons as growth substrates. However, the relative inertness of hydrocarbons renders their degradation by microorganisms a challenging task. Aromatic compounds are relatively resistant to biodegradation because of the stability of the aromatic ring resulting from the resonance energy (150 kJ/mol) (Bugg and Winfield, 1998). The dissociation energy of the C–H bonds in hydrocarbons is relatively high, ranging from 350 to more than 550 kJ/mol. The highest C–H bond dissociation energies are found among unsaturated hydrocarbons such as aliphatic compounds with double bonds and aromatic compounds such as benzene and naphthalene (460–475 kJ/mol). In alkynes, such as acetylene, it can reach 556 kJ/mol (Boll et al., 2020a). Saturated alkanes occupy the second place in terms of C–H bond dissociation energies, having values between 400 and 417 kJ/mol in linear alkanes and cycloalkanes, up to 439 kJ/mol in methane (Luo, 2003: Thauer and Shima, 2008). Interestingly, hydrocarbon-degrading bacteria are equipped with diverse catabolic strategies that enable them to overcome the inherent inertness of hydrocarbons under both oxic and anoxic conditions.

In general, microbial degradation of aromatic compounds starts with the so-called channeling reactions. These reactions serve to funnel the great variety of aromatic compounds into a few key intermediates which are substrates for further catabolism by central or core reactions. The latter transforms the key intermediates into metabolites of the central metabolic network (Fuchs, 1999). Under aerobic conditions, the channeling reactions produce catechol, protocatechuate and gentisate (Fig. 9). These aromatic diols are then attacked by ring-fission dioxygenases to cleave the aromatic ring. Subsequent reactions transform the non-aromatic ring cleavage products to metabolites of the TCA cycle (Harwood and Parales, 1996). Under anoxic conditions, the channeling reactions generate mainly benzoyl-coenzyme A (benzoyl-CoA), the most common central intermediate in the anaerobic metabolism of aromatic compounds (Heider and Fuchs, 1997). The benzene ring of benzoyl-CoA is initially reduced to a cyclic diene followed by hydrolytic cleavage of the resulting alicyclic ring. Eventually, a modified β-oxidation system transforms the non-aromatic products into central metabolites such as acetyl-CoA (Boll et al., 2020b).

Obviously, aerobic microorganisms employ an oxygen-dependent catabolic strategy to utilize aromatic compounds. The key reactions are the activation of the aromatic ring by hydroxylation and the oxidative ring cleavage. On the contrary, in the absence of oxygen, anaerobes depend on CoA thioesterification as an activation alternative, as well as reductive mechanisms to overcome the aromaticity of the substrates. Moreover, anaerobic microorganisms cleave the ring via a hydrolytic, oxygen-independent, reaction.

A third catabolic route for aromatic compounds was discovered three decades ago. The new pathway operates mainly in facultative anaerobes such as *P. putida*, *E. coli* and *Azoarcus evansii* when they grow under oxic conditions. The so-called "hybrid pathway" encompasses reactions common to the classical aerobic pathways (ring hydroxylation) as well as reactions known to operate only in the absence of oxygen (CoA thioester formation and oxygen-independent ring cleavage) (Fuchs et al., 2011; Ismail and Gescher, 2012).

6.3.1 Aerobic degradation of aromatic compounds

The process of aromatic compounds degradation or catabolism can be broadly divided into two main phases based on the objective or purpose of each phase: the initial (preparatory) phase and the central

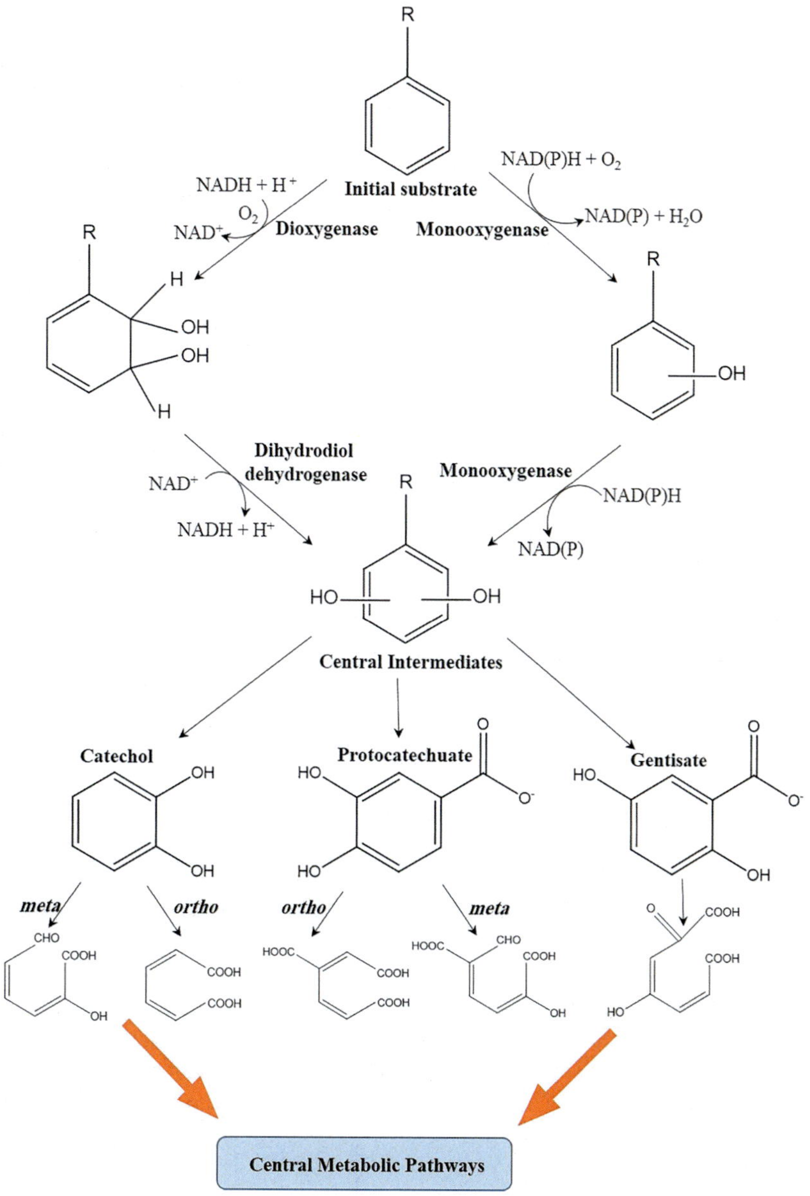

FIG. 9

Initial hydroxylation reactions and modes of aromatic ring cleavage under aerobic conditions.

(core) phase. During the initial stage of the degradation process, the aromatic substrates are prepared for ring cleavage that takes place during the second phase of degradation. This is accomplished by a number of ring modification reactions which are also called channeling (peripheral) or funneling reactions because they lead to the production of a few central intermediates from the huge array of the aromatic substrates. All channeling reactions have in common a hydroxylation step. Hydroxylation of the aromatic ring is catalyzed by monooxygenases and dioxygenases that require NAD(P)H as an electron donor and oxygen as a co-substrate (Mason and Cammack, 1992; Butler and Mason, 1997; Widdel and Musat, 2019). As shown in Fig. 9, monooxygenases (mixed function oxygenases) catalyze the incorporation of one oxygen atom of dioxygen into the aromatic ring and reduce the other oxygen atom to water, whereas dioxygenases introduce both atoms into the aromatic nucleus, thus producing a nonaromatic *cis*-dihydrodiol. The latter rearomatizes through the elimination of 2H, a reaction that is catalyzed by dihydrodiol dehydrogenase (Pérez-Pantoja et al., 2019). Other modification reactions are oxidation or removal of substituents at the aromatic ring (Ribbons and Eaton, 1982). Eventually, the aromatic substrates are transformed into a few central intermediates that are subjected to ring cleavage. The most common central intermediates are catechol (1, 2-dihydroxy benzene), protocatchuate (3, 4-dihydroxybenzoate) and gentisate (2, 5-dihydroxybenzoate) which share the presence of two hydroxyl groups either on two adjacent or on opposing carbons of the aromatic ring.

The central phase of the degradation process aims at breaking the aromatic ring of the key intermediates as well as transforming the ring cleavage products to Krebs cycle metabolites. Ring fission is catalyzed by non-heme iron-containing ring cleavage dioxygenases which introduce both atoms of dioxygen into the aromatic ring (Harayama et al., 1992). Two main modes of aromatic ring fission operate when the aromatic ring bears two adjacent hydroxyl groups (catechol and protocatechuate).

The first mode of cleavage is the *ortho* cleavage, where the aromatic ring is cleaved between the two hydroxyl groups (intradiol cleavage) (Fig. 9). The other ring fission mode, the *meta* cleavage, proceeds via cleavage adjacent to one of the hydroxyl groups (extradiol cleavage). A third mode of cleavage, which occurs with gentisate, comes into play when the two hydroxyl groups on the aromatic ring are *para* to each other. In this case, the cleavage takes place between the carboxyl-substituted carbon and the adjacent hydroxylated carbon (Fig. 9) (Pérez-Pantoja et al., 2019). Fig. 10 shows the degradation pathways for toluene and naphthalene, representing monocyclic and polycyclic aromatic compounds, respectively, that are common components of crude oil.

The most common central pathway for the microbial degradation of aromatic compounds is the β-ketoadipate pathway (Harwood and Parales, 1996; Pérez-Pantoja et al., 2008; Seo et al., 2009). Both branches of the bacterial β-ketoadipate pathway are shown in Fig. 11. Catechol and protocatechuate are subjected to intradiol *(ortho)* cleavage and the nonaromatic products, muconate and β-carboxymuconate, are transformed to the corresponding lactones. Both intermediates are then transformed to β-ketoadipate enol-lactone which is the convergence point of both branches of the β-ketoadipate pathway. The β-ketoadipate enol-lactone is hydrolyzed to β-ketoadipate, the key intermediate of the *ortho* cleavage pathway, which is transformed to the corresponding thioester, β-ketoadipyl-CoA. Thiolytic cleavage of the latter yields the Krebs cycle metabolites, succinyl-CoA and acetyl-CoA.

In eukaryotic microorganisms, the protocatechuate branch of the pathway exhibits two unique reactions that are not found in the bacterial version of the pathway (Cain, 1988). The first reaction deals with the cyclization of β-carboxy-*cis, cis*-muconate to give β-carboxymuconolactone instead of γ-carboxymuconolactone seen in the prokaryotic pathways. The second reaction is the direct conversion of β-carboxymuconolactone to β-ketoadipate, bypassing the formation of β-ketoadiapte enol-lactone.

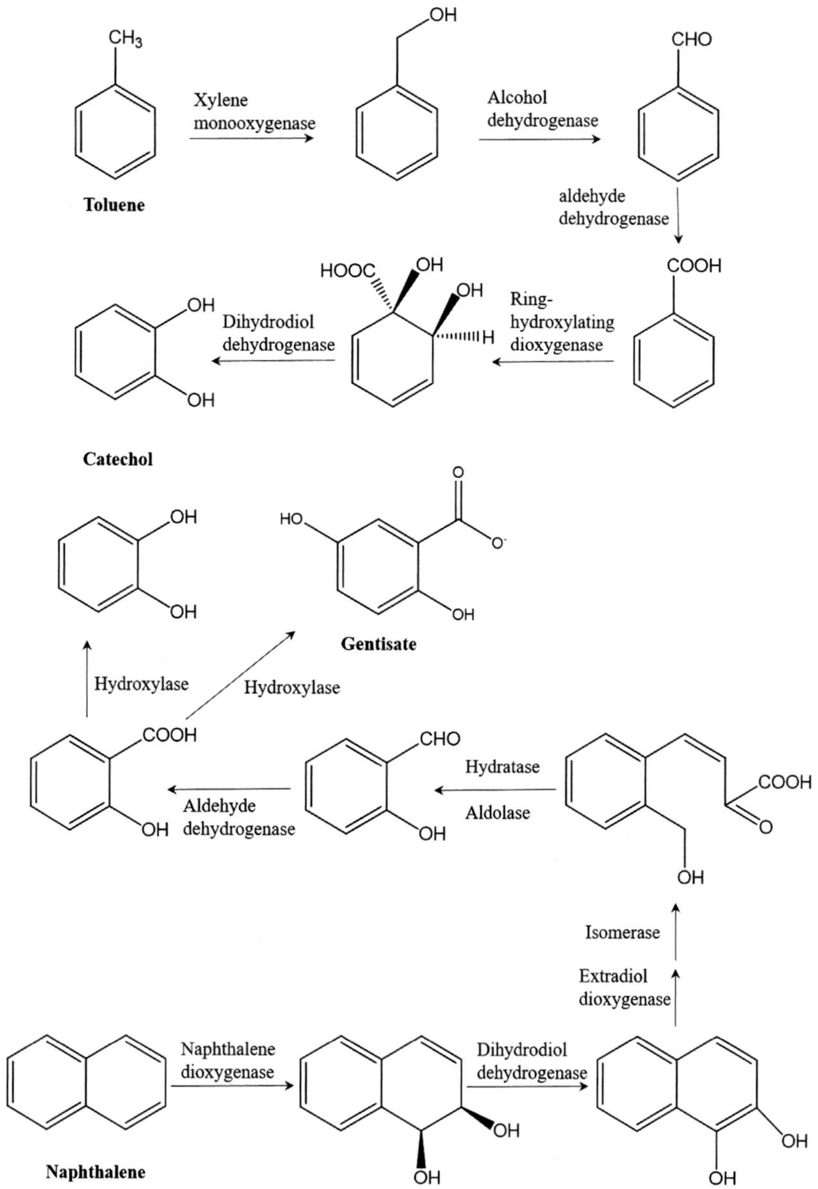

CH3
Toluene
Xylene monooxygenase
OH
Alcohol dehydrogenase
CHO
aldehyde dehydrogenase
COOH
Ring-hydroxylating dioxygenase
HOOC
OH
OH
H
Dihydrodiol dehydrogenase
OH
OH
Catechol
OH
OH
HO
O
O-
OH
Gentisate
Hydroxylase
Hydroxylase
COOH
OH
Aldehyde dehydrogenase
CHO
OH
Hydratase
Aldolase
COOH
O
OH
Isomerase
Extradiol dioxygenase
Naphthalene
Naphthalene dioxygenase
OH
OH
Dihydrodiol dehydrogenase
OH
OH

FIG. 10

Aerobic pathways for degradation of toluene and naphthalene (Fuentes et al., 2014).

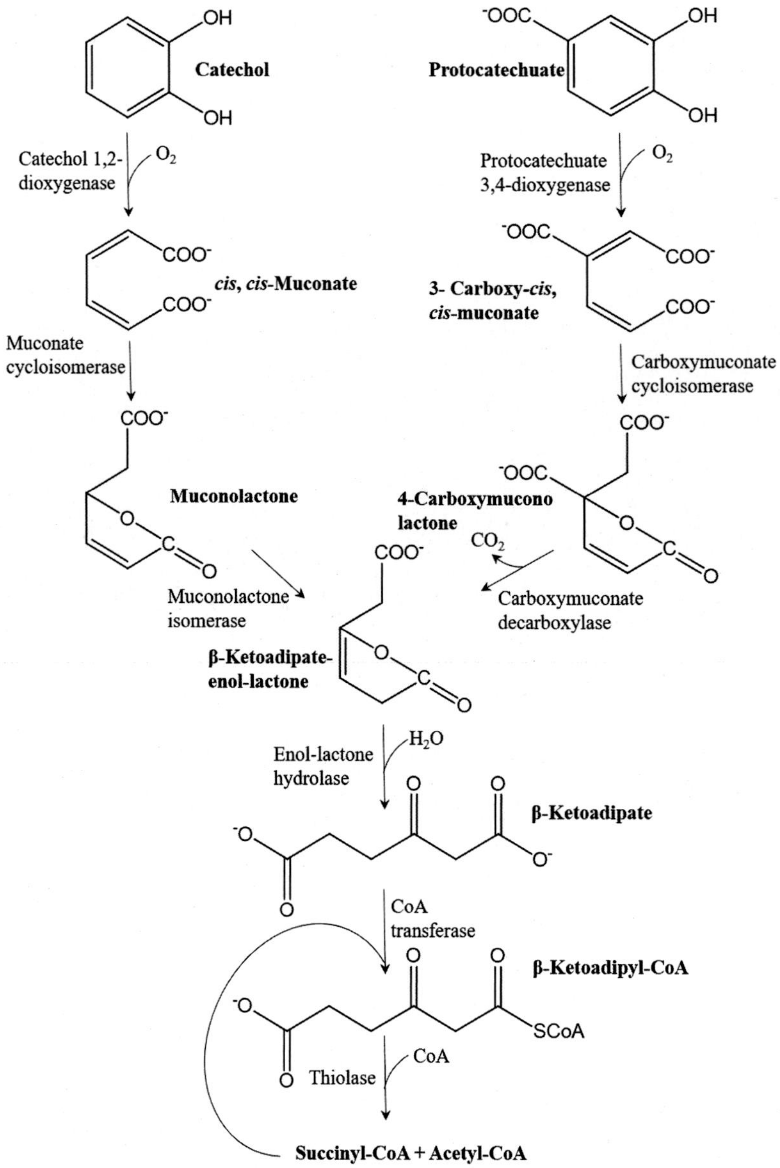

OH
Catechol
OH
-OOC
OH
Protocatechuate
OH
Catechol 1,2-dioxygenase
O2
Protocatechuate 3,4-dioxygenase
O2
COO-
cis, cis-Muconate
COO-
-OOC
COO-
3- Carboxy-cis, cis-muconate
COO-
Muconate cycloisomerase
Carboxymuconate cycloisomerase
COO-
Muconolactone
O
C
O
COO-
-OOC
4-Carboxymucono lactone
O
C
O
CO2
COO-
Muconolactone isomerase
Carboxymuconate decarboxylase
β-Ketoadipate-enol-lactone
O
C
O
Enol-lactone hydrolase
H2O
O
O
β-Ketoadipate
-O
O-
O
CoA transferase
O
O
β-Ketoadipyl-CoA
-O
SCoA
O
Thiolase
CoA
Succinyl-CoA + Acetyl-CoA

FIG. 11

The β-ketoadipate pathway of aerobic degradation of aromatic compounds. *CoA*, Coenzyme A (Harwood and Parales, 1996; Pérez-Pantoja et al., 2019).

6.3.2 Anaerobic degradation of aromatic compounds

It was long perceived that microbial metabolism of aromatic compounds obligatorily depends on oxygen. The situation remained disputable until the observation that the benzene ring of many aromatic compounds was completely decomposed to CO_2 and CH_4 when incubated under anoxic conditions with sewage sludge (Tarvin and Busswell, 1934). Over the many years, it became evident that aromatic compounds can be metabolized by microorganisms in the absence of oxygen (Evans and Fuchs, 1988; Heider and Fuchs, 1997). A growing number of anaerobic bacteria representing various physiological groups were found to utilize a multitude of aromatic compounds as a carbon and energy source. These bacteria comprise denitrifying, sulfate-reducing, metal-reducing, fermenting, as well as photosynthetic members (Lovley et al., 1993; Rahalkar et al., 1993; Anders et al., 1995; Boll et al., 2020b).

Despite the fact that anaerobic metabolism of aromatic compounds has been known for several decades, the novel enzymes and reactions inherent to that field of research have only been unraveled during the past three decades. We have seen in the previous section how oxygen is an indispensable cosubstrate for the catabolism of aromatic compounds under oxic conditions. A similar reactive cosubstrate is not accessible under anoxic conditions. To that end, one should ask how microorganisms degrade aromatic compounds without oxygen? Obviously, anaerobic bacteria have to rely on alternative mechanisms for activating and funneling diverse compounds into catabolic pathways and dearomatizing reactions (Boll et al., 2020a). In contrast to the aerobic pathways, which are essentially oxidative, anaerobic catabolism of aromatic compounds proceeds via reductive modifications of the substrates. Particularly remarkable is the reduction of the aromatic ring to nonaromatic cyclohexane derivatives (Boll et al., 2020a).

The anaerobic pathways, in common with the aerobic ones, start with the so-called channeling or funneling reactions. The latter converts the multitude of aromatic substrates to a few key intermediates, followed by ring reduction and opening. Currently, there are five known strategies for oxygen-independent initial attacks on hydrocarbons, depending on the substrate. The first involves anaerobic hydroxylation reactions of the alkyl chains of alkyl benzenes and secondary or tertiary carbon atoms of aliphatic hydrocarbons. These reactions are catalyzed by periplasmic molybdenum- or flavin-containing hydroxylases and typically activate hydrocarbons with low C–H bond dissociation energies (350–400 kJ/mol) such as *p*-cresol, *p*-ethylbenzene, alkanes, sterols, etc. The second major activation mechanism proceeds via fumarate addition, leading to succinate adducts of the hydrocarbon substrate, which is catalyzed by specialized glycyl radical-containing enzymes. Toluene, substituted toluene, 2-methylnaphthalene and some alkanes are typically activated via this mechanism. The anaerobic degradation of compounds with very high C–H bonds dissociation energies such as benzene and naphthalene is proposed to proceed via initial carboxylation reactions catalyzed by unique UbiD enzyme family members. The fourth strategy applies to acetylene and involves water addition to the C=C double bond. Methane is degraded via the unique pathway of reverse methanogenesis, which occurs mostly in syntrophic consortia consisting of certain types of Archaea and sulfate-reducing bacteria (Boll et al., 2020a).

For monocyclic aromatic compounds, the channeling reactions or pathways yield four central intermediates, namely; benzoyl-CoA, resorcinol, phloroglucinol and hydroxyhydroquinone (Fuchs et al., 1994; Gibson and Harwood, 2002; Boll et al., 2020a). Three of those key intermediates, phloroglucinol, resorcinol and hydroxyhydroquinone, possess *meta*-positioned hydroxyl groups which weaken the aromaticity of these substrates and facilitate the exergonic two-electron reduction of the aromatic ring with common physiological reducing agents such as NAD(P)H and ferredoxin. On the contrary, benzoyl-CoA, which is the most common central intermediate of the anaerobic pathways, has a fully developed aromatic character. Therefore, it is not feasible to be reduced simply by physiological reductants.

Organic chemists usually use solvated electrons to reduce the benzene ring in alternate electron and proton transfer steps (Birch reduction). By applying solvated electrons it becomes possible to overcome the very low redox barrier of the first electron transfer to the benzene ring (−3 V for benzene, −1.9 V for the benzoyl-CoA analogue S-ethylbenzoic acid thiol ester) (Boll, 2005; Fuchs, 2008).

Anaerobic bacteria solved this problem by using the low potential electron donor ferredoxin and coupling the reduction of the aromatic ring to stoichiometric hydrolysis of ATP. The enzyme that catalyzes benzoyl-CoA reduction is benzoyl-CoA reductase (class I) that is best studied in the denitrifying *Thauera aromatica*. The enzyme catalyzes the transfer of two electrons from reduced ferredoxin to benzoyl-CoA accompanied by the hydrolysis of two moles of ATP to produce a cyclic diene derivative. Another type of aryl-CoA reductases, which are ATP-independent, was discovered in obligate anaerobic bacteria (class II benzoyl-CoA reductases) (Loffler et al., 2011; Boll et al., 2020b). During the final phase of degradation, the reduced ring is hydrolytically cleaved and the products of ring cleavage are degraded via β-oxidation-like and conventional β-oxidation reactions, thus yielding acetyl-CoA and CO_2 (Fuchs, 2008; Fuchs et al., 2011) (Fig. 12).

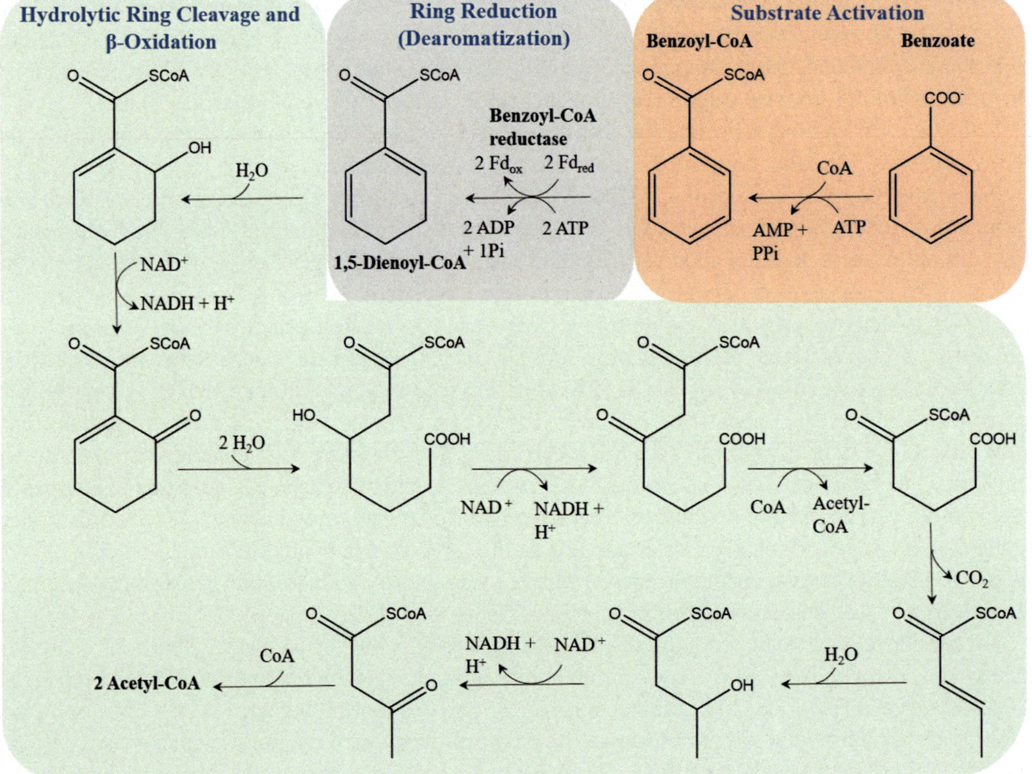

FIG. 12

The benzoyl-CoA pathway for anaerobic degradation of aromatic compounds. Benzene ring reduction (dearomatization) is catalyzed by the ATP-dependent benzoyl-CoA reductase. *Fd,* Ferredoxin (Boll et al., 2020b).

In contrast to monocyclic aromatics, the anaerobic degradation of polycyclic aromatic hydrocarbons is much less characterized. To date, naphthalene and methylnaphthalene are the sole polycyclic aromatic hydrocarbons whose anaerobic degradation pathways have been described. However, common principles are proposed to apply for the higher molecular weight substrates such as: initial carboxylation, formation of CoA thioester derivatives, reductive dearomatization of the ring system and further metabolism via modified β-oxidation reactions to produce the central metabolite acetyl-CoA. These are the same biochemical processes that are involved in the anaerobic degradation of monocyclic aromatic hydrocarbons via the benzoyl-CoA pathway (Boll and Estelmann, 2020). Fig. 13 shows the initial and ring reduction reactions for the anaerobic degradation of naphthalene and 2-methylnaphthalene. The naphthalene pathway starts with a carboxylation reaction yielding 2-naphthoate, followed by CoA thioesterification to produce 2-naphthoyl-CoA. The activation of 2-methylnaphthalene proceeds via fumarate addition in a process analogous to that reported for toluene. Both pathways converge at naphthoyl-CoA which is further degraded through the central 2-naphthoyl-CoA degradation pathway (Annweiler et al., 2000; Selesi et al., 2010; Boll and Estelmann, 2020). The dearomatization reactions involve three consecutive two-electron transfers to the naphthyl ring system. A member of the flavin-containing old yellow enzyme family reduces the non-substituted ring of 2-naphthoyl-CoA, followed by a second reduction reaction catalyzed by another old yellow enzyme. A class I (ATP-dependent) benzoyl-CoA reductase homolog catalyzes the dearomatization of the substituted ring of naphthoyl-CoA. The subsequent cleavage of the nonaromatic hexahydronaphthoyl-CoA ring system via modified β-oxidation reactions produces cyclohexane-containing intermediates.

The complete anaerobic degradation of naphthalene, methylnaphthalene and phenanthrene was frequently reported in sulfate-reducing bacteria, mainly members of the Desulfobacteriaceae. Degradation of naphthalene was also reported by a Fe(III)-respiring enrichment culture N49, whose members belong to the Peptococcaceae (Gram-positive bacteria) (Kleemann and Meckenstock, 2011). There is also an evidence for anaerobic naphthalene degradation under methanogenic and denitrifying conditions (Rockne et al., 2000; Chang et al., 2006).

6.3.3 Degradation of aromatic compounds via the hybrid pathways

A novel catabolic strategy for the aerobic degradation of aromatic compounds has been elucidated in some bacteria for the biodegradation of benzoate, phenylacetate and anthranilate (Ferrández et al., 1998; Olivera et al., 1998; Mohamed et al., 2002; Ismail et al., 2003; Ismail and Gescher, 2012). This pathway involves hydroxylation of the aromatic ring, metabolism of CoA thioester intermediates, epoxide formation, isomerization of the epoxide to an oxepin, as well as hydrolytic (oxygen-independent) ring cleavage. Accordingly, modules from both the aerobic and anaerobic catabolic pathways are combined, hence the designation "hybrid pathway" that reveals its mixed catabolic nature.

Phenylacetate is initially transformed to phenylacetyl-CoA, a reaction that takes place only during the anaerobic degradation of aromatic compounds and is catalyzed by a phenylacetate-CoA ligase (Fig. 14) (Ismail and Gescher, 2012; Teufel et al., 2010). The subsequent formation of ring-1, 2-epoxide is catalyzed by a novel multi-component oxygenase which is the prototype of a new group of bacterial di-iron multi-component oxygenases acting on CoA-thioesters. A member of the crotonase superfamily, which is highly similar to Δ^3, Δ^2-enoyl-CoA isomerases, produces a seven-membered O-heterocyclic enol-ether (oxepin), followed by hydrolytic ring cleavage that is catalyzed by a fusion protein (PaaZ) consisting of a C-terminal MaoC-like domain and an N-terminal aldehyde dehydrogenase domain.

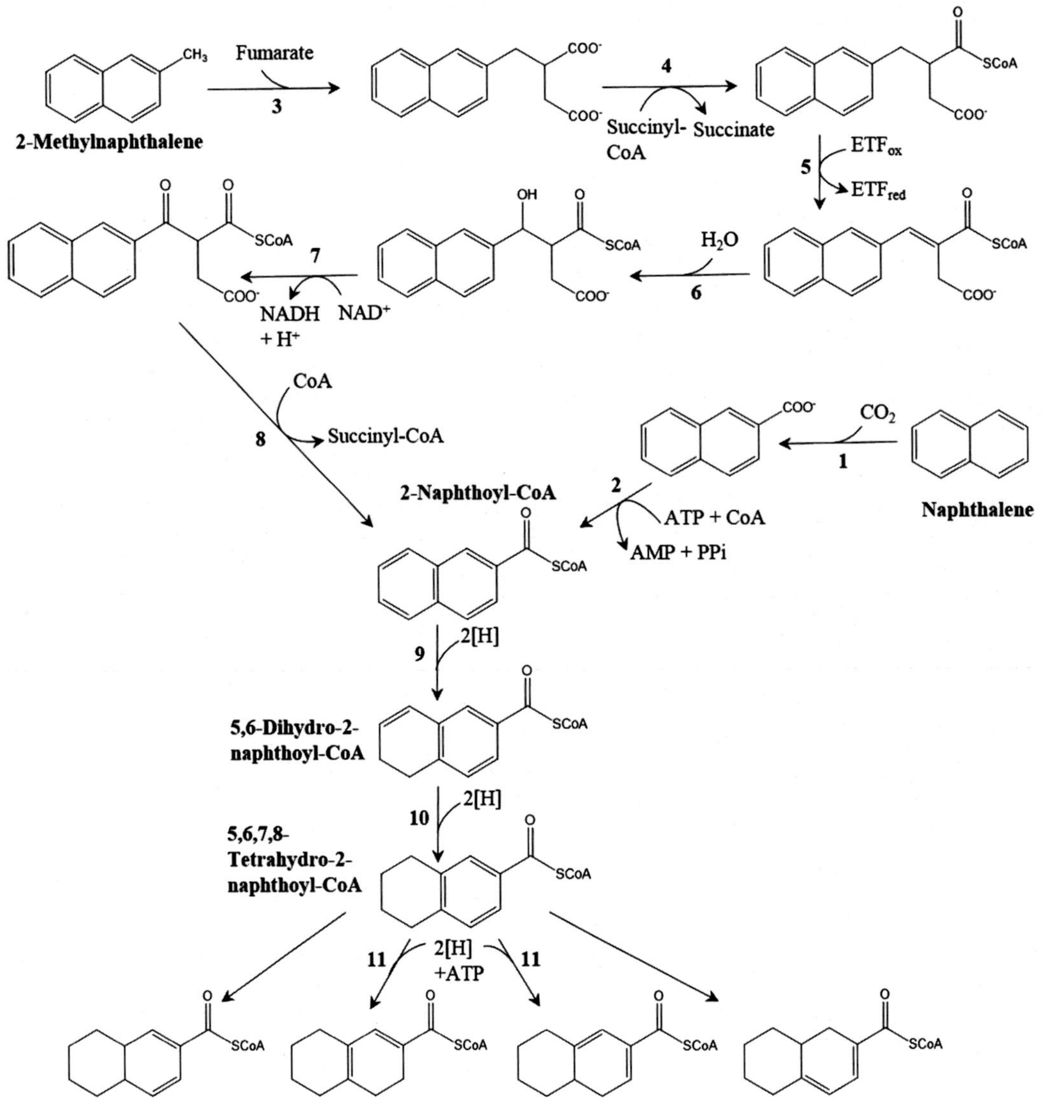

FIG. 13

Anaerobic degradation of naphthalene and 2-methylnaphthalene. Initial activation of naphthalene proceeds via a carboxylation reaction, while 2-methylnaphthalene is activated via addition to fumarate. Both pathways converge at 2-naphthoyl-CoA which is then further degraded via the central naphthoyl-CoA pathway. Reduction of tetrahydronaphthoyl-CoA may lead to different possible isomers. *CoA*, coenzyme A; *ETF*, electron transfer flavoprotein; 1: naphthalene carboxylase, 2: 2-naphthoate-CoA ligase, 3: naphthyl-2-methylsuccinate synthase, 4: CoA transferase; 5: naphthyl-2-methylsuccinyl-CoA dehydrogenase; 6: naphthyl-2-methylenesuccinyl-CoA hydratase, 7: naphthyl-2-hydroxymethylsuccinyl-CoA dehydrogenase, 8: naphthyl-2-oxomethylsuccinyl-CoA thiolase, 9: naphthoyl-CoA reductase (ATP-independent), 10: dihydronaphthoyl-CoA reductase (ATP-independent), 11: tetrahydronaphthoyl-CoA reductase (ATP-dependent) (Boll and Estelmann, 2020).

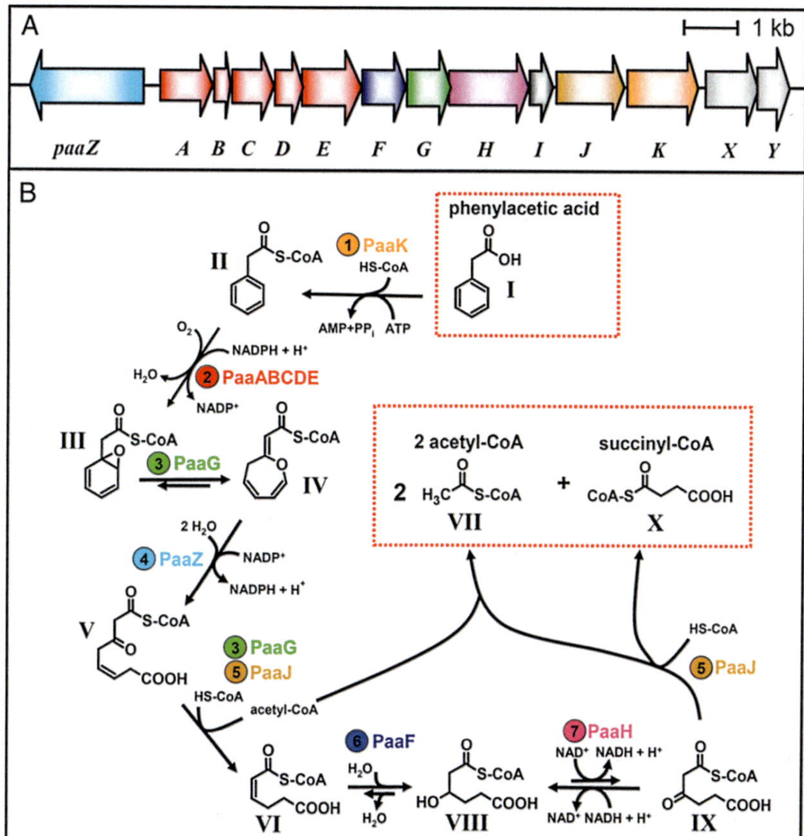

FIG. 14

Aerobic phenylacetate catabolic pathway. (A) Catabolic gene cluster for phenylacetate degradation in *Escherichia coli* K12. (B) Reactions and intermediates of the pathway as studied in *Escherichia coli* K12 and *Pseudomonas* sp. strain Y2. Enzyme names: 1: phenylacetate-CoA ligase (AMP forming), 2: ring 1,2-phenylacetyl-CoA epoxidase (NADPH), 3: ring 1,2 epoxyphenylacetyl-CoA isomerase (oxepin-CoA forming), postulated 3,4-dehydroadipyl-CoA isomerase, 4: oxepin-CoA hydrolase/3-oxo-5,6-dehydrosuberyl-CoA semialdehyde dehydrogenase (NADP+), 5: 3-oxoadipyl-CoA/3-oxo-5,6-dehydrosuberyl-CoA thiolase, 6: 2,3 dehydroadipyl-CoA hydratase; 7: 3-hydroxyadipyl-CoA dehydrogenase (NAD^+) (probably (*S*)-3-specific). Compounds: I: phenylacetate, II: phenylacetyl-CoA, III: ring 1,2-epoxyphenylacetyl-CoA, IV: 2-oxepin-2(3H)-ylideneacetyl-CoA, V: 3-oxo-5,6-dehydrosuberyl-CoA, VI: 2,3-dehydroadipyl-CoA, VII, acetyl-CoA, VIII: 3-hydroxyadipyl-CoA, IX: 3-oxoadipyl-CoA, X: succinyl-CoA.

Reprinted with permission from Teufel, R., Mascaraque, V., Ismail, W., Voss, M., Perera, J., Eisenreich, W., et al. 2010. Bacterial phenylalanine and phenylacetate catabolic pathway revealed. Proc. Natl. Acad. Sci. U. S. A. 32, 14390–14395. Copyright (2010) Proceedings of the National Academy of Sciences USA.

The aldehyde group of the C_8 ring opening product is further oxidized to a carboxylic acid by the aldehyde dehydrogenase domain of PaaZ. Eventually, β-oxidation-like reactions convert the ring opening product to metabolites of the central metabolic pathways.

From the above discussion, it can be reconciled that hydrocarbonoclastic bacteria are endowed with diverse catabolic machineries for aromatic compounds that can be exploited to introduce structural changes in the different crude oil fractions. Of particular interest are reactions which can functionalize asphaltenes by introducing hydroxyl, oxo, or carboxyl groups, in addition to cleavage and hydrogenation of aromatic rings. These types of bioconversions might provoke destabilization of aromatic stacking and lead to dissociation of asphaltenes aggregates into smaller clusters, which would potentially result in viscosity reduction. Furthermore, aromatic ring opening and degradation could reduce aromatic stacking and the apparent molecular weight of asphaltenes and resins. Oxygenated compounds can also act as surfactants enabling the formation of crude oil-water emulsions which could be another route toward viscosity reduction. Since the splitting of stacked polycyclic aromatic molecules contributes to viscosity reduction, pathways of polycyclic aromatic hydrocarbon degradation merit more attention, particularly the anaerobic aromatic ring reduction and cleavage. These reactions might reduce the π–π stacking in the aromatic structures and induce dissociation of the asphaltenes into lighter fragments. As presented above, hydrogenation of the aromatic ring occurs during the anaerobic biodegradation of monocyclic and polycyclic aromatic compounds via the benzoyl-CoA and naphthoyl-CoA pathways, respectively. Moreover, aromatic ring hydrogenation has been reported for the aerobic biodegradation of trinitrotoluene and trinitrophenol via hydride transfer reactions catalyzed by hydride transferases using NADPH as the hydride donor and coenzyme F420 (Vorbeck et al., 1994, 1998). However, to date the applicability of these biochemical reactions for specific transformations of aromatic structures in heavy crudes has not been addressed.

6.3.4 Biodegradation of aliphatic compounds

From bioupgrading perspective, molecular weight reduction can be achieved via targeted splitting or cracking of the alkyl bridges linking the high-molecular weight structures. Furthermore, there is a potential to deal with wax and asphaltene deposition through the introduction of functional groups and biodegradation of the problematic components. Indeed, many bacteria are known for their alkane degradation capabilities that can be exploited to provoke structural changes in the heavy crudes and improve their physicochemical properties. Biodegradation of *n*-alkanes has been thoroughly characterized at the biochemical and molecular levels under both oxic and anoxic conditions (Fuentes et al., 2014; Moreno and Rojo, 2019a,b; Wilkes and Rabus, 2020). Here we present an overview of the various strategies adopted by alkane-degrading microorganisms to activate and degrade *n*-alkanes.

Similar to the aerobic degradation of aromatic compounds, the aerobic degradation of aliphatic compounds is initiated via hydroxylation reactions. The latter may take place at the terminal, subterminal, or diterminal carbons and is catalyzed by monooxygenases (hydroxylases) (Solano-Serena et al., 2008). In the terminal and diterminal degradation pathways the initial hydroxylation produces primary alcohols which are sequentially oxidized to the corresponding carboxylic acids. These are then metabolized via β-oxidation or incorporated into cellar lipids.

In the sub-terminal oxidation pathways, a secondary carbon is oxidized to produce an alcohol that is subsequently dehydrogenated to the corresponding ketone. Various microorganisms are known to convert alkanes into ketones (Solano-Serena et al., 2008). For instance, *P. aeruginosa* produces several alcohols and the corresponding ketones from decane. Sub-terminal oxidation of alkanes was also reported in fungi of the genus *Mucor*. However, in this case, the aliphatic chain is ruptured due to a Baeyer-Villiger oxygenation of the ketone (Solano-Serena et al., 2008).

In contrast to branched and cycloalkanes, *n*-alkanes are the subclass of saturated hydrocarbons whose biodegradation under anoxic conditions has been most extensively studied. The anaerobic degradation of *n*-alkanes has been reported for C_3 up to C_{50} substrates and *n*-alkane-degrading bacteria exist mostly among the sulfate reducers and denitrifying bacteria (Kaiya et al., 2012; Mohapatra et al., 2017; Wilkes and Rabus, 2020). In addition, recent studies have shown that anaerobic alkane degradation may occur under electrophototrophy/electrode reduction conditions (Venkidusamy and Megharaj, 2016a,b). Syntrophic utilization of *n*-alkanes under anoxic conditions was also reported, where archaeal and bacterial members may be involved in the initial activation of the hydrocarbon substrate and consumption of the reducing equivalents produced by their syntrophic partners (Gieg et al., 2014). Methanogenic degradation of hydrocarbons is also gaining increasing momentum due to the notion that anaerobic degradation of hydrocarbons by methanogenic communities in petroleum reservoirs produces huge amounts of methane (Milkov, 2018).

Currently, four metabolic strategies are known to play a role in the utilization of *n*-alkanes under anoxic conditions, namely; addition to fumarate, anaerobic hydroxylation, transformation to alkyl-coenzyme M and "intra-aerobic" oxidation. Kropp et al. (2000) reported the first evidence for anaerobic activation of *n*-alkanes via addition to fumarate, in analogy with toluene transformation to benzylsuccinate by the glycyl radical enzyme benzylsuccinate synthase. Subsequent biochemical and molecular investigations provided compelling evidence for *n*-alkanes addition to fumarate (Wilkes and Rabus, 2020). In case of *n*-hexane, this initial activation produces 1-methylpentylsuccinate, a reaction catalyzed by the glycyl radical enzyme 1-methylalkylsuccinate synthase (Grundmann et al., 2008). Thioesterification transforms the initial adduct to 1-methylpentylsuccinyl-CoA which is then rearranged to 2-methylhexylmalonyl-CoA followed by decarboxylation to yield (*R*)-4-methyloctanoyl-CoA (Wilkes et al., 2002; Jarling et al., 2012). β-Oxidation reactions produce acetyl-CoA and (*R*)-2-methylhexanoyl-CoA, and after epimerization two more rounds of β-oxidation produce two molecules of acetyl-CoA and one molecule of propionyl-CoA (Wilkes et al., 2002, 2016).

Dehydrogenation/anaerobic hydroxylation was proposed for long-chain *n*-alkanes. Hydroxylation produces alkan-2-ol followed by dehydrogenation to the corresponding alkane-2-one that is carboxylated at C3 to produce 2-acetylalkanoate. The latter is transformed a CoA thioester and further degraded via β-oxidation (Heider et al., 2016). Initial activation of *n*-alkanes via transformation to alkyl-coenzyme M derivatives was proposed for the anaerobic degradation of *n*-butane by a syntrophic consortium (Laso-Pérez et al., 2016). During this mode of degradation, an archaeal member of the syntrophic association activates and completely oxidizes *n*-butane, whereas a sulfate-reducing bacterial member functions as a sink for the resulting reducing equivalents. The initial activation reaction for *n*-butane produces 2-(butylsulfanyl)ethane-1-sulfonate (1-butyl-coenzyme M).

The capability of some anaerobic bacteria to produce oxygen via dismutation reactions of alternative electron acceptors during *n*-alkane degradation laid the foundation for "intra-aerobic" oxidation as a novel mode of hydrocarbon degradation by anaerobic bacteria (Wilkes and Rabus, 2020). This mechanism was proposed for methane activation in a nitrite-reducing Candidatus *Methylomirabilis oxyfera*, where dismutation of NO produces O_2 that serves as the cosubstrate of particulate methane monooxygenases (Ettwig et al., 2010). Furthermore, *n*-alkanes degradation by the chlorate-respiring *Pseudomonas chloritidismutans* AW-1^T was reported to involve oxygen produced via chlorate reduction/chlorite dismutation (Mehboob et al., 2009).

7 Conclusions and research needs

Despite many studies on asphaltenes biotransformation and unconventional oil bioupgrading/bioconversion, it is still a long way ahead to reach the commercialization phase. Apparently, we have just scratched the surface of a very interesting and promising, albeit challenging, field which needs intensive research and development work. Given the well-characterized and astonishing microbial catabolic activities on a wide array of hydrocarbons and heterocyclic compounds, hydrocarbonoclastic bacteria constitute a resourceful niche that can be further developed and adapted to applications in heavy crudes upgrading. Obviously, the field is facing major obstacles that can be potentially overcome via multidisciplinary, systematic and problem-oriented research and development efforts. Many of these obstacles deal directly with the biological component of the system, the microbial biocatalyst, while others emanate from the complex and heterogeneous nature of the chemical feedstock, the heavy (unconventional) crude oil.

Although the large body of literature and accumulating knowledge on microbial biodegradation of crude oil in various polluted ecosystems shed light on analogous interventions for bioupgrading, the objective of the two processes can be fundamentally different in several cases. Bioupgrading of the crude does not require complete biodegradation, albeit targeted or specific bioconversions/biotransformations are more beneficial. However, this requires thorough and deep understanding of the bioupgrading mechanisms in terms of not only the involved biochemical pathways, but also the physiological and metabolic adaptations and regulatory cascades/networks elicited by the heavy crude substrate. Genome sequencing and systems biology investigations (proteomics, transcriptomics and metabolomics) can provide a holistic view and in-depth understanding of the asphaltenes biodegradation/biotransformation machinery and mechanisms, as well as the underlying physiological adaptations. These together with functional genomics via gene knockout and complementation studies can enable the identification of metabolic engineering hotspots that can be manipulated to tailor more efficient recombinant biocatalysts. Furthermore, considering the structural complexity of unconventional oils in general, and asphaltenes in particular, it is tempting to pay more attention to microbial consortia. These are known to be more efficient than axenic cultures, particularly with complex and challenging substrates like asphaltenes and other high-molecular weight fractions found in unconventional oils.

Biodesulfurization is an integral process in oil upgrading and deserves more in-depth investigations. However, it is important to pay more attention to the basic science to gain insights into the physiological and metabolic adaptations that take place under biodesulfurization conditions. Systems biology approaches can enable better understanding of the physiology and metabolism of fuel-biodesulfurizing bacteria. The ultimate goal of these studies is to identify metabolic bottlenecks that can be manipulated to overcome or bypass the natural limits of sulfur requirements of native biodesulfurizers, thus producing more catalytically efficient recombinants.

To date, nothing is known about asphaltenes degradation or transformation products, catabolic pathways, underlying genes, enzymes and regulation cascades, to name a few aspects. It is of utmost importance to have a deeper understanding of the structural changes that take place in the asphaltene fraction as a result of microbial attack. This is one of the prohibiting tasks in bioupgrading studies due to the difficulties associated with the detailed analysis of asphaltenes. Therefore, advances on the analytical toolbox for asphaltenes including NMR and high-resolution massspectrometric techniques can provide more realistic elucidation of the bioupgrading mechanisms and enable fine-tuning of the involved pathways.

It is obvious from the literature that many of the heavy crude bioupgrading studies suffer from experimental design issues due to the use of aspheltenes fraction precipitated from crude oil and the presence of additional carbon sources together with the oil or aspheltene substrate. Although the asphaltene fractions or model compounds can be used for the initial enrichment and isolation of microbial cultures, the presence of heavy crude oil as the sole carbon and energy source in the bioupgrading cultures is more conclusive. Moreover, the heave crude substrate can be used not only as the sole source of carbon, but also as the sole source of sulfur and/or nitrogen in minimal medium completely lacking these nutrients. This medium composition might enable the enrichment of microbial cultures capable of more efficient attack on the heavy components to access these essential nutrients, thus causing deeper structural alterations. In addition, it is not appropriate to depend on estimating the quantity of asphaltenes lost during biological treatments by gravimetry as the sole tool to assess the bioupgrading potential of microbial cultures. A combination of various analytical techniques will make the experimental design more robust and provide a comprehensive view of the physicochemical changes in the treated oil. In this context, proper control cultures must be included to assess abiotic alterations.

The long incubation time of the bioupgrading cultures, mostly in the range of 15–60 days, is one of the current limitations and is not commercially applicable. The toxicity and structural complexity of the substrate can delay the onset of the biocatalytic activity due to microorganisms that need a long time of acclimatization to such complex substrates (Gudiña and Teixeira, 2017). Furthermore, the heavy components of unconventional crude oils have very low solubility and mass transfer rates impede uptake across the cell membrane (León and Kumar, 2005; Gudiña and Teixeira, 2017). Therefore, the application of mixed cultures with relevant and complementary activities might overcome these problems. Solvent-tolerant and biosurfactants/bioemulsifiers-producing microorganisms can be beneficial in this matter. Pretreatment of high-molecular weight components of unconventional crude oils with biosurfactants and bioemulsifiers will enable better solubility and bioavailability via surface tension reduction, depolymerization and emulsification. Moreover, solvent-tolerant enzymes (native or engineered) are catalytically active in the presence of a minimal amount of water, and thus represent an advantage as compared to whole-cell biocatalysts.

References

Abbad-Andaloussi, S., Lagnel, C., Warzywoda, M., Monot, F., 2003. Multi-criteria comparison of resting cell activities of bacterial strains selected for biodesulfurization of petroleum compounds. Enzyme Microb. Technol. 32 (3–4), 446–454.

Abbasian, F., Lockington, L., Megharaja, M., Naidua, R., 2016. Identification of a new operon involved in the desulfurization of dibenzothiophenes using a metagenomic study and cloning and functional analysis of the genes. Enzyme Microb. Technol. 87–88, 24–28.

Abin-Fuentes, A., Mohamed, M.E.S., Wang, D., Prather, K., 2013. Exploring the mechanism of biocatalyst inhibition in microbial desulfurization. Appl. Environ. Microbiol. 79 (24), 7807–7817.

Aggarwal, S., Karimi, I., Ivan, G., 2013. In silico modeling and evaluation of *Gordonia alkanivorans* for biodesulfurization. Mol. Biosyst. 9 (10), 2530–2540.

Aggarwal, S., Karimi, I., Kilbane, J., Lee, D., 2012. Roles of sulfite oxidoreductase and sulfite reductase in improving desulfurization by *Rhodococcus erythropolis*. Mol. Biosyst. 8 (10), 2724–2732.

Akbarzadeh, K., Hammami, A., Kharrat, A., Zhang, D., Allenson, S., Creek, J., et al., 2007. Asphaltenes-problematic but rich in potential. Oilfield Rev. 19 (2), 22–43.

Alaei, M., Bazmi, M., Rashidi, A., Rahimi, A., 2017. Heavy crude oil upgrading using homogenous nanocatalyst. J. Petrol. Sci. Eng. 158, 47–55.

Al-Bahry, S.N., Al-Wahaibi, Y.M., Al-Hinai, B., Joshi, S.J., Elshafie, A.E., Al-Bemani, A.S., et al., 2016. Potential in heavy oil biodegradation via enrichment of spore forming bacterial consortia. J. Pet. Explor. Prod. Technol. 6 (4), 787–799.

Al-Mailem, D., Eliyas, M., Khanafer, M., Radwan, S., 2014. Culture-dependent and culture-independent analysis of hydrocarbonoclastic *microorganisms* indigenous to hypersaline environments in Kuwait. Microb. Ecol. 67 (4), 857–865.

Al-Sayegh, A., Al-Wahaibi, Y., Joshi, S., Al-Bahry, S., Elshafie, A., Al-Bemani, A., 2016. Bioremediation of heavy crude oil contamination. Open Biochem. J. 10 (1), 301–311.

Al-Sulaimani, H., Joshi, S., Al-Wahaibi, Y.M., Al-Bahry, S., Elshafie, A., Al-Bemani, A., 2011. Microbial biotechnology for enhancing oil recovery: current developments and future prospects. Invited Rev. Biotechnol. Bioinformatics Bioeng. 1 (2), 147–158.

Alves, L., Melo, M., Mendonça, D., Simóes, F., Matos, J., Tenreiro, R., et al., 2007. Sequencing, cloning and expression of the *dsz* genes required for dibenzothiophene sulfone desulfurization from *Gordonia alkanivorans* strain 1B. Enzyme Microb. Technol. 40 (6), 1598–1603.

Ancheyta, J., Trejo, F., Rana, M.S., 2009. Asphaltenes: Chemical Transformation during Hydroprocessing of Heavy Oils. CRC Press. Taylor and Francis Group, Boca Raton, https://doi.org/10.1201/9781420066319.

Anders, A., Kaetzke, A., Kaempfer, P., Ludwig, W., Fuchs, G., 1995. Taxonomic position of aromatic degrading denitrifying pseudomonad strains K172 and KB740 and their description as new members of the genera *Thauera*, *T. aromatica* sp. nov., and *Azoarcus*, *A. evansii* sp. nov., respectively, members of the β-subclass of proteobacteria. Int. J. Syst. Bacteriol. 45, 327–333.

Annweiler, E., Materna, A., Safinowski, M., Kappler, A., Richnow, H.H., Michaelis, W., et al., 2000. Anaerobic degradation of 2-methylnaphthalene by a sulfate-reducing enrichment culture. Appl. Environ. Microbiol. 66 (12), 5329–5333.

Assadi, M.M., Tabatabaee, M.S., 2010. Biosurfactants and their use in upgrading petroleum vacuum distillation residue. Int. J. Environ. Sci. 4 (4), 549–572.

Ayala, M., Vazquez-Duhalt, R., Morales, M., Le Borgne, S., 2017. Application of *microorganisms* to the processing and upgrading of crude oil and fractions. In: Lee, S.Y. (Ed.), Consequences of Microbial Interactions with Hydrocarbons, Oils, and Lipids: Production of Fuels and Chemicals. Handbook of Hydrocarbon and Lipid Microbiology, Springer, Cham, Switzerland, pp. 705–740.

Banat, I.M., Makkar, R.S., Cameotra, S.S., 2000. Potential commercial applications of microbial surfactants. Appl. Microbiol. Biotechnol. 53 (5), 495–508.

Banat, I.M., Satpute, S.K., Cameotra, S.S., Patil, R., Nyayanit, N.V., 2014. Cost effective technologies and renewable substrates for biosurfactants' production. Front. Microbiol. 5, 697. https://doi.org/10.3389/fmicb.2014.00697.

Bhatia, S., Sharma, D.K., 2010. Mining of genomic databases to identify novel biodesulfurizing *microorganisms*. J. Ind. Microbiol. Biotechnol. 37 (4), 425–429.

Biria, D., Maghsoudi, E., Roostaazad, R., 2013. Application of biosurfactants to wettability alteration and IFT reduction in enhanced oil recovery from oil-wet carbonates. Pet. Sci. Technol. 31 (12), 1259–1267.

Boll, M., 2005. Key enzymes in the anaerobic aromatic metabolism catalysing Birch-like reductions. Biochim. Biophys. Acta 1707, 34–50.

Boll, M., Estelmann, S., 2020. Catabolic pathways and enzymes involved in the anaerobic degradation of polycyclic aromatic hydrocarbons. In: Boll, M. (Ed.), Anaerobic Utilization of Hydrocarbons, Oils, and Lipids. Handbook of Hydrocarbon and Lipid Microbiology, Springer, Cham, Switzerland, pp. 135–150.

Boll, M., Estelmann, S., Heider, J., 2020a. Anaerobic degradation of hydrocarbons: mechanisms of hydrocarbon activation in the absence of oxygen. In: Boll, M. (Ed.), Anaerobic Utilization of Hydrocarbons, Oils, and Lipids. Handbook of Hydrocarbon and Lipid Microbiology, Springer, Cham, Switzerland, pp. 3–30.

Boll, M., Estelmann, S., Heider, J., 2020b. Catabolic pathways and enzymes involved in the anaerobic degradation of monocyclic aromatic compounds. In: Boll, M. (Ed.), Anaerobic Utilization of Hydrocarbons, Oils, and Lipids. Handbook of Hydrocarbon and Lipid Microbiology, Springer, Cham, Switzerland, pp. 85–134.

Boniek, D., Figueiredo, D., dos Santos, A.F.B., de Resende Stoianoff, M.A., 2014. Biodesulfurization: a mini review about the immediate search for the future technology. Clean Techn. Environ. Policy 17 (1), 29–37.

Bordoloi, N., Rai, S., Chaudhuri, M., Mukherjee, A., 2014. Deep-desulfurization of dibenzothiophene and its derivatives present in diesel oil by a newly isolated bacterium *Achromobacter* sp. to reduce the environmental pollution from fossil fuel combustion. Fuel Process. Technol. 119, 236–244.

BPE Outlook, 2019. BP Statistical Review of World Energy: Total Proved Reserves, sixty eighth annual ed.

Britton, L.N., 1984. Microbial degradation of aliphatic hydrocarbons. In: Microbial Degradation of Organic Compounds. Marcel Dekker, New York, pp. 89–129.

Brown, L.R., Vadie, A.A., Stephens, J.O., 2000. Slowing production decline and extending the economic life of an oil field: new MEOR Technology. In: Society of Petroleum Engineers, SPE/DOE Improved Oil Recovery Symposium. Tulsa, Oklahoma., https://doi.org/10.2118/59306-MS.

Bugg, T.D., Winfield, C.J., 1998. Enzymatic cleavage of aromatic rings: mechanistic aspects of the catechol dioxygenases and the later enzymes of the bacterial oxidative cleavage pathways. Nat. Prod. Rep. 15, 513–530.

Butler, C.S., Mason, J.R., 1997. Structure-function analysis of the bacterial ring-hydroxylating dioxygenases. Adv. Microb. Physiol. 38, 47–84.

Cain, R.P., 1988. Aromatic metabolism by mycelial organisms: actinomycete and fungal strategies. In: Hagedorn, S.R., Hanson, R.S., Kunz, D.A. (Eds.), Microbial Metabolism and the Carbon Cycle. Harwood Academic Publishers, New York, pp. 101–144.

Caro, A., Boltes, K., Leton, P., García-Calvo, E., 2007. Dibenzothiophene biodesulfurization in resting cell conditions by aerobic bacteria. Biochem. Eng. J. 35 (2), 191–197.

Castorena-Cortés, G., Roldán-Carrillo, T., Reyes-Avila, J., Zapata-Peñasco, I., Mayol-Castillo, M., Olguín-Lora, P., 2012. Coreflood assay using extremophile *microorganisms* for recovery of heavy oil in Mexican oil fields. J. Biosci. Bioeng. 114 (4), 440–445.

Castro, L.V., Vázquez, F., 2009. Fractionation and characterization of Mexican crude oils. Energy Fuel 23 (3), 1603–1609.

Chaillan, F., Le Flèche, A., Bury, E., Phantavong, Y., Grimont, P., Saliot, A., Oudot, J., 2004. Identification and biodegradation potential of tropical aerobic hydrocarbon-degrading *microorganisms*. Res. Microbiol. 155 (7), 587–595.

Chang, W., Um, Y., Holoman, T.R., 2006. Polycyclic aromatic hydrocarbon (PAH) degradation coupled to methanogenesis. Biotechnol. Lett. 28 (6), 425–430.

Chen, H., Zhang, W., Chen, J., Cai, Y., Li, W., 2008. Desulfurization of various organic sulfur compounds and the mixture of DBT+4,6-DMDBT by *Mycobacterium* sp. ZD-19. Bioresour. Technol. 99, 3630–3634.

Chen, S., Zhao, C., Liu, Q., Zang, M., Liu, C., Zhang, Y., 2018. Thermophilic biodesulfurization and its application in oil desulfurization. Appl. Microbiol. Biotechnol. 102 (21), 9089–9103.

Choi, K.-H., Al-majnoiuni, K.A., Al-shareef, A., 2018. Process to upgrade highly waxy crude oil by hot pressurized water. Patent No.: US 10,010,839 B2.

Christofi, N., Ivshina, I.B., 2002. Microbial surfactants and their use in field studies of soil remediation. J. Appl. Microbiol. 93 (6), 915–929.

Chronopoulou, P.-M., Sanni, G.O., Silas-Olu, D.I., van der Meer, J.R., Timmis, K.N., Brussaard, C.P.D., et al., 2014. Generalist hydrocarbon-degrading bacterial communities in the oil-polluted water column of the North Sea. J. Microbial. Biotechnol. 8 (3), 434–447.

Clemente, J., 2015. How Much Oil Does the World Have Left? Forbes. Retrieved from https://www.forbes.com/sites/judeclemente/2015/06/25/how-much-oil-does-the-world-have-left/2/. (Accessed 5 December 2017).

Constanti, M., Giralt, J., Bordons, A., 1996. Degradation and desulfurization of dibenzothiophene sulfone and other sulfur compounds by *Agrobacterium* MC501 and mixed culture. Enzyme Microb. Technol. 19 (3), 214–219.

Das, N., Chandran, P., 2011. Microbial degradation of petroleum hydrocarbon contaminants: an overview. Biotechnol. Res. Int. 2011, 941810.

De Sena, M.F., Rosa, L.P., Szklo, A., 2013. Will Venezuelan extra-heavy oil be a significant source of petroleum in the next decades? Energy Policy 61, 51–59.

Dehghani, A., Sattarin, M., Bridjanian, H., Mohamadbeigy, K.H., 2009. Investigation on effectiveness parameters in residue upgrading methods. Pet. Coal 51 (4), 229–236.

Denis-Larose, C., Labbe, D., Bergeron, H., Jones, A., Greer, C., Al-Hawari, J., et al., 1997. Conservation of plasmid–encoded dibenzothiophene desulfurization genes in several *Rhodococci*. Appl. Environ. Microbiol. 63 (7), 2915–2919.

Denome, S.A., Oldfield, C., Nash, L.J., Young, K.D., 1994. Characterization of the desulfurization genes from *Rhodococcus* sp. strain IGTS8. J. Bacteriol. 176 (21), 6707–6716.

Denome, S.A., Olson, E.S., Young, K.D., 1993. Identification and cloning of genes involved in specific desulfurization of dibenzothiophene by *Rhodococcus* sp. strain IGTS8. Appl. Environ. Microbiol. 59 (9), 2837–2843.

Desando, M., Ripmeester, J., 2002. Chemical derivatization of Athabasca oil sand asphaltene for analysis of hydroxyl and carboxyl groups via nuclear magnetic resonance spectroscopy. Fuel 81 (10), 1305–1319.

Doman, L., 2016. International Energy Outlook 2016, DOE/EIA-0484. U. S. Energy Information Administration, Washington, DC, p. 20585.

Dorado-Morales, P., Martínez, I., Rivero-Buceta, V., Díaz, E., Bähre, H., Lasa, I., Solano, C., 2020. Elevated c-di-GMP levels promote biofilm formation and biodesulfurization capacity of *Rhodococcus erythropolis*. Microb. Biotechnol. 14 (3), 923–937.

Duarte, G.F., Rosado, A.S., Seldin, L., de Araujo, W., van Elsas, J.D., 2001. Analysis of bacterial community structure in sulfurous-oil-containing soils and detection of species carrying dibenzothiophene desulfurization (*dsz*) genes. Appl. Environ. Microbiol. 67 (3), 1052–1062.

Elraies, K.A., Tan, I.M., 2012. The application of a new polymeric surfactant for chemical EOR. In: Romero-Zerón, L. (Ed.), Introduction to Enhanced Oil Recovery (EOR), Processes and Bioremediation of Oil-contaminated Sites. InTech Europe, Rijeka, Croatia, pp. 45–70.

Elshafie, A.E., Joshi, S.J., Al-Wahaibi, Y.M., Al-Bemani, A.S., Al-Bahry, S.N., Al-Maqbali, D., et al., 2015. Sophorolipids production by *Candida bombicola* ATCC 22214 and its potential application in microbial enhanced oil recovery. Front. Microbiol. 6, 1324.

Ettwig, K.F., Butler, M.K., Le Paslier, D., Pelletier, E., Mangenot, S., Kuypers, M.M., et al., 2010. Nitrite-driven anaerobic methane oxidation by oxygenic bacteria. Nature 464 (7288), 543–548.

Evans, W.C., Fuchs, G., 1988. Anaerobic degradation of aromatic compounds. Annu. Rev. Microbiol. 42, 289–317.

Evdokimov, I.N., 2005. Bifurcated correlations of the properties of crude oils with their asphaltene content. Fuel 84 (1), 13–28.

Ferrández, A., Miñambres, B., García, B., Olivera, E.R., Luengo, J.M., García, J.L., et al., 1998. Catabolism of phenylacetic acid in *Escherichia coli*. Characterization of a new aerobic hybrid pathway. J. Biol. Chem. 273 (40), 25974–25986.

Ferreira, P., Sousa, S., Fernandes, P., Ramos, M., 2017. Improving the catalytic power of the DszD enzyme for the biodesulfurization of crude oil and derivatives. Chem. A Eur. J. 23 (68), 17231–17241.

Fracchia, L., Ceresa, C., Franzetti, A., Cavallo, M., Gandolfi, I., Van Hamme, J., et al., 2015. Industrial applications of biosurfactants. In: Kosaric, N., Vardar-Sukan, F. (Eds.), Biosurfactants: Production and Utilization-Processes, Technologies, and Economics. CRC Press, New York, pp. 245–268.

Fuchs, G., 1999. Diversity of metabolic pathways. In: Lengler, J.V., Drews, G., H.G. Schlegel HG. (Eds.), Biology of the Prokaryotes. Thieme, Stuttgart, pp. 215–221.

Fuchs, G., 2008. Anaerobic metabolism of aromatic compounds. Ann. N. Y. Acad. Sci. 11 (25), 82–99.

Fuchs, G., Boll, M., Heider, J., 2011. Microbial degradation of aromatic compounds-from one strategy to four. Nat. Rev. Microbiol. 9 (11), 803–816.

Fuchs, G., Mohammed, M., Altenschmidt, U., Koch, J., Lack, A., Brackmann, R., et al., 1994. Biochemistry of anaerobic biodegradation of aromatic compounds. In: Ratledge, C. (Ed.), Biochemistry of Microbial Degradation. Kluwer Academic Publishers, Dordrechts, pp. 513–553.

Fuentes, S., Méndez, V., Aguila, P., Seeger, M., 2014. Bioremediation of petroleum hydrocarbons: catabolic genes, microbial communities, and applications. Appl. Microbiol. Biotechnol. 98 (11), 4781–4794.

Gallagher, J.R., Olson, E.S., Stanley, D.C., 1993. Microbial desulfurization of dibenzothiophene: a sulfur-specific pathway. FEMS Microbiol. Lett. 107 (1), 31–36.

Gao, H., Zhang, J., Lai, H., Xue, Q., 2017. Degradation of asphaltenes by two *Pseudomonas aeruginosa* strains and their effects on physicochemical properties of crude oil. Int. Biodeter. Biodegr. 122, 12–22.

García-Arellano, H., Buenrostro-Gonzalez, E., Vazquez-Duhalt, R., 2004. Biocatalytic transformation of petroporphyrins by chemical modified cytochrome C. Biotechnol. Bioeng. 85 (7), 790–798.

Gibson, J., Harwood, C.S., 2002. Metabolic diversity in aromatic compound utilization by anaerobic microbes. Annu. Rev. Microbiol. 56, 345–369.

Gieg, L.M., Fowler, S.J., Berdugo-Clavijo, C., 2014. Syntrophic biodegradation of hydrocarbon contaminants. Curr. Opin. Biotechnol. 27, 21–29.

Gray, K.A., Pogrebinsky, O.S., Mrachko, G.T., Xi, L., Monticello, D.J., Squires, C.H., 1996. Molecular mechanisms of biocatalytic desulfurization of fossil fuels. Nat. Biotechnol. 14 (13), 1705–1709.

Grundmann, O., Behrends, A., Rabus, R., Amann, J., Halder, T., Heider, J., et al., 2008. Genes encoding the candidate enzyme for anaerobic activation of *n*-alkanes in the denitrifying bacterium, strain HxN1. Environ. Microbiol. 10 (2), 376–385.

Gudiña, E.J., Teixeira, J.A., 2017. Biological treatments to improve the quality of heavy crude oils. In: Heimann, K., Karthikeyan, O.P., Muthu, S.S. (Eds.), Biodegradation and Bioconversion of Hydrocarbons. Springer Science + Business Media, Singapore, pp. 337–351.

Gunam, I., Iqbal, M., Arnata, I., Antara, N., Anggreni, A., Setiyo, Y., et al., 2017. Biodesulfurization of dibenzothiophene by a newly isolated *Agrobacterium tumefaciens* LSU20. Appl. Mech. Mater. 885, 143–149.

Gupta, N., Roychoudhury, P., Deb, J., 2005. Biotechnology of desulfurization of diesel: prospects and challenges. Appl. Microbiol. Biotechnol. 66, 356–366.

Hao, R., Lu, A., 2009. Biodegradation of heavy oils by halophilic bacterium. Prog. Nat. Sci. 19 (8), 997–1001.

Hao, R., Lu, A., Zeng, Y., 2004. Effect on crude oil by thermophilic bacterium. J. Petrol. Sci. Eng. 43, 247–258.

Harayama, S., Kok, M., Neidle, E.L., 1992. Functional and evolutionary relationships among diverse oxygenases. Annu. Rev. Microbiol. 446, 565–601.

Harner, N.K., Richardson, T.L., Thompson, K.A., Best, R.J., Best, A.S., Trevors, J.T., 2011. Microbial processes in the Athabasca oil sands and their potential applications in microbial enhanced oil recovery. J. Ind. Microbiol. Biotechnol. 38 (11), 1761–1775.

Harwood, C.S., Parales, R.E., 1996. The β-ketoadipate pathway and the biology of self identity. Annu. Rev. Microbiol. 50, 553–590.

Hausmann, R., Syldatk, C., 2015. Types and classification of microbial surfactants. In: Kosaric, N., Vardar-Sukan, F. (Eds.), Biosurfactants: Production and Utilization-Processes, Technologies, and Economics. CRC Press, New York, pp. 3–18.

Hazen, T.C., Prince, R.C., Mahmoudi, N., 2016. Marine oil biodegradation. Environ. Sci. Technol. 50 (5), 2121–2129.

He, L., Lin, F., Li, X., Sui, H., Xu, Z., 2015. Interfacial sciences in unconventional petroleum production: from fundamentals to applications. Chem. Soc. Rev. 44 (15), 5446–5494.

Heider, J., Fuchs, G., 1997. Microbial anaerobic aromatic metabolism. Anaerobe 3, 1–22.

Heider, J., Szaleniec, M., Sünwoldt, K., Boll, M., 2016. Ethylbenzene dehydrogenase and related molybdenum enzymes involved in oxygen-independent alkyl chain hydroxylation. J. Mol. Microbiol. Biotechnol. 26 (1–3), 45–62.

Heinberg, R., 2005. The Party's Over: Oil, War and the Fate of Industrial Societies. https://archive.org/stream/fe_The_Partys_Over-Oil_War_and_the_Fate_of_Industrial_Societies/The_Partys_Over-Oil_War_and_the_Fate_of_Industrial_Societies_djvu.txt. (Accessed 19 December 2019).

Hernández-López, E.L., Ayala, M., Vazquez-Duhalt, R., 2015b. Microbial and enzymatic biotransformations of asphaltenes. J. Pet. Sci. Technol. 33 (9), 1017–1029.

Hernández-López, E.L., Ramírez-Puebla, S.T., Vazquez-Duhalt, R., 2015a. Microarray analysis of *Neosartorya fischeri* using different carbon sources, petroleum asphaltenes and glucose-peptone. Genom. Data 5, 235–237.

Hirschler, A., Carapito, C., Maurer, L., Zumsteg, J., Villette, C., Heintz, D., et al., 2021. Biodesulfurization induces reprogramming of sulfur metabolism in *Rhodococcus qingshengii* IGTS8: proteomics and untargeted metabolomics. Microbiol. Spectr., e0069221.

Höök, M., Hirsch, R., Aleklett, K., 2009. Giant oil field decline rates and their influence on world oil production. Energy Policy 37 (6), 2262–2272.

Ismail, W., Gescher, J., 2012. Epoxy-coenzyme A thioester pathways for degradation of aromatic compounds. Appl. Environ. Microbiol. 78 (15), 5043–5051.

Ismail, W., Mohamed, M.E.-S., Awadh, M.N., Obuekwe, C., El Nayal, A.M., 2017. Simultaneous valorization and biocatalytic upgrading of heavy vacuum gas oil by the biosurfactant-producing *Pseudomonas aeruginosa* AK6U. J. Microbial. Biotechnol. 10 (6), 1628–1639.

Ismail, W., Mohamed, M.E.-S., Wanner, B.L., Datsenko, K.A., Eisenreich, W., Rhodich, F., et al., 2003. Functional genomics by NMR spectroscopy: phenylacetate catabolism in *Escherichia coli*. Eur. J. Biochem. 270, 3047–3054.

Jahromi, H., Fazaelipoor, M.A., Ayatollahi, S., Niazi, A., 2014. Asphaltenes biodegradation under shaking and static conditions. Fuel 117, 230–235.

Jarling, R., Sadeghi, M., Drozdowska, M., Lahme, S., Buckel, W., Rabus, R., et al., 2012. Stereochemical investigations reveal the mechanism of the bacterial activation of *n*-alkanes without oxygen. Angew. Chem. Int. Ed. Engl. 51 (6), 1334–1338.

Javadli, R., de Klerk, A., 2012. Desulfurization of heavy oil. Appl. Petrochem. Res. 1, 3–19.

Jiang, X., Yang, S., Li, W., 2014. Biodesulfurization of model compounds and de–asphalted bunker oil by mixed culture. Appl. Biochem. Biotechnol. 172 (1), 62–72.

Joshi, S.J., Al-Wahaibi, Y., Al-Bahry, S., 2019. Biotransformation of heavy crude oil and biodegradation of oil pollution by arid zone bacterial strains. In: Arora, P.K. (Ed.), Microbial Metabolism of Xenobiotic Compounds. Springer, Singapore, pp. 103–122.

Kaiya, S., Rubaba, O., Yoshida, N., Yamada, T., Hiraishi, A., 2012. Characterization of *Rhizobium naphthalenivorans* sp. nov. with special emphasis on aromatic compound degradation and multilocus sequence analysis of housekeeping genes. J. Gen. Appl. Microbiol. 58 (3), 211–224.

Kang, J., Myint, A.A., Sim, S., Kim, J., Kong, W.B., Lee, Y.-W., 2018. Kinetics of the upgrading of heavy oil in supercritical methanol. J. Supercrit. Fluids 133 (1), 133–138.

Kaufman, E.N., Harkins, J.B., Borole, A.P., 1998. Comparison of batch stirred and electrospray reactors for biodesulfurization of dibenzothiophene in crude oil and hydrocarbon feedstocks. Appl. Biochem. Biotechnol. 73, 127–144.

Kertesz, M.A., 1999. Riding the sulfur cycle-metabolism of sulfonates and sulfate esters in Gram-negative bacteria. FEMS Microbiol. Rev. 24 (2), 135–175.

Khire, M., 2010. Bacterial biosurfactants and their role in microbial enhanced oil recovery (MEOR). In: Sen, R. (Ed.), Biosurfactants. Advances in Experimental Medicine and Biology. Springer, New York, pp. 146–157.

Kilbane, J., Stark, B., 2016. Biodesulfurization: a model system for microbial physiology research. World J. Microbiol. Biotechnol. 32 (137), 1–9.

Kilbane, J.J., 1990. Sulfur-specific microbial metabolism of organic compounds. Resour. Conserv. Recycl. 3 (2–3), 69–79.

Kilbane, J.J., 2006. Microbial biocatalyst developments to upgrade fossil fuels. Curr. Opin. Biotechnol. 17 (3), 305–314.

Kilbane, J.J., Jackowski, K., 1992. Biodesulfurization of water-soluble coal-derived material by *Rhodococcus rhodochrous* IGTS8. Biotechnol. Bioeng. 40 (9), 1107–1114.

Kirimura, K., Furuya, T., Nishii, Y., Ishii, Y., Kino, K., Usami, S., 2001. Biodesulfurization of dibenzothiophene and its derivatives through the selective cleavage of carbon-sulfur bonds by a moderately thermophilic bacterium *Bacillus subtilis* WU-S2B. J. Biosci. Bioeng. 91 (3), 262–266.

Kirkwood, K.M., Foght, J.M., Gray, M.R., 2004. Prospects for biological upgrading of heavy oils and asphaltenes. In: Vazquez-Duhalt, R., Quintero-Ramirez, R. (Eds.), Petroleum Biotechnology, Developments and Perspectives. Studies in Surface Science and Catalysis, vol 151. Elsevier, Amsterdam, pp. 113–144.

Kleemann, R., Meckenstock, R.U., 2011. Anaerobic naphthalene degradation by Gram-positive, iron-reducing bacteria. FEMS Microbiol. Ecol. 78 (3), 488–496.

Kok, M.V., Gul, K.G., 2013. Thermal characteristics and kinetics of crude oils and SARA fractions. Thermochim. Acta 569, 66–70.

Kotlar, H.K., 2012. Extreme to the fourth power! Oil-, high temperature-, salt- and pressure-tolerant *microorganisms* in oil reservoirs. What secrets can they reveal? In: Anitori, R.P. (Ed.), Extremophiles-Microbiology and Biotechnology. Caister Academic Press, Norfolk, pp. 159–182.

Kropp, K.G., Davidova, I.A., Suflita, J.M., 2000. Anaerobic oxidation of *n*-dodecane by an addition reaction in a sulfate-reducing bacterial enrichment culture. Appl. Environ. Microbiol. 66 (12), 5393–5398.

Kropp, K.G., Fedorak, P.M., 1998. A review of the occurrence, toxicity, and biodegradation of condensed thiophenes found in petroleum. Can. J. Microbiol. 44 (7), 605–622.

Lacotte, D., Mille, G., Acquaviva, M., Berttand, J., 1996. Arabian light 150 asphaltene biotransformation with *n*-alkanes as Co-substrates. Chemosphere 32 (9), 1755–1761.

Laso-Pérez, R., Wegener, G., Knittel, K., Widdel, F., Harding, K.J., Krukenberg, V., et al., 2016. Thermophilic archaea activate butane via alkyl-coenzyme M formation. Nature 539 (7629), 396–401.

Lavania, M., Cheema, S., Lal, B., 2015. Potential of viscosity reducing thermophilic anaerobic bacterial consortium TERIB#90 in upgrading heavy oil. Fuel 144, 349–357.

Lavania, M., Cheema, S., Sarma, P.M., Mandal, A.K., Lal, B., 2012. Biodegradation of asphalt by *Garciaella petrolearia* TERIG02 for viscosity reduction of heavy oil. Biodegradation 23 (1), 15–24.

Le Borgne, S., Quintero, R., 2003. Biotechnological processes for the refining of petroleum. Fuel Process. Technol. 81 (2), 155–169.

Lee, W.C., Ohshiro, T., Matsubara, T., Izumi, Y., Tanokura, M., 2006. Crystal structure and desulfurization mechanism of 2-hydroxybiphenyl-2-sulfinic acid desulfinase. J. Biol. Chem. 281 (43), 32534–32539.

Lei, B., Tu, S., 1996. Gene overexpression, purification, and identification of a desulfurization enzyme from *Rhodococcus* sp. strain IGTS8 as a sulfide/sulfoxide monooxygenase. J. Bacteriol. 178 (19), 5699–5705.

León, V., Cordova, J., Munoz, S., De Sisto, A., Naranjo, L., 2007. Process for the Upgrading Of Heavy Crude Oil, Extra-Heavy Crude Oil or Bitumens Through the Addition of a Biocatalyst. United States patent no US2007/0231870 Kind code: A1.

León, V., Kumar, M., 2005. Biological upgrading of heavy crude oil. Biotechnol. Bioprocess Eng. 10, 471–481.

Linguist, L.K., Pacheco, M.A., 1999. Enzyme-based diesel desulfurization process offers energy, CO_2 advantages. Oil Gas J. 97 (8), 45–48.

Loffler, C., Kuntze, K., Vazquez, J.R., Rugor, A., Kung, J.W., Böttcher, A., et al., 2011. Occurrence, genes and expression of the W/Se-containing class II benzoyl-coenzyme A reductases in anaerobic bacteria. Environ. Microbiol. 13 (3), 696–709.

Lovley, D.R., Giovannoni, S.J., White, D.C., Champine, J.E., Phillips, E.J.P., Gorby, Y.A., et al., 1993. *Geobacter metallireducens* gen. nov. sp. nov., a microorganism capable of coupling the complete oxidation of organic compounds to the reduction of iron and other metals. Arch. Microbiol. 159, 336–344.

Luo, Y.R., 2003. Handbook of Bond Dissociation Energies in Organic Compounds, first ed. CRC Press, Boca Raton.

Madden, P.B., Morawski, J.D., 2011. The future of the Canadian oil sands: engineering and project management advances. Energy Environ. 22 (5), 579–596.

Martín-Cabello, G., Terrón-González, L., Ferrer, M., Santero, E., 2020. Identification of a complete dibenzothiophene biodesulfurization operon and its regulator by functional metagenomics. Environ. Microbiol. 22 (1), 91–106.

Martínez, I., Mohamed, M.E.-S., Santos, V.E., García, J.L., García-Ochoac, F., Díaz, E., 2017. Metabolic and process engineering for biodesulfurization in Gram-negative bacteria. J. Biotechnol. 262, 47–55.

Mason, J.R., Cammack, R., 1992. The electron transport proteins of hydroxylating bacterial dioxygenases. Annu. Rev. Microbiol. 46, 277–305.

Mason, O., Scott, N., Gonzalez, A., Robbins-Pianka, A., Baelum, J., Kimbrel, J., et al., 2014. Metagenomics reveals sediment microbial community response to Deepwater Horizon oil spill. ISME J. 8 (7), 1464–1475.

Mawad, A.M.M., Hassanein, M., Aldaby, E.S., Yousef, N., 2021. Desulphurisation kinetics of thiophenic compound by sulphur oxidizing *Klebsiella oxytoca* SOB-1. J. Appl. Microbiol. 130 (4), 1181–1191.

Mehboob, F., Junca, H., Schraa, G., Stams, A.J., 2009. Growth of *Pseudomonas chloritidismutans* AW-1(T) on *n*-alkanes with chlorate as electron acceptor. Appl. Microbiol. Biotechnol. 83 (4), 739–747.

Milkov, A.V., 2018. Secondary microbial gas. In: Wilkes, H. (Ed.), Hydrocarbons, Oils, and Lipids: Diversity, Origin, Chemistry and Fate. Handbook of Hydrocarbon and Lipid Microbiology, Springer, Cham, pp. 613–622.

Mirchi, A., Hadian, S., Madani, K., Rouhani, O.M., Rouhani, A.M., 2012. World energy balance outlook and OPEC production capacity: implications for global oil security. Energies 5 (6), 2626–2651.

Mohamed, M.E.-S., Al-Yacoub, Z., Vedakumar, J., 2015. Biocatalytic desulfurization of thiophenic compounds and crude oil by newly isolated bacteria. Front. Microbiol. 6, 112.

Mohamed, M.E.-S., Ismail, W., Heider, J., Fuchs, G., 2002. Aerobic metabolism of phenylacetic acids in *Azoarcus evansii*. Arch. Microbiol. 178, 180–192.

Mohapatra, B., Sarkar, A., Joshi, S., Chatterjee, A., Kazy, S.K., Maiti, M.K., et al., 2017. An arsenate-reducing and alkane-metabolizing novel bacterium, *Rhizobium arsenicireducens* sp. nov., isolated from arsenic-rich groundwater. Arch. Microbiol. 199 (2), 191–201.

Mohebali, G., Ball, A.S., 2016. Biodesulfurization of diesel fuels—past, present and future perspectives. Int. Biodeter. Biodegr. 110, 163–180.

Monticello, D.J., 2000. Biodesulfurization and the upgrading of petroleum distillates. Curr. Opin. Biotechnol. 11 (6), 540–546.

Monticello, D.J., Finnerty, W.R., 1985. Microbial desulfurization of fossil fuels. Annu. Rev. Microbiol. 39 (1), 371–389.

Montiel, C., Quintero, R., Aburto, J., 2009. Petroleum biotechnology: technology trends for the future. Afr. J. Biotechnol. 8 (25), 2653–2666.

Morales, M., Ayala, M., Vazquez-Duhalt, R., Le Borgne, S., 2010. Application of *microorganisms* to the processing and upgrading of crude oil and fractions. In: Timmis, K.N. (Ed.), Handbook of Hydrocarbon and Lipid Microbiology. Springer-Verlag, Berlin, Heidelberg, pp. 2767–2785, https://doi.org/10.1007/978-3-540-77587-4_205.

Moreno, R., Rojo, F., 2019a. Enzymes for aerobic degradation of alkanes in bacteria. In: Rojo, F. (Ed.), Aerobic Utilization of Hydrocarbons, Oils, and Lipids. Handbook of Hydrocarbon and Lipid Microbiology, Springer, Cham, pp. 117–142.

Moreno, R., Rojo, F., 2019b. Genetic features and regulation of *n*-alkane metabolism in bacteria. In: Rojo, F. (Ed.), Aerobic Utilization of Hydrocarbons, Oils, and Lipids. Handbook of Hydrocarbon and Lipid Microbiology, Springer, Cham, pp. 521–542.

Muhammad, I., Tijjanii, N., Dioha, I.J., Musa, A., Sale, H., Lawal, A., 2013. SARA separation and determination of concentration levels of some heavy metals in organic fractions of Nigerian crude oil. Chem. Mater. Res. 3 (4), 7–14.

Mulligan, C.N., 2005. Environmental applications for biosurfactants. Environ. Pollut. 133 (2), 183–198.

Mullins, O.C., Sheu, E.Y., Hammami, A., Marshall, A.G. (Eds.), 2007. Asphaltenes, Heavy Oils, and Petroleomics. Springer-Verlag, USA, New York, https://doi.org/10.1007/0-387-68903-6.

Naranjo, L., Urbina, H., De Sisto, A., León, V., 2007. Isolation of autochthonous non-white rot fungi with potential for enzymatic upgrading of Venezuelan extra-heavy crude oil. Biocatal. Biotransformation 25 (2–4), 341–349.

Naranjo-Briceño, L., Pernía, B., Guerra, M., Demey, J.R., DeSisto, A., Inojosa, Y., et al., 2013. Potential role of oxidative exoenzymes of the extremophilic fungus *Pestalotiopsis palmarum* BM-04 in biotransformation of extra-heavy crude oil. J. Microbial. Biotechnol. 6 (6), 720–730.

Natural Gas Intelligence, 2019. Canadian Oil Sands Facts and Information. https://www.naturalgasintel.com/canadian-oil-sands. (Accessed 2 February 2020).

Ni'matuzahroh, N.R., Silvia, R.A., Nurhariyati, T., Surtiningsih, T., 2015. Effectiveness in enhancing oil recovery through combination of biosurfactant and lipases bacteria. J. Appl. Environ. Biol. Sci. 5 (6), 83–87.

Offshore Technology, 2019. Safaniya Field Upgrade. Persian Gulf. https://www.offshore-technology.com/projects/safaniya-upgrade-persian-gulf/. (Accessed 2 February 2020).

Ohshiro, T., Izumi, Y., 1999. Microbial desulfurization of organic sulfur compounds in petroleum. Biosci. Biotechnol. Biochem. 63 (1), 1–9.

Olivera, E.R., Miñambres, B., García, B., Muñiz, C., Moreno, M.A., Ferrández, A., et al., 1998. Molecular characterization of the phenylacetic acid catabolic pathway in *Pseudomonas putida* U: the phenylacetyl-CoA catabolon. Proc. Natl. Acad. Sci. U. S. A. 95 (11), 6419–6424.

Omajali, J.B., Hart, A., Walker, M., Wood, J., Macaskie, L.E., 2017. In-situ catalytic upgrading of heavy oil using dispersed bionanoparticles supported on gram-positive and gram-negative bacteria. Appl. Catal. Environ. 203, 807–819.

OPEC, 2017. http://www.opec.org/opec_web/flipbook/MOMRNovember2017/MOMRNovember2017.html#1/z. (Accessed 4 December 2017).

Oudot, J., Dupont, J., Haloui, S., Roquebert, M.F., 1993. Biodegradation potential of hydrocarbon-assimilating tropical fungi. Soil Biol. Biochem. 25 (9), 1167–1173.

Oxford Business Groups, 2017. Kuwaiti Government Focusing Investment in Heavy Oil. https://oxfordbusinessgroup.com/analysis/heavy-duty-government-increasingly-focusing-investment-heavy-oil. (Accessed 19 December 2017).

Pacheco, M.A., Lange, E.A., Pienkos, P.T., Yu, L.Q., Rouse, M.P., Lin, Q., et al., 1999. Recent advances in biodesulfurization of diesel fuel. In: NPRA AM-99-27, 1999, National Petrochemical and Refiners Association, Annual Meeting, 21–23 March, San Antonio, Texas, pp. 1–26.

Pan, Y., Liao, Y., Zheng, Y., 2015. Effect of biodegradation on the molecular composition and structure of asphaltenes: clues from quantitative Py-GC and THM-GC. Org. Geochem. 86, 32–44.

Parveen, S., Akhtar, N., Ghauri, M.A., Akhtar, K., 2020. Conventional genetic manipulation of desulfurizing bacteria and prospects of using CRISPR-Cas systems for enhanced desulfurization activity. Crit. Rev. Microbiol. 46 (3), 300–320.

Patel, J., Borgohain, S., Kumar, M., Rangarajan, V., Somasundaran, P., Sen, R., 2015. Recent developments in microbial enhanced oil recovery. Renew. Sustain. Energy Rev. 52, 1539–1558.

Pérez-Pantoja, D., De la Iglesia, R., Pieper, D.H., González, B., 2008. Metabolic reconstruction of aromatic compounds degradation from the genome of the amazing pollutant-degrading bacterium *Cupriavidus necator* JMP134. FEMS Microbiol. Rev. 32 (5), 736–794.

Pérez-Pantoja, D., González, B., Pieper, D.H., 2019. Aerobic degradation of aromatic compounds. In: Rojo, F. (Ed.), Aerobic Utilization of Hydrocarbons, Oils, and Lipids. Handbook of Hydrocarbon and Lipid Microbiology, Springer, Cham, pp. 157–200.

Perfumo, A., Rancich, I., Banat, I.M., 2010. Possibilities and challenges for biosurfactants use in petroleum industry. In: Sen, R. (Ed.), Biosurfactants. Springer, New York, pp. 135–145.

Pineda-Flores, G., Boll-Arguello, G., Lira-Galeana, C., Mesta-howard, A.M., 2004. A microbial consortium isolated from a crude oil sample that uses asphaltenes as a carbon and energy source. Biodegradation 15 (3), 145–151.

Pineda-Flores, G., Mesta-Howard, A.M., 2001. Petroleum asphaltenes: generated problematic and possible biodegradation mechanisms. Rev. Latinoam. Microbiol. 43 (3), 143–150.

Porto, B., Maass, D., Oliveira, J.V., de Oliveira, D., Yamamoto, C.I., Ulson de Souza, A.A., et al., 2018. Heavy gas oil biodesulfurization using a low-cost bacterial consortium. J. Chem. Technol. Biotechnol. 93 (8), 2359–2363.

Pourfakhraei, E., Badraghi, J., Mamashli, F., Nazari, M., Saboury, A.A., 2018. Biodegradation of asphaltene and petroleum compounds by a highly potent *Daedaleopsis* sp. J. Basic Microbiol. 58 (7), 609–622.

Premuzic, E.T., Lin, M.S., Bohenek, M., Zhou, W.M., 1999. Bioconversion reactions in asphaltenes and heavy crude oils. Energy Fuel 13 (2), 297–304.

Rahalkar, S.B., Joshi, S.R., Shivaraman, N., 1993. Photometabolism of aromatic compounds by *Rhodopseudomonas palustris*. Curr. Microbiol. 26, 1–9.

Rana, M.S., Sámano, V., Ancheyta, J., Diaz, J., 2007. A review of recent advances on process technologies for upgrading of heavy oils and residua. Fuel 86 (9), 1216–1231.

Riazi, M.R., Eser, S., 2013. Properties, specifications, and quality of crude oil and petroleum products. In: ASTM Manual Series MNL. vol. 58, pp. 79–100.

Ribbons, D.W., Eaton, R.W., 1982. Chemical transformation of aromatic hydrocarbons that support the growth of *microorganisms*. In: Chakrabarty, A.M. (Ed.), Biodegradation and Detoxification of Environmental Pollutants. CRC Press Inc., Boca Raton, Florida.

Rockne, K.J., Chee-Sanford, J.C., Sanford, R.A., Hedlund, B.P., Staley, J.T., Strand, S.E., 2000. Anaerobic naphthalene degradation by microbial pure cultures under nitrate-reducing conditions. Appl. Environ. Microbiol. 66 (4), 1595–1601.

Rojas-Avelizapa, N.G., Cervantes-Gonzalez, E., Cruz-Camarillo, R., Rojas Avelizapa, L.I., 2002. Degradation of aromatic and asphaltenic fractions by *Serratia liquefasciens* and *Bacillus* sp. Bull. Environ. Contam. Toxicol. 69 (6), 835–842.

Rontani, J.F., Bosser-Joulak, F., Rambeloarisoa, E., Bertrand, J.C., Giusti, G., Faure, R., 1985. Analytical study of Asthart crude oil asphaltenes biodegradation. Chemosphere 14 (9), 1413–1422.

Sadare, O., Obazu, F., Daramola, O., 2017. Review: biodesulfurization of petroleum distillates: current status, opportunities and future trends. Environments 4 (85), 1–20.

Sanchez-Minero, F., Ancheyta, J., Silva-Oliver, G., Flores-Valle, S., 2013. Predicting SARA composition of crude oil by means of NMR. Fuel 110, 318–321.

Santos, D., Rufino, R., Luna, J., Santos, V., Sarubbo, L., 2016. Biosurfactants: multifunctional biomolecules of the 21st century. Int. J. Mol. Sci. 17 (3), 401.

Santos, R., Loh, W., Bannwart, A., Trevisan, O., 2014. An overview of heavy oil properties and its recovery and transportation methods. Braz. J. Chem. Eng. 31 (3), 571–590.

Sar, T., Chen, Y., Bai, Y., Liu, B., Agarwal, P., Stark, B.C., et al., 2021. Combining co-culturing of *Paenibacillus* strains and *Vitreoscilla* hemoglobin expression as a strategy to improve biodesulfurization. Lett. Appl. Microbiol. 72 (4), 484–494.

Schenk, C., Cook, T., Charpentier, R., Pollastro, R., Klett, T., Tennyson, M., et al., 2009. An Estimate of Recoverable Heavy Oil Resources of the Orinoco Oil Belt. U.S. Geological Survey, Venezuela. https://pubs.usgs.gov/fs/2009/3028/pdf/FS09-3028.pdf. (Accessed 24 February 2020).

Schlumberger, 2010. Heavy Crude Oil. Library | Schlumberger. http://www.slb.com/~/media/Files/industry_challenges/heavy_oil/Resource. (Accessed 7 December 2017).

Schuler, B., Meyer, G., Peña, D., Mullins, O.C., Gross, L., 2015. Unraveling the molecular structures of asphaltenes by atomic force microscopy. J. Am. Chem. Soc. 137 (31), 9870–9876.

Selesi, D., Jehmlich, N., von Bergen, M., Schmidt, F., Rattei, T., Tischler, P., et al., 2010. Combined genomic and proteomic approaches identify gene clusters involved in anaerobic 2-methylnaphthalene degradation in the sulfate-reducing enrichment culture N47. J. Bacteriol. 192 (1), 295–306.

Sen, R., 2008. Biotechnology in petroleum recovery: the microbial EOR. Prog. Energy Combust. Sci. 34 (6), 714–724.

Seo, J.-S., Keum, Y.-S., Li, Q.X., 2009. Bacterial degradation of aromatic compounds. Int. J. Environ. Res. Public Health 6, 278–309.

Shahebrahimi, Y., Fazlali, A., Motamedi, H., Kord, S., Mohammadi, A.H., 2020. Effect of various isolated microbial consortiums on the biodegradation process of precipitated asphaltenes from crude oil. ACS Omega 5 (7), 3131–3143.

Shi, Q., Zhao, S., Zhou, Y., Gao, J., Xu, C., 2019. Development of heavy oil upgrading technologies in China. Rev. Chem. Eng. 36 (1), 1–19.

Shibulal, B., Al-Bahry, S., Al-Wahaibi, Y., Elshafie, A., Al-Bemani, A., Joshi, S., 2018. Microbial-enhanced heavy oil recovery under laboratory conditions by *Bacillus firmus* BG4 and *Bacillus halodurans* BG5 isolated from heavy oil Fields. Colloids Interfaces 2 (1), 1.

Shibulal, B., Al-Bahry, S., Al-Wahaibi, Y.M., Elshafie, A.E., Al-Bemani, A.S., Joshi, S.J., 2014. Microbial enhanced heavy oil recovery by the aid of inhabitant spore-forming bacteria: an insight review. Sci. World J. 2014, 309159.

Shibulal, B., Al-Bahry, S.N., Al-Wahaibi, Y.M., Elshafie, A.E., Al-Bemani, A.S., Joshi, S.J., 2017. The potential of indigenous *Paenibacillus ehimensis* BS1 for recovering heavy crude oil by biotransformation to light fractions. PLoS One 12 (2), e0171432.

Siddiqui, M.N., 2003. Infrared study of hydrogen bond types in asphaltenes. Pet. Sci. Technol. 21 (9–10), 1601–1615.

Silva, R., Almeida, D.G., Rufino, R.D., Luna, J.M., Santos, V.A., Sarubbo, L.A., 2014. Applications of biosurfactants in the petroleum industry and the remediation of oil spills. Int. J. Mol. Sci. 15 (7), 12523–12542.

Singh, R.N., Bahuguna, A., Munjal, A., 2012. Biodesulfurization of dibenzothiophene: a molecular approach: bioremediation of PASH by *Lysinibacillus sphaericus* DMT-7. Lambert Academic Publishing, Saarbrucken, Germany, pp. 14–50.

Solano-Serena, F., Marchal, R., Vandecasteele, J.-P., 2008. Biodegradation of aliphatic and alicyclic hydrocarbons. In: Vandecasteele, J.-P. (Ed.), Petroleum Microbiology: Concepts, Environmental Implications, Industrial Applications. vol. 1. Editions Technip, Paris, France, pp. 173–240.

Speight, J.G., 2006. History and terminology. In: The Chemistry and Technology of Petroleum, fourth ed. CRC Press, Boca Raton.

Strubinger, A., Ehrmann, U., León, V., DeSisto, A., González, M., 2015. Changes in Venezuelan Orinoco belt crude after different biotechnological approaches. J. Petrol. Sci. Eng. 127, 421–432.

Tanaka, Y., Yoshikawa, O., Maruhashi, K., Kurane, R., 2002. The cbs mutant strain of *Rhodococcus erythropolis* KA2-5-1 expresses high levels of Dsz enzymes in the presence of sulfate. Arch. Microbiol. 178, 351–357.

Tang, X.-D., Zhou, T.-D., Li, J.-J., Deng, C.-L., Qin, G.-F., 2019. Experimental study on a biomass-based catalyst for catalytic upgrading and viscosity reduction of heavy oil. J. Anal. Appl. Pyrolysis 143, 104684.

Tarvin, D., Busswell, A.M., 1934. The methane fermentation of organic acids and carbohydrates. J. Am. Chem. Soc. 56, 1751–1755.

Tavassoli, T., Mousavi, S.M., Shojaosadati, S.A., Salehizadeh, H., 2012. Asphaltene biodegradation using *microorganisms* isolated from oil samples. Fuel 93, 142–148.

Teufel, R., Mascaraque, V., Ismail, W., Voss, M., Perera, J., Eisenreich, W., et al., 2010. Bacterial phenylalanine and phenylacetate catabolic pathway revealed. Proc. Natl. Acad. Sci. U. S. A. 32, 14390–14395.

Thauer, R.K., Shima, S., 2008. Methane as fuel for anaerobic *microorganisms*. Ann. N. Y. Acad. Sci. 1125, 158–170.

The Economist Intelligence Unit, 2019. Kuwait Presses Ahead with Heavy Oil Production Expansion. http://www.eiu.com/industry/article/717944455/kuwait-presses-ahead-with-heavy-oil-production-expansion/2019-04-25. (Accessed 2 March 2020).

Thompson, D., Cognat, V., Goodfellow, M., Koechler, S., Heintz, D., Carapito, C., et al., 2020. Phylogenomic classification and biosynthetic potential of the fossil fuel-biodesulfurizing *Rhodococcus* strain IGTS8. Front. Microbiol. 11, 1417. https://doi.org/10.3389/fmicb.2020.01417.

Tverberg, G., 2007. Peak oil overview: discussions about energy and our future. In: The Oil Drum. http://theoildrum.com/node/2693. (Accessed 20 December 2018).

Uribe-Álvarez, C., Ayala, M., Perezgasga, L., Naranjo, L., Urbina, H., Vázquez-Duhalt, R., 2011. First evidence of mineralization of petroleum asphaltenes by a strain of *Neosartorya fischeri*. J. Microbial. Biotechnol. 4 (5), 663–672.

US Environmental Protection Agency, 2017. Integrated Science Assessment for Sulfur Oxides–Health Criteria. EPA/600/R–17/451 https://www.epa.gov/isa.

Van Hamme, J.D., Fedorak, P.M., Foght, J.M., Gray, M.R., Dettman, H.D., 2004. Use of a novel fluorinated organosulfur compound to isolate bacteria capable of carbon-sulfur bond cleavage. Appl. Environ. Microbiol. 70 (3), 1487–1493.

Vazquez-Duhalt, R., Quintero-Ramirez, R., 2004. Petroleum Biotechnology: Developments and Perspectives. Studies in Surface Science and Catalysis, Elsevier, Amsterdam.

Venkateswaran, K., Hoaki, T., Kato, M., Maruyama, T., 1995. Microbial degradation of resins fractionated from Arabian light crude oil. Can. J. Microbiol. 41 (4–5), 418–424.

Venkidusamy, K., Megharaj, M., 2016a. Identification of electrode respiring, hydrocarbonoclastic bacterial strain *Stenotrophomonas maltophilia* MK2 highlights the untapped potential for environmental bioremediation. Front. Microbiol. 7, 1965.

Venkidusamy, K., Megharaj, M., 2016b. A Novel electrophototrophic bacterium *Rhodopseudomonas palustris* strain RP2, exhibits hydrocarbonoclastic potential in anaerobic environments. Front. Microbiol. 7, 1071.

Vorbeck, C., Lenke, H., Fischer, P., Knackmuss, H.J., 1994. Identification of a hydride-Meisenheimer complex as a metabolite of 2,4,6-trinitrotoluene by a *Mycobacterium* strain. J. Bacteriol. 176 (3), 932–934. https://doi.org/10.1128/jb.176.3.932-934.1994.

Vorbeck, C., Lenke, H., Fischer, P., Spain, J.C., Knackmuss, H.J., 1998. Initial reductive reactions in aerobic microbial metabolism of 2,4,6-trinitrotoluene. Appl. Environ. Microbiol. 64 (1), 246–252. https://doi.org/10.1128/AEM.64.1.246-252.1998.

Widdel, F., Musat, F., 2019. Diversity and common principles in enzymatic activation of hydrocarbons: an introduction. In: Rojo, F. (Ed.), Aerobic Utilization of Hydrocarbons, Oils, and Lipids. Handbook of Hydrocarbon and Lipid Microbiology, Springer, Cham, pp. 3–32.

Wilkes, H., Buckel, W., Golding, B.T., Rabus, R., 2016. Metabolism of hydrocarbons in *n*-alkane-utilizing anaerobic bacteria. J. Mol. Microbiol. Biotechnol. 26 (1–3), 138–151.

Wilkes, H., Rabus, R., 2020. Catabolic pathways involved in the anaerobic degradation of saturated hydrocarbons. In: Boll, M. (Ed.), Anaerobic Utilization of Hydrocarbons, Oils, and Lipids. Handbook of Hydrocarbon and Lipid Microbiology, Springer, Cham, pp. 61–84.

Wilkes, H., Rabus, R., Fischer, T., Armstroff, A., Behrends, A., Widdel, F., 2002. Anaerobic degradation of *n*-hexane in a denitrifying bacterium: further degradation of the initial intermediate (1-methylpentyl)succinate via C-skeleton rearrangement. Arch. Microbiol. 177 (3), 235–243.

Xia, W., Tong, L., Jin, T., Hu, C., Zhang, L., Shi, L., et al., 2021. N,S-Heterocycles biodegradation and biosurfactant production under CO_2/N_2 conditions by *Pseudomonas* and its application on heavy oil recovery. Chem. Eng. J. 413, 128771.

Xu, P., Ma, C., Li, F., Tong, M., Zeng, Y., Wang, S., et al., 2002. Preparation of microbial desulfurization catalysts. Chin. Sci. Bull. 47 (13), 1077–1081.

Xu, P., Yu, B., Li., F.L., Cai, X.F., Ma, C.Q., 2006. Microbial degradation of sulfur, nitrogen and oxygen heterocycles. Trends Microbiol. 14 (9), 398–405. https://doi.org/10.1016/j.tim.2006.07.002.

Yernazarova, A., Kayirmanova, G., Baubekova, A., Zhubanova, A., 2016. Microbial enhanced oil recovery. In: Romero-Zerón, L. (Ed.), Chemical Enhanced Oil Recovery (cEOR)—A Practical Overview. InTechOpen, Rijeka, pp. 147–167, https://doi.org/10.5772/64805.

Zargar, A.N., Kumar, A., Sinha, A., Kumar, M., Skiadas, I., Mishra, S., Srivastava, P., 2021. Asphaltene biotransformation for heavy oil upgradation. AMB Express 11 (1), 127.

Zhang, J.H., Xue, Q.-H., Gao, H., Ma, X., Wang, P., 2015. Degradation of crude oil by fungal enzyme preparations from *Aspergillus* spp. for potential use in enhanced oil recovery. J. Chem. Technol. Biotechnol. 91 (4), 865–875.

Zhou, J.-F., Gao, P.-K., Dai, X.-H., Cui, X.-Y., Tian, H.-M., Xie, J.-J., et al., 2018. Heavy hydrocarbon degradation of crude oil by a novel thermophilic *Geobacillus stearothermophilus* strain A-2. Int. Biodeter. Biodegr. 126, 224–230.

CHAPTER

Microbial approaches for amino acids production

7

Ani M. Paloyan[a], Lusine H. Melkonyan[b,c], and Gayane Ye. Avetisova[b,c]

[a]*Laboratory of Protein Technologies, Scientific and Production Center "Armbiotechnology" of the National Academy of Sciences of Armenia, Yerevan, Armenia,*

[b]*Laboratory of Strain-Producers of BAS and Biosynthesis, Scientific and Production Center "Armbiotechnology" of the National Academy of Sciences of Armenia, Yerevan, Armenia,*

[c]*Institute of Pharmacy, Yerevan State University, Yerevan, Armenia*

Abstract

Amino acids are the most important primary metabolites with constantly increasing market capacity requirements. They are biochemical building blocks having a number of industrial applications, such as ingredients being used in pharmaceutical, medical, cosmetic, food, and feed industries.

In many other applied biotechnological fields, the role of bacteria for amino acid production is sufficiently large. The industrial microbial ways of amino acids production are fermentation and enzymatic process (biotransformation).

Currently, from the economic and environmentally friendly point of view, fermentation is the most used process for manufacturing; moreover, the application of new genetic engineering techniques makes it more attractive and preferable. Almost all amino acids can be produced by fermentation using strain-producers and inexpensive carbon sources. One of the main downranks for this process is purification costs. Construction of bacterial strains with improved productivity for desirable amino acids together with decreased by-product synthesis is important for reducing production expenses. Overall, for the production of L-amino acids having high consumption and demand, fermentation is the best way of choice.

At small-scale production, it is advantageous to use an enzymatic mean, which is based on a single or some enzymes. This process required substrates which are generally produced by chemical synthesis and usually can be expensive. Therefore, several techniques, for the improvement of physicochemical and biochemical characteristics of biocatalysts as well as their immobilization have been applied to make the biotransformation process more attractive.

Keywords: Amino acids synthesis, Fermentative way, Enzymatic way, Strain-producers, Enzymes

1 Introduction

Amino acids are structural units of proteins. They have a number of applications. Thus, from a pharmaceutical point of view the amino acids are used as direct pharmaceutical products as well as starting materials for synthesizing other products, in dietary they can serve as food supplements to compensate amino acid deficiencies or as a component for functional food and beverage products, in the food industry they are used as a flavor enhancer or sweetener, in feed industry the essential amino acids are

Microbial Syntrophy-mediated Eco-enterprising. https://doi.org/10.1016/B978-0-323-99900-7.00010-9

used as additives or source of nutrients. Amino acids are also used in cosmetics, agriculture, polymer industries, etc. In other words, the demand for amino acids is growing day by day which also leads to the emergence of new, more profitable technologies that will meet the demand for different amino acids in the market (Ikeda, 2003; Yokota and Ikeda, 2017).

Amino acids are mainly obtained in three ways: extraction from protein hydrolysates, chemical production, and microbial mean—fermentative and enzymatic ways with the aid of microorganisms (Subhashini et al., 2017). The benefits of each method hinge on a number of facts such as process economics, availability of the raw materials, market sizes, environmental regulations, etc. (Ikeda, 2003; D'Este et al., 2018).

In the initial period when amino acid production technologies start to develop the extraction method was dominated. Even now it is still profitable for a few types of amino acids. Particularly, this method is being applied for those amino acids (such as L-tyrosine, L-leucine, L-cysteine, and L-asparagine) which have rather small markets and available protein sources for extraction. Given the fact that the demand for amino acids is growing, the extraction method cannot alone meet the market requirements and in parallel other technologies for that amino acid production are being applied (Ikeda, 2003; D'Este et al., 2018).

Chemical synthesis can produce only the racemic forms of amino acids. The resulting racemate must be then subjected to optical division by using expensive chemicals and technologies; moreover the concept of being "chemically synthesized" often is not acceptable for consumers. While the asymmetric chemical synthesis method which has developed recently led to the formation of one isomer of an amino acid (Ikeda, 2003; Saghyan and Lange, 2016; D'Este et al., 2018; Parpart et al., 2018).

The fermentative way which is mainly applicable for the L-amino acids industry is widely used to produce most L-amino acids. Notwithstanding the large application prospect fermentation is nonviable for a few kinds of L-amino acids conditioned with low-production yields of the process, and/or the existence of other profitable methods for the production.

The success of this method mainly depends not only on efficient strain producers but also on the cost of the fermentation medium where the main part is occupied by the expenses of the carbon source, as well as available and cost-effective methods for isolation and purification of amino acid from cultivation broth which is the expensive part of the overall process. Continual process improvement has made the fermentation way the most economic method, especially for the amino acids large-scale production (Ikeda, 2003; Kumagai, 2006; D'Este et al., 2018).

The enzymatic way has a number of advantages, the most important of which is that D- and L-amino acids can be obtained with absolute optical purity in higher concentrations. This kind of process offering enantiopurity of amino acids and fewer by-products formation at the end greatly simplify the downstream processing. The choice of this method for large-scale production is mainly conditioned by the cost and manufacturing of the substrate. The scenario is rapidly changing with the development of stereospecific biocatalytic methods, by using highly enantioselective enzymes. During the past three decades, applications of biotechnological methods that rely on the help of free or immobilized cells/enzymes as catalysts and enzymatic production of chiral amino acids have replaced chemical methods. It is a unique strategy that has great potential. New enzymes are being identified, engineered, as well as new technologies are being developed and applied regularly to generate the process cost-effectively and to design processes at the low environmental impact.

In this chapter, L-tryptophan, L-valine, L-alanine production by fermentative method, including producing strains and approaches to their improvements, as well as L- and D-amino acids production based on aminoacylases and hydantoins carbamoylases by the enzymatic method have been reviewed.

2 Amino acids production by the fermentative way

2.1 Approaches to amino acids-producing strains improvements

Fermentation is a well-known method for L-amino acids production. Since 1957, after the discovery of glutamic acid-producing *Corynebacterium glutamicum* this method has been applied for numerous L-amino acids production. One of the advantages of this process is stereospecificity which makes this method preferred among other methods. The key point for fermentation is the availability of the corresponding strain producers, which are normally developed by classical or modern genetic engineering methods. At an early stage amino acids producing strains as by classical mutagenesis as overproducers screening and selection have been improved. The strain-producers of amino acids are divided into four groups: wild-type strains, auxotrophic strains, regulatory strains, auxotrophic regulatory strains. Especially, the most useful are auxotrophic strains (Figs. 1 and 2). Succeeding mutations guarantee that substances are canaliculated to the target products (El-Mansi et al., 2012).

These modifications include as break feedback controls as increase formation of pathway precursors and intermediates. This way for improvement of commercial strain-producers of amino acids has been used (El-Mansi et al., 2012).

Genetic engineering ensures the highest production of amino acids by gene regulation. This recombinant DNA technique is more rational, and it gives a chance to assemble the appropriate characteristics. This is important for the strains which have complex regulatory systems. Production of amino acids by deregulated organisms may have some restrictions as they cannot produce enough biosynthetic enzymes to achieve high productivity. This problem can be solved by increasing the copy number of structural genes for these enzymes by recombinant DNA technology. These two strategies often are being applied in combination. The recombinant plasmid carrying the biosynthetic genes transformed in the cell which replicates and forms multiple copies in the cell. For the increasing frequency of transcription, the recombinant plasmid carries only structural genes without promoter

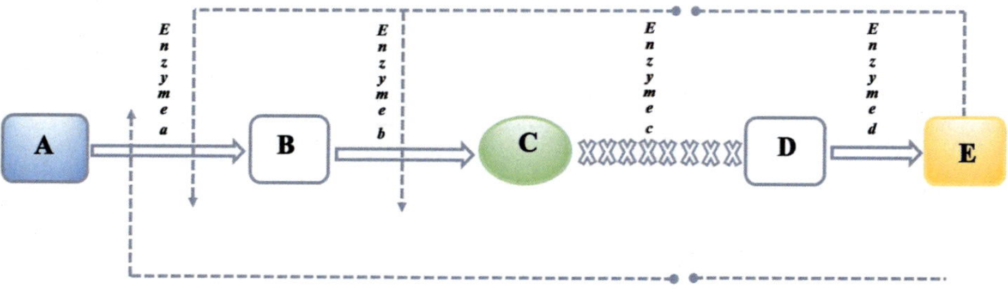

FIG. 1

Overproduction of an intermediate of a linear primary metabolic pathway. Feedback inhibition by the end product inhibits the activity of enzyme and feedback repression represses the formation of enzymes a and b. By making a genetic block (mutation) at enzyme c, an auxotrophic mutant is made, which cannot grow unless the metabolite E is added to the medium. As long as the amount of E present is not excessive, there will be no feedback effects and the metabolite C will be overproduced.

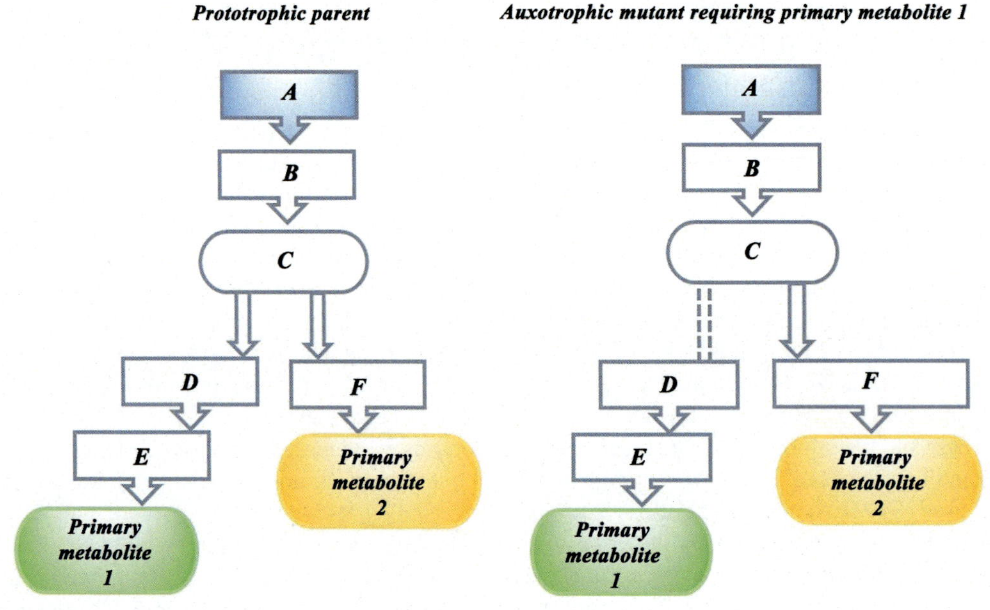

FIG. 2

Use of auxotrophic mutation in a branched pathway. The auxotrophic mutant will require primary metabolite 1 for growth. When the concentration of metabolite 1 is not excessive to biosynthetic demands, primary metabolite 2 will be overproduced.

and operator. The lofty plasmid would contain a regulatory region of the preferable phenotype. These genetic approaches give a chance to obtain a better strain which has only mutations important for hyperproduction (El-Mansi et al., 2012).

2.2 L-tryptophan production

Tryptophan ($C_{11}H_{12}N_2O_2$; MW 204.23) is a proteinogenic amino acid. It is necessary to produce metabolites, most notably kynurenine, neurotransmitter, serotonin, etc. Tryptophan can relax nerves, decrease carbohydrate cravings, ease premenstrual symptoms, and improve relaxation, sleep, etc. It is used widely in the pharmaceutical and food industries. It was first isolated from the milk protein casein by Hopkins and Cole (1901), Wexler (2007), and Jenkins et al. (2016).

In general, aromatic amino acids (AAA) biosynthesis is taking place the same way in most organisms (unicellular organisms, fungi, plants) (Fig. 3). This is named the shikimate pathway as well. Two molecules of phosphoenolpyruvate, one molecule of erythrose-4-phosphate as well as reducing equivalents ATP are necessary for AAA pathway, to form a precursor of chorismate by seven steps for three aromatic amino acids (Yokota and Ikeda, 2017).

Overall, L-tryptophan is obtained via chemical synthesis or protein hydrolysis. These two means are not the greatest of choice for large production of this amino acid, because as there are limited amounts of raw materials available as well as chemical synthesis has a negative ecological impact. Moreover, the complex processes are not viable from an economic point of view. The alternative way

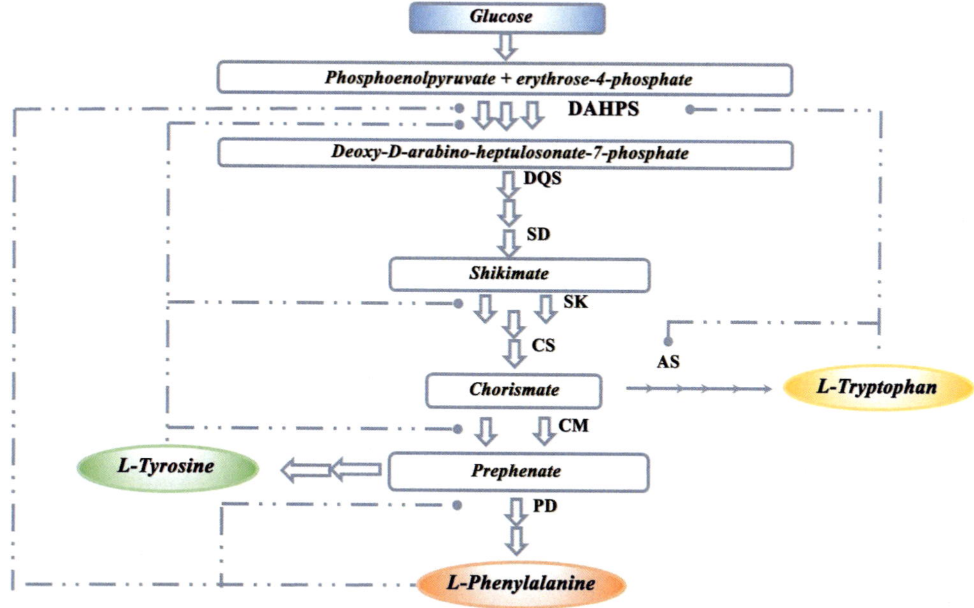

FIG. 3

Biosynthetic pathways for AAA productions in *Corynebacterium glutamicum* ssp. flavum (*Brevibacterium flavum*) 3-deoxy-D-arabino-heptulosonate 7-phosphate synthase (DAHPS), dehydroquinate synthase (DQS), shikimate dehydrogenase (SD), shikimate kinase (SK), chorismate synthase (CS), anthranilate synthase (AS), chorismate mutase (CM), and prephenate dehydratase (PD).

for obtaining L-tryptophan is microbial production, in particular, microbial transformation, enzymatic synthesis, and direct fermentation. Due to some advantages is preferable to direct fermentation by microorganisms (Yokota and Ikeda, 2017; Niu et al., 2019).

L-tryptophan can produce a wide number of bacteria including *Brevibacterium*, *Corynebacterium*, *Escherichia*, *Bacillus*, etc. (Ikeda, 2003). Currently, *C. glutamicum* and *Escherichia coli* are the main producers of L-tryptophan (Niu et al., 2019).

Obtaining strains able to produce L-tryptophan was done by classical mutagenesis and appropriate screening mutants resistant to aromatic amino acids analogs. For instance, strains resistant to p-fluorophenylalanine and m-fluorophenylalanine from wild-type strain *Brevibacterium flavum* ATCC 14067[a] were obtained (Fig. 4).

The ability of resistant strains to produce L-tryptophan during short-time tube fermentation conditions was studied. The synthetic activity of the selected *Br. flavum* 27, *Br. flavum* 8 (p-FP-r), *Br. flavum* 59, and *Br. flavum* 18 strains (m-FP-r) using L-tryptophan-auxotrophic strain as a microbiological test was evaluated. As controls 14,067 strain and isoleucine-auxotrophic strain (ile^-) were used (Fig. 5).

[a]*Br. flavum* ATCC 14067 has renamed as *C. glutamicum* ATCC 14067 (Lv et al., 2012).

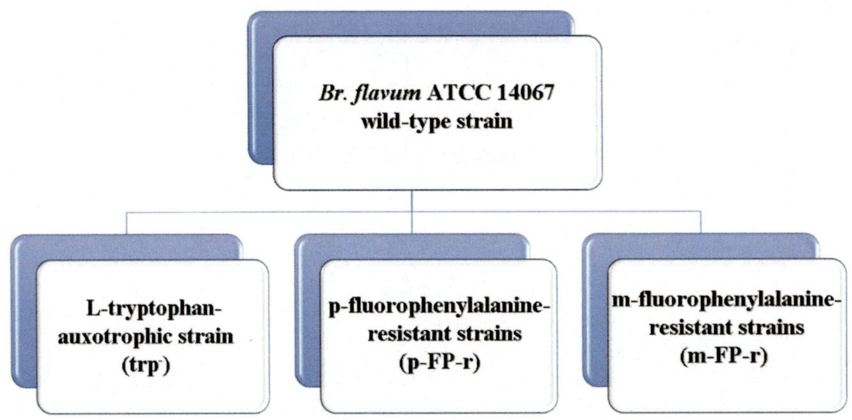

FIG. 4

Obtaining schema of L-tryptophan-producing strains.

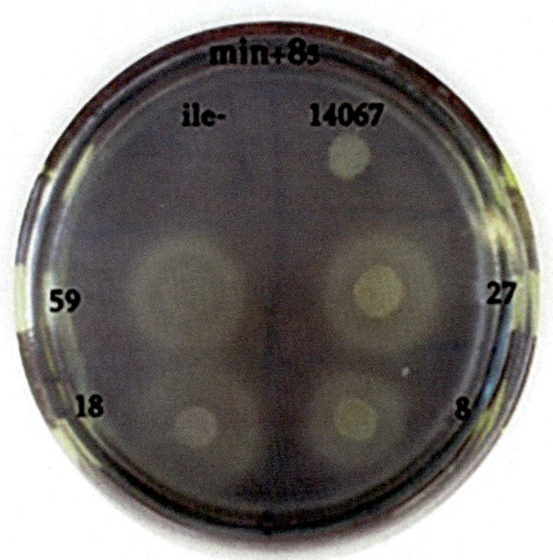

FIG. 5

Evaluation of *Br. flavum* strains L-tryptophan-producing ability by microbiological test L-tryptophan-auxotrophic strain containing synthetic medium (min + 8_5).

Br. flavum 27 and *Br. flavum* 18 strains (Fig. 6) as the most active were selected. These strains were cultivated in submerged fermentation and produced up to 2 g/L L-tryptophan from sucrose within 96 h. There is scope for improvement of these strains by classical mutagenesis.

C. glutamicum ssp. flavum strain resistant to sulfaguanidine was able to produce up to 19 g/L L-tryptophan. Genetic engineering allowed to increase L-tryptophan yield up to 43 g/L (El-Mansi et al., 2012). Other *C. glutamicum* KY9218 strain carrying pKW9901 plasmid for L-tryptophan overproduction was constructed. This strain up to 50 g/L L-tryptophan from sucrose within 80 h in fed-batch cultivation, without antibiotic pressure, was produced. Further genetic improvement allowed to increase the strain L-tryptophan yield up to 58 g/L (Ikeda and Katsumata, 1999).

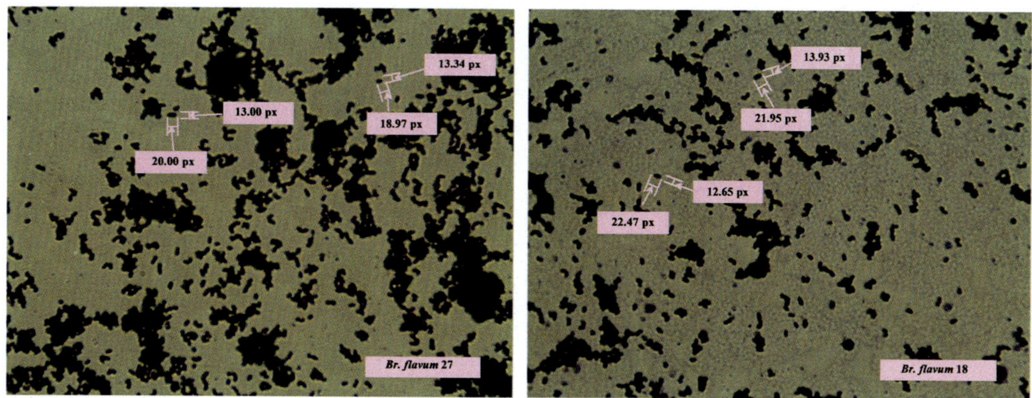

FIG. 6

Cells of strains under the microscope (×10,000). Microscopic images by Leica DM500 trinocular microscope and Digital Camera EC3 Leica Microsystem software were obtained.

A genetically engineered *E. coli* NT1259/pF112*aro*FBL$_{kan}$ during 68 h from glycerol up to 14.3 g/L L-tryptophan was produced (Tröndle et al., 2020).

Various production processes for L-tryptophan with *E. coli* have been described, most of them using glucose as a carbon source. A strain of *E. coli* was engineered to overproduce L-tryptophan. It was produced up to 44 g/L L-tryptophan from glucose in a fed-batch process (Dodge and Gerstner, 2002). In the framework of other research, from A genetic engineered *E. coli* W3110 *trpAE1 trpR tna* was obtained SGIII1032 sulphaguanidine-resistant strain, which produced up to 54.6 g/L L-tryptophan from glucose within 78 h (Azuma et al., 1993).

Further improvement of L-tryptophan-producing strains is an important aspect of a knowledge-based bioeconomy.

2.3 L-valine production

Valine ($C_5H_{11}NO_2$; MW 117.15) is an essential amino acid that is not synthesized in the human body but in main time has fundamental role in the metabolism and physiology of mammals. Cited amino acid is used in pharmaceuticals, in cosmetics, in agriculture. It was first isolated from casein by Hermann Emil Fischer in 1901 (Vickery and Schmidt, 1931; Oldiges et al., 2014; Yokota and Ikeda, 2017).

Biosynthesis pathways of branched-chain amino acids (BCAAs) are densely bound. Almost all living forms of the synthesis of this amino acid occur such as in *C. glutamicum* summarized in Fig. 7 (Yokota and Ikeda, 2017).

Acetohydroxy acid synthase (AHAS) enzyme catalyzes the initial step of the biosynthesis of BCAAs. The enzyme of *C. glutamicum* is encoded by thc *ilvBN* gene This enzyme is responsible for the production of 2-acetolactate and 2-aceto-2-hydroxybutyrate which are intermediate substrates for the biosynthesis of BCAAs. The reaction starts from pyruvate and 2-ketobutyrate. Pyruvate is supplied from the glycolytic pathway and 2-ketobutyrate is supplied from L-threonine dehydration.

The only one isoform of AHAS has been reported in *C. glutamicum*, whereas other microorganisms have some isoforms of this enzyme. In the case of *E. coli*, they are three—AHAS I, II, and III. *ilvBN*, *ilvGM*, and *ilvIH* genes are encoded mentioned enzymes (Yokota and Ikeda, 2017).

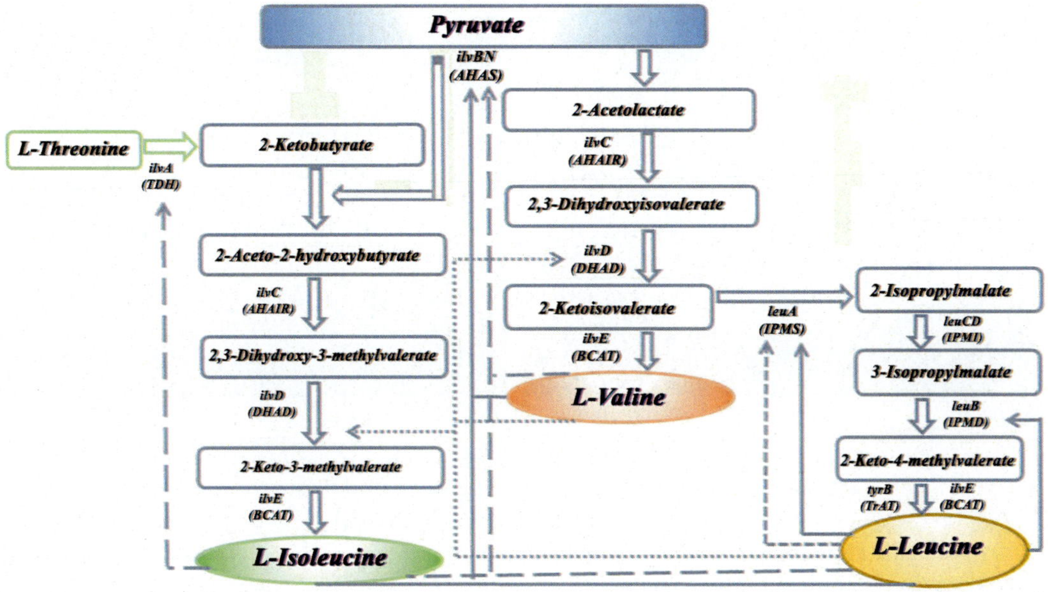

FIG. 7

The BCAAs biosynthetic pathways in *C. glutamicum* acetohydroxy acid isomeroreductase (AHAIR), acetohydroxy acid synthase (AHAS), aspartate kinase (AK), aspartate semialdehyde dehydrogenase (ASADH), branched-chain amino acid aminotransferase (BCAT), dihydroxy acid dehydratase (DHAD), homoserine dehydrogenase (HDH), homoserine kinase (HK), isopropylmalate dehydrogenase (IPMD), isopropylmalate isomerase (IPMI), isopropylmalate synthase (IPMS), threonine dehydratase (TDH), tyrosine-repressible transaminase (TrAT), and threonine synthase (TS).

Formerly, the chemo enzymatic method has been used for BCAAs, and later the enantiomers of chemically synthesized BCAAs were separated enzymatically. However, with the increased demand for BCAAs, the fermentative way is the best choice from the economic and the ecological point of view. *Br. flavum*, subspecies of *C. glutmicum*, one of the most used corynebacteria in industrial fermentation, has been widely used to produce BCAAs and other amino acids (Park and Lee, 2008; Hou et al., 2012).

For BCAAs production, most strains were obtained by classical mutagenesis and selection method. *Br. flavum* AA53 (VKPM[b] B-3714) and AA54 (VKPM B-3715) strain-producers (Fig. 8) isoleucine-auxotrophic and resistant to non-specific analog of valine—L-arginine hydroxamate, as well as to the analog of valine—DL-α-aminobutyric acid in the existence of L-leucine were obtained for L-valine production (Azizian et al., 1990a, b, 1991).

Under optimized conditions the most active strain *Br. flavum* AA53 synthesized up to 55 g/L L-valine at a yield of 0.35 mol mol of glucose^{-1}, with low concomitant amino acids formation (Avetisova, 2002; Avetisova et al., 2003a, b). It was shown that high production of L-valine results from significant de-repression of AHAS and partial removal of enzyme inhibition by the target product (Avetisova et al., 2003a, b).

[b]VKPM—Russian National Collection of Industrial Microorganisms.

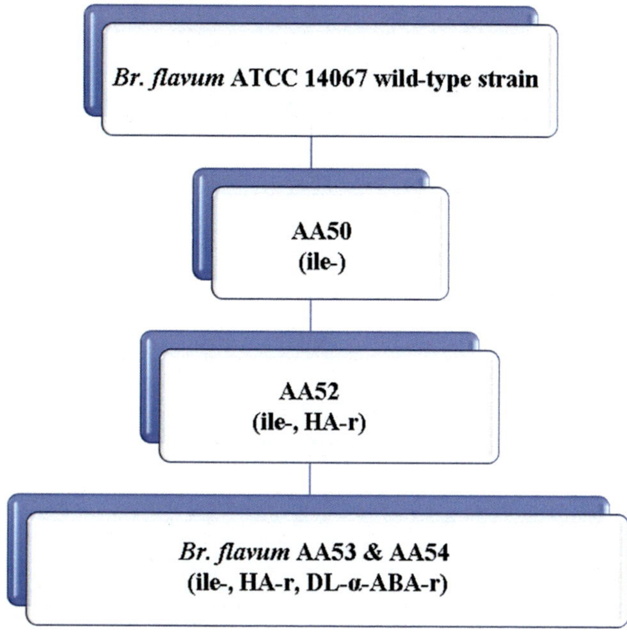

FIG. 8

Obtaining stages of L-valine producing strains ile^{-}—auxotrophic to isoleucine, HA-r—resistant to L-arginine hydroxamate, DL-α-ABA-r—resistant to DL-α-aminobutyric acid.

Based on *Br. flavum* AA53 strain the microbiological process for L-valine production was developed at the pilot plant.

Currently, genetic engineering methods give a great opportunity to obtain highly productive strains. The overexpression of biosynthetic genes of L-valine synthesis ensures desirable results. A clear example of this is genetically modified L-valine producing strains improved in the main by *C. glutamicum*. There are also other strain-producers of this amino acid, designed based on other bacteria including *E. coli*.

Modified *C. glutamicum* strain with *ldhA* gene knockout and *ilvBNCDE* overexpression produced L-valine under low oxygen conditions. It has been reported that the main drawback during this synthesis pathway is a noticeable imbalance of cofactor production and consumption. To solve this problem, the coenzyme for L-valine production was changed from NADPH to NADH by modified acetohydroxy acid isomeroreductase and *Lysinibacillus sphaericus* leucine dehydrogenase insertion instead of endogenous transaminase B. The engineered producer, *C. glutamicum* BN^{GE}C™DLD/ΔLDH, accordingly produced 1470 mM (172.2 g/L) after 24 h, with a yield of 0.63 mol mol of glucose^{-1}, and 1940 mM (227.3 g/L) after 48 h (Hasegawa et al., 2012). The strain synthesized by-product succinate, which had a negative effect on L-valine production, therefore there is still room for improvement. The final improved strain, named *C. glutamicum* Val-9 (BN^{GE}C™DLD/ΔLP_ΔAc + GP_ilvN^{GE}C™_ΔAla), produced 1280 mM (150 g/L) L-valine at a yield of 0.88 mol mol of glucose^{-1} after 24 h under low oxygen conditions (Hasegawa et al., 2013).

The development of genetically engineered strains of *E. coli* for L-valine production is more difficult and less successful compared with *C. glutamicum*. As it has extra complex regulatory mechanisms for biosynthesis of this amino acid (Yokota and Ikeda, 2017). Despite the mentioned complications number of results concerning modified *E. coli* strains have been reported. Thus, *E. coli* VAMF (pK-BRilvBN^{mut}CED) (pTrc184ygaZHlrp) strain with AHAS I mutation was generated. Under optimized conditions, the strain synthesized up to 32.3 g/L target amino acid, with 0.58 g/L/h productivity (Park et al., 2011a, b).

Further, *E. coli* W was metabolically engineered for L-valine production. The engineered *E. coli* W (Δ*lacI*Δ*ilvA*) strain overexpressing the *ilvBN^{mut}*, *ilvCED*, *ygaZH*, and *lrp* genes was able to synthesize 60.7 g/L L-valine within 29.5 h, with 2.06 g/L/h productivity (Park et al., 2011a, b).

Thereby, improved *E. coli* and *C. glutamicum* strain-producers for L-valine production can have industrial-scale application (Oldiges et al., 2014).

2.4 L-alanine production

Alanine ($C_3H_7NO_2$; MW 89.09) is a non-polar, aliphatic amino acid. This amino acid is non-essential to humans and can be synthesized metabolically. For the nervous system, this amino acid plays an important role. In 1850 Schutzenberger and Bourgeois extracted it from the hydrolyzate of fibroin for the first time. Alanine has also been found in almost all bacteria. L-alanine is important for the synthesis of proteins as a building block, and D-alanine is integrated into the cell wall of microorganisms (Wendisch, 2007; Melkonyan et al., 2013a, b; Sidiq et al., 2019).

L-alanine is extensively used in the medical, pharmaceutical, chemical, and food industries.

It can be synthesized in bacteria by several enzymes. In bacteria, there are three major biosynthetic pathways for alanine metabolism: reversible conversion of pyruvate to alanine catalyzed by alanine dehydrogenase (AlaD), beta-decarboxylation of L-aspartate catalyzed by aspartate-4-decarboxylase (AspD), transfer of amino group from amino acids and amines to pyruvate catalyzed by aminotransferases (ATs) such as glutamate-pyruvate AT (AlaB) and valine-pyruvate AT (AvtA) (Fig. 9). In *Br. flavum, C. glutamicum* L-alanine is produced from pyruvate by AlaB and AvtA enzymes (Ambartsumyan and Bezirdzhyan, 1994a, b; Van der Kaaij et al., 2004; Wendisch, 2007; Melkonyan et al., 2013a, b; Avetisova et al., 2014a, b; Jojima et al., 2014).

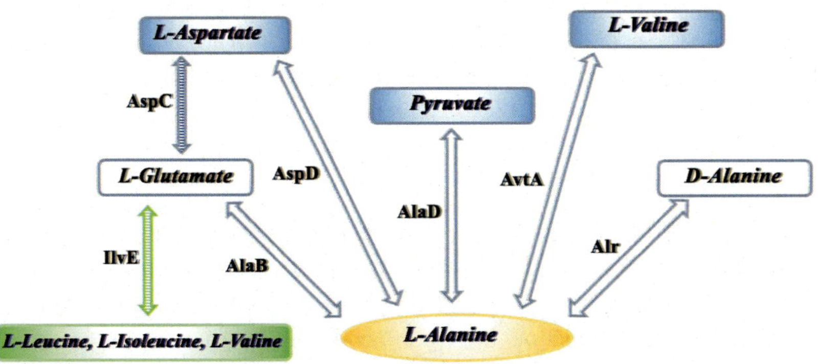

FIG. 9

Alanine metabolic pathways L-alanine dehydrogenase (AlaD), glutamate-pyruvate AT (AlaB), valine-pyruvate AT (AvtA), alanine racemase (Alr), branched-chain amino acid AT (IlvE), aspartate AT (AspC), and aspartate 4-decarboxylase (AspD).

L-alanine can be produced as fermentative as enzymatic ways. It is synthesized from L-aspartate by a one-step enzymatic method action of AspD. L-alanine production by fermentation is difficult as microorganisms have an Alr enzyme (Kumagai, 2006).

L-alanine production by the fermentative way with *Br. flavum*, *C. glutamicum* and *E. coli* higher yields have been reported (Azizian et al., 1992; Avetisova et al., 2014a, b; Jojima et al., 2014; Zhou et al., 2016).

Alr-deficient and resistant to amino acids structural analogs *Br. flavum* strains were obtained by classical mutagenesis and selection method. *Br. flavum* AA5 initial strain up to 43.8 g/L of L-alanine at a yield of 0.29 mol mol of glucose^{-1} in flask fermentation conditions was synthesized (Azizian et al., 1992; Vardanyan et al., 2008; Melkonyan et al., 2013a, b). First, the activity of initial strain key enzymes involved in L-alanine synthesis was studied. It has been shown that DL-α-ABA, L-cycloserine, and β-Cl-L-alanine were inhibitors for AlaB and AvtA enzymes. β-Cl-L-alanine was the strongest inhibitor for AlaB (Melkonyan et al., 2013a, b).

The new strains resistant to β-Cl-L-alanine from initial strain *Br. flavum* AA5 were obtained (Fig. 10). *Br. flavum* GL1 (MDC[c] 11,841) and GL18 (MDC 11842) strains produced up to 53.7 and 60.5 g/L target amino acid at a yield of 0.36 and 0.40 mol mol of glucose^{-1}, respectively (Avetisova et al., 2012; Melkonyan, 2013; Melkonyan et al., 2013a, b; Avetisova et al., 2014a, b).

To reveal the mechanism of L-alanine overproduction in GL1 and GL18 strains, the influence of β-Cl-L-alanine on AlaB enzyme was studied. It has been shown that 96 and 5.0 mM β-Cl-L-alanine was inhibited AlaB activity of GL1 and GL18 by 50%, respectively. L-alanine and D-alanine had a significant derepressive effect on AlaB synthesis of GL18. On the other hand, DL-α-ABA and β-Cl-L-alanine (in the case of AvtA) had a strong derepressive effect on the enzymes synthesis of GL1 strain. No activity of AvtA was observed in GL18 strain compared to which all investigated substances exhibited a derepressive effect (Melkonyan et al., 2013a, b).

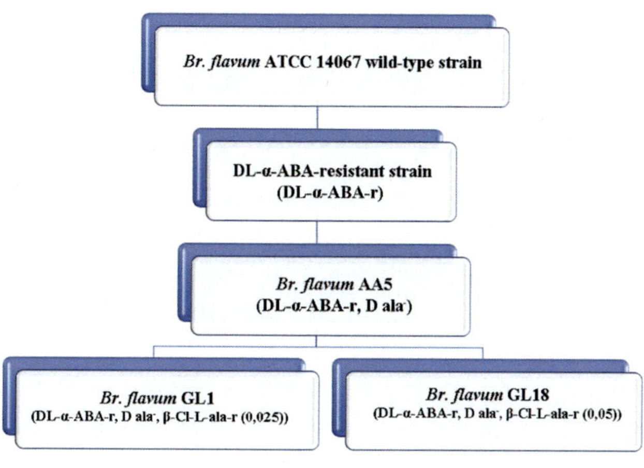

FIG. 10

Obtaining stages of L-alanine-producing strains D ala^{-}—auxotrophic to D-alanine, DL-α-ABA-r—resistant to DL-α-aminobutyric acid (DL-α-ABA), β-Cl-L-ala-r (0.025)—resistant to 0.025 mg/mL of β-Cl-L-alanine, β-Cl-L-ala-r (0.05)—resistant to 0.05 mg/mL of β-Cl-L-alanine.

[c]MDC—Microbial Depository Center (Armenia).

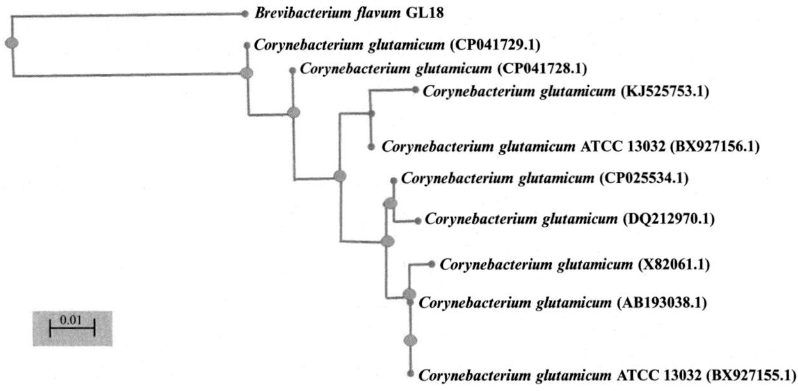

FIG. 11

The phylogenetic three *of Br. flavum* GL18 constructed by Neighbor-joining method.

Based on GL18 strain the technology for L-alanine biosynthesis in the "Biostat-S" bioreactor was developed. By the optimization of pH, temperature, and aeration profiles, as well as sugar and nitrogen fed-batch, the effective fermentation process was developed for L-alanine production. Final strain GL18 produced up to 62.8 g/L L-alanine with 1.08 $gL^{-1}h^{-1}$ productivity within 58 h (Avetisova et al., 2014a, b).

For GL18 the 16S rRNA gene sequence was carried out. It has been shown that the strain belongs to the species *Corynebacterium glutamicum* (Fig. 11).

A genetically engineered *C. glutamicum* producer for L-alanine production was also obtained. To provide the highest yield of target amino acid, the synthesis of lactate and succinate needs to be suppressed. For that purpose, Jojima and co-authors inactivated *ldhA* and *ppc* genes. The engineered producer synthesized up to 98 g/L L-alanine within 32 h in mineral salts medium. The *gapA* gene encoding 3-phosphate dehydrogenase is playing a key role in bacterial growth on glucose. Overexpression of this gene in the alanine-producing strain may improve glucose consumption rates, which will enhance the productivity of alanine. The final engineered strain GLY3/pCRD500 produced 216 g/L alanine within 48 h (Jojima et al., 2010, 2014).

Another strain-producer, recombinant *E. coli* B0016-060 BCE under optimized growth conditions produced up to 120.8 g/L L-alanine in glucose medium (Zhou et al., 2016).

Thereby, the fermentative way is the best choice for L-alanine large-scale production.

3 Amino acids production by the enzymatic way

3.1 L- and D-amino acids production: Aminoacylases

Aminoacylases catalyze the hydrolysis of an N-acyl group from N-α-acyl amino acids to form corresponding amino acids. Depending on the stereospecificity to L- or D-isomers of N-α-acyl amino acids aminoacylases can be divided into two groups—L- and D-aminoacylases. There are different approaches concerning the role of these enzymes. Thus, L-aminoacylase participates in detoxication of xenobiotics and some endogenous compounds by conversion to the corresponding conjugates.

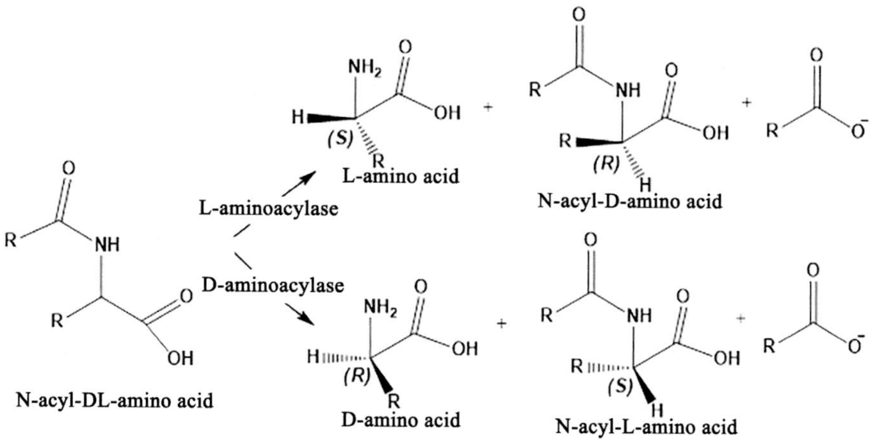

L-aminoacylase
D-aminoacylase
N-acyl-DL-amino acid
L-amino acid
N-acyl-D-amino acid
D-amino acid
N-acyl-L-amino acid

FIG. 12

Biocatalytic routes for amino acid preparation using aminoacylase.

In contrast to L-aminoacylase, the role of the D-aminoacylases is unknown. Despite the fact that these two enzymes catalyze the same type of reaction they do not have sequence identity as well as structure homology. Presumably, the hydrolytic reaction mechanism catalyzes by L-aminoacylase and D-aminoacylases is different. Based on the crystal structures of mononuclear (D-aminoacylase) (Liaw et al., 2003) and binuclear (aminoacylase-1, ACY1) (Lindner et al., 2003) aminoacylases they are classified as a member of different superfamilies. L-aminoacylase belongs to the amidohydrolase and D-aminoacylases belong to zinc-hydrolase superfamilies (Hernick and Fierke, 2010). L-aminoacylase have used to produce L-amino acids for quite a long time. Moreover, immobilized enzyme technology has been successfully applied for this enzyme and immobilized biocatalyst has been used for large-scale production of L-amino acids. Unlike L-amino acids which have developed technologies, such as fermentation for production, large-scale production of D-amino acids mainly has been accomplished by enzymatic methods.

Since L- and D-aminoacylases show substrate specificity toward N-acyl-L- and N-acyl-D-amino acids, respectively, they can be used to separate L- or D-enantiomers. After the reaction, amino acid can be easily separated from the unreacted N-acyl-amino acids (Sato and Tosa, 2010). Schematically the process is presented in Fig. 12.

3.1.1 L-aminoacylases

Since 1950s after the characterization of pig kidney L-aminoacylase (EC 3.5.1.14. N-acyl-L-amino acid amidohydrolase), these enzymes have been applied for the production of L-amino acids by optical resolution (Greenstein, 1957) and is the first enzyme to be used industrially. In 1969 Chibata and co-workers were successful for the first time in the industrial application of the fungal immobilized L-aminoacylase to produce a number of L-amino acids such as methionine, phenylalanine, valine, etc. (Chibata et al., 1972).

Number of L-aminoacylase from different sources: mammals (Lindner et al., 2000), fungal (Solov'eva and Stepanov, 1983), bacteria (Sakanyan et al., 1993; Kempf and Bremer, 1996; Cho et al., 1987; Curley and Sinderen, 2000; Yang et al., 1994; Aganyants et al., 2013a, b) Achaea (Tanimoto et al., 2008; Story et al., 2001; Toogood et al., 2002a, b), have been isolated and characterized. The first L-aminoacylase obtained from microbial sources was isolated from *Aspergillus oryzae*. The enzyme in free and immobilized form has been applied for asymmetric hydrolyses of a number of *N*-acetyl L-amino acids to the corresponding amino acid (Yuan et al., 2002; Chibata et al., 1972; He et al., 1992). Most of the characterized L-aminoacylases, including of Archaea origin, are homotetrameric with subunits molecular weights varying in the range of 41.5–43 kDa (Cho et al., 1987; Sakanyan et al., 1993; Story et al., 2001; Tanimoto et al., 2008; Toogood et al., 2002a, b), however, L-aminoacylases from *Aspergillus oryzae* as well as of bacterial origin with dimeric composition have been characterized as well (Weiss et al., 1995; Yang et al., 1994; Gentzen et al., 1980; Curley et al., 2003; Aganyants et al., 2013a, b).

The microbial L-aminoacylases are a zinc-containing metalloenzymes. These enzymes are very sensitive to metal chelating agents, but this type of inactivation is completely reversible. After incubation with divalent metal ions such as Zn^{2+}, Mn^{2+}, and Ni^{2+} (Gentzen et al., 1980; Weiss et al., 1995; Story et al., 2001; Tanimoto et al., 2008). L-aminoacylases restore the initial activity. Divalent metals can also show activating action on the enzyme. L-aminoacylase can be significantly activated by Co^{2+} ions (Gentzen et al., 1980; Weiss et al., 1995), Mn^{2+} and Cd^{2+} also act as an activator in some cases (Cho et al., 1987; Weiss et al., 1995). Some divalent metals (Ni^{2+}, Cd^{2+}, Fe^{2+}, Cd^{2+}, Cu^{2+}, Ba^{2+}) have shown inhibitory effects on enzyme activity (Cho et al., 1987; Toogood et al., 2002a, b; Gentzen et al., 1980).

The L-aminoacylase has a broad substrate specificity. Fungal enzyme preferers *N*-acetyl- methionine as a favorable substrate (Solov'eva and Stepanov, 1983) while the enzymes of bacterial or Archaeal origin differ substantially by substrate preference. Characterized L-aminoacylase of *Bacillus* genius prefers *N*-acetyl-aromatic amino acids and its derivatives (Sakanyan et al., 1993) or *N*-chloroacetyl amino acids and shows specificity to di- and tripeptides (Cho et al., 1987), while the same enzyme characterized in *Lactococcus* shows the highest activity to *N*-acetyl alanine (Curley and Sinderen, 2000), the enzyme of *Alcaligenes* shows higher activity toward *N*-acetyl-derivative of hydrophobic L-amino acids, *N*-acetyl-L-valine is the favorite substrate for this enzyme (Liaw et al., 2003). Only three representatives of this enzyme, having an archaeal origin, have been isolated and characterized up to date, which is quite different by its substrate specificity. Thus, purified L-aminoacylase of *Pyrococcus horikoshii* showed the ability to hydrolyze *N*-acetyle groups of *N*-acetyl-L-methionine, *N*-acetyl-L-glutamine, and *N*-acetyl-L-leucine to yield corresponding L-amino acids (Tanimoto et al., 2008). *Pyrococcus furiosus* L-aminoacylase hydrolyzed nonpolar N-acylated L amino acids (Story et al., 2001) and *Thermococcus litoralis* L-aminoacylase has a broad substrate specificity preferring the amino acids: phenylalanine ≫ Methionine > Cysteine > Alanine ≃ Valine > Tyrosine > Propargylglycine > Trypyophane > Proline > Arginine (Toogood et al., 2002a, b).

The microbial L-aminoacylases have temperature optimum varying from 30 °C to 100 °C and pH optimum varying 6.5–8.5 (Gentzen et al., 1980; Weiss et al., 1995; Sakanyan et al., 1993; Solov'eva and Stepanov, 1983; Tanimoto et al., 2008; Story et al., 2001; Toogood et al., 2002a, b; Yang et al., 1994; Curley and Sinderen, 2000).

Different methods, such as recombinant protein technologies, immobilization, have been applied for the development of biocatalyst applicable in industrial biotransformation reactions. In a number of reports, the successful results are presented, concerning the use of immobilization techniques allowing to reduce the amount of enzyme needed for obtaining the unit product. A method for obtaining L- and D-methionine simultaneously from *N*-acetyl D,L-methionine by using recombinant L-aminoacylase of *E. coli* has been developed (Mkrtchyan et al., 2005), as well as the possibility of production of enantiometrically pure amino acids using thermophilic L-aminoacylase of *Geobacillus stearotermophilus* immobilized on silochrome C-80 has been shown by our group (Aganyants et al., 2013a, b).

Table 1 L-aminoacylases and their sources being used or having the potential to be used in amino acid production.

Organism	Example of amino acid produced	References
Aspergillus	L-methionine, L-phenylalanine, L-tryptophan, L-valine	Chibata et al. (1975)
Geobacillus	L-methionine	Mkrtchyan et al. (2005), Adenan et al. (2018), Cho et al. (1987), and Aganyants et al. (2013a, b)
E. coli		
Bacillus		
Bacillus	L-alanine, L-tyrosine, L-phenylalanine, etc.	Sakanyan et al. (1993) and Weiss et al. (1995)
Lactococcus	L-alanine	Curley et al. (2003)
Thermococcus	L-tryptophane	Toogood et al. (2002a, b) and Ngamsom et al. (2010)
Alcaligenes	L-valine, L-alanine	Yang et al. (1994)

Among the immobilization methods, the covalent binding to glyoxyl-Sepharose and Amberlite XAD7 were the most successful methods for this enzyme. By application of these methods, it was possible to bind 15 and 80 mg protein (calculated for per gram of support), respectively, with nearly 80% of activity recovery in both cases (Toogood et al., 2002a, b). Moreover, micro-reactors containing a monolith-immobilized enzyme have been developed which has a great potential to be used for high throughput biotransformations (Ngamsom et al., 2010). Two companies (Chirotech/Dow Pharma and Chirotech/Dr. Reddy's) are using the *Thermococcus* L-aminoacylase enzyme for large-scale production of L-amino acids as well as their analogs.

Table 1 summarizes the L-aminoacylases and their sources being used or having the potential to be used in amino acid production.

3.1.2 D-aminoacylases

D-aminoacylase (N-acyl-D-amino acid amidohydrolase EC 3.5.1.81) catalyzed the hydrolytic reaction of N-acyl-D-amino acids to liberate corresponding D-amino acids. The enzymes are common among a variety of microorganisms and have been found in a number of strains of *Alcaligenes* (Moriguchi and Ideta, 1988; Moriguchi et al., 1993a; Tsai et al., 1988; Yang et al., 1991), *Pseudomonas* (Sakai et al., 1991a, b; Kubo et al., 1980), *Variovorax* (Lin et al., 2002), *Stenotrophomonas* (Muniz-Lozano et al., 1998), *Streptomyces* (Sugie and Suzuki, 1978, 1980; Arima et al., 2013), *Bordetella* (Cummings et al., 2009), *Defluvibacter* (Kumagai et al., 2004), *Microbacterium* (Liu et al., 2005, 2012) and *Achromobacter* (Wang et al., 2013) genius. By our group *Rhodococcus armeniensis* AM6.1 strain with D-aminoacylase activity have been isolated from soil samples. This enzyme was purified and its physico-chemical parameters, substrate specificity profile have been studied (Mkhitaryan, 2013; Hambardzumyan et al., 2016).

Based on the substrate specificity *N*-acyl-D-amino acid amidohydrolases have been classified into three types. D-aminoacylase isolated and characterized from a number of bacterial strains (*Pseudomonas*, *Streptomyces*, *Alcaligenes*, etc.) has a brought substrate spectrum, but has no ability to hydrolyze N-acyl-D-aspartate. *N*-acyl-D-aspartate amidohydrolase was isolated and characterized only from *Alcaligenes xylosoxydans* subsp. *xylosoxydans* A-6 (*Alcaligenes* A-6) shows activity to *N*-acyl-D-aspartate. The *N*-acyl-D-glutamate amidohydrolase characterized in *Pseudomonas* sp. 5f-1 and *Alcaligenes* A-6 shows activity to *N*-acyl-D-glutamate (Wakayama and Moriguchi, 2001). The D-aminoacylases from *Alcaligenes*

(Yang et al., 1991, 1992; Moriguchi et al., 1993a; Sakai et al., 1991a, b), *Pseudomonas* (Kubo et al., 1980; Kameda et al., 1978), and *Streptomyces* (Sugie and Suzuki, 1978, 1980) and fungi (Mitsuhashi et al., 2003) have high stereospecificity and broad substrate specificity while *N*-acetyl-derivatives of D-methionine, D-leucine, and *D*-phenylalanine were the preferred substrates for all of the mentioned enzymes. D-aminoacylase of *Rhodococcus armeniensis* AM6.1 shows the highest activity to *N*-acetyl-D-methionine, as well as to aromatic and hydrophobic *N*-acetyl-amino acids. The specific activity toward the basic substrates is the week (Hambardzumyan et al., 2016). Characterized *N*-acyl-D-glutamate amidohydrolase was highly specific to N-acyl derivatives of D-glutamate, formyl-D-glutamate was the most preferred substrate. The enzyme shows an ability to hydrolyze dipeptides such as glycyl-D-glutamate (Sakai et al., 1991a, b). *N*-acyl-D-aspartate amidohydrolase was specific for N-acyl derivatives of D-aspartate (Wang et al., 2013).

Characterized D-aminoacylases are inducible enzymes being induced by D- or N-acyl-D-amino acids. *N*-acetyl-DL-leucine is the good inducer for *Alcaligenes denitrificans* subsp. *xylosoxydans* MI-4 (Moriguchi and Ideta, 1988).

D-aminoacylases, as well as *N*-acyl-D-aspartate amidohydrolase and *N*-acyl-D-glutamate amidohydrolase, were monomeric, with the molecular weight varying from 45 to 100 kDa (Kubo et al., 1980; Kameda et al., 1978; Wang et al., 2013; Arima et al., 2013; Wakayama and Moriguchi, 2001; Yang et al., 1991, 1992; Sakai et al., 1991a, b; Moriguchi et al., 1993a).

Known D-aminoacylases are zinc-containing enzymes (Wakayama et al., 2003; Seibert and Raushel, 2009; Cummings et al., 2009) and some of them are known to be strongly inhibited by Zn^{2+} and metal chelating agents (Wakayama et al., 1995; Sakai et al., 1991a, b; Lin et al., 2002; Yang et al., 1992).

D-aminoacylases of *M. natoriense* TNJL143–2 have been inhibited by Zn^{2+}. This enzyme as well as D-aminoacylases from *Defluvibacter sp.* A131–3 shows full activity in the presence of EDTA (Kumagai et al., 2004). Hg^{2+} has been found to be a strong inhibitor for most of characterized D-aminoacylases, as well as for *N*-acyl-D-glutamate amidohydrolase and for *N*-acyl-D-aspartate amidohydrolase (Yang et al., 1992; Moriguchi et al., 1993a, b; Lin et al., 2002; Wang et al., 2013; Sakai et al., 1991a, b). The activator effect of metal ions on D-aminoacylases was not observed (Kubo et al., 1980; Kameda et al., 1978; Moriguchi et al., 1993a; Yang et al., 1992; Sakai et al., 1991a, b). While Co^{2+} acts as a protector for D-aminoacylase of *Streptomyces olivaceus* to preserve from thermal denaturation, although the metal content has not been reported (Sugie and Suzuki, 1978). *N*-acyl-D-glutamate amidohydrolase of *Pseudomonas* sp. 5f-1 is metalloenzyme. Incubation of EDTA treated *N*-acyl-D-glutamate amidohydrolase of *Pseudomonas* sp. 5f-1 in the presence of Zn^{2+} restored the enzymatic activity. This result indicates the important role of Zn^{2+} for this enzyme, whereas like other metalloenzymes Co^{2+} also restore the activity (Wakayama and Moriguchi, 2001; Sakai et al., 1991a, b). Most of the known D-aminoacylases have temperature optimum in a range from 40 °C to 50 °C (Yang et al., 1992; Wakayama et al., 1995; Moriguchi et al., 1993a; Liu et al., 2012; Wang et al., 2013; Mkhitaryan, 2013) while thermolabile representative has been characterized as well (Arima et al., 2013). All of the characterized enzymes showed optimal activity in the natural or alkaline pH, ranging from 7.0 to 8.5 (Yang et al., 1992; Lin et al., 2002; Wang et al., 2013; Mkhitaryan, 2013).

To be applicable for large-scale applications in industrial harsh conditions number of methods such as different optimization methods, protein engineering and immobilization technologies have been applied to enhance enzyme activity, yield, and stability (Hsu et al., 2002; Wakayama et al., 2003; Yano et al., 2011; Hambardzumyan et al., 2017). Table 2 summarizes the D-aminoacylases and their sources being used or having the potential to be used in amino acid production.

Table 2 D-aminoacylases and their sources being used or having the potential to be used in amino acid production.

Organism	Example of amino acid produced	References
Sebekia	D-methionine, D-valine, D-tryptophan, D-asparagine, *D*-phenylalanine, D-alanine, D-leucine	Tokuyama (2000)
Streptomyces	D-phenylglycine	Sugie and Suzuki (1980)
Defluvibacter	D-valine	Kumagai et al. (2004)
Alcaligenes	D-methionine, D-tryptophan	Tsai et al. (1992) and Nakajima and Yamamoto (2003)
Streptomyces	D-methionine, D-valine, D-tryptophan, D-phenylalanine, D-alanine, D-leucine	Tokuyama and Matsuyama (2005)
Rhodococcus	D-methionine, D-tyrosine, D-allylglycine, D-oxyvaline, D-tryptophane	Hambardzumyan et al. (2017)

3.2 L- and D-amino acids production: Hydantoinase-carbamoilase

One of the most established enzymatic bioconversion methods of obtaining optically pure amino acids is the "Hydantoinase Process." Racemic 5-monosubstituted hydantoins are the starting products of the "Hydantoinase Process" and with the involvement of three enzymes it yields to L- or D-amino acids (Wilms et al., 2001). In this cascade of reactions, the DL-5-monosubstituted hydantoin ring obtained by chemical synthesis is first enantiospecifically hydrolyzed by L- or D-hydantoinase resulting in N-carbamoyl-amino acid. Further, the product of the first reaction hydrolyzed to corresponding free amino acid by highly enantiospecific L- or D-carbamoylase. At the same time, there is also hydantoin racemase enzyme involved in the racemization process of 5-monosubstituted hydantoin from L- or D-form, or vice versa (Fig. 13). It is possible to prepare a wide range of optically pure D- as well as L-amino acids by this method.

From the beginning of the discovery of the "Hydantoinase Process," when enzymes of this process have not been characterized yet and whole-cell systems were used it was the opinion that only a single enzyme catalyzed the conversion of hydantoins to corresponding amino acids. Later in 1978 it has been shown that the "Hydantoinase Process" is a multi-enzymatic system. Yamada and co-authors obtained the intermediate carbamoyl product of the reaction catalyzed by hydantoinase which led to conclude the involvement of more than one enzyme in the process (Yamada et al., 1978). Other facts such as enzymes isolation and characterization, came out later indicating that several steps/reactions catalyzed by different enzymes are involved in the process. After the immobilization of the "enzyme" extract of microorganisms the researchers of Ajinomoto Co. Inc. noted that there was more than one enzyme involved in the conversion (Nakamori et al., 1980).

For the cell, the biological function of these enzymes still remains unclear, although it has been shown the role of some D-hydantoinases being involved in the reductive pathway of pyrimidine catabolism (Vogels and van der Drift, 1976), as well as it has been proven role of L-N-carbamoylases in the synthesis of L-cysteine (Ohmachi et al., 2002). A number of studies indicate that three enzymes (hydantoinase, hydantoin racemase, and carbamoylase) associated with this process are constituents of one operon (Wiese et al., 2001; Grifanti et al., 1998). Hils et al. describe the operon structure for the three genes of D-hydantoinase, D-carbamoylase and putative hydantoin racemase

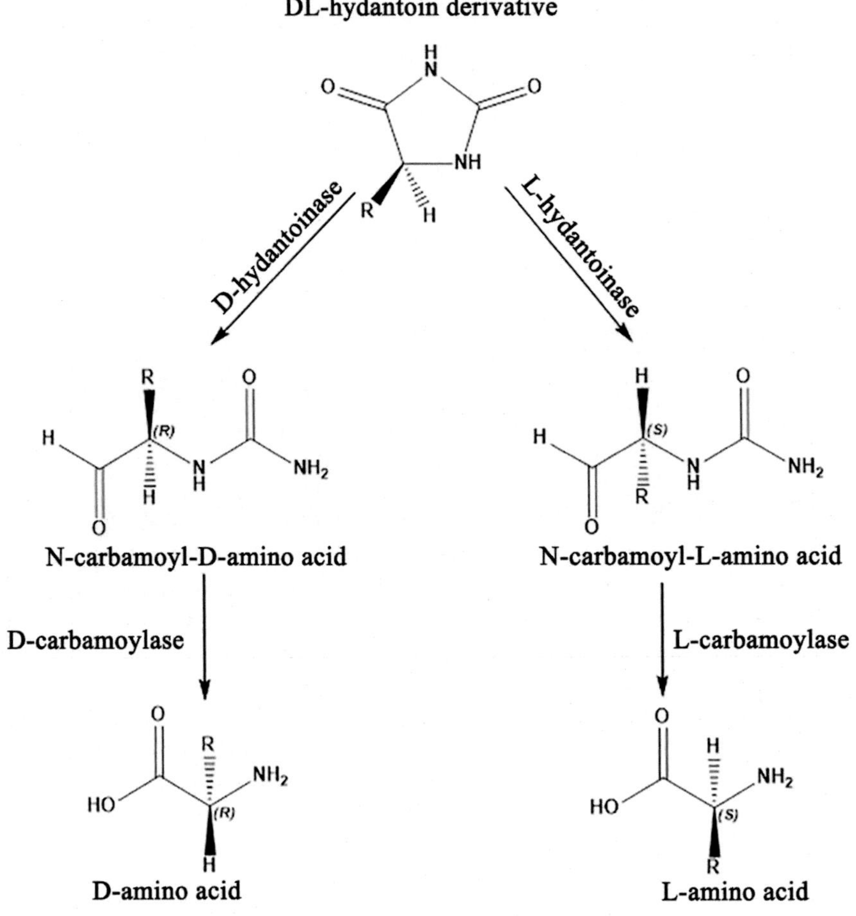

FIG. 13

"Hydantoinase Process."

in *Agrobacterium sp* IP I-671. They have shown that hydantoin utilization genes are localized on the 190 kb plasmid. The fragment of 7125 pb hold the ORF of D-carbamoylase gene (hyuC), at the C-terminal end of operon, following the ORF encoding a hydantoin racemase (hyuA) in the same orientation and the ORF of D-hydantoinase (hyuH) located in the opposite orientation (Hils et al., 2001). D-hydantoinase gene of *Pseudomonas putida* DSM 84 was cloned and characterized (LaPointe et al., 1994). The hydantoinase process is well established in industry and used for the synthesis of optically pure amino acids. It is cost effective, viable and eco-friendly process with number of advantages compared to chemical methods.

With the time in addition to L-form D-form of "Hydantoinase Process" based on microorganisms/enzymes/whole-cell systems was developed and nowadays being applied industrially. This is because it is challenging to obtain D-amino acids from natural sources or by fermentation, but at the same time the demand for D-amino acids and its derivatives is growing.

3.2.1 Hydantoinases

Hydantoinase/dihydropyrimidinase, are the first enzyme of "Hydantoinase Process" which catalyzes the hydrolysis DL-5-mono-substituted hydantoins to corresponding carbamoyl amino acid. These enzymes are widely distributed in microorganisms, plants and animals (LaPointe et al., 1994). In the literature, it is mentioned that hydantoinases (EC 3.5.2.2) were, probably are the microbial equivalent of eukaryotic dihydropyrimidinases. Moreover, often these two terms are used synonymously in EC nomenclature (Engel et al., 2012).

With this in mind, we will not separate dihydropyrimidinase and hydantoinase within this review as we are viewing them in terms of biotechnological importance. Based on the substrate specificity L-, D-, and DL-configuration, hydantoinase can be classified into three types (Syldatk et al., 1999). However, the classification of hydantoinases on the basis of enantioselectivity is arguable as there are literature data showing that, substrate-dependent changes of enantioselectivity. Thus, enantioselectivity of hydantoinases can strongly depend on the substrates tested.

According to the literature data, the same enzyme showing specificity to the L-form of the substrate shows D-selective for another substrate (Yokozeki et al., 1987). This fact indicates that L-hydantoinase can also be D-hydantoinase if other substrate is being tested. Moreover, stereospecificity of hydantoinase is not a key point of hydantoinase process as a ring-opening hydrolysis of 5-substituted hydantoins to corresponding N-carbamyl-amino acids is reversible (Watabe et al., 1992), as well as the spontaneous racemization of 5-substituted hydantoin occurred in the reaction mixture (Yamashiro et al., 1988). This is the reason, that stereoselectivity of the hydantoinase would probably not have a significant influence on the biotransformation reaction. Various microorganisms such as *Pseudomonas* sp. NS671 (Watabe et al., 1992), *P. putida* (Chien et al., 1998), *Agrobacterium* sp. IP I-671 (Hils et al., 2001), *Arthrobacter aurescens* DSM 3745 (Wiese et al., 2001), *Arthrobacter crystallopoietes* DSM 20117 (Slomka et al., 2017), *Ochrobactrum anthropi* (Pozo et al., 2002), *Burkholderia pickettii* (Xu et al., 2003), *Bacillus stearothermophilus* (Cheon et al., 2002), *Geobacillus stearothermophilus* (Weigel et al., 2013; Aganyants et al., 2020), etc., have been isolated and characterized as a potential source of this enzyme. All characterized hydantoinases are either homodimers or homotetramers. The molecular weight of subunits varies between 50 and 60 kDa. Homodimer form of enzymes from *B. stearothermophilus SD-1* (Lee et al., 1995), *A. crystallopoietes* DSM 20117 (Slomka et al., 2017), *B. pickettii* (Xu et al., 2003), *Pseudomonas stutzeri* (Xu and West, 1994) and homotetramer enzymes from *Thermus* sp. (Abendroth et al., 2002) and *B. stearothermophilus* (Cheon et al., 2002) have been characterized.

Hydantoinases require metal ions for their activity possessing a binuclear metal-binding site. Zn^{2+} is cofactor of these enzymes but other divalent metals such as Mn^{2+} or Co^{2+} can play the role of a cofactor. Thus, in the active site of L-hydantoinase from Arthobacter aurescens DSM 3745 has zinc in the active center (Abendroth et al., 2002), wears *B. pickettii* contains Cd2+ ion (Xu et al., 2003). The authors also indicated that the enzymatic activity of hydantoinase can be stimulated by various metal ions, including Co^{2+}, Mn^{2+}, Zn^{2+}, Fe^{2+}, and Ni^{2+}, Co^{2+} being the most effective. Thermostability is very important for this enzyme as higher temperatures increase the solubility of hydantoins and enhanced the rate of their racemization. Recently a thermophile hydantoinase has been characterized from *G. stearothermophilus* ATCC 31783 (Weigel et al., 2013; Aganyants et al., 2020) showing high activity in the pH range from 8.0 to 9.2 after incubation at 65 °C for 20 min as well as from *Pseudomonas aeruginosa* (MCM B-887) showing optimal activity at 42 °C and pH 9.0 (Engineer et al., 2020).

Characterization of D- and L-specific hydantoinases of different microorganisms and other natural sources, the discovery of highly enantioselectivity and a broad spectrum of substrate specificity of

Table 3 Hydantoinases and their sources being used or having the potential to be used in amino acid production.

Organism	Example of amino acid produced	References
Arthrobacter, Flavobacterium	L-tryptophan	Miyoshi et al. (1985), Syldatk et al. (1987), Yokozeki et al. (1987b), Nishida et al. (1987), and Wilms et al. (2001)
Arthrobacter	L-methionine	Wagner et al. (1996)
Nocardia	L-leucine	Klages et al. (1988)
Bacillus, Pseudomonas, Aerobacter, Micorcoccus, Flavobacterium	L-glutamic acid	Tsugawa et al. (1966)
Agrobacterium	D-p-OH-phenylglycine	Olivieri et al. (1979)
Arthrobacter	D-serine, D-phenylalanine	Slomka et al. (2017)
Bacillus	D-phenylglycine	Lee et al. (1995) and Slomka et al. (2017)

hydantoinases, as well as the possibility of easy racemization of the hydantoin derivatives led to renewed interest toward hydantoinases, and characterization of novel enzymes and sources with new properties remains in the spotlight of researchers. Table 3 summarise the hydantoinases and their sources being used or having the potential to be used in amino acid production.

3.2.2 Carbamoilases

Carbamoylase (N-carbamoyl-amino acid amidohydrolases), included in E.C. 3.5.1, catalyzes the conversion of N-carbamoyl-amino acid into the free amino acid, forming ammonia and carbon dioxide as secondary products. Unlike hydantoinase this enzyme has absolute stereospecificity. Based on the substrate specificity there are two types of N-carbamoyl-amino acid amidohydrolases, L-carbamoylase (E.C. 3.5.1.77) and D-carbamoylase (E.C. 3.5.1.87). N-carbamoyl β-alanine amidohydrolase which is a key enzyme for pyrimidine degradation, have been characterized as a L-carbamoylase given the fact that it has brought substrate spectrum and enantiospecificity toward N-carbamoyl-L-amino acids (Martínez-Gómez et al., 2011). Carbamoylase attracted the attention as an industrially applicable enzyme due to its 100% stereospecificity and substrate promiscuity (May et al., 2002; Clemente-Jiménez et al., 2008).

3.2.3 L-carbamoylases

L-N-carbamoylases has been characterized among microorganisms belonging to diverse genera, such as *Arthrobacter* (Wilms et al., 1999; Pietzsch et al., 2000), *Alcaligenes* (Ogawa et al., 1995), *Bacillus* (Batisse et al., 1997), *Blastobacter* (Yamanaka et al., 1997), *Flavobacterium* (Yokozeki et al., 1987), *Microbacterium* (Suzuki et al., 2005), and *Sinorhizobium* (Martínez-Rodríguez et al., 2005). L-carbamoylases obtained from *Bacillus* genious are the most studied of these enzymes (Martínez-Rodríguez et al., 2010). N-carbamoyl β-alanine amidohydrolase with the L-carbamoylase activity has been has been characterized from microbial sources (Martínez-Gómez et al., 2009, 2011).

L-carbamoylases are described as homodimers. Purified enzyme of *Alcaligenes xylosoxidans* has 135 kDa molecular weight consisting of two identical subunits with 65 kDa molecular weight (Ogawa et al., 1995) while L-carbamoylases characterized from *Arthrobacter aurescens* DSM3747 (Wilms et al., 1999), *Pseudomonas* sp. NS671 (Ishikawa et al., 1996), and *Geobacillus stearotherm* (Martínez-Rodríguez et al., 2008) have smaller molecular weights consisting of two identical subunits with 45 kDa or 44 kDa molecular weights. L-carbamoylases from *Pseudomonas* sp. ON-4 is the only representative which has a homotetrameric structure having four 45 kDa monomers (Ohmachi et al., 2004).

L-carbamoylases are metal-containing enzymes, several divalent metals, such as Mn^{2+} and Co^{2+}, increase their activity (Martínez-Rodríguez et al., 2005). Chelating agents such as EDTA inhibit enzyme activity, metal ions restore the enzyme activity. This effect has been reported for *B. kaustophilus* CCRC 11223, *A. xylosoxidans* and *P. putida* IFO 12996L-carbamoylases as well as for N-carbamoyl-L-cysteine amidohydrolase from *Pseudomonas* sp. ON-4a. (Hu et al., 2003; Ogawa and Shimizu, 1994; Ogawa et al., 1995; Ohmachi et al., 2004).

According to date reported, the substrate specificity of L-N-carbamoylases characterized from different bacterial sources is varying. *Bacillus fordii* MH602 prefers aromatic (such as carbamoyl-phenylalanine, -phenylglycine and -tryptophan) or very short chain substrates (Mey et al., 2008), *P. putida* only yields aliphatic L-amino acids (Buchanan et al., 2001), the corresponding enzymes of *Bacillus stearothermophilus* NS1122A (Ishikawa et al., 1993) and *Arthrobacter* sp. DSM7330 (Wagner et al., 1996) shows specificity only toward carbamoyl-L-methionine, whereas the enzymes from *Arthrobacter* sp. DSM3747 (Gross et al., 1990) and *Flavobacterium* sp. (Nishida et al., 1987) yields L-tryptophan.

Based on the substrate specificity authors tried to classify the three types of L-N-carbamoylases: aliphatic, aromatic, or no clear preference for aliphatic/aromatic substrates. However, with the characterization of new representatives of carbamoylases classification becomes more difficult due to unclear substrate type tendency (Martínez-Rodríguez et al., 2010). The studied L-carbamoylases present low thermostability in general. Thus, the maximum specific activity of *Sinorhizobium meliloti* CECT4114 was obtained at pH 8.0 and 60 °C; however, enzyme was unstable at 60 °C (Martínez-Rodríguez et al., 2005). Similar picture has been obtained for L-N-carbamoylase of *Pseudomonas* sp. NS 671 (Ishikawa et al., 1993). On the contrary, *A. aurescens* DSM 3747, *A. xylosoxidans* and *P. putida* IFO 12996 L-N-carbamoylases are thermostable (Wilms et al., 1999; Ogawa et al., 1995; Ogawa and Shimizu, 1994). The specific activity L-N-carbamoylases of *B. kaustophilus* CCRC 11223 increased by only 10%–20% after incubation at 50 °C for 20 min (Hu et al., 2003).

3.2.4 D-carbamoylase

D-carbamoylases have been found from several Gram-positive and Gram-negative bacteria. This enzyme is selective for D-forms of carbamoyle substrates and catalyzed its enzymatic conversion to corresponding D-amino acid. The interest toward D-carbamoylase has been grown especially after finding its application in the multienzymatic synthesis of semisynthetic antibiotics such as penicillins and cephalosporins (ampicillin and amoxicillin), where *D*-phenylglycine and D-p-hydroxyphenylglycine have been used for the formulation of side chains. The chemical production of these compounds has many disadvantages such as low yield and waste management problems. Therefore, biocatalytic process is a great alternative for the chemical process.

The first microbial source of D-carbamoylase was *Agrobacterium radiobacter* (Olivieri et al., 1979) and later the enzyme has been characterized from number of microorganisms such as *Arthrobacter* (Möller et al., 1988), *Pseudomonas* (Ikenaka et al., 1998), *Sinorhizobium* (Wu et al., 2005), *Comamonas* (Ogawa et al., 1993), *Burkholderia* (Xu et al., 2002), amd *Bradyrhizobium* (Bellini et al., 2019). The molecular weight of the native D-carbamoylase varies from 67 to 150 kDa. The enzyme from different sources has dimeric, trimeric, and tetrameric quaternary structures. D-carbamoylase monomers calculated by amino acid sequence or experimentally range from 32 to 40 kDa. Thus, it has been shown that the enzymes isolated from *A. radiobacter* is a homodimer with 34 kDa subunit molecular weight (Grifantini et al., 1996; Wang et al., 2001), the enzymes of *Blastobacter* spp. and *Comamonas* spp. were identified as a homotrimer with subunit molecular mass approximately 40 kDa (Ogawa et al., 1993, 1994), as well as enzymes with homotetrameric structure with 38 kDa subunit molecular weight, has been characterized from *Sinorhizobium morelens* S-5 (Wu et al., 2006). Nevertheless, the crystallographic structure of D-carbamoylase was initially reported to be homotetrameric (Wang et al., 2001; Nakai et al., 2000; Chiu et al., 2006).

D-carbamoylases exhibit broad substrate specificity toward different N-carbamoyl-D-amino acids. Thus, *N*-carbamoyl-DL-methionine was found to be the best substrate for D-carbamoylase of *A. tumefaciens* AM 10 (Sareen et al., 2001) and D-carbamoylases of other *Agrobacterium* strains (Louwrier and Knowles, 1996; Nanba et al., 1998). The purified D-carbamoylase of *Blastobacter* sp. preferred N-carbamoyl-D-amino acids having an aromatic or long alkyl group. The enzyme seems to prefer long aliphatic (N-carbamoyl-D-methionine and N-carbamoyl-D-norleucine) and aromatic derivatives (N-carbamoyl-*D*-phenylglycine and N-carbamoyl-hydroxy-phenylglycine), whereas charged/polar amino acids were not good substrates (Nanba et al., 1998; Wu et al., 2005). The enzyme characterized by *Arthrobacter crystallopoietes* is different from D-carbamoylases of *Sinorhizobium morelens* S-5, *Flavobacterium* sp. AJ11199. It preferred aromatic to aliphatic substituted *N*-carbamoyl-amino acids, displaying the highest specific activity toward *N*-carbamoyl-D-tryptophan and no detectable activity toward all tested aliphatic substituted *N*-carbamoyl-amino acids (Liu et al., 2018).

Characterized D-carbamoylases have a broad spectrum of pH optimum varying from the range of 7.0 to 9.0 and are generally thermostable. The optimum temperature of the reaction is found to be varying from 40 °C to 73 °C (Ogawa et al., 1994; Grifantini et al., 1996; Nozaki et al., 2005; Wu et al., 2006; Louwrier and Knowles, 1996). D-carbamoylases are not affected by chelating agents which means that cations are not necessary for enzymatic activity (Nozaki et al., 2005; Wu et al., 2006). Table 4 summarizes the carbamoylase and their sources being used or having the potential to be used in amino acid production.

Table 4 Carbamoylase and their sources being used or having the potential to be used in amino acid production.

Organism	Example of amino acid produced	References
Arthrobacter, Flavobacterium	L-tryptophan	Wilms et al. (2001) and Yokozeki et al. (1987b)
Pseudomonas	L-cysteine, L-alanine, L-valine, L-norleucine	Ohmachi et al. (2002) and Buchanan et al. (2001)
Arthrobacter	L-methionine	Slomka et al. (2017) and Olivieri et al. (1979)
Agrobacterium	*D*-hydroxyphenylglycine	Jiwaji et al. (2009)
Arthrobacter	D-serine D-phenylglycine *D*-phenylalanine	
Agrobacterium	D-valine	Battilotti and Barberini (1988)
Arthrobacter	D-tryptophan	Liu et al. (2018)
Agrobacterium	D-parahydroxy-phenylglycine, *D*-phenylglycine	Olivieri et al. (1982)

3.2.5 Whole-cell systems for hydantoinase process

Whole-cell biocatalyst gives a unique asset to develop sustainable processes and has been widely used for the efficient biosynthesis of value-added products, as well as pharmaceutically active ingredients providing reliable outcomes. Nowadays research is going to the direction of producing amino acids using whole-cell system, which allows to design the process rationally. Thus, a number of whole-cell biocatalyst systems were designed and applied for L-amino acids synthesis from hydantoins. L-hydantoinase, L-*N*-carbamoylase, and hydantoin racemase encoding genes from *A. aurescens* were cloned and co-expressed in *E. coli*. The productivity of the system was more than 6-fold higher than the activity achieved with *Arthrobacter* enzymes. The multienzyme system has been applied to produce L-tryptophan from the corresponding hydantoin (Wilms et al., 2001).

An L-stereoselective whole-cell system was used to produce a number of aromatic L-amino acids as well as their derivatives. The whole-cell system was developed based on L-hydantoinase, L-carbamoylase, and racemase enzymes of *A. aurescens* DSM 3745 and DSM 3747 (Matcher et al., 2012).

A fully enzymatic process employing D-hydantoinase of *B. stearothermophilus* SD1and N-carbamoylase of *A. tumefaciens* NRRL B11291 was developed for the production of D-amino acid from 5′-monosubstituted hydantoin. Co-expressed *E. coli* host cells were applied for production of *D*-hydroxyphenylglycine (Park et al., 2000).

Nozaki and co-authors have used whole-cell system containing three enzymes. With the use of *E. coli* containing co-expressed D-specific hydantoinase, *N*-carbamoyl-D-amino acid amidohydrolase and hydantoin racemase. With this system, it was possible to produce D-forms of eight amino acids, such as *D*-phenylalanine, D-tyrosine, D-tryptophan, *O*-benzyl-D-serine, D-valine, D-norvaline, D-leucine and D-norleucine from the corresponding DL-5-monosubtituted hydantoins (Nozaki et al., 2005).

Recombinant *E. coli* strain with one plasmid harboring three genes encoding D-hydantoinase and D-carbamoylase obtained from *A. tumefaciens* BQL9 and hydantoin racemase obtained from *A. tumefaciens* C58, co-expressed by the same promoter, was applied to produce optically pure D-amino acids and their derivatives from the corresponding hydantoin racemic mixture (Martinez-Gomez et al., 2007).

4 Conclusion

The demand for amino acids is growing, which mostly is due to the expansion of their application in different fields, including as precursors for the synthesis of many biologically important compounds. In parallel with the growing market amino acid production technology has made large progress. Currently, almost all L-amino acids are produced by microbial approaches—fermentative or enzymatic ways.

Due to the number of advantages, the fermentation way of amino acid production is a widely used method at an industrial scale. It is applied to produce a number of L-amino acids except for a few kinds. The creation of microbial strains with improved amino acid productivity and lower by-product formation is essential to reduce the overall process expenses. Basic requirements for strain producers include also non-pathogenicity and growth on cheap substrates. Nowadays by application of genetic engineering tools such as genetic recombination, recombinant DNA technologies let eliminate the metabolic regulatory/control processes and develop highly efficient amino acid-producing strains by mutagenesis and screening techniques. The major advantages of the fermentative way are that it produces mainly the L-forms although, this way requires sterility and mixing that impact on operation costs.

The enzymatic way is based on the action of an enzyme or multienzyme complex which catalyze the production of the target amino acids. Not only fermentation way, but also enzymatic way are continuously improving, especially after the development of metagenome and metatranscriptome technologies allowing to identify novel enzymes. By this way, it can be produced as optically pure D- and as well as L-amino acids, but it is especially popular for the production of D-amino acids and nonproteinogenic L-amino acids. With this whey it is possible to obtain products with higher yields and with an extremely decreased by-products contamination. However, this method has a number of disadvantages such as price, possible product inhibitions, limited stability of enzyme which required strong regulations of reaction conditions. All these are the main drawbacks of the enzymatic way.

Thus, the choice of the most advantageous method for amino acids synthesis depends on lots of factors, such as available sustainable technologies, market size operation costs as well as environmental impact.

References

Abendroth, J., Niefind, K., Schomburg, D., 2002. X-ray structure of a dihydropyrimidinase from thermus sp. at 1.3 Å resolution. J. Mol. Biol. 320 (1), 143–156.

Adenan, S., et al., 2018. Isolation, identification and molecular characterization of thermophilic aminoacylase from Geobacillus sp. strain SZN. Int. J. Sci. Environ. Technol. 7 (5), 1483–1494.

Aganyants, H.A., et al., 2013a. Biochemical characteristics of recombinant L-aminoacylase of *Escherichia coli*. Biol. J. Armenia Supplement 1 (65), 29–30.

Aganyants, H.A., et al., 2013b. Immobilization of recombinant L-aminoacylase from *Geobacillus stearotermophilus* and characteristics of obtained preparations. Chem. Biol. 1, 32–35.

Aganyants, H., et al., 2020. Rational engineering of the substrate specificity of a thermostable D-hydantoinase (dihydropyrimidinase). High-Throughput 9 (1), 5.

Ambartsumyan, A., Bezirdzhyan, K., 1994a. Catalytic properties of valine:pyruvate aminotransferase from Brevibacterium flavum. Biochemistry (Mosc.) 59 (9), 1027–1032.

Ambartsumyan, A., Bezirdzhyan, K., 1994b. Isolation and preliminary characterization of valine: pyruvate aminotransferase from Brevibacterium flavum. Biochemistry (Mosc.) 59 (9), 1021–1026.

Arima, J., Isoda, Y., Hatanaka, T., Mori, N., 2013. Recombinant production and characterization of an N-acyl-D-amino acid amidohydrolase from Streptomyces sp. 64E6. World J. Microbiol. Biotechnol. 29 (5), 899–906.

Avetisova, G., 2002. Study of physiological properties of the L-valine producing strain. Biol. J. Armenia 54 (3–4), 204–209.

Avetisova, G., Hambardzumyan, A., Kocharyan, S., Saghyan, A., 2003a. The calculation of maximum productivity and other coefficients for valine biosynthesis by Brevibacterium flavum. Proc. Yerevan State Univ. (2), 94–99.

Avetisova, G., Pogosyan, N., Azizian, A., Hambardzumyan, A., 2003b. Study on the properties of acetolactate synthase in Brevibacterium flavum strains with overproduction of L-valine. J. Biotechnol. 1, 39–43.

Avetisova, G., Melkonyan, L., Chakhalyan, A. & Saghiyan, A., 2012. Method of L-alanine production. RA, Patent No. 2691 A.

Avetisova, G., et al., 2014a. Selection of new highly active L-alanine producer strains of Brevibacterium flavum and comparison of their activity in alanine synthesis. Russ. J. Genet. Appl. Res. 4 (1), 23–26.

Avetisova, G., et al., 2014b. Development of L-Alanine Biosynthesis Technology. Yerevan, Armenia, s.n, p. 48.

Azizian, A., Avetisova, G., Arushanian, A. & Kocharian, S., 1990a. Method for production of L-valine. United Kingdom, Patent No. GB2199323B.

Azizian, A., Avetisova, G., Arushanian, A. & Kocharian, S., 1990b. Procede de Preparation de L-Valine. France, Patent No. FR 2609047-B1.

Azizian, A., Avetisova, G., Arushanian, A. & Kocharian, S., 1991. Method for production of L-valine. USSR, Patent No. SU 1675327 A1.

Azizian, A., Ambartsumian, A., Ananikian, M. & Kocharian, S., 1992. Method for preparing L-alanine. USA, Patent No. US5124257A.

Azuma, S., et al., 1993. Hyper-production of L-trytophan via fermentation with crystallization. Appl. Microbiol. Biotechnol. 39, 471–476.

Batisse, N., Weigel, P., Lecocq, M., Sakanyan, V., 1997. Two amino acid amidohydrolase genes encoding L-stereospecific carbamoylase and aminoacylase are organized in a common operon in *Bacillus stearothermophilus*. Appl. Environ. Microbiol. 63, 763–766.

Battilotti, M., Barberini, U., 1988. Preparation of D-valine from DL-5-isopropylhydantoin by stereoselective biocatalysis. J. Mol. Catal. 43, 343–352.

Bellini, R., et al., 2019. Structural analysis of a novel N-carbamoyl-d-amino acid amidohydrolase from a Brazilian *Bradyrhizobium japonicum* strain: in silico insights by molecular modelling, docking and molecular dynamics. J. Mol. Graph. Model. 86, 35–42.

Buchanan, K., Burton, S., Dorrington, R., 2001. A novel *Pseudomonas putida* strain with high levels of hydantoin-converting activity, producing L-amino acids. J. Mol. Catal. B: Enzym. 11, 397–406.

Cheon, Y., et al., 2002. Crystal structure of D-hydantoinase from *Bacillus stearothermophilus*: insight into the stereochemistry of enantioselectivity. Biochemistry 41, 9410–9417.

Chibata, I., et al., 1972. Preparation and industrial application of immobilized aminoacylases. In: Proc. IV IFS: Ferment Technology Today, pp. 383–389.

Chibata, I., et al., 1975. Applications of immobilized enzymes and immobilized microbial cells for L-amino acid production. In: Immobilized Enzyme Technology. Springer, Boston.

Chien, H., Jih, Y.-L., Yang, W.-Y., Hsu, W., 1998. Identification of the open reading frame for the *Pseudomonas putida* D-hydantoinase gene and expression of the gene in *Escherichia coli*. Biochim. Biophys. Acta 1395, 68–77.

Chiu, W., et al., 2006. Structure–stability–activity relationship in covalently cross-linked N-carbamoyl d-amino acid amidohydrolase and N-acylamino acid racemase. J. Mol. Biol. 359, 741–753.

Cho, H., Tanizawa, K., Tanaka, H., Soda, K., 1987. Thermostable aminoacylase from *Bacillus thermoglucosidius*: purification and characterization. Agric. Biol. Chem. 51 (10), 2793–2800.

Clemente-Jiménez, J., Martínez-Rodríguez, S., Rodríguez-Vico, F., Las Heras-Vázquez, F., 2008. Optically pure alpha-amino acids production by the "Hydantoinase Process". Recent Pat. Biotechnol. 2, 35–46.

Cummings, J., et al., 2009. Annotating enzymes of uncertain function: the deacylation of D-amino acids by members of the amidohydrolase superfamily. Biochemistry 48, 6469–6481.

Curley, P., Sinderen, D., 2000. Identification and characterisation of a gene encoding aminoacylase activity from *Lactococcus lactis* MG1363. FEMS Microbiol. Lett. 183 (1), 177–182.

Curley, P., van der Does, C., Driessen, A.J., van Sinderen, J.D., 2003. Purification and characterisation of a lactococcal aminoacylase. Arch. Microbiol. 179, 402–408.

D'Este, M., Alvarado-Morales, M., Angelidaki, I., 2018. Amino acids production focusing on fermentation technologies—a review. Biotechnol. Adv. 36 (1), 14–25.

Dodge, T., Gerstner, J., 2002. Optimization of the glucose feed rate profile for the production of tryptophan from recombinant E. coli. J. Chem. Technol. Biotechnol. 77 (11), 1238–1245.

El-Mansi, E., et al., 2012. Fermentation Microbiology and Biotechnology, third ed. CRC Press, Boca Raton.

Engel, U., Syldatk, C., Rudat, J., 2012. Novel amidases of two *Aminobacter* sp. strains: biotransformation experiments and elucidation of gene sequences. AMB Express 2 (33), 1–13.

Engineer, A., Yadav, K., Kshirsagar, P., Dhakephalkar, P., 2020. A novel, enantioselective, thermostable recombinant hydantoinase to aid the synthesis of industrially valuable non-proteinogenic amino acids. Enzyme Microb. Technol. 138, 109554.

Gentzen, I., Löffler, H.G., Schneider, F., 1980. L-Aminoacylase from *Aspergillus oryzae*. Comparison with the pig kidney enzyme. Z. Naturforsch. C: Biosci. 35 (7–8), 544–550.

Greenstein, J.P., 1957. Methods in Enzymology. Academic Press, New York.

Grifanti, R., et al., 1998. Efficient conversion of 5-substituted hydantoins to D-α-amino acids using recombinant *Escherichia coli* strains. Microbiology 144, 947–954.

Grifantini, R., Pratesi, C., Galli, G., Grandi, G., 1996. Topological mapping of the cysteine residues of N-carbamyl-D-amino-acid amidohydrolase and their role in enzymatic activity. J. Biol. Chem. 271, 9326–9331.

Gross, C., Syldatk, C., Mackowiak, V., 1990. Cell growth and enzyme synthesis of a mutant of *Arthrobacter* sp. (DSM3747) used for the production of L-amino acids from DL-5-monosubstituted hydantoins. Biotechnology 14, 363–376.

Hambardzumyan, A.A., et al., 2016. Catalytic properties of aminoacylase of strain *Rhodococcus armeniensis* AM6.11. Appl. Biochem. Microbiol. 52 (3), 250–255.

Hambardzumyan, A.A., Mkhitaryan, A.V., Paloyan, A.M., Dadayan, S.A., 2017. Covalent immobilization of D-aminoacylase of strain *Rhodococcus armeniensis* AM6.1 and the characteristics of the biocatalyst. Appl. Biochem. Microbiol. 53 (1), 20–24.

Hasegawa, S., et al., 2012. Improvement of the redox balance increases L-valine production by *Corynebacterium glutamicum* under oxygen deprivation conditions. Appl. Environ. Microbiol. 78 (3), 865–875.

Hasegawa, S., et al., 2013. Engineering of corynebacterium glutamicum for high-yield L-valine production under oxygen deprivation conditions. Appl. Environ. Microbiol. 79 (4), 1250–1257.

He, B., Li, M., Wang, D., 1992. Studies on aminoacylase from *Aspergillus oryzae* immobilized on polyacrylate copolymer. React. Polym. 17 (3), 341–346.

Hernick, M., Fierke, C., 2010. Mechanisms of metal-dependent hydrolases. In: Liu, L.M.H. (Ed.), *Comprehensive Natural Products II Chemistry and Biology.* s.l. Elsevier Science, pp. 562–563.

Hils, M., et al., 2001. Cloning and characterization of genes from *Agrobacterium* sp. IP I-671 involved in hydantoin degradation. Appl. Microbiol. Biotechnol. 57, 680–688.

Hopkins, F., Cole, S., 1901. A contribution to the chemistry of proteids: part I. A preliminary study of a hitherto undescribed product of tryptic digestion. J. Physiol. 27 (4–5), 418–428.

Hou, X., et al., 2012. Improvement of L-valine production at high temperature in *Brevibacterium flavum* by overexpressing ilvEBNrC genes. J. Ind. Microbiol. Biotechnol. 39, 63–72.

Hsu, C.-S., Lai, W.-L., Chang, W.-W., Liaw, S.-H., Tsai, Y.-C., 2002. Structural based mutational analysis of D-aminoacylase from *Alcaligenes faecalis* DA1. Protein Sci. 11, 2545–2550.

Hu, H., Hsu, W., Chien, H., 2003. Characterization and phylogenetic analysis of a thermostable N-carbamoyl-L-amino acid amidohydrolase from *Bacillus kaustophilus* CCRC11223. Arch. Microbiol. 179, 250–257.

Ikeda, M., 2003. Amino acid production processes. In: Faurie, R., et al. (Eds.), Microbial Production of L-Amino Acids. Advances in Biochemical Engineering/Biotechnology. Springer, Berlin, Heidelberg, Berlin, Heidelberg, pp. 1–35.

Ikeda, M., Katsumata, R., 1999. Hyperproduction of tryptophan by *Corynebacterium glutamicum* with the modified pentose phosphate pathway. Appl. Environ. Microbiol. 65 (6), 2497–2502.

Ikenaka, Y., et al., 1998. Screening, characterization, and cloning of the gene for N-carbamyl-D-amino acid amidohydrolase from thermotolerant soil bacteria. Biosci. Biotechnol. Biochem. 62 (5), 882–886.

Ishikawa, T., et al., 1993. Microbial conversion of DL-5-substituted hydantoins to the corresponding L-amino acids by Pseudomonas sp. strain NS671. Biosci. Biotechnol. Biochem. 57, 982–986.

Ishikawa, T., Watabe, K., Mukohara, Y., Nakamura, 1996. N-carbamyl-L-amino acid amidohydrolase of *Pseudomonas* sp. strain NS671: purification and some properties of the enzyme expressed in *Escherichia coli.* Biosci. Biotechnol. Biochem. 60, 612–615.

Jenkins, A., Nguyen, J., Polglaze, K., Bertrand, P., 2016. Influence of tryptophan and serotonin on mood and cognition with a possible role of the gut-brain axis. Nutrients 8 (1), 56.

Jiwaji, M., et al., 2009. Enhanced hydantoin-hydrolyzing enzyme activity in an *Agrobacterium tumefaciens* strain with two distinct N-carbamoylases. Enzyme Microb. Technol. 44, 203–209.

Jojima, T., et al., 2010. Engineering of sugar metabolism of *Corynebacterium glutamicum* for production of amino acid L-alanine under oxygen deprivation. Appl. Microbiol. Biotechnol. 87 (1), 159–165.

Jojima, T., Vertès, A., Inui, M., Yukawa, H., 2014. Development of growth—arrested bioprocesses with corynebacterium glutamicum for cellulosic ethanol production from complex sugar mixtures. In: Biorefineries Integrated Biochemical Processes for Liquid Biofuels. Elsevier B.V., Amsterdam, pp. 121–139.

Kameda, Y., Hase, H., Kanamoto, S., Kita, Y., 1978. Studies on acylase activity and micro-organisms. XXVI. purification and properties of D-acylase (N-Acyl-D-amino-acid Amidohydrolase) from AAA 6029 (*Pseudomonas* sp.). Chem. Pharm. Bull.(Tokyo) 26, 2698–2704.

Kempf, B., Bremer, E., 1996. A novel amidohydrolase gene from *Bacillus subtilis* cloning: DNA-sequence analysis and map position of amhX. FEMS Microbiol. Lett. 141, 129–137.

Klages, U., Weber, A. & Wilschowitz, L., 1988. Verfahren zur Herstellung von L-Aminosäuren. German patent, Patent No. 3702384 A1.

Kubo, K., Ishikura, T., Fukagawa, Y., 1980. Deacetylation of PS-5, a new beta-lactam compound. II. Separation and purification of L- and D-amino acid acylases from *Pseudomonas* sp. 1158. J. Antibiot. 33 (6), 550–555.

Kumagai, H., 2006. Amino acid production. In: The Prokaryotes. Springer Science+Business Media, Inc, New York, pp. 756–765.

Kumagai, S., et al., 2004. A new d-aminoacylase from *Defluvibacter* sp. A 131-3. J. Mol. Catal. B: Enzym. 30, 159–165.

LaPointe, G., et al., 1994. Cloning, sequencing, and expression in *Escherichia coli* of the D-hydantoinase gene from *Pseudomonas putida* and distribution of homologous genes in other microorganisms. Appl. Environ. Microbiol. 60 (3), 888–895.

Lee, S.-G., et al., 1995. Ther- mostable D-hydantoinase from thermophilic *Bacillus stearothermophilus* SD-1: characteristics of the purifed enzyme. Appl. Microbiol. Biotechnol. 43, 270–276.

Liaw, S., et al., 2003. Crystal structure of D-aminoacylase from Alcaligenes faecalis DA1. A novel subset of amidohydrolases and insights into the enzyme mechanism. J. Biol. Chem. 278, 4957–4962.

Lin, P., Su, S., Tsai, Y., Lee, C., 2002. Identification and characterization of a new gene from Variovorax paradoxus Iso1 encoding N-acyl-D-aminoacid amidohydrolase responsible for D-amino acid production. Eur. J. Biochem. 269, 4868–4878.

Lindner, H., et al., 2000. The distribution of aminoacylase I among mammalian species and localization of the enzyme in porcine kidney. Biochimie 82, 129–137.

Lindner, H.A., et al., 2003. Essential roles of zinc ligation and enzyme dimerization for catalysis in the aminoacylase-1/M20 family. J. Biol. Chem. 278, 44496–44504.

Liu, J., et al., 2005. *Microbacterium natoriense* sp. nov., a novel D-aminoacylaseproducing bacterium isolated from soil in Natori, Japan. Int. J. Syst. Evol. Microbiol. 55, 661–665.

Liu, J., et al., 2012. Purification, characterization, and primary structure of a novel N-acyl-D-amino acid amidohydrolase from *Microbacterium natoriense* TNJL143-2. J. Biosci. Bioeng. 114, 391–397.

Liu, Y., et al., 2018. Identification of D-carbamoylase for biocatalytic cascade synthesis of D-tryptophan featuring high enantioselectivity. Bioresour. Technol. 249, 720–728.

Louwrier, A., Knowles, C., 1996. The purification and characterization of a novel D(-)-specific carbamoylase enzyme from an *Agrobacterium* sp. Enzyme Microb. Technol. 19, 562–571.

Lv, Y., et al., 2012. Genome sequence of *Corynebacterium glutamicum* ATCC 14067, which provides insight into amino acid biosynthesis in coryneform bacteria. J. Bacteriol. 194 (3), 742–743.

Martinez-Gomez, A., et al., 2007. Recombinant polycistronic structure of hydantoinase process genes in *Escherichia coli* for the production of optically pure D-amino acids. Appl. Environ. Microbiol. 73 (5), 1525–1531.

Martínez-Gómez, A., et al., 2009. Potential application of N-carbamoyl-β-alanine amidohydrolase from *Agrobacterium tumefaciens* C58 for β-amino acid production. Appl. Environ. Microbiol. 75, 514–520.

Martínez-Gómez, A., et al., 2011. N-Carbamoyl-β-alanine amidohydrolase from *Agrobacterium tumefaciens* C58: a promiscuous enzyme for the production of amino acids. J. Chromatogr. B Analyt. Technol. Biomed. Life Sci. 879 (29), 3277–3282.

Martínez-Rodríguez, S., Clemente-Jiménez, J.M., Rodríguez-Vico, F., Heras-Vázquez, F.J.L., 2005. Molecular cloning and biochemical characterization of L-N-carbamoylase from *Sinorhizobium meliloti* CECT4114. Mol. Microbil. Biotechnol. 9, 16–25.

Martínez-Rodríguez, S., et al., 2008. Crystallization and preliminary crystallographic studies of the recombinant L-N-carbamoylase from *Geobacillus stearothermophilus* CECT43. Acta Crystallogr. F64, 1135–1138.

Martínez-Rodríguez, S., et al., 2010. Carbamoylases: characteristics and applications in biotechnological processes. Appl. Microbiol. Biotechnol. 85, 441–458.

Matcher, G., Dorrington, A., Burton, S., 2012. Enzymatic production of enantiopure amino acids from mono-substituted hydantoin substrates. Methods Mol. Biol. 794, 37–54.

May, O., Verseck, S., Bommarius, A., Drauz, K., 2002. Development of dynamic kinetic resolution processes for biocatalytic production of natural and nonnatural L-amino acids. Org. Process Res. Dev. 6 (4), 452–457.

Melkonyan, L., 2013. Optimization of L-alanine biosynthesis parameters by *Brevibacterium flavum* strain-producers. Biol. J. Armenia 60 (1), 77–84.

Melkonyan, L., et al., 2013a. Study of regulation of some key enzymes of L-alanine biosynthesis by *Brevibacterium flavum* producer strains. Appl. Biochem. Microbiol. 49 (2), 120–124.

Melkonyan, L., et al., 2013b. Characteristics of material and energy balance of L-alanine biosynthesis by *Brevibacterium flavum*. J. Biotechnol. 29 (3), 47–50.

Mey, Y., He, B., Ouyang, P., 2008. Enzymatic production of L-amino acids from the corresponding DL-5-substituted hydantoins by *Bacillus fordii* MH602. World J. Microbiol. Biotechnol. 24, 375–381.

Mitsuhashi, K., Yamamoto, H., Matsuyama, A. & Tokuyama, S., 2003. D-aminoacylase, method for producing the same, and method for producing D-amino acids using the same. United States, Patent No. 6514742B1.

Miyoshi, T., Kitagawa, H., Kato, M. & Chiba, S., 1985. Process for production of A-amino acids. Eur patent, Denki Kagaku Kogyo Kabushiki, Patent No. 0159866.

Mkhitaryan, A.V., 2013. Isolation and identification of strain with D-aminoacylase activity: preliminary characterization of the enzyme. Biol. J. Armenia 4 (65), 58–63.

Mkrtchyan, G.M., et al., 2005. D- and L-methionine preparation from N-acetyl-DL-methionine by the means of bacterial L-aminoacylase. Biotechnology (Rus.) 1, 37–41.

Möller, A., Syldatk, C., Schulze, M., Wagner, F., 1988. Stereo- and substratespecificity of a D-hydantoinase and a D-N-carbamyl-amino acid amidohydrolase of *Arthrobacter crystallopoietes* AM 2. Enzyme Microb. Technol. 10 (10), 618–625.

Moriguchi, M., Ideta, K., 1988. Production of D-aminoacylase from *Alcaligenes denitrificans* subsp. xylosoxydans MI-4. Appl. Environ. Microbiol. 54, 2767–2770.

Moriguchi, M., Sakai, K., Miyamoto, Y., Wakayama, M., 1993a. Production, purification, and characterization of D-aminoacylase from *Alcaligenes xylosoxydans* subsp. *xylosoxydans* A-6. Biosci. Biotechnol. Biochem. 57, 1149–1152.

Moriguchi, M., et al., 1993b. Purification and characterization of novel N-acyl-D-aspartate amidohydrolase from *Alcaligenes xylosoxydans* subsp. *xylosoxydans* A-6. Biosci. Biotechnol. Biochem. 57, 1145–1148.

Muniz-Lozano, F., Dominguez-Sanchez, G., Diaz-Viveros, Y., Barradas-Dermitz, D., 1998. D-aminoacylase from a novel producer: *Stenotrophomonas maltophilia* ITV-0595. J. Ind. Microbiol. Biotechnol. 21, 296–299.

Nakai, T., et al., 2000. Crystal structure of N-carbamyl-D-amino acid amidohydrolase with a novel catalytic framework common to amidohydrolases. Structure 8 (7), 729–737.

Nakajima, T. & Yamamoto, H., 2003. D-aminoacylase mutants from alcaligenes denitrificans for improved D-amino acid production. Europe, Patent No. 1435388B1.

Nakamori, S. et al., 1980. Method for producing D-α-amino acid. United States, Patent No. US4211840A.

Nanba, H., et al., 1998. Isolation of *Agrobacterium* sp. strain KNK 712 that produces N-carbamoyl-D-amino acid amidohydrolase, cloning of the gene for this enzyme, and properties of the enzyme. Biosci. Biotechnol. Biochem. 62 (5), 875–881.

Ngamsom, B., et al., 2010. Development of a high throughput screening tool for biotransformations utilising a thermophilic L-aminoacylase enzyme. J. Mol. Catal. B: Enzym. 63 (1–2), 81–86.

Nishida, Y., Nkamichi, K., Nabe, K., Tosa, T., 1987. Enzymatic production of L-tryptophan from DL-5-indolmethylhydantoin by *Flavobacterium* sp. Enzyme Microb. Technol. 9, 721–725.

Niu, H., et al., 2019. Metabolic engineering for improving L-tryptophan production in *Escherichia coli*. J. Ind. Microbiol. Biotechnol. 46 (1), 55–65.

Nozaki, H., et al., 2005. D-amino acid production by *E. coli* co-expressed three genes encoding hydantoin racemase, D-hydantoinase and coexpressed D-hydantoinase and N-carbamoylase. J. Mol. Catal. B: Enzym. 32 (5–6), 213–218.

Ogawa, J., Shimizu, S., 1994. Beta-ureidopropionase with N-carbamoyl-L-amino acid amidohydrolase activity from an aerobic bacterium, *Pseudomonas putida* IFO 12996. Eur. J. Biochem. 223, 625–630.

Ogawa, J., Shimizu, S., Yamada, H., 1993. N-carbamoyl-D-amino acid amidohydrolase from *Comamonas* sp. E222c purification and characterization. Eur. J. Biochem. 212 (3), 685–691.

Ogawa, J., Chung, M., Hida, S., Yamada, H., 1994. Thermostable N-carbamoyl-D-amino acid amidohydrolase: screening, purification and characterization. J. Biotechnol. 38, 11–19.

Ogawa, J., Miyake, H., Shimizu, S., 1995. Purification and characterization of N-carbamoyl-L-amino acid amidohydrolase with broad substrate specificity from *Alcaligenes xylosoxidans*. Appl. Microbiol. Biotechnol. 43, 1039–1043.

Ohmachi, T., et al., 2002. Identifcation, cloning, and sequencing of the genes involved in the conversion of D,L-2-amino-D2-thiazoline-4-carboxylic acid to L-cysteine in *Pseudomonas* sp. strain ON-4a. Biosci. Biotechnol. Biochem. 66, 1097–1104.

Ohmachi, T., et al., 2004. A novel N-carbamoyl-L-amino acid amidohydrolase of *Pseudomonas* sp. strain ON-4a: purification and characterization of N-carbamoyl-L-cysteine amidohydrolase expressed in *Escherichia coli*. Appl. Microbiol. Biotechnol. 65, 686–693.

Oldiges, M., Eikmanns, B., Blombach, B., 2014. Application of metabolic engineering for the biotechnological production of L-valine. Appl. Microbiol. Biotechnol. 98 (13), 5859–5870.

Olivieri, R., Fascetti, E., Angelini, L., Degen, L., 1979. Enzymatic conversion of N-carbamoyl-D-amino acids to D-amino acids. Enzyme Microb. Technol. 1 (3), 201–204.

Olivieri, R. et al., 1982. Enzymic microbiological process for producing optically active aminoacids starting from hydantoins and/or racemic carbamoyl derivatives. United States, Patent No. US4312948.

Park, J., Lee, S., 2008. Towards systems metabolic engineering of microorganisms for amino acid production. Curr. Opin. Biotechnol. 19 (5), 454–460.

Park, J., Kim, G., Kim, H., 2000. Production of D-amino acid using whole cells of recombinant *Escherichia coli* with separately and coexpressed D-hydantoinase and N-carbamoylase. Biotechnol. Prog. 16 (4), 564–570.

Park, J., Jang, Y.-S., Lee, J., Lee, S., 2011a. *Escherichia coli* W as a new platform strain for the enhanced production of L-valine by systems metabolic engineering. Biotechnol. Bioeng. 108 (5), 1140–1147.

Park, J., Kim, T., Lee, K., Lee, S., 2011b. Fed-batch culture of *Escherichia coli* for L-valine production based on in silico flux response analysis. Biotechnol. Bioeng. 108 (4), 934–946.

Parpart, S., et al., 2018. Synthesis of optically pure (S,E)-2-amino-5-arylpent-4-enoic acids by heck reactions of nickel complexes. Synlett 29, A–F.

Pietzsch, M., et al., 2000. Purification of recombinant hydantoinase and L-N-carbamoylase from *Arthrobacter aurescens* expressed in *Escherichia coli*: comparison of wild-type and genetically modified proteins. J. Chromatogr. B 737, 179–186.

Pozo, C., Rodelas, B., de la Escalera, S., Gonzalez-Lopez, J., 2002. D,L-hydantoinase activity of an *Ochrobactrum anthropi* strain. J. Appl. Microbiol. 92, 1028–1034.

Saghyan, A., Lange, P., 2016. Asymmetric Synthesis of Non-Proteinogenic Amino Acids. Wiley-VCH, Weinheim.

Sakai, K., et al., 1991a. Purification and characterization of N-acyl-D-glutamate deacylase from *Alcaligenes xylosoxydans* subsp. *xylosoxydans* A-6. FEBS Lett. 289, 44–46.

Sakai, K., Oshima, K., Moriguchi, M., 1991b. Production and characterization of N-acyl-D-glutamate amidohydrolase from *Pseudomonas* sp. strain 5f-1. Appl. Environ. Microbiol. 57, 2540–2543.

Sakanyan, V., et al., 1993. Gene cloning, sequence analysis, purification, and characterization of a thermostable aminoacylase from *Bacillus stearothermophilus*. Appl. Environ. Microbiol. 59 (11), 3878–3888.

Sareen, D., Sharma, R., Rakesh, S., Vohra, R., 2001. Two-step purification of D(2)-specific carbamoylase from *Agrobacterium tumefaciens* AM 10. Protein Expr. Purif. 21, 170–175.

Sato, T., Tosa, T., 2010. L-amino acids production by aminoacylase. In: Flickinger, M.C. (Ed.), Encyclopedia of Industrial Biotechnology: Bioprocess, Bioseparation, and Cell Technology. Wiley, pp. 1–20.

Seibert, C., Raushel, F., 2009. Structural and catalytic diversity within the amidohydrolase superfamily. Biochemistry 44, 6383–6391.

Sidiq, K., Chow, M., Zhao, Z., Daniel, R., 2019. Alanine Metabolism in *Bacillus subtilis*. *bioRxiv*, p. 562850.

Slomka, C., et al., 2017. Toward a cell-free hydantoinase process: screening for expression optimization and one-step purification as well as immobilization of hydantoinase and carbamoylase. AMB Express 7, 122.

Solov'eva, T.A., Stepanov, V.M., 1983. Isolation and purification of an A minoacylase from *Aspergillus oryzae*. Chem. Nat. Compd. 19, 582–586.

Story, S.V., Grunden, A.M., Adams, M.W., 2001. Characterization of an aminoacylase from the hyperthermophilic archaeon *Pyrococcus furiosus*. J. Bacteriol. 183 (14), 4259–4268.

Subhashini, D.V., Singh, R.P., Manchanda, G., 2017. OMICS Approaches: Tools to Unravel Microbial Systems. https://books.google.co.in/books?id=vSaLtAEACAAJ.

Sugie, M., Suzuki, H., 1978. Purification and properties of D-aminoacylase of *Streptomyces olivaceus*. Agric. Biol. Chem. 42, 107–113.

Sugie, M., Suzuki, H., 1980. Optical resolution of DL-amino acids with D-aminoacylase of *Streptomyces*. Agric. Biol. Chem. 44 (5), 1089–1095.

Suzuki, S., Takenaka, Y., Onishi, N., Yokozeki, K., 2005. Molecular cloning and expression of the *hyu* genes from *Microbacterium liquefaciens* AJ 3912, responsible for the conversion of 5-substituted hydantoins to α-amino acids, in *Escherichia coli*. Biosci. Biotechnol. Biochem. 69, 1473–1482.

Syldatk, C., et al., 1987. Substrate- and stereospecificity, induction and metallo-dependence of a microbial hydantoinase. Biotechnol. Lett. 9, 25–30.

Syldatk, C., et al., 1999. Microbial hydantoinases—industrial enzymes from the origin of life? Appl. Microbiol. Biotechnol. 51, 293–309.

Tanimoto, K., et al., 2008. Characterization of thermostable aminoacylase from hyperthermophilic archaeon *Pyrococcus horikoshii*. FEBS J. 275 (6), 1140–1149.

Tokuyama, S., 2000. D-aminoacylase. United States, Patent No. 6030823.

Tokuyama, S. & Matsuyama, A., 2005. Heat-stable D-aminoacylase. United States, Patent No. 6902915B2.

Toogood, H.S., et al., 2002a. A thermostable L-aminoacylase from *Thermococcus litoralis*: cloning, overexpression, characterization, and applications in biotransformations. Extremophiles 6, 111–122.

Toogood, H.S., et al., 2002b. Immobilisation of the thermostable L-aminoacylase from *Thermococcus litoralis* to generate a reusable industrial biocatalyst. Biocatal. Biotransformation 20 (4), 241–249.

Tröndle, J., et al., 2020. Metabolic control analysis of L-tryptophan production with *Escherichia coli* based on data from short-term perturbation experiments. J. Biotechnol. 307, 15–28.

Tsai, Y., Tseng, C., Hsiao, K., Chen, L., 1988. Production and purification of D-aminoacylase from *Alcaligenes denitrificans* and taxonomic study of the strain. Appl. Environ. Microbiol. 54 (4), 984–988.

Tsai, Y., et al., 1992. Production and immobilization of D-aminoacylase of *Alcaligenes faecalis* D AI for optical resolution of N-acyl-DL-amino acids. Enzyme Microb. Technol. 14 (5), 384–389.

Tsugawa, R., Okumura, S., Ito, T., Katsuga, N., 1966. Production of L-glutamic acid from DL-hydantoin-5-propionic acid by microoganisms. Agric. Biol. Chem. 30, 27–34.

Van der Kaaij, H., Desiere, F., Mollet, B., Germond, J., 2004. L-alanine auxotrophy of *Lactobacillus johnsonii* as demonstrated by physiological, genomic, and gene complementation approaches. Appl. Environ. Microbiol. 70 (3), 1869–1873.

Vardanyan, A. et al., 2008. Method of L-alanine production. RA, Patent No. 2239 A.

Vickery, H., Schmidt, C., 1931. The history of the discovery of the amino acids. Chem. Rev. 9 (2), 169–318.

Vogels, G., van der Drift, C., 1976. Degradation of purines and pyrimidines by microorganisms. Bacteriol. Rev. 40 (2), 403–468.

Wagner, T., Hantke, B., Wagner, F., 1996. Production of L-methionine from DL-5-(2-methylthioethyl)hydantoin by resting cells of a new mutant strain of *Arthrobacter* sp. DSM7330. Biotechnology 46, 63–68.

Wakayama, M., Moriguchi, M., 2001. Comparative biochemistry of bacterial N-acyl-D-amino acid amidohydrolase. J. Mol. Catal. B: Enzym. 12, 15–25.

Wakayama, M., et al., 1995. Cloning and sequencing of a gene encoding D-aminoacylase from *Alcaligenes xylosoxydans* subsp. *xylosoxydans* A-6 and expression of the Gene in *Escherichia coli*. Biosci. Biotechnol. Biochem. 59, 2115–2119.

Wakayama, M., Yoshimune, K., Hirose, Y., Moriguchi, M., 2003. Production of D-amino acids by N-acyl-D-amino acid amidohydrolase and its structure and function. J. Mol. Catal. B: Enzym. 23, 71–85.

Wang, W.-C., Hsu, W.-H., Chien, F.-T., Chen, C.-Y., 2001. Crystal structure and site-directed mutagenesis studies of N-carbamoyl-D-amino-acid amidohydrolase from *Agrobacterium radiobacter* reveals a homotetramer and insight into a catalytic cleft. J. Mol. Biol. 306, 251–261.

Wang, W., et al., 2013. Cloning, expression and characterization of D-aminoacylase from *Achromobacter xylosoxidans* subsp. *denitrificans* ATCC 15173. Microbiol. Res. 168 (6), 360–366.

Watabe, K., Ishikawa, T., Mukohara, Y., Nakamura, H., 1992. Cloning and sequencing of the genes involved in the conversion of 5-substituted hydantoins to the corresponding L-amino acids from the native plasmid of *Pseudomonas* sp. strain NS671. J. Bacteriol. 174 (3), 962–969.

Weigel, P., Marc, F., Aganyants, H., Sakanyan, V., 2013. Characterization of thermostable hydantoinases cloned from *Geobacillus stearothermophilus*. Rep. NAS RA 113, 92–98.

Weiss, H.M., Palm, G.J., Röhm, K.H., 1995. Thermostable aminoacylase from *Bacillus stearothermophilus*: significance of the metal center for catalysis and protein stability. Biol. Chem. Hoppe Seyler 376 (11), 643–649.

Wendisch, V., 2007. Amino Acid Biosynthesis—Pathways, Regulation and Metabolic Engineering. Springer-Verlag Berlin Heidelberg, New York.

Wexler, B., 2007. Tryptophan: Powerful Serotonin Booster, illustrated ed. Woodland Pub, Woodland.

Wiese, A., Syldatk, C., Mattes, R., Altenbuchner, J., 2001. Organization of genes responsible for the stereospecific conversion of hydantoins to α-amino acids in *Arthrobacter aurescens* DSM 3747. Arch. Microbiol. 176, 187–196.

Wilms, B., et al., 1999. Cloning, nucleotide sequence and expression of a new L-N-carbamoylase gene from *Arthrobacter aurescens* DSM 3747 in *E. coli*. J. Biotechnol. 68, 101–113.

Wilms, B., Wiese, A., Syldatk, C., Mattes, R., 2001. Development of an *Escherichia coli* whole cell biocatalyst for the production of L-amino acids. J. Biotechnol. 86, 19–30.

Wu, S., et al., 2005. Enzymatic production of D-P-hydroxyphenylglycine from DL-5-phydroxyphenylhydantoin by *Sinorhizobium morelens* S-5. Enzyme Microb. Technol. 36 (4), 520–526.

Wu, S., et al., 2006. Thermostable D-carbamoylase from *Sinorhizobium morelens* S-5: purification, characterization and gene expression in *Escherichia coli*. Biochimie 88, 237–244.

Xu, Z., Jiang, W.-H., Jiao, R.-S., Yang, Y.-L., 2002. Cloning, sequencing and high expression in *Escherichia coli* of D-hydantoinase gene from *Burkholderia pickettii*. Sheng Wu Gong Cheng Xue Bao 18 (2), 149–154.

Xu, Z., et al., 2003. Crystal structure of D-hydantoinase from *Burkholderia pickettii* at a resolution of 2.7 angstroms: insights into the molecular basis of enzyme thermostability. J. Bacteriol. 185 (14), 4038–4049.

Yamada, H., Takahashi, S. & Yoneda, K., 1978. Process for preparing D-(-)-N-carbamoyl-2-(phenyl or substituted phenyl)glycines. United States, Patent No. US4094741A.

Xu, G., West, Th., 1994. Characterization of dihydropyrimidinase from *Pseudomonas stutzeri*. Archives of Microbiology 161, 70–74.

Yamanaka, H., Kawamoto, T., Tanaka, A., 1997. Efficient preparation of optically active *p*-trimethylsilylphenylalanine by using cell-free extract of *Blastobacter* sp. A17p-4. J. Ferment. Bioeng. 84, 181–184.

Yamashiro, A., Kubota, K., Yokozaki, K., 1988. Mechanism of stereospecific production of L-amino acids from the corresponding 5-substituted hydantoins by *Bacillus brevis*. Agric. Biol. Chem. 52 (11), 2857–2863.

Yang, Y., et al., 1991. Purification and characterization of D-aminoacylase from *Alcaligenes faecalis* DA1. Appl. Environ. Microbiol. 57 (4), 1259–1260.

Yang, Y., et al., 1992. Characterization of D-aminoacylase from *Alcaligenes denitrificans* DA181. Biosci. Biotechnol. Biochem. 56, 1392–1395.

Yang, Y., et al., 1994. Purification and characterization of L-aminoacylase from *Alcaligenes denitrificans* DA181. Biosci. Biotechnol. Biochem. 58 (1), 204–205.

Yano, S., et al., 2011. Engineering the substrate specificity of *Alcaligenes* D-aminoacylase useful for the production of D-amino acids by optical resolution. J. Chromatogr. B 879, 3247–3252.

Yokota, A., Ikeda, M., 2017. Amino Acid Fermentation. Springer, Tokyo, Tokyo.

Yokozeki, K., Hirose, Y., Kubota, K., 1987. Mechanism of asymmetric production of L-aromatic amino acids from the corresponding hydantoins by *Flavobacterium* sp. Agric. Biol. Chem. 51, 737–746.

Yokozeki, K., et al., 1987b. Enzymatic production of L-tryptophan from DL-5-indolylmethylhydantoin by mutants of *Flavobacterium* sp. T-523. Agric. Biol. Chem. 51, 819–825.

Yuan, Y., Wang, S., Song, Z., Gao, R., 2002. Immobilization of an L-aminoacylase-producing strain of *Aspergillus oryzae* into gelatin pellets and its application in the resolution of D,L-methionine. Biotechnol. Appl. Biochem. 35 (2), 107–113.

Zhou, L., et al., 2016. Efficient L-alanine production by a thermo-regulated switch in *Escherichia coli*. Appl. Biochem. Biotechnol. 178 (2), 324–337.

CHAPTER

Pseudomonas for sustainable agricultural ecosystem

Pooja Misra[a], Archana[b], Shikha Uniyal[c], and Atul Kumar Srivastava[c]

[a]*Crop Protection Division, CSIR-Central Institute of Medicinal and Aromatic Plants, Lucknow, Uttar Pradesh, India,*
[b]*Department of Endocrinology, Sanjay Gandhi Post Graduate Institute of Medical Sciences, Lucknow, Uttar Pradesh, India,*
[c]*Research and Development Department, Uttaranchal University, Dehradun, Uttarakhand, India*

Abstract

Pseudomonas is a gram-negative bacterium that belongs to the family Psedomonadaceae, found in a wide range of niches. *Pseudomonas* is an essential member of the soil microbial community and plays a significant role in the regulation of the agricultural soil ecosystem. For plant growth and soil health, *Pseudomonas* is an illustrious feature of plant growth-promoting bacteria (Siderophores, IAA, NH3, HCN, and P solubilization), and bioremediation as well. In addition, *Pseudomonas* also produces active secondary metabolites, act against the pesticide to make it a significant biological control agent. In 21st century, researchers were curious to know the biosynthetic pathways involve in the synthesis of secondary metabolites by using advanced techniques. Thus, the *Pseudomonas* successively regulates the sustainable agricultural soil ecosystem. It can also modulate the soil ecosystem functioning with the help of the application, advanced biosynthetic pathways, and plant growth-promoting factors.

Keywords: *Pseudomonas*, PGPR, Sustainable agriculture, Bioremediation

1 Introduction

Pseudomonas is one of the most studied bacterial groups. It was first identified by Migula (1894). *Pseudomonas* is gram-negative, slender, rod-shaped, polar-flagellated bacteria. It is an aerobic bacterium that belongs to the family of Pseudomonadaceae. The morphological feature of *Pseudomonas* is more similar to other bacteria of prokaryote genera (Lysenko, 1961). The advanced nucleic acid-based method can easily differentiate these bacteria from a similar group of genera. *Pseudomonas* are ubiquitous pathogen can be found in soil, water, plant, and animal tissue. Different species of *Pseudomonas* affect both humans and plants.

Continuous planting of similar crops on land brings yield loss with time. To grow the crop farmer use a huge amount of pesticides and fertilizers in the field, that deteriorates the land fertility and effect on biological diversity of local flora and fauna. The use of agricultural bio-product such as biofertilizer stimulators is an attractive choice for sustainable development (Singh et al., 2021). Even during the storage period of crop, pest causes heavy loss of the food grain. Besides this many environmental and social problems arise due to modern agriculture. It poses pressure on natural resources. The most common problems are soil erosion, groundwater pollution, depletion in the quality of nearby water

Microbial Syntrophy-mediated Eco-enterprising. https://doi.org/10.1016/B978-0-323-99900-7.00012-2

bodies, and soil infertility. The social problems like unemployment of the farm laborers due to the use of heavy machines also rise. An alternative for the adverse impacts of modern agriculture is sustainable agriculture.

Sustainable agriculture is defined as "a system of ecological farming practices, which is based on scientific innovations through which it is possible to produce healthy foods with respect for the land, air, water, and farmers' health and rights." In other words, sustainable agriculture uses sustainable farm techniques which are economically viable to produce food, fibers; animal products and protects the environmental entities, natural resources, public health, and welfare of both humans and animals.

2 Methods of sustainable agriculture

Sustainable agriculture can be carried out in various ways. It not only helps the soil fertility to enhance but is also economically viable. The various methods of sustainable agriculture are mentioned below. Farmers must choose the method according to the topographical conditions, soil conditions, local climatic conditions, pests, and most importantly farmer's goal.

1. **Mixed or diverse cropping:**

   In a mixed or diverse cropping system two or more crops are grown at the same time and in the same field. Mostly the first crop grown is of long duration accompanied by the short duration crop to maintain the fertility of the soil and both the crop can get nutrition at the maturity time. To increase the fertility of the field leguminous plant species are grown with the main crop. Leguminous species can fix the atmospheric nitrogen and are known for increasing the fertility of the soil. For example, carbohydrate-rich cereal can be grown with a legume as carbohydrate cereals use the nitrogen from the soil while leguminous species put nitrogen back to the soil by fixing the atmospheric nitrogen (St-Martin et al., 2017).

2. **Crop rotation:**

   It is defined as "the practice of alternating the annual crops grown on a specific field in a planned pattern or sequence in successive crop years so that crops of the same species or family are not grown repeatedly without interruption on the same field." Crop rotation is considered as a significant part of organic farming as it helps in retaining soil fertility and crop diversity (Sumner, 2018). Grasses and legumes are preferably grown during the rotation of crops. Table 1 describes the advantages and disadvantages related to crop rotation.

3. **Mixed Farming:**

   Farming of the crops with the livestock in the same field is known as mixed farming. The farming of the mixed crops along with the livestock at the same time has many advantages. Crops are grown at same leveled land while forages and pastures along with grasses are grown at the steeper slopes. It lessens soil erosion and when pastures and forages are grown in rotation it enhances the fertility of the soil (Vereijken, 2020). The manure that comes from livestock farming is used as the manure in the field. If recent technology of vermicomposting is used, the fertility of the soil increases at a higher scale with very less financial input. At the time of poor climatic conditions or low rainfall or failure of the crop, livestock farming works as the cushion for the farmers. It strengthens the financial condition of the farmer.

Table 1 Advantages and disadvantages of crop rotation.

Advantages	Disadvantages
• Soil quality gets improved	• Initial significant investments
• Soil structure does not deplete	• Crop rotation is laborious
• Less water is required	• Experience is required to work
• It is a sustainable practice	• Unexpected weather conditions may harm sensitive plants
• Nutrients in the soil get depleted quite slowly	• Lack of knowledge in crop rotation
• Higher crop yield as time passes	• Its efficiency depends on the geographical factors
• It is a natural remedy against pests and weeds	• Short term follow may prevent advantages of crop rotation
• Assurance of the uninterrupted food supply to the locals	• It is not a magic wand to solve all the problems related to agriculture

3 Plant growth-promoting rhizobacteria

The overall problem of agriculture farming can be solved to some extent by plant growth-promoting rhizobacteria (PGPR). It is an eco-friendly approach and a safe choice for plant growth and disease management. PGPR are the rhizosphere bacteria are found in conjugation with the plant. They enhance the plant growth by a broad range of mechanisms like solubilization of phosphate, siderophore production, Nitrogen fixation, aminocyclopropane-1 carboxylate deaminase (ACC) production, prevent the formation of biofilms, promote phytohormone production, exhibiting the volatile organic compounds (VOCs), inhibit the growth of pathogen by the production of the toxin, etc. (Singh et al., 2016). Rhizosphere microorganisms produce a large amount of growth-promoting substances which indirectly promotes the overall growth of the plant. Based on the association of roots with the plant, PGPRs are classified as intracellular plant growth-promoting rhizobacteria (iPGPR) and extracellular plant growth-promoting rhizobacteria (ePGPR) (Martínez-Viveros et al., 2010). The iPGPR lies in the specialized structure of root cells (Verma et al., 2010) whereas PGPR lies between the cells of the cortex. Endophyte and Frankia are Ipgpr whereas *Pseudomonas* is the ePGPR (Gray and Smith, 2005). *Pseudomonas* is an indispensable part of rhizosphere biota (Glick, 1995; Kaymak, 2010) exhibiting successful colonization (Lugtenberg et al., 2001).

PGPR is used to control insect pests (Ruiu, 2020) plant disease (Saravanakumar et al., 2019) to alleviate abiotic stress (Dimkpa et al., 2009) to allow the nutrient to the plant. The main function of PGPR is to protect the plant from the hyperosmotic condition, accumulation of NaCl concentration, salinity tolerance. PGPR could eliminate the negative impact of salinity on the plant by two mechanisms. First, activate the stress response system in the plant after exposure to salt stress conditions. Second promoting synthesis of anti-stress biochemical such as AEs, NEAs, and Os that are used to remove reactive oxygen species (ROS) (Lata et al., 2018). PGPR reduce salt stress symptom by producing Na^+ binding exopolyassachiride (EPS) and decreasing ethylene hormone production by synthesizing phytohormones (Ashraf et al., 2004; Upadhyay et al., 2011; Atouei et al., 2019). PGPR possesses a different mechanism to maintain plant growth by producing ACC deaminase thus decreases the production of ethylene hormone, hence maintain plant growth. ACC act as the nitrogen source in a plant as nitrogen uptake is

Table 2 Plant growth accelerating substances of *Pseudomonas* species.

Pseudomonas species	Plant growth accelerating substances	References
Pseudomonas putida	IAA, Siderophotes, HCN, Ammonia, *exo*-polysaccharides, phosphate solubilization, Antifungal activity, ACC deaminase	Ahemad and Khan (2012a,c) and Pandey et al. (2006)
Pseudomonas aeruginosa	IAA, siderophores, HCN, ammonia, exo-polysaccharides, phosphate solubilization, siderophore, ACC deaminase	Ahemad and Khan (2012b), Naik and Dubey (2011), Ganesan (2008), and Ma et al., 2011
Pseudomonas fluorescens	ACC deaminase, phosphate solubilization, Induced systemic resistance, antifungal activity, IAA, siderophores	Shaharoona et al. (2008), Saravanakumar et al. (2007), Gupta et al. (2005), Dey et al. (2004), and Jeon et al. (2003)
Pseudomonas jessenii	ACC deaminase, IAA, siderophore, heavymetal solubilization, phosphate solubilization	Rajkumar and Freitas (2008)
Pseudomonas chlororaphis	Antifungal activity	Liu et al. (2007)
Pseudomonas sp.	Phosphate solubilization, IAA, siderophore, HCN, biocontrol potentials, ACC deaminase, heavy metal solubilization	Tank and Saraf (2009) and Rajkumar and Freitas (2008)
Pseudomonas argentinesis	ACC deaminase	Phour and Sindhu (2020)
Pseudomonas azotoformans	ACC deaminase	Phour and Sindhu (2020)

suppressed under salt condition (Fuertes-Mendizábal et al., 2020). PGPR mainly *Pseudomonas putida* UW4 (Cheng et al., 2012), *Pseudomonas argentinesis*, *Pseudomonas azotoformans* (Phour and Sindhu, 2020) can produce ACC deaminase. PGPR can change the root system architecture by producing auxin and controlling ethylene level by ACC deaminase. *Pseudomonas PSO1* inhibits the primary root formation and triggers lateral root formation and root hair development (Table 2).

4 Biosynthesis of *Pseudomonas*

4-Hydroxy-2-Alkylquinolins (HAQs). HAQ signal molecules produced in the stationary phase of growth play an important role during the bacterial population under stress. N-acyl-homoserine lactones, 4-hydroxy-2-alkylquinoline. *P. aeruginosa* produces a compound called pycoins and heterocyclic compounds such as phenazines, quinolines, and phenylpyrole. These compounds are known to eliminate microorganisms. *P. aeruginosa* produces a compound with bactericide activity that controls Multi-Drug Resistant (MDR) bacteria. *P. aeruginosa* also produces metallic compounds with inhibitory activity against MDR bacteria.

5 Bioremediation activity

Bioremediation is a process by the use of microorganism's complete transformation of organic pesticides to harmless product such as CO_2 and H_2O. The release of waste from industry without treatment possesses a significant negative impact on public health. During the time it accumulates in the food chain. Removal of metal from the waste by the use of microorganisms is an effective strategy to remove contaminants with lower toxicity, high efficiency, and ecofriendly (Rajendran et al., 2003; Wasi et al., 2011a,b). Microorganism has a mechanism to remove metal and radionuclides from water (Ji and Silver, 1995). Detoxification of metal by bacteria includes several processes such as oxidation-reduction, complexation, methylation, and reaction involving bioemulsification and sidophores (Wasi et al., 2008; Alam and Ahmad, 2011). Markandey et al. (2002) reviewed the removal of Pb and Cr metal through bioremediation. Bioremediation activity is also influenced by environmental factors. Degradation of pesticides by microorganisms extensively reviewed by Nawaz et al. (2011). Degradation of phenoxy acetate herbicides is available in the literature (Mai et al., 2001; Müller et al., 2001). Bacteria, algae, fungi, and yeast are reported to carry out degradation of phenolic compounds example *Pseudomonas* (Kumar et al., 2005; Yap et al., 1999). Aerobic degradation of HCH in pesticides contaminant by a *Pseudomonas strain* (Matsumura et al., 1976) and its degradation achieved by the *Pseudomonas paucimobilis* (Wada et al., 1989). Biodegradation of g-HCN by *Pseudomonas* species was well established by Wasi (Wasi et al., 2011a,b). *P. aeruginosa* was well reported for removal of Cadmium (Wang et al., 1997). Detoxification of Cd by *P. putida* (Lee et al., 2001) and *P. putida* strain precipitate the copper, thus act as a good candidate for bioremediation of copper. Morever *P. putida*, *Pseudomonas* sp. *Bu34* degrade phenols (Chung et al., 2004; Basha et al., 2010). Immobilized bacteria are a good candidate for bioremediation.

6 How *Pseudomonas* directly benefitted agriculture

Agriculture is a broad term as it includes all the facets of crop production including food and fiber crops, livestock farming, and also fish farming. Agriculture farming is the primary need of human farming for their survival whereas a growing population increases the demand for food products obtained/made from the field crops. In the last two decades alteration in the ecological climate damages agriculture farming and plant, pathogen molest increases, which results diminish the health of crops and that's why the yield and quality of crops reduces.

Due to the modernization in the current agriculture farming, as we discuss above the quality and productivity-increasing day by day and reduces the extra manpower. Modern equipment of agriculture farming (like Stroke Sprayer Pump, Combine Harvester, Roto Seed Drill, Manure Spreader, Rotary Tiller, etc.) replaces the farmer manual work into mechanical work, which gives comfort to farmers for living their life. Biopesticides and biofertilizer is the biological aspect in agriculture to make a better life of farmers that increase the irrigation facilities and help to the improvement of quality and quantity of crops. Though modern agriculture came up with many positive effects but sooner the deleterious effects of modern agriculture were sighted.

Plants are fully covered by the beneficial and non-beneficial microbial environment from root cap to shoot tip. Soil is the primary habitat of microbes and also a nutrient source of plants. But due to the demand for food and greed for good crops, we pour chemicals for improving the fertility of the soil and inadvertently harm the health of the soil. This unhealthy soil directly or indirectly affects the agricultural crops growing by us. Beneficial microbes that serve as plant growth developing agents associated with roots are known as plant growth-promoting rhizobacteria (PGPR). Rhizobacteria are the bacteria that symbiotically associated with the rhizospheric (originate from Greek word *rhiza* meaning root) region of plants. Rhizobacterial group directly benefit to plant because they work as a biofertilizer and

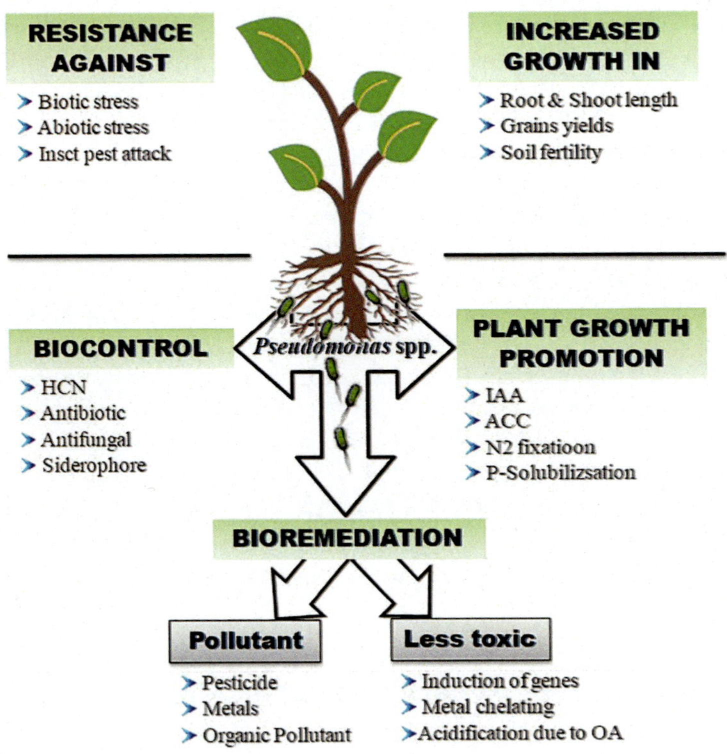

FIG. 1

Showing role of pseudomonas in above and below the ground in plant.

fix around 65% of environmental gaseous nitrogen into ammonia making it an available nutrient to plants. Plenty of nutrients including phosphorus found in the soil but it is unreachable to plant because of its insolubility. PGPR enhance the availability of nutrient present in soil by the solubilization and by the production of siderophore that facilitates the nutrient transport (Reference).

Pseudomonas spp., *Azospirillum* spp., and nitrogen-fixing bacteria (*Rhizobium*, *Azorhizobium*, *Bradyrhizobium,* and *Allorhizobium*) are included in the PGPR bacteria (Willey et al., 2011). Here, we mainly describe the role of *Pseudomonas* in the plant growth-promoting activity. *Pseudomonas* is found to be the best plant growth-promoting bacteria (PGPB) directly influences the plant and is mainly found in the soil habitat (Fig. 1).

7 Plant growth-promoting activity of *Pseudomonas*

Pseudomonas spp. found everywhere in diverse environmental conditions. *Pseudomonas aeruginosa* is the best-studied species that behave as an opportunistic human pathogen and *Pseudomonas syringae* identifies as the soil-born plant pathogen. Most of the species of *Pseudomoans* species are surrounded by the rhizospheric region and play an important role in plant growth promotion. *P. fluorescens*,

P. lini, *P. migulae*, *P. smigulae*, and *P. graminis*are the best studied plant growth-promoting species of *Pseudomonas* (Padda et al., 2018, 2019). *Pseudomonas* species enhance plant growth and development through nitrogen fixation, increase essential metal and phosphate solubilization, lowering ethylene, and enhanced the production of plant hormones.

In the management of plant disease *Pseudomonas* spp. play an important role in the inhibition or suppression of pathogenic microbes by producing antibiotics and generating inhibitory mechanisms against them. List of antibiotic produces by *Pseudomonas* spp. are shown here.

1. Antifungal antibiotics	11. Antibacterial antibiotics
2. Phenazines	12. Pseudomonic acid
3. Phenazine-1-carboxylic acid	13. Azomycin
4. Phenazine-1-carboxamide	14. Antitumor antibiotics
5. Pyrrolnitrin	15. FR901463
6. Pyoluteorin	16. Cepafungins
7. Cepaciamide A	17. Antiviral antibiotic
8. Oomycin A	18. Karalicin
9. Vscosinamide	
10. Pyocyanin	

Pseudomnas spp. are not only known for their growth promotions but they also activate some of the responses in stress conditions of the plant to maintain the growth condition in a stressed environment. Some mechanisms of growth promotion include growth hormones (phytohormones) production, mineral solubilization, mineral and phosphate solubilization, siderophore production along with protection against biotic and abiotic stresses as ACC-deaminase, chitinase, osmolyte, and exopolysaccharides production. Kandasamy in 2019 studied three different proteins of *Pseudomonas fluorescens* involved in growth promotion. The growth hormones i.e. phytohormones produced by *Pseudomonas* spp. (Ahmad et al., 2005; Hariprasad and Niranjana, 2009; Somers et al., 2004; Pallai et al., 2012) and the growth regulators like cytokinins and auxin protect the plant from stress (Yao et al., 2010; Malik and Sindhu, 2011). In the same way siderophores of *Pseudomonas* increase the iron availability for plants and make them unavailable for pathogens (Saharan and Nehra, 2011). In stress conditions, the enzymes like ACC-deaminase and chitinase produce by *Pseudomonas* spp. and act as bioinoculant. The elevated level of ACC by microbes protects the plant from abiotic stresses like salinity, drought, heavy metals, and temperature and ACC-deaminase also dilutes the byproducts of ACC into ammonia and α-ketobutyrate (Glick et al., 1998) and consequently reduce the ethylene level. Some *Pseudomonas* species produce exopolysaccharides to enhance the growth in salinity as well as in drought (Ashraf et al., 2004; Sandhya et al., 2009, 2010). In salinity, these exopolysaccharides bind to Na^+ ion and minimize their toxicity and in drought, it protects the plant from desiccation by maintaining its growth (Sandhya et al., 2009). The *Pseudomonas* spp. was also reported producing antioxidant enzymes (catalases, peroxidases, glutathion, and ascarbate) that help plants in stress conditions (Mittler, 2002; Yan et al., 2010). Some *Pseudomonas* spp. is also well known for providing protection against pathogens to plants, i.e., act as biocontrol agents either by competition or by antibiosis and induce systemic resistance. In competition, bacteria reduce the availability of particular nutrients to restrict the growth of pathogens like iron chelation by siderophore-producing bacteria (Subba, 1993). Antibiosis comprises the production of such molecule or compound that reduces the pathogen growth like production of antifungal metabolites as HCN, phenazines, and pyrrolnitrin by *Pseudomonas* spp. (Beneduzi et al., 2012; Bhattacharyya and Jha, 2012; Sivasakthi et al., 2014). In induced systemic resistance bacteria enhance the resistance of plants by changing the host-plat vulnerability (Van Loon, 2007). Likewise, the chitinase produced by bacteria degrade the fungal cell wall and thereby decrease the susceptibility of the plant (Shanmugam and Kanoujia, 2011). Some antifungal compounds like flavonoids act indirectly and protect the host plant. Garcia-Seco et al. (2015) suggested that *Pseudomonas* improve the quality of fruits by modifying flavonoid content (Table 3).

Table 3 PGPR activity and related mechanism of *Pseudomonas* species.

S.No.	*Pseudomonas* spp.	PGPR activity	Mechanism of action	References
1.	*Pseudomonas fluorescens*	# Increased shoot-root length and root diameter. Improved wheat yield, #Increased root-shoot growth, enhanced uptake of N, P, and K, # Enhanced root elongation and root weight. Increased number of tillers, 1000-grain weight, and grain yield	#Root colonization, #Production of plant growth regulators, #ACC-deaminase activity, auxin production, phosphate solubilization	Gamalero et al. (2009) and Egamberdiyeva and Höflich (2003)
2.	*P. putida*	# Increased root/shoot weight, inhibit the growth of various fungal pathogens # Improved plant height and root length. Improved crop yield # Increases in total dry weights of root and shoot inhibit fungal growth	# Production of metabolites and IAA # Auxin and siderophores production, phosphate solubilization # IAA production and antifungal metabolites	Abbas-Zadeh et al. (2010) and Mehnaz and Lazarovits (2006)
3.	*P. jessenii*	Enhanced root-shoot length and yield, and tolerance against foliar pathogens	Interactive effect of PGPR with cow dung by acting as biopesticide agent	Srivastava et al. (2010)
4.	*P. corrugata*	#Enhanced grain yield of maize, high root-shoot ratio	#Root colonization and stimulation of indigenous microflora	Kumar et al. (2007)
5.	*P. entomophila*	Reduced disease severity on pepper plants and enhanced root length	ACC deaminase, production of secondary metabolites and enzymes	Kamala-Kannan et al. (2010)
6.	*P. chlororaphis*	Improved seed emergence, seedling fresh weight, and yield of carrot and onion	Production of phytohormones and secondary metabolites	Bennett et al. (2009)
7.	*P. cepacia*	Significantly higher P availability improved P uptake, and increased plant biomass	Phosphate solubilization and IAA production	Katulanda and Rajapaksha (2012)
8.	*P. chlororaphis*	Significant increase in plant height, root length, and number of grains per spike	Production of siderophores and phytohormones, phosphate solubilization	Carlier et al. (2008)

8 Bioremediation by pseudomonas

Bioremediation is the process in which the contaminants of soil, water, or sediments which affect directly or indirectly to public health detoxify or removed by microorganisms (Talley, 2005; Wasi et al., 2008). In this type of process, the harmful toxic organic pesticides or chemicals are transformed into harmless end products like CO_2 and H_2O. In the same way, microbes also transform the inorganic pollutants into the compound which has less solubility, mobility, and toxicity as well (Kamaludeen et al., 2003; Wasi et al., 2008). Nowadays by natural and anthropogenic activities, heavy metals, pesticide and phenolic like contaminants adding continuously into the environment which is a global concern. Without proper treatment, these heavy metals and pollutants are easily incorporated into our food chain because of their persistence, accumulation, and biomagnification. The microbes have the ability to immobilize, remove or detoxify these contaminants through various mechanisms (Ji and Silver, 1995). Several processes are known by which microorganisms detoxify the metal contaminants such as oxidation-reduction, methylation, complexation, and by use of biosurfactants and siderophores (Wasi et al., 2008; Alam and Ahmad, 2011). Several organic contaminants are transformed by the microbial enzymatic attack and the end product may differ drastically (Alexander, 1999; Kunamneni et al., 2008; Rao et al., 2010). Yeast, fungi algae, and bacteria all have a property to degrade the phenolic compounds among them, *Pseudomonas* (Kumar et al., 2005; Agarry and Solomon, 2008), *Bacillus* sp. (Gurujeyalakshmi and Oriel, 1989; Kim and Oriel, 1995), *Alcaligenes* sp. (Nair et al., 2008), *Streptomyces* sp. (Antai and Crawford, 1983), *Trichosporon* sp. (Basha et al., 2010), *Candida* sp. (Tsai et al., 2005), *Ochromonas* sp. (Semple et al., 1999), and the most widely studied are the *Pseudomonas* species (Yap et al., 1999; Mollaeia et al., 2010; Wasi et al., 2011a,b). When *Pseudomonas putida*, immobilize with Ca-alginate then its phenol degradation capacity increased (Bettmann and Rehm, 1984). *Pseudomonas fluorescens* SM1strain is reported for its bioremediation of major toxicants viz. phenols, pesticides, and heavy metals in immobilizing stage (Wasi et al., 2011a; Asthana et al., 1995). Ferschl et al. (1991) found that *Pseudomonas acidovorans* CA28 degrades 3-chloroaniline by calcium alginate immobilized cells. *P. putida* US2 has the ability to degrade 2-chloroethanol.

9 Advantages of sustainable agriculture

- Sustainable agriculture protects the earth's natural resources. It focuses on preserving the soil as it lessens soil erosion and enhances the fertility of the soil.
- It reduces the farmer's dependency on costly fertilizers and pesticides thereby strengthening the economic condition of farmers.
- It does not require higher irrigation facilities, as many crops used in crop rotation and mixed farming do not require exclusive irrigation facilities.
- It is beneficial to humans because the ultimate consumers of the crops are humans. Chemical fertilizers and pesticides harm human health because many synthetic fertilizers and pesticides contain carcinogens. It does not pollute the groundwater nor are the nearby aquatic bodies polluted by agriculture run-off.
- Sustainable agriculture acts as a natural pesticide against pests. With very little input, manure received in mixed farming can be converted into vermicompost which in turn increases the fertility of the soil.

10 Drawbacks of sustainable agriculture

- Products of sustainable agriculture are expensive and less attractive due to their production method. Chemicals that gives shiny luster to the eatable products are not used in sustainable agriculture.
- It is quite laborious as it uses the common farm techniques of agriculture and mechanization is omitted.
- It has a shorter shelf life.

11 Conclusion

Pseudomonas is a gram-negative bacterium useful in sustainable agriculture proof by the different researchers by the expatriation of its illustrious feature of plant growth-promoting bacteria (Siderophores, IAA, NH3, HCN, and P solubilization), and bioremediation as well. In addition, *Pseudomonas* also produces active secondary metabolites, act against the pesticide to make it a significant biological control agent. In 21st century, researchers were curious to know the biosynthetic pathways involve in the synthesis of secondary metabolites by using advanced techniques. Thus, the *Pseudomonas* successively regulates the sustainable agricultural soil ecosystem. It can also modulate the soil ecosystem functioning with the help of the application, advanced biosynthetic pathways, and plant growth-promoting factors.

References

Abbas-Zadeh, P., Saleh-Rastin, N., Asadi-Rahmani, H., Khavazi, K., Soltani, A., Shoary-Nejati, A.R., Miransari, M., 2010. Plant growth-promoting activities of fluorescent pseudomonads, isolated from the Iranian soils. Acta Physiol. Plant. 32 (2), 281–288. https://doi.org/10.1007/s11738-009-0405-1.

Agarry, S.E., Solomon, B.O., 2008. Kinetics of batch microbial degradation of phenols by indigenous Pseudomonas fluorescence. Int. J. Environ. Sci. Technol. 5 (2), 223–232. https://doi.org/10.1007/BF03326016.

Ahemad, M., Khan, M.S., 2012a. Alleviation of fungicide-induced phytotoxicity in greengram [Vigna radiata (L.) Wilczek] using fungicide-tolerant and plant growth promoting Pseudomonas strain. Saudi J. Biol. Sci. 19 (4), 451–459. https://doi.org/10.1016/j.sjbs.2012.06.003.

Ahemad, M., Khan, M.S., 2012b. Effect of fungicides on plant growth promoting activities of phosphate solubilizing Pseudomonas putida isolated from mustard (Brassica compestris) rhizosphere. Chemosphere 86 (9), 945–950. https://doi.org/10.1016/j.chemosphere.2011.11.013.

Ahemad, M., Khan, M.S., 2012c. Evaluation of plant-growth-promoting activities of rhizobacterium Pseudomonas putida under herbicide stress. Ann. Microbiol. 62 (4), 1531–1540. https://doi.org/10.1007/s13213-011-0407-2.

Ahmad, F., Ahmad, I., KHAN, M.S., 2005. Indole acetic acid production by the indigenous isolates of Azotobacter and fluorescent Pseudomonas in the presence and absence of tryptophan. Turk. J. Biol. 29 (1), 29–34.

Alam, M.Z., Ahmad, S., 2011. Chromium removal through biosorption and bioaccumulation by bacteria from tannery effluents contaminated soil. Clean: Soil, Air, Water 39 (3), 226–237. https://doi.org/10.1002/clen.201000259.

Alexander, M., 1999. Biodegradation and Bioremediation. Gulf Professional Publishing.

Antai, S.P., Crawford, D.L., 1983. Degradation of phenol by Streptomyces setonii. Can. J. Microbiol. 29, 142–143.

Ashraf, M., Hasnain, S., Berge, O., Mahmood, T., 2004. Inoculating wheat seedlings with exopolysaccharide-producing bacteria restricts sodium uptake and stimulates plant growth under salt stress. Biol. Fertil. Soils 40 (3), 157–162. https://doi.org/10.1007/s00374-004-0766-y.

Asthana, R.K., Chatterjee, S., Singh, S.P., 1995. Investigations on nickel biosorption and its remobilization. Process Biochem. 30 (8), 729–734.

Atouei, M.T., Pourbabaee, A.A., Shorafa, M., 2019. Alleviation of salinity stress on some growth parameters of wheat by exopolysaccharide-producing bacteria. Iran. J. Sci. Technol. Trans. A: Sci. 43 (5), 2725–2733. https://doi.org/10.1007/s40995-019-00753-x.

Basha, K.M., Rajendran, A., Thangavelu, V., 2010. Recent advances in the biodegradation of phenol: a review. Asian J. Exp. Biol. Sci. 1 (2), 219–234.

Beneduzi, A., Ambrosini, A., Passaglia, L.M., 2012. Plant growth-promoting rhizobacteria (PGPR): their potential as antagonists and biocontrol agents. Genet. Mol. Biol. 35 (4), 1044–1051. https://doi.org/10.1590/S1415-47572012000600020.

Bennett, A.J., Mead, A., Whipps, J.M., 2009. Performance of carrot and onion seed primed with beneficial microorganisms in glasshouse and field trials. Biol. Control 51 (3), 417–426. https://doi.org/10.1016/j.biocontrol.2009.08.001.

Bettmann, H., Rehm, H.J., 1984. Degradation of phenol by polymer entrapped microrganisms. Appl. Microbiol. Biotechnol. 20, 285–290.

Bhattacharyya, P.N., Jha, D.K., 2012. Plant growth-promoting rhizobacteria (PGPR): emergence in agriculture. World J. Microbiol. Biotechnol. 28 (4), 1327–1350. https://doi.org/10.1007/s11274-011-0979-9.

Carlier, E., Rovera, M., Jaume, A.R., Rosas, S.B., 2008. Improvement of growth, under field conditions, of wheat inoculated with Pseudomonas chlororaphis subsp. aurantiaca SR1. World J. Microbiol. Biotechnol. 24 (11), 2653–2658. https://doi.org/10.1007/s11274-008-9791-6.

Cheng, Z., Woody, O.Z., McConkey, B.J., Glick, B.R., 2012. Combined effects of the plant growth-promoting bacterium Pseudomonas putida UW4 and salinity stress on the Brassica napus proteome. Appl. Soil Ecol. 61, 255–263. https://doi.org/10.1016/j.apsoil.2011.10.006.

Chung, T.P., Wu, P.C., Juang, R.S., 2004. Process development for degradation of phenol by Pseudomonas putida in hollow-fiber membrane bioreactors. Biotechnol. Bioeng. 87 (2), 219–227. https://doi.org/10.1002/bit.20133.

Dey, R.K.K.P., Pal, K.K., Bhatt, D.M., Chauhan, S.M., 2004. Growth promotion and yield enhancement of peanut (Arachis hypogaea L.) by application of plant growth-promoting rhizobacteria. Microbiol. Res. 159 (4), 371–394. https://doi.org/10.1016/j.micres.2004.08.004.

Dimkpa, C., Weinand, T., Asch, F., 2009. Plant–rhizobacteria interactions alleviate abiotic stress conditions. Plant Cell Environ. 32 (12), 1682–1694. https://doi.org/10.1111/j.1365-3040.2009.02028.x.

Egamberdiyeva, D., Höflich, G., 2003. Influence of growth-promoting bacteria on the growth of wheat in different soils and temperatures. Soil Biol. Biochem. 35 (7), 973–978. https://doi.org/10.1016/S0038-0717(03)00158-5.

Ferschl, A., Loidl, M., Ditzelmüller, G., Hinteregger, C., Streichsbier, F., 1991. Continuous degradation of 3-chloroaniline by calciumalginate-entrapped cells of Pseudomonas acidovorans CA28: influence of additional substrates. Appl. Microbiol. Biotechnol. 35, 544–550.

Fuertes-Mendizábal, T., Bastías, E.I., González-Murua, C., González-Moro, M., 2020. Nitrogen assimilation in the highly salt-and boron-tolerant ecotype Zea mays L. Amylacea. Plants 9 (3), 322. https://doi.org/10.3390/plants9030322.

Gamalero, E., Lingua, G., Berta, G., Lemanceau, P., 2009. Methods for studying root colonization by introduced beneficial bacteria. Sustain. Agric., 601–615. https://doi.org/10.1007/978-90-481-2666-8_37.

Ganesan, V., 2008. Rhizoremediation of cadmium soil using a cadmium-resistant plant growth-promoting rhizopseudomonad. Curr. Microbiol. 56 (4), 403–407. https://doi.org/10.1007/s00284-008-9099-7.

Garcia-Seco, D., Zhang, Y., Gutierrez-Mañero, F.J., Martin, C., Ramos-Solano, B., 2015. Application of Pseudomonas fluorescens to blackberry under field conditions improves fruit quality by modifying flavonoid metabolism. PLoS One 10 (11). https://doi.org/10.1371/journal.pone.0142639, e0142639.

Glick, B.R., 1995. The enhancement of plant growth by free-living bacteria. Can. J. Microbiol. 41 (2), 109–117. https://doi.org/10.1139/m95-015.

Glick, B.R., Penrose, D.M., Li, J., 1998. A model for the lowering of plant ethylene concentrations by plant growth-promoting bacteria. J. Theor. Biol. 190 (1), 63–68. https://doi.org/10.1006/jtbi.1997.0532.

Gray, E.J., Smith, D.L., 2005. Intracellular and extracellular PGPR: commonalities and distinctions in the plant–bacterium signaling processes. Soil Biol. Biochem. 37 (3), 395–412. https://doi.org/10.1016/j.soilbio.2004.08.030.

Gupta, A., Rai, V., Bagdwal, N., Goel, R., 2005. In situ characterization of mercury-resistant growth-promoting fluorescent pseudomonads. Microbiol. Res. 160 (4), 385–388. https://doi.org/10.1016/j.micres.2005.03.002.

Gurujeyalakshmi, G., Oriel, P., 1989. Isolation of phenoldegrading Bacillus stearothermophilus BR219 and partial characterization of the phenol hydroxylase. Appl. Environ. Microbiol. 55, 500–502.

Hariprasad, P., Niranjana, S.R., 2009. Isolation and characterization of phosphate solubilizing rhizobacteria to improve plant health of tomato. Plant Soil 316 (1), 13–24. https://doi.org/10.1007/s11104-008-9754-6.

Jeon, J.S., Lee, S.S., Kim, H.Y., Ahn, T.S., Song, H.G., 2003. Plant growth promotion in soil by some inoculated microorganisms. J. Microbiol. 41 (4), 271–276.

Ji, G., Silver, S., 1995. Bacterial resistance mechanisms for heavy metals of environmental concern. J. Ind. Microbiol. 14 (2), 61–75. https://doi.org/10.1007/BF01569887.

Kamala-Kannan, S., Lee, K.J., Park, S.M., Chae, J.C., Yun, B.S., Lee, Y.H., Oh, B.T., 2010. Characterization of ACC deaminase gene in Pseudomonas entomophila strain PS-PJH isolated from the rhizosphere soil. J. Basic Microbiol. 50 (2), 200–205. https://doi.org/10.1002/jobm.200900171.

Kamaludeen, S.P.B., Arunkumar, K.R., Ramasamy, K., 2003. Bioremediation of Chromium Contaminated Environments. NISCAIR.

Katulanda, P., Rajapaksha, C.P., 2012. Response of maize grown in an alfisol of SriLanka to inoculants of plant growth promoting rhizobacteria. J. Plant Nutr. 35 (13), 1984–1996. https://doi.org/10.1080/01904167.2012.716891.

Kaymak, H.C., 2010. Potential of PGPR in agricultural innovations. In: Plant Growth and Health Promoting Bacteria. Springer, pp. 45–79. https://doi.org/10.1007/978-3-642-13612-2_3.

Kim, I.C., Oriel, P.J., 1995. Characterization of the Bacillus stearothermophilus BR219 phenol hydroxylase gene. Appl. Environ. Microbiol. 61 (4), 1252–1256.

Kumar, A., Kumar, S., Kumar, S., 2005. Biodegradation kinetics of phenol and catechol using Pseudomonas putida MTCC 1194. Biochem. Eng. J. 22 (2), 151–159. https://doi.org/10.1016/j.bej.2004.09.006.

Kumar, B., Trivedi, P., Pandey, A., 2007. Pseudomonas corrugata: a suitable bacterial inoculant for maize grown under rainfed conditions of Himalayan region. Soil Biol. Biochem. 39 (12), 3093–3100. https://doi.org/10.1016/j.soilbio.2007.07.003.

Kunamneni, A., Camarero, S., García-Burgos, C., Plou, F.J., Ballesteros, A., Alcalde, M., 2008. Engineering and applications of fungal laccases for organic synthesis. Microb. Cell Factories 7 (1), 1–17. https://doi.org/10.1186/1475-2859-7-32.

Lata, R., Chowdhury, S., Gond, S.K., White Jr., J.F., 2018. Induction of abiotic stress tolerance in plants by endophytic microbes. Lett. Appl. Microbiol. 66 (4), 268–276. https://doi.org/10.1111/lam.12855.

Lee, S.W., Glickmann, E., Cooksey, D.A., 2001. Chromosomal locus for cadmium resistance in Pseudomonas putida consisting of a cadmium-transporting ATPase and a MerR family response regulator. Appl. Environ. Microbiol. 67 (4), 1437. https://doi.org/10.1128/AEM.67.4.1437-1444.2001.

Liu, H., He, Y., Jiang, H., Peng, H., Huang, X., Zhang, X., Xu, Y., 2007. Characterization of a phenazine-producing strain Pseudomonas chlororaphis GP72 with broad-spectrum antifungal activity from green pepper rhizosphere. Curr. Microbiol. 54 (4), 302–306. https://doi.org/10.1007/s00284-006-0444-4.

Lugtenberg, B.J., Dekkers, L., Bloemberg, G.V., 2001. Molecular determinants of rhizosphere colonization by Pseudomonas. Annu. Rev. Phytopathol. 39 (1), 461–490. https://doi.org/10.1146/annurev.phyto.39.1.461.

Lysenko, O., 1961. Pseudomonas—an attempt at a general classification. Microbiology 25 (3), 379–408. https://doi.org/10.1099/00221287-25-3-379.

Ma, Y., Rajkumar, M., Luo, Y., Freitas, H., 2011. Inoculation of endophytic bacteria on host and non-host plants—effects on plant growth and Ni uptake. J. Hazard. Mater. 195, 230–237. https://doi.org/10.1016/j.jhazmat.2011.08.034.

Mai, P., Jacobsen, O.S., Aamand, J., 2001. Mineralization and co-metabolic degradation of phenoxyalkanoic acid herbicides by a pure bacterial culture isolated from an aquifer. Appl. Microbiol. Biotechnol. 56 (3), 486–490. https://doi.org/10.1007/s002530000589.

Malik, D.K., Sindhu, S.S., 2011. Production of indole acetic acid by Pseudomonas sp.: effect of coinoculation with Mesorhizobium sp. Cicer on nodulation and plant growth of chickpea (Cicer arietinum). Physiol. Mol. Biol. Plants 17 (1), 25–32. https://doi.org/10.1007/s12298-010-0041-7.

Markandey, D.K., Maiti, S.K., Makhijani, S.D., Rajvaida, M., 2002. In: Markandey, D.K., Markandey, N.R. (Eds.), Microorganisms in Bioremediation. Capital, New Delhi.

Martínez-Viveros, O., Jorquera, M.A., Crowley, D.E., Gajardo, G.M.L.M., Mora, M.L., 2010. Mechanisms and practical considerations involved in plant growth promotion by rhizobacteria. J. Soil Sci. Plant Nutr. 10 (3), 293–319. https://doi.org/10.4067/S0718-95162010000100006.

Matsumura, F., Benezet, H.J., Patil, K.C., 1976. Factors affecting microbial metabolism of y-BHC. J. Pestic. Sci. 1 (1), 3–8. https://doi.org/10.1584/jpestics.1.3.

Mehnaz, S., Lazarovits, G., 2006. Inoculation effects of Pseudomonas putida, Gluconacetobacter azotocaptans, and Azospirillum lipoferum on corn plant growth under greenhouse conditions. Microb. Ecol. 51 (3), 326–335. https://doi.org/10.1007/s00248-006-9039-7.

Migula, W., 1894. Über ein neues System der Bakterien. Arbeiten aus dem Bakteriologischen Institut der Technischen Hochschule zu Karlsruhe.

Mittler, R., 2002. Oxidative stress, antioxidants and stress tolerance. Trends Plant Sci. 7 (9), 405–410. https://doi.org/10.1016/S1360-1385(02)02312-9.

Mollaeia, M., Abdollahpoura, S., Atashgahia, S., Abbasia, H., Masoomi, F., Rad, I., et al., 2010. Enhanced phenol environ Monit assess degradation by Pseudomonas sp. SA01: gaining insight into the novel single and hybrid immobilizations. J. Hazard. Mater. 175 (1–3), 284–292.

Müller, R.H., Kleinsteuber, S., Babel, W., 2001. Physiological and genetic characteristics of two bacterial strains utilizing phenoxypropionate and phenoxyacetate herbicides. Microbiol. Res. 156 (2), 121–131. https://doi.org/10.1078/0944-5013-00089.

Naik, M.M., Dubey, S.K., 2011. Lead-enhanced siderophore production and alteration in cell morphology in a Pb-resistant Pseudomonas aeruginosa strain 4EA. Curr. Microbiol. 62 (2), 409–414. https://doi.org/10.1007/s00284-010-9722-2.

Nair, C.I., Jayachandran, K., Shashidhar, S., 2008. Biodegradation of phenol. Afr. J. Biotechnol. 7 (25), 4951–4958.

Nawaz, K., Hussain, K., Choudary, N., Majeed, A., Ilyas, U., Ghani, A., Lashari, M.I., 2011. Eco-friendly role of biodegradation against agricultural pesticides hazards. Afr. J. Microbiol. Res. 5 (3), 177–183. https://doi.org/10.5897/AJMR10.375.

Padda, K.P., Puri, A., Chanway, C.P., 2018. Isolation and identification of endophytic diazotrophs from lodgepole pine trees growing at unreclaimed gravel mining pits in central interior British Columbia, Canada. Can. J. For. Res. 48 (12), 1601–1606. https://doi.org/10.1139/cjfr-2018-0347.

Padda, K.P., Puri, A., Chanway, C., 2019. Endophytic nitrogen fixation–a possible 'hidden'source of nitrogen for lodgepole pine trees growing at unreclaimed gravel mining sites. FEMS Microbiol. Ecol. 95 (11). https://doi.org/10.1093/femsec/fiz172, fiz172.

Pallai, R., Hynes, R.K., Verma, B., Nelson, L.M., 2012. Phytohormone production and colonization of canola (Brassica napus L.) roots by Pseudomonas fluorescens 6-8 under gnotobiotic conditions. Can. J. Microbiol. 58 (2), 170–178. https://doi.org/10.1139/w11-120.

Pandey, A., Trivedi, P., Kumar, B., Palni, L.M.S., 2006. Characterization of a phosphate solubilizing and antagonistic strain of Pseudomonas putida (B0) isolated from a sub-alpine location in the Indian Central Himalaya. Curr. Microbiol. 53 (2), 102–107. https://doi.org/10.1007/s00284-006-4590-5.

Phour, M., Sindhu, S.S., 2020. Amelioration of salinity stress and growth stimulation of mustard (Brassica juncea L.) by salt-tolerant Pseudomonas species. Appl. Soil Ecol. 149, 103518. https://doi.org/10.1016/j.apsoil.2020.103518.

Rajendran, P., Muthukrishnan, J., Gunasekaran, P., 2003. Microbes in Heavy Metal Remediation. http://nopr.niscair.res.in/handle/123456789/17153.

Rajkumar, M., Freitas, H., 2008. Effects of inoculation of plant-growth promoting bacteria on Ni uptake by Indian mustard. Bioresour. Technol. 99 (9), 3491–3498. https://doi.org/10.1016/j.biortech.2007.07.046.

Rao, M.A., Scelza, R., Scotti, R., Gianfreda, L., 2010. Role of enzymes in the remediation of polluted environments. J. Soil Sci. Plant Nutr. 10 (3), 333–353. https://doi.org/10.4067/S0718-95162010000100008.

Ruiu, L., 2020. Plant-growth-promoting Bacteria (PGPB) against insects and other agricultural pests. Agronomy 10 (6), 861. https://doi.org/10.3390/agronomy10060861.

Saharan, B.S., Nehra, V., 2011. Plant growth promoting rhizobacteria: a critical review. Life Sci. Med. Res. 21 (1), 30.

Sandhya, V.Z.A.S., Grover, M., Reddy, G., Venkateswarlu, B., 2009. Alleviation of drought stress effects in sunflower seedlings by the exopolysaccharides producing Pseudomonas putida strain GAP-P45. Biol. Fertil. Soils 46 (1), 17–26. https://doi.org/10.1007/s00374-009-0401-z.

Sandhya, V.S.K.Z., Ali, S.Z., Grover, M., Reddy, G., Venkateswarlu, B., 2010. Effect of plant growth promoting Pseudomonas spp. on compatible solutes, antioxidant status and plant growth of maize under drought stress. Plant Growth Regul. 62 (1), 21–30. https://doi.org/10.1007/s10725-010-9479-4.

Saravanakumar, D., Vijayakumar, C., Kumar, N., Samiyappan, R., 2007. PGPR-induced defense responses in the tea plant against blister blight disease. Crop Prot. 26 (4), 556–565. https://doi.org/10.1016/j.cropro.2006.05.007.

Saravanakumar, D., Thomas, A., Banwarie, N., 2019. Antagonistic potential of lipopeptide producing Bacillus amyloliquefaciens against major vegetable pathogens. Eur. J. Plant Pathol. 154 (2), 319–335. https://doi.org/10.1007/s10658-018-01658-y.

Semple, K.T., Cain, R.B., Stefan, S., 1999. Biodegradation of aromatic compounds by microalgae. FEMS Microbiol. Lett. 170 (2), 291–300.

Shaharoona, B., Naveed, M., Arshad, M., Zahir, Z.A., 2008. Fertilizer-dependent efficiency of Pseudomonads for improving growth, yield, and nutrient use efficiency of wheat (Triticum aestivum L.). Appl. Microbiol. Biotechnol. 79 (1), 147–155. https://doi.org/10.1007/s00253-008-1419-0.

Shanmugam, V., Kanoujia, N., 2011. Biological management of vascular wilt of tomato caused by fusarium oxysporum f. sp. lycospersici by plant growth-promoting rhizobacterial mixture. Biol. Control 57 (2), 85–93. https://doi.org/10.1016/j.biocontrol.2011.02.001.

Singh, R.P., Manchanda, G., Singh, R.N., Srivastava, A.K., Dubey, R.C., 2016. Selection of alkalotolerant and symbiotically efficient chickpea nodulating rhizobia from North-West Indo Gangetic Plains. J. Basic Microbiol. 56 (1), 14–25. https://doi.org/10.1002/jobm.201500267. 26377641.

Singh, R.P., Handa, R., Manchanda, G., 2021. Nanoparticles in sustainable agriculture: an emerging opportunity. J. Control Release 329, 1234–1248. https://doi.org/10.1016/j.jconrel.2020.10.051. 33122001.

Sivasakthi, S., Usharani, G., Saranraj, P., 2014. Biocontrol potentiality of plant growth promoting bacteria (PGPR)-Pseudomonas fluorescens and Bacillus subtilis: a review. Afr. J. Agric. Res. 9 (16), 1265–1277. https://doi.org/10.5897/AJAR2013.7914.

Somers, E., Vanderleyden, J., Srinivasan, M., 2004. Rhizosphere bacterial signalling: a love parade beneath our feet. Crit. Rev. Microbiol. 30 (4), 205–240. https://doi.org/10.1080/10408410490468786.

Srivastava, R., Aragno, M., Sharma, A.K., 2010. Cow dung extract: a medium for the growth of pseudomonads enhancing their efficiency as biofertilizer and biocontrol agent in rice. Indian J. Microbiol. 50 (3), 349–354. https://doi.org/10.1007/s12088-010-0032-y.

St-Martin, A., Vico, G., Bergkvist, G., Bommarco, R., 2017. Diverse cropping systems enhanced yield but did not improve yield stability in a 52-year long experiment. Agric. Ecosyst. Environ. 247, 337–342. https://doi.org/10.1016/j.agee.2017.07.013.

Subba, R., 1993. Biofertilizers in agriculture and forestry. In: Biofertilizers in Agriculture and Forestry, third ed. CABI.

Sumner, D.R., 2018. Crop rotation and plant productivity. In: CRC Handbook of Agricultural Productivity. CRC Press, pp. 273–314.

Talley, J., 2005. Introduction of recalcitrant compounds. In: Bioremediation of Recalcitrant Compounds. Taylor and Francis, pp. 1–9.

Tank, N., Saraf, M., 2009. Enhancement of plant growth and decontamination of nickel-spiked soil using PGPR. J. Basic Microbiol. 49 (2), 195–204. https://doi.org/10.1002/jobm.200800090.

Tsai, S.-C., Tsai, L.-D., Li, Y.-K., 2005. An isolated candida albicans TL3 capable of degrading phenol at large concentration. Biosci. Biotechnol. Biochem. 69 (12), 2358–2367.

Upadhyay, S.K., Singh, J.S., Singh, D.P., 2011. Exopolysaccharide-producing plant growth-promoting rhizobacteria under salinity condition. Pedosphere 21 (2), 214–222. https://doi.org/10.1016/S1002-0160(11)60120-3.

Van Loon, L.C., 2007. Plant responses to plant growth-promoting rhizobacteria. In: New Perspectives and Approaches in Plant Growth-Promoting Rhizobacteria Research. Springer, Dordrecht, pp. 243–254. https://doi.org/10.1007/978-1-4020-6776-1_2.

Vereijken, P., 2020. Research on integrated arable farming and organic mixed farming in the Netherlands. In: Sustainable Agricultural Systems. CRC Press, pp. 287–296.

Verma, J.P., Yadav, J., Tiwari, K.N., Lavakush, S., Singh, V., 2010. Impact of plant growth promoting rhizobacteria on crop production. Int. J. Agric. Res. 5 (11), 954–983.

Wada, H., Senoo, K., Takai, Y., 1989. Rapid degradation of 2,-HCH in upland soil after multiple applications. Soil Sci. Plant Nutr. 35 (1), 71–77. https://doi.org/10.1080/00380768.1989.10434738.

Wang, C.L., Michels, P.C., Dawson, S.C., Kitisakkul, S., Baross, J.A., Keasling, J.D., Clark, D.S., 1997. Cadmium removal by a new strain of Pseudomonas aeruginosa in aerobic culture. Appl. Environ. Microbiol. 63 (10), 4075–4078.

Wasi, S., Jeelani, G., Ahmad, M., 2008. Biochemical characterization of a multiple heavy metal, pesticides and phenol resistant Pseudomonas fluorescens strain. Chemosphere 71 (7), 1348–1355. https://doi.org/10.1016/j.chemosphere.2007.11.023.

Wasi, S., Tabrez, S., Ahmad, M., 2011a. Detoxification potential of Pseudomonas fluorescens SM1 strain for remediation of major toxicants in Indian water bodies. Water Air Soil Pollut. 222 (1), 39–51. https://doi.org/10.1007/s11270-011-0802-0.

Wasi, S., Tabrez, S., Ahmad, M., 2011b. Suitability of immobilized Pseudomonas fluorescens SM1 strain for remediation of phenols, heavy metals, and pesticides from water. Water Air Soil Pollut. 220 (1), 89–99. https://doi.org/10.1007/s11270-010-0737-x.

Willey, J.M., Sherwood, L.M., Woolverton, C.J., 2011. Chapter 29: Microorganisms in terrestrial ecosystems. In: Prescott's Microbiology. McGraw-Hill, ISBN: 978-0-07-131367-4, pp. 703–706.

Yan, L.G., Xu, Y.Y., Yu, H.Q., Xin, X.D., Wei, Q., Du, B., 2010. Adsorption of phosphate from aqueous solution by hydroxy-aluminum, hydroxy-iron and hydroxy-iron–aluminum pillared bentonites. J. Hazard. Mater. 179 (1–3), 244–250. https://doi.org/10.1016/j.jhazmat.2010.02.086.

Yao, L., Wu, Z., Zheng, Y., Kaleem, I., Li, C., 2010. Growth promotion and protection against salt stress by *Pseudomonas putida* Rs-198 on cotton. Eur. J. Soil Biol. 46 (1), 49–54. https://doi.org/10.1016/j.ejsobi.2009.11.002.

Yap, L.F., Lee, Y.K., Poh, C.L., 1999. Mechanism for phenol tolerance in phenol-degrading Comamonas testosteroni strain. Appl. Microbiol. Biotechnol. 51 (6), 833–840. https://doi.org/10.1007/s002530051470.

CHAPTER

9

Relationship between organic matter and microbial biomass in different vegetation types

Emre Babur[a], Turgay Dindaroğlu[a], Rana Roy[b], Mahmoud F. Seleiman[c,d], Ekrem Ozlu[e], Martin L. Battaglia[f], and Ömer Suha Uslu[g]

[a]*Department of Soil Science and Ecology, Faculty of Forestry, Kahramanmaras Sutcu Imam University, Kahramanmaraş, Turkey,*

[b]*Department of Agroforestry & Environmental Science, Sylhet Agricultural University, Sylhet, Bangladesh,*

[c]*Plant Production Department, College of Food and Agriculture Sciences, King Saud University, Riyadh, Saudi Arabia,*

[d]*Department of Crop Sciences, Faculty of Agriculture, Menoufia University, Shibin El-Kom, Egypt,*

[e]*Great Lakes Bioenergy Research Center, W.K. Kellogg Biological Station, Michigan State University, Hickory Corners, MI, United States,*

[f]*Department of Animal Science, Cornell University, Ithaca, NY, United States,*

[g]*Field Crops Department, Agriculture Faculty, Kahramanmaras Sutcu Imam University, Kahramanmaraş, Turkey*

Abstract

Organic matter (OM) is composed of organic debris of plant mass (forest floor and dead wood) and animals. Vegetation cover is one of the most important sources for greater organic matter accumulation in different ecosystems like agriculture, forests, and grasslands. It is well known that forests have higher soil organic carbon at surface soil depth due to the higher amount of litter return and organic matter input. Organic matter has direct and indirect critical impacts on soil structure and health because it is a source of nutrients such as nitrogen, sulfur, phosphorus, and many of the micronutrients, and it helps to enhance microbial community compositions and activities. However, the availability of these substances depends on the presence of microbial biomass (MB) in the environment. The OM and MB are sensitive indicators for ecosystem productivity and health and are recognized by ecosystem quality monitoring programs. Both OM and MB also influence carbon sequestration rate, dynamics of atmospheric C, and climate change regulations in terrestrial ecosystems. This is where microorganisms show their crucial role, especially that in the decomposition of OM from different vegetation, and for nutrient availability. Therefore, it is necessary to determine the microbial biomass to understand the biological activities of organic matter. This chapter reports an overview of the understanding of the relationship between organic matter and microbial biomass in different vegetation types.

Keywords: Climate change, Organic matter, Microbial biomass, Soil health, Vegetation types

Microbial Syntrophy-mediated Eco-enterprising. https://doi.org/10.1016/B978-0-323-99900-7.00005-5

1 Introduction

Forest vegetation cover is an important component of several functions related to ecosystem stability. These functions decrease the effects of water and wind erosion, increase the soil water holding capacity and soil organic carbon (C) sequestration, moderate the impact of extreme temperatures and moisture contents on the soil surface, and protect the soil particles against the effects of intensive rainfall events (Gallardo et al., 1998; Babur et al., 2021a,c). The contribution of soil microorganisms to C mineralization, nutrient cycling, and other microbial processes including microbial respiration and biomass growth depends on both substrate quality and local climatic conditions (Manzoni et al., 2012; Spohn and Chodak, 2015; Adnan et al., 2020). The main component of the C mineralization process is the microbial C use efficiency, defined as the ratio between the microbial C used for growth and the total C involved in the process. This ratio shows the difference between C used for building biomass (anabolism), and C utilized for respiration to obtain the energy necessary for maintenance, enzyme synthesis, growth, and nutrient accession (catabolism) (Sinsabaugh et al., 2013; Spohn et al., 2016). However, differences in the biochemical composition of soil organic matter (SOM) may affect the rate and overall extent of litter decomposition, thus contributing to increased structural and fertility heterogeneity of surface soils in forests.

Litterfall is the main soil C input in forest ecosystems. Consequently, to maintain soil fertility and productivity, the continuous input of fresh organic matter (OM) through litter addition is important. In agricultural ecosystems, management practices like plant residue addition, animal manure applications, and food or animal wastes, and green manure can be used to add OM into the soil (Ozlu and Kumar, 2018; Babur and Dindaroglu, 2020; Battaglia et al., 2021a,b). Regardless of the ecosystem, soil microbial communities are the main decomposers of this organic litter. However, the activity and presence of microbes are constantly influenced by the physical, chemical, and biological properties of soils in different ecosystems. Tree species is another factor impacting the microbial diversity and the biochemical properties in different ecosystems, mainly as a result of the ample variations in the quantity and quality of litter production across tree species (Bauhus and Pare, 1998; Kara et al., 2016). In addition, seasonal changes in microbial activity due to changes in temperature, precipitation, and substrate availability (Basiliko et al., 2005; Lepcha and Devi, 2020) play a paramount role in plant growth, nutrient availability and plant nutrient uptake, microbial biomass, and metabolic quotient (qCO_2). The impacts of strong seasonal changes on microbial C use efficiency may decrease qCO_2 as decomposition proceeds (Maraun and Scheu, 1995).

The SOM, arguably one of the most studied soil properties (Battaglia et al., 2021b), has a profound impact on soil health and soil structure by maintaining nutrient supply and C mineralization, and sequestration rates (Babur and Dindaroglu, 2020). Soil microbes and enzymatic activities control the decomposition of litter and act as a reservoir of soil nutrients and energy (Jenkinson and Ladd, 1981). Although a myriad of previous studies have documented the biochemical mechanisms and properties of a large number of soils around the globe, the dynamics of soil microbial communities and their diversity are still unclear in different ecosystems (Torsvik and Øvreås, 2007).

This chapter aims to understand the impact that differences in forest soils and different land-use types have on SOM, soil microbial biomass C, microbial respiration, and microbial activity. Here, we focus on the current knowledge related to soil development under different tree species in forest ecosystems. Since soil development in forest ecosystems is continuously affected by the interaction of different ecological factors, soil types, and soil microorganisms, changes in soil properties and their interactions are also constant. These interactions, therefore, should be accounted for when assessing the properties of soils in different climate regimes, topographies, and vegetation types.

2 Factors affecting soil microbial properties

Microbial biomass is considered an early indicator of changes in soil organic C (SOC), which indicates that SOC response to changes in management practices is generally slower than changes in microbial biomass C. These findings reflect the strong and continuous relations among plants, soils, and microorganisms (Fig. 1). Kara et al. (2016) reported that positive significant changes in microbial biomass and other chemical properties such as pH, SOC with afforestation application in the northern region of Turkey. Previous studies evaluate the influence of retaining standing crop residues after harvest vs burning crop residues on short-term gross N transformation rates in Western Australia. Hoyle et al. (2006) found short-term microbial biomass C increases between 100 and 150 kg C ha^{-1} by the short-term impacts of planting. Since soils are generally dry and poor in terms of organic matter and, thus, they are also deficient in microbial biomass. In all cases, the amount of unstable SOC is particularly important, as it provides a source of readily available C, which represents the energy source for the microbial activity (Hoyle et al., 2006). Thus, soils with more labile C tend to have higher microbial biomass. Among the sources of labile C, fresh organic residues and root exudates are the most important and readily available energy sources for soil microorganisms (Babur and Dindaroglu, 2020).

Soil microbial properties are generally influenced by environmental and anthropogenic factors such as climate (moisture and temperature), topography, vegetation, land use, management practices, geological and geomorphological properties, soil type, nutrient availability, and quality of substrates, etc. (Babur and Dindaroglu, 2020). Even though chemical properties of soil such as clay content, soil pH and SOC impact microbial biomass (Figs. 2 and 3), factors such as soil depth, temperature, and moisture content are the limiting factors for microbial biomass in forest soils of the Mediterranean

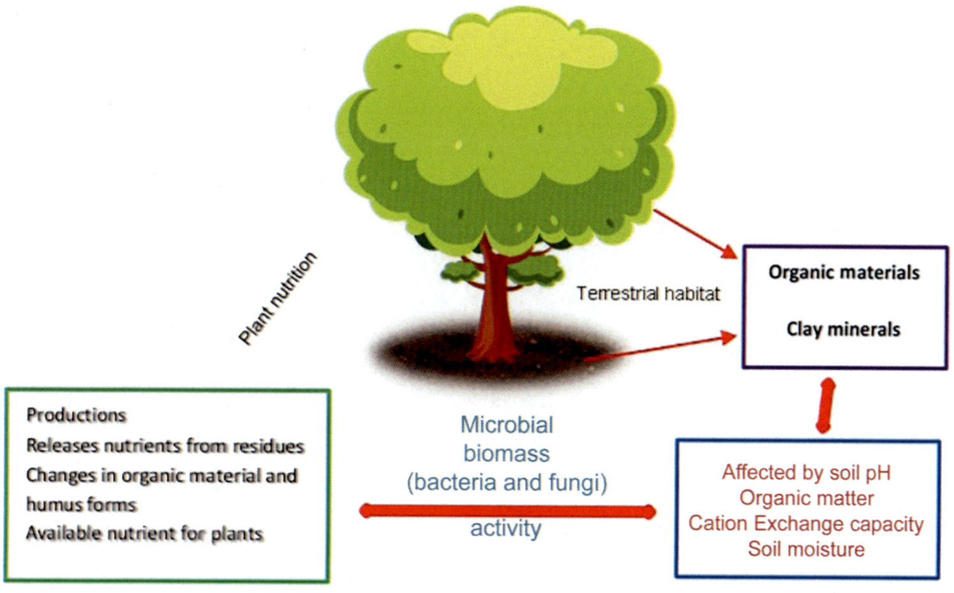

FIG. 1

The interaction among the main soil properties, microbial biomass, and plants.

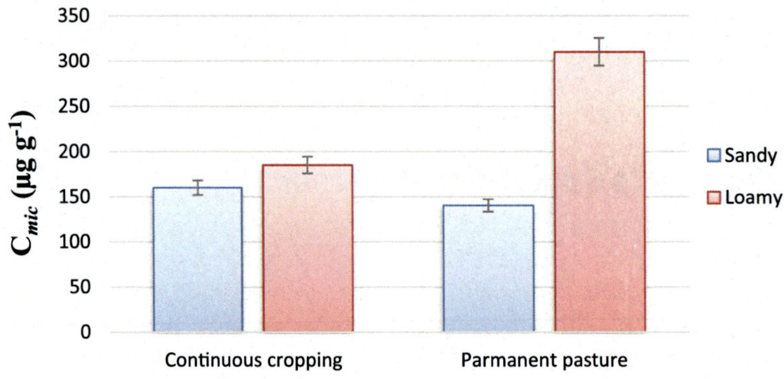

FIG. 2

Different clay contents and management practices significantly affect soil microbial biomass in topsoil (0–10 cm) (Anonymous, 2021).

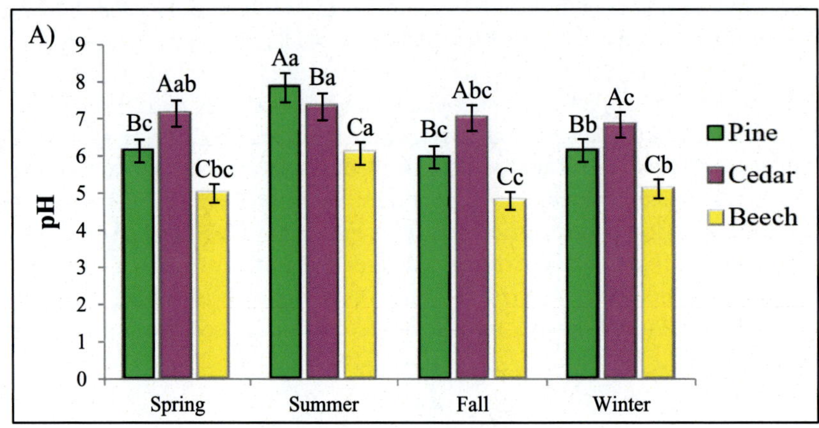

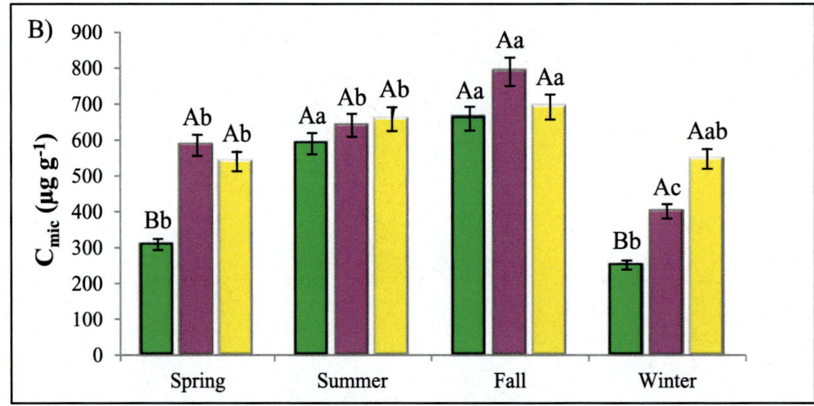

FIG. 3

The effect of season and tree species on (A) pH and (B) microbial biomass C (C_{mic} in µg g^{-1}) and means followed by different capital letters show differences across tree species for the same season (between column comparisons). Different lower-case letters show differences for a given tree species across seasons (within column comparisons). All differences are set at 0.05 probability level † (Babur et al., 2021a).

region (Figs. 4 and 5). Although N release was higher than microbial biomass N at 0–10 and 30–40 cm depth, these soils had greater microbial biomass N compared to N release at 10–20, 20–30, and 40–50 cm depths.

Soils rich in clay content and SOC, generally hold more water and hence have higher microbial biomass. A soil pH around neutral (6.5–7.5) is most suitable for the microbial biomass (Fig. 3).

The management activities, such as the management of plant residues, affect the microbial biomass and enzyme activities because post-production residues are the source of SOC, and nutrients used by the microbial communities (Yang et al., 2019). Plant residue removal or even burning residue is a common practice among farmers around the world. However, the plant residue is an organic plant material and source of SOM and a food source for microbial organisms (Table 1). Returning these residues to soils, therefore, provides a practical way to increase the soil microbial biomass SOM (Fig. 6).

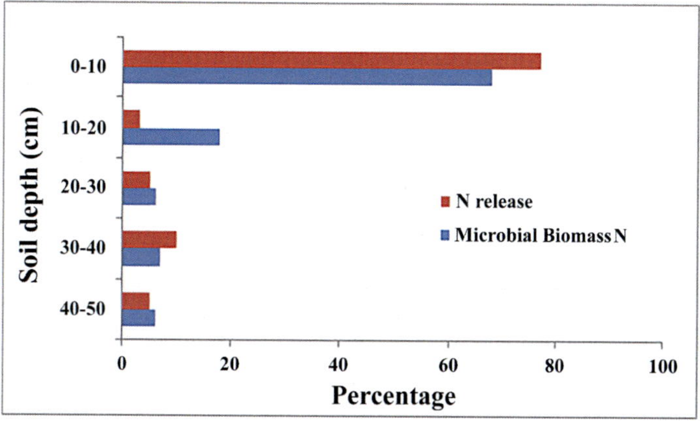

FIG. 4

The microbial biomass N and N release decrease deeper into the soil (Murphy et al., 1998).

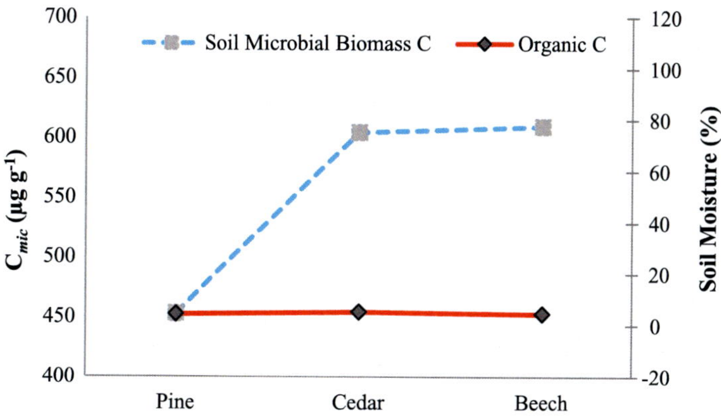

FIG. 5

Microbial biomass C and soil moisture content in different tree species Eastern Mediterranean Forests (Babur et al., 2021a).

Table 1 The effect of retaining (17 years) or burning stubble on microbial biomass carbon at different soil depths at Merredin, WA (Hoyle et al., 2006).

	Microbial biomass carbon (kg/ha)	
Soil depth (cm)	**Stubble retained**	**Stubble burnt**
0–10	229	165
10–20	112	93
20–30	69	58

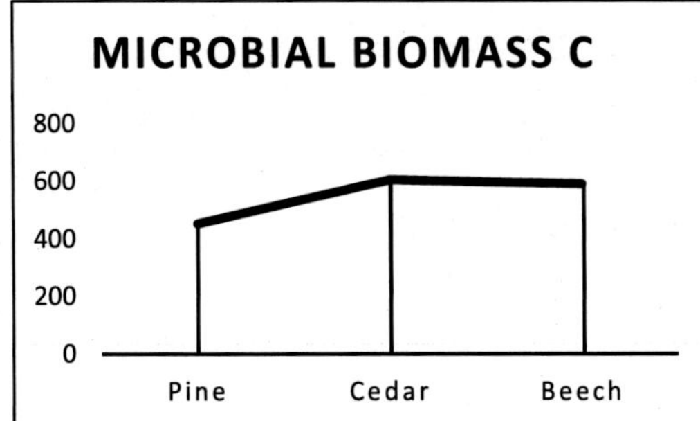

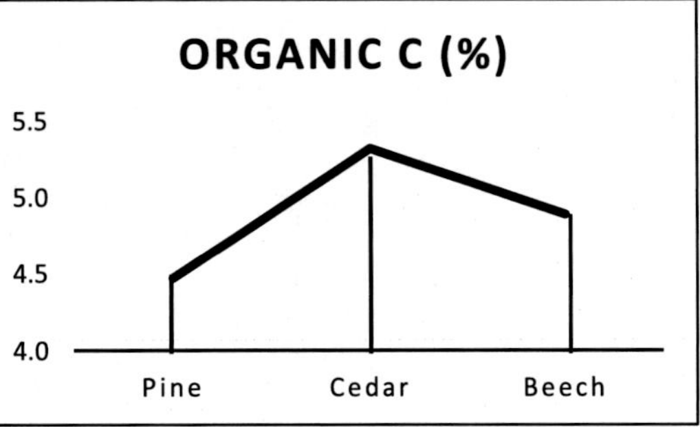

FIG. 6

Microbial biomass carbon of soils is directly proportional to organic soil carbon (Babur et al., 2021a).

Minimizing soil disturbance preserves soil aggregates (Ozlu and Arriaga, 2021) formed by fungal or mycorrhizal networks and hence increases microbial biomass (Fig. 9). The pore spaces in soil aggregates are an important habitat for soil microorganisms. Moreover, more destructive combined applications can cause rapid depletion of SOC and microbial biomass C in the surface soil. Direct drilling increased microbial biomass in the surface soil layer (0–5 cm) compared to other tillage applications (Fig. 7). Stubble incorporation causes residues to migrate deeper into the soil, which can adversely affect aeration in the soil, limiting decomposition and thus microbial activity and microbial biomass C (Pankhurst et al., 2002). However, disturbing aggregates may cause negative effects on stable SOC in smaller aggregates (Ozlu, 2000).

The residues of leguminous can increase microbial biomass and their activities due to their greater nitrogen content (Fig. 8). In addition, longer pasture rotations generally increase soil microbial biomass under reduced disturbances and increased SOM levels (Fig. 2).

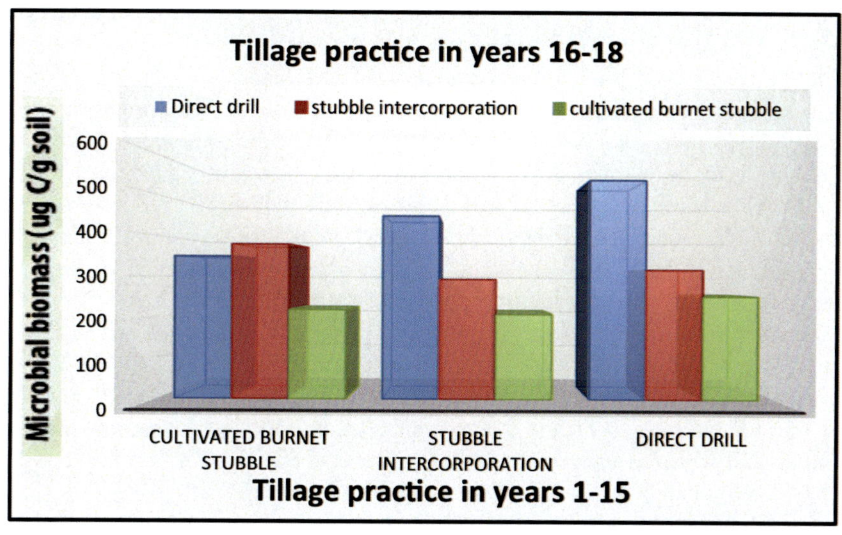

FIG. 7

Rapid changes occur to microbial biomass in topsoil (0–5 cm) after tillage practice (Pankhurst et al., 2002).

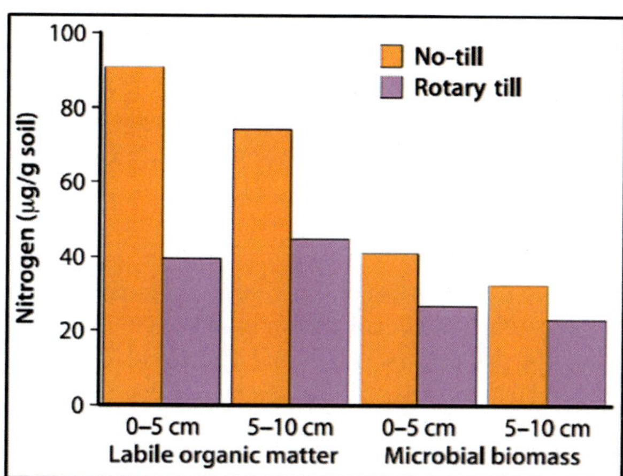

FIG. 8

Relationship between nitrogen content and the microbial biomass with no-till compared to rotary till at Wongan Hills, WA (Cookson et al., 2008).

3 Effects of vegetation types, species, and afforestation

The variability in vegetation types in a particular ecosystem has significant effects on the SOC stocks and C dynamics (Vesterdal et al., 2002; Hansson, 2011; Babur et al., 2021a,b,c; Laganière et al., 2010). Laganière et al. (2010) also found that the broad-leaved species store 25% more C than the coniferous species. In addition, naturally grown or mixed-species forests may have significantly different decomposition rates and the amount of C stored in the litter layer and soil itself (Sarıyıldız et al., 2005; Tolunay and Çömez, 2008). Shi and Cui (2010) studied SOC stocks in plantation areas in China and found that more exchangeable Cat 0–40 cm was represented in broadleaf species (5.92 g $Cm^{-2}year^{-1}$) and mixed forests (4.88 g $Cm^{-2}year^{-1}$) than coniferous forests (−0.90 g $Cm^{-2}year^{-1}$). In another study conducted in broad and coniferous afforestation areas, it was documented that the soils formed under broadleaf types store more C whereas the C loss was higher in coniferous soils (Guo and Gifford, 2002).

The root systems of trees help to develop soil structure and SOC content. The broad-leaved forests store more C throughout the soil profile due to their long and deep root structures and root biomass (Lorenz and Lal, 2005; Strong and La Roi, 1985), whereas coniferous forests store less C because of shallow root systems (Jandl et al., 2007). However, other studies reported greater SOC in the litter for coniferous species compared to those of broad-leaved species (Vesterdal and Raulund-Rasmussen, 1998; Vesterdal et al., 2008). The reason behind this difference is the slower decomposition in coniferous forests. Therefore, studies that aim to determine the accumulated sequestration rate in soil and litter need to consider the effects of different tree species. Besides the quality and quantity of litter and SOC, tree species significantly affect stand productivity, stand cover, and nitrogen accumulation. Moreover, the canopy was also among the factors that affect the amount of C stored in forests. Since climate change differentiates vegetation types in specific ecosystems, including tree species in forests (Koca et al., 2006), these changes in vegetation type may also impact SOC and microbial properties of soil. For instance, Tate et al. (1991) researched the C stocks in the litter layer and soil profile, and their relationships with respiration in natural old beech forests in the South of New Zealand. From this study, Tate et al. (1991) documented that the amount of total C in the litter layer was 3 kg m^{-2} significantly lower than that of mineral soil (15.8 kg m^{-2}) while the annual average respiratory rate of the litter was 0.65 kg CO_2-C m^{-2}.

There are important relationships between the chemical composition of litters and the physicochemical properties of soils, and their microbial biomass and activity (Parr and Papendick, 1997). Since soil microbial activity has a direct effect on ecosystem productivity, microbial biomass provides prompt and precise information about soil quality (Smith et al., 1992). In addition, seasonal changes regulate nutrient availability as they accelerate microbial decomposition.

The metabolic quotient (qCO_2) was measured as the rate of basal respiration (μg CO_2-C h^{-1}) per milligram of microbial biomass C (Dilly and Munch, 1998). The qCO_2 has been widely used as an indicator to assess soil substrate quality, stress response, and land-use practices (Anderson and Domsch, 1993; Martín-Lammerding et al., 2015). The higher qCO2 is generally related to a recalcitrant substrate and environmental stress (Spohn, 2015). Babur et al. (2021a,b,c) found that the high metabolic quotient in both the winter and pine forest land indicated a low efficiency of the substrate utilization by the microbial community composition (Fig. 9). In general, soils with high microbial diversity are better able to utilize the substrate leading to lower metabolic quotient values (Babur and Dindaroglu, 2020; Maková et al., 2011). Fig. 9 indicates that high microbial diversity under different forest soil is energetically more efficient with lower qCO2.

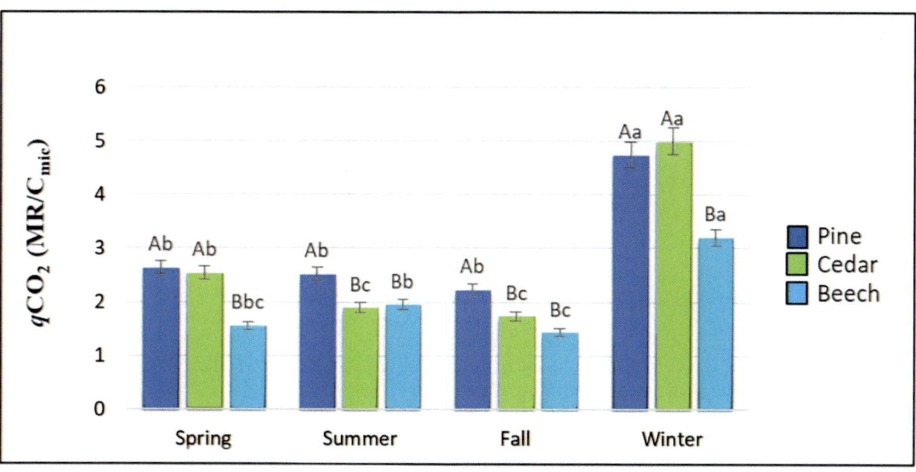

FIG. 9

The metabolic quotient is the ratio of microbial respiration to biomass and is indicated as a microbial stress indicator and interpreted as "microbial efficiency" (Babur et al., 2021a).

Martinez et al. (2019) conducted a study on how tree species affected SOM and soil biochemical properties in trace element contaminated soils. They found no differences between directions (North and South), and afforestation increased soil microbial biomass compared to tree-less areas with higher microbial activity rates in poplar and pine forests (Fig. 10).

Scheu and Parkinson (1995) conducted a study to determine biochemical properties of the litter of the aspen (*Populustremuloides* Michx.) and Contarta pine (*Pinuscontorta* Loud.) species in Alberta. As

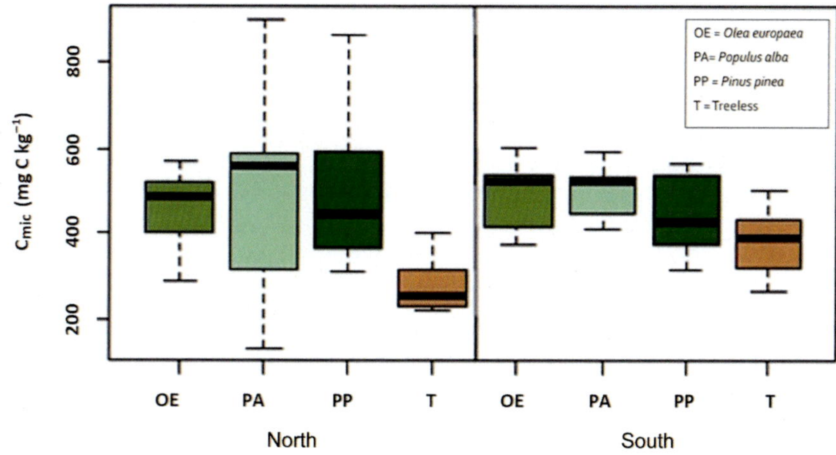

FIG. 10

The effect of direction and vegetation type on microbial biomass C (C_{mic} in $\mu g\ g^{-1}$). All differences are set at 0.05 probability level † OE: *Olea europaea*, PA: *Populus alba*, PP: *Pinuspinea*, T: treeless (bare space) (Martinez et al., 2019).

a result of this research, the microbial biomass C content and microbial quotient (C_{mic}/C_{org}) in the litter of aspen forest were found higher than those in a pine forest. Priha and Smolander (1997) investigated the microbial properties and activities of litter at 0–10 cm soil depth in forested areas with scotch pine (*Pinus sylvestris* L.), European spruce (*Piceaabies* L.), and birch (*Betula pendula* L.) species in two different locations in Malahati. There was no statistically significant difference between species regarding soil microbial activity and microbial properties since they are not very old stands. In addition, with respect to the C_{org}/N_{total} ratio, 30 was found in spruce leaves, 69 in pine leaves, and 54 in birch leaves. Furthermore, Nmic and net N mineralization in spruce leaves were higher than other species. Microbial properties of litter and soils in forests consisting of different tree species were compared with southern boreal forests of Canada by Bauhus and Pare (1998). The C_{mic}/C_{org} and N_{mic}/N_{total} ratios increased in parallel with the stand age. In particular, lower C_{mic}/C_{org} (1.9%) and N_{mic}/N_{total} (7.5%) ratios were reported for litter layers of the coniferous species compared to those of broadleaf species (2.2% and 8.6% in birch and 2.4% and 9.2% in poplar). Chen et al. (2000) investigated the amount of microbial biomass C in the litter (L) and fermented (F) layers of litter layers of mountain beech (*Nothofagussolandri* var. *cliffortioides*) stand in New Zealand. While the mean Cmic value was found at 2481 $\mu g\, g^{-1}$ in the litter layer, it was found at 5388 $\mu g\, g^{-1}$ in the fermentation layer. In the same study, the microbial characteristics of litter, fermented and humic (H) layers of birch (*Betula pendula* Roth), and European spruce (*Piceaabies* (L.) Karst) and Scot's pine (*Pinus sylvestris* L.) stands were investigated.C_{mic}/C_{org} percentages were found at 4.1% in L, 2.1% in F, and 1.4% in H layers of European spruce, while 3.9% in L, 2.6% in F, and 1.6% in H layers of birch trees. Basal respiration values of the litter layers were found as 14.62 $\mu g\, CO_2$-C g^{-1} litter h^{-1} in L and 37.00 $\mu g\, CO_2$-C g^{-1} litter h^{-1} in the F layer, respectively. The metabolic quotient (qCO2) was found at 5.89 mg CO_2-C $g^{-1} C_{mic}\, h^{-1}$ in the L layer and 6.87 mg CO_2-C $g^{-1} C_{mic}\, h^{-1}$ in the F layer.

Kanerva and Smolander (2007) also investigated the microbial properties of the three litter layers (L, F, and H) in the birch (*Betula pendula* Roth), European spruce (*Piceaabies* (L.) Karst) and Scot's pine (*Pinus sylvestris* L.) stands. As a result of the study, C_{mic}/C_{org} percentages were found to be 9.6% in L, 6.2% in F, and 6.9% in H of birch, 12.5% in L, 6.4% in F, and 6.7% in H of European spruce, and 16.3% in L, 7.8% in F, and 6.9% in H of scots pine forests. Regarding qCO_2 ratio, no difference was found in the L layer across different tree species. However, significant differences were found in the F and H layers of birch, European spruce, and Scotch pine forests. Another study reported that tree species affected soil microbial properties, the mean N_{mic} value of birch species was higher than other species, and birch organic materials decomposed more slowly than spruce species (Kiikkilä et al., 2014). On the other hand, Ravindran and Yang (2014) noticed that the soil microbial biomass C and N were higher than those of the grassland soils.

The OM compositions such as lignin, cellulose, and hemicellulose may differ in C_{mic} amount and activities (Sparling, 1997). The relationships between the chemical composition of litter and the biochemical properties of soils were significant (Parr and Papendick, 1997). In addition, temperature and humidity play a critical role in Cmic and its activity (Bauhus and Khanna, 1999).

4 Effects of land-use on soil organic carbon and microbial biomass

Land use and management, serving as anthropogenic human activities, have strong influences on soil ecosystem dynamics and functions (Singh et al., 2016). Therefore, it is necessary to identify soil processes under different land uses, to preserve, maintain, reclaim, and renovate the ecosystem services. Among ample research, one, for example, showed that the change in SOC was directly related to different land-use patterns (Zhang et al., 2011). Crucial factors affecting SOC stocks in soils were land use and relief. The C stocks at 0–10 cm soil depth in the karstic ecosystems were found to be more in depressed areas (48.71 Mg C ha^{-1}) than in non-depressed areas (27.22 Mg C ha^{-1}). In addition, SOC stocks were investigated with respect to changes in land use in depression and forest areas. According to the study, an average loss of 55% in SOC was observed in the transition from fertile forest areas to agricultural areas in depressed areas, and 80% in the transition from fertile forest areas to agricultural areas in non-depressed areas in the karst ecosystem. Thus, the importance of geomorphological units in the storage of SOC was identified in the karst ecosystem (Dindaroglu et al., 2019). Similar relations were discovered in another study in which SOC was analyzed by collecting surface soil samples at 0–10 and 10–20 cm in different land use. In general, the average soil organic carbon stocks in natural and mixed forest land (29.62 ± 1.95 Mg C ha^{-1}) were observed more than other land-uses. The SOC stocks in forest land are 36.14%, 28.36%, and 27.63% higher than those of cultivated areas, open shrubland, and Eucalyptus cultivated areas. The lower input of plant litter and decomposition might be the reason behind lower SOC stocks (Amanuel et al., 2018). The greater C sequestration through photosynthesis and the more plant biomass, and the more SOC stocks are presented in the soil (Matamala et al., 2003). The SOC stocks at 0–30 cm depth were significantly higher in forest soils (average 112.0 Mg ha^{-1}) than those of continuous agricultural fields (on average 108.8 Mg ha^{-1}) (Chaplot et al., 2010; Ajami et al., 2016). Tesfaye et al. (2016) also reported afforestation with different tree species in cultivated lands had 56.4% more SOC storage. In tropical regions, the conversion of forest lands to agricultural managements reduced SOC stocks by an average of 16% while the conversion of forest lands to pastures and pastures into secondary forest lands increased SOC stocks. Planting perennial trees in previously cultivated soils increases SOC stocks but transforming unmanaged degraded forests or pastures into agriculture did not affect SOC stocks (Powers et al., 2011).

Managed farmlands, grazed fields (controlled grazing), forests (functional management practices on tree species, afforestation, stand management, species selection), wetland/peat restoration, and degraded soils have great potential to increase SOC sequestration (Lal, 2014) with proper management. For instance, to provide sustainable C sequestration, natural vegetation should be protected and lower disturbance by conservation tillage methods might be recommended to increase the SOC and total N (Smith, 2004). No-till systems due to lower disturbance and field traffic increase the formation of macroaggregates, aggregate stability, (Ozlu and Arriaga, 2021) soil moisture, improved water infiltration, increased ventilation but reduces mass density, surface runoff, and effects of erosion (Batey, 2009). Soil management practices provide an important planning tool to increase and maintain actual SOC stocks and the potential for C sequestration. However, the effects of land-use change on these properties might be necessary. However, especially in peat areas where the accumulation of OM is high, significant amounts of Care present deeper in the profile and these SOCstocks tend to deform (Fontaine et al., 2007). One priority for sustainable soil C management can be to protect soils with high SOC stocks such as peatlands, rangelands, and natural forests from intensive disturbance and/or over-fertilization. For instance, decreasing the density of grazing in natural rangelands can increase SOC stocks, but the sustainability of these practices is linked with planned grazing practices (Conant et al., 2005).

Previously numerous studies have focused on the dynamics of total C in soils (Zhang et al., 2016; Han et al., 2018), but only a few have examined the impacts of management practices on dissolved organic carbon (DOC), easily oxidizable organic carbon (EOC), and microbial biomass carbon (MBC) together (Yang et al., 2009; Sheng et al., 2015). Similarly, the conversion of forests to agriculture led to decreased SOCby 40% over 40–50 years (Belay et al., 2018). Microbial biomass C is a measure of C contained within the living and active component of SOC pools, which is generally lower than 5% of the total C and N in soils, and responsible for OM decomposition, nutrient contents, and the biogeochemical processes in terrestrial ecosystems (Gregorich et al., 2000; Haney et al., 2001). The microbial biomass generally consists of bacteria and fungi, which directly play a role in the decomposition of plant residues and OM in the soil. The importance of microorganisms in the dynamics and functioning of the ecosystem increases the attention to investigate soil microbial biomass (Azam et al., 2003; Babur et al., 2021b). Moreover, microbial biomass and other biochemical properties are the best features to better understand and estimate variability related to land use impacts in different terrestrial ecosystems (Sharma et al., 2004). Soil biochemical mechanisms lead by soil microbes are sensitive to land use due to impacts of vegetation type, root exudates and depth, and litter content and composition (Hooper et al., 2005; Stenberg, 1999).

Different ranges of microbial biomass C in different tropical forest soils were reported by different studies such as those by Vance et al. (1987) between 61 and 2000 $\mu g\, g^{-1}$, and by Henrot and Robertson (1994) between 102 and 2073 $\mu g\, g^{-1}$. In addition, others showed a range of 97 to 1539 $\mu g\, g^{-1}$ in different soil types, land uses, stand ages, and vegetation types (Bauhus and Pare, 1998; Sharma et al., 2004; Tracy and Frank, 1998; Arunachalam et al., 1999) between 726.70 and 1529.14 $\mu g\, g^{-1}$in the wasteland and 806.1 $\mu g\, g^{-1}$ for grassland, and Kara and Bolat (2008) between 588 and 1310 $\mu g\, g^{-1}$ in the forest, from 429 to 1418 $\mu g\, g^{-1}$ in pasture and from to 682 $\mu g\, g^{-1}$ in agricultural soils. Further, the microbial biomass N ranged from 18 to 463 $\mu g\, g^{-1}$ under impacts of different land-use (Sharma et al., 2004; Diaz-Ravina et al., 1988; Martikainen and Palojärvi, 1990; Tracy and Frank, 1998; Garcia and Rice, 1994; Kara and Bolat, 2008). The key controlling factors in these differences generally are ground cover vegetation, the number of roots, land use type, and management application (Vásquez-Murrieta et al., 2007; Babur et al., 2021b). Generally, SOC, total N, C_{mic}, and N_{mic} concentrations are significantly greater in the forest soils than those under pasture and agriculture (Sharma et al., 2004; Kara and Bolat, 2008; Van Leeuwen et al., 2017; Lepcha and Devi, 2020). Thus, SOC and TN are significant factors for the soil microbial biomass and microbial activities (Diaz-Ravina et al., 1988; Babur et al., 2021a,b,c). The greater accumulation of litter and fine roots under pasture and forest can contribute to the growth of microbial populations and increase microbial biomass.

5 Importance of organic litter layer

The litter layer is a surface layer of predominantly dead organic plant matter. In a forest, the litter layer is made up of undecayed to decayed leaves and twigs from the forest canopy (Fig. 11). The litter layer is essential to the ecological health of terrestrial ecosystems because it provides OM to soil and acts as a primary location for several biological processes. It is also a major energy supplier in soil and therefore crucial for the preservation of organisms and their wide biodiversity. Because of its physical and nutritional properties, this habitat is frequented by higher trophic level predators. Moreover, several other species which is very important in soil fertility like the earthworm also depend on litter for their feeding purposes. About 120 billion tons of SOC are generated every year on land, of which only a small amount is consumed by herbivores (Cebrián and Duarte, 1995) or sequestered (Battle et al., 2000). Around 90% of the OM that is produced globally is decomposed,

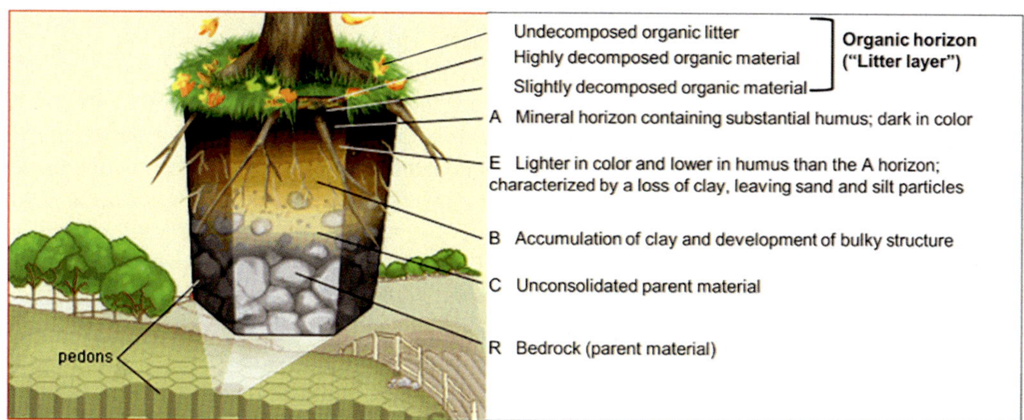

FIG. 11

Soil profile including strata in the litter layer after Encyclopaedia Britannica (© 1999, Encyclopaedia Britannica, Inc).

leading to the recycling of C and other nutrients (Cebrian, 1999; Gessner et al., 2010). SOC and plant nutrient cycling are generally measured in terms of litter production. Besides, plant heights and diameters, litter is also considered as a mark of prime production in the forest ecosystem (Vitousek, 1982). The littering of the plants was used to prevent soil freezes and erosion, to protect weed infestation, to improve reclamation of degraded soil, to buffer soil hydrotherapy, and to enhance forest ecosystem function (Cornwell et al., 2008; Sayer, 2006). Litterfall is an essential part of nutritional cycles in forest ecosystems that govern SOM buildup, nutrient uptake and output, nutrient recovery, the preservation of biodiversity, and other ecosystem processes. These inputs refill SOM inventories, but at the same time also promote decomposition of less labile SOM components (Kuzyakov et al., 2000; Prescott, 2010).

Decomposition happens through three primary processes (Giebelmann et al., 2011): (i) litter fragmentation into smaller pieces, (ii) leaching of soluble chemicals, and (iii) catabolism by decomposer organisms. Swift et al. (1979) stated that the literal quality, surrounding environment and the decomposing organisms are the main agents and influence liter breakdown. Generally, plant leaves with low C/N content and a low phenolic and lignin content degrade faster than other plants and are regarded as good quality litter (Berg and Laskowski, 2006). This is because they are quickly colonized by microorganisms and more appealing to detritivores and microbial grazers (Giebelmann et al., 2011).

Litter breakdown strongly depends on the organic forest topsoil's high microbial decomposer populations and microclimatic conditions, which encourage stand-specific decomposition (Hayes and Holl, 2003). In addition, concentrations of soil pH, temperature, and NH_4 impact the litter decomposition rate considerably. In addition, other microclimate conditions might impact the litter decomposition rate. For instance, Krishna and Mohan (2017) reported that decomposition rates were slow during the winter and rapid during the rainy season, and that increased moisture, rainfall, and a high level of microfungal populations are factors for a faster rate of litter decomposition in the rainy season. Kumar et al. (2010) reached the same conclusion that rainfall and microbial load result in high rates of litter breakdown during wet seasons.

Variation in leaf quality might be partially attributed to the variability in leaf lifespans. A very long-lived plant has a lower specific leaf area (which is typically linked to the plant's toughness) (Chapman and Koch, 2007) whereas long-living leaf types have a lot of lignin and tannin (Giebelmann et al., 2011). In addition, inter- and intra-plant litter variations have a major influence on nitrogen input and loss (Seta et al., 2018). The variations in the quality of the leaf litter have a significant influence on the pace of decline and mineralization of leaf litter in a forest (Giebelmann et al., 2011; Scheer, 2009). Genetic differences across species may explain the variance in litter quality among species (Rawat et al., 2009). Moreover, litter decomposition is faster at its origin. For instance, Broadleaf decays faster in broadleaf environments than in conifer ones (Rawat et al., 2009). Litter decays at lower altitudes faster than at higher elevations, depending on plant species (Veen et al., 2015). Sometimes neighboring species also influence the rate of decomposition (Mélo et al., 2013). Thus, the mixed species degrade more quickly than monocultures, which suggests the lower beneficial influence on non-additives in mixes. This also means that the strong lignified leaf tissue can hamper the decomposition of a leaf litter due to its high structural stability in some monoculture species (Mishra et al., 2004). When nutrient levels fluctuate between species, litters disintegrate fast in forests with a great diversity of trees (Clark et al., 2001).

The effect of litter mixture on decomposition rate is highly dependent on the litter quality and species composition (Hättenschwiler and Jørgensen, 2010). Decomposition is much quicker on litter mixture with a similar texture, but not on litter combination with different leaf textures (Seta et al., 2018). After mixing the recalcitrant litters, decomposition is enhanced, since the transfer of litters from high-quality to low-quality encourages the proliferation of micronutrients. Management of crops and the age of the stands can also affect the breakdown of litter. For instance, Tutua et al. (2002) discovered that decomposition and N release in apple orchards is coupled to management techniques for tree lines. Moreover, as ecosystems grow and mature the rate of plant protection product breakdown in litter rises (Walters, 1999). A more diverse microbial community is included in decomposition as tree species richness rises, resulting in increased leaf breakdown (Chapman and Koch, 2007).

The decomposers in the soil are very varied and functionally diverse (Crawford, 1988; Schinner et al., 2012). The soil fauna is important in preparing the litter and promoting microbial activity, while soil microorganisms are the primary drivers for decomposition (Coleman and Crossley, 1996). In particular, algae, actinomycetes, fungus, and bacteria (McCarthy, 1987) are connected with litter decomposition and the organization where the depth of these soil faunas and microbial communities influence litter breakup (McCarthy, 1987; Crawford, 1988). The functions of the decomposers influence the pace of litter decomposition and enhance soil fertility status (Swift et al., 1979; McCarthy, 1987). In addition, the decomposition activities that decompose leaf litter and convert OM to nutrients and emit $CO^2$to the environment are significantly impacted by the makeup of the decomposer populations (Dilly et al., 2004). Fungi and bacteria are the primary engines in these processes, even if their roles and methods differ (McCarthy, 1987). The fungus may colonize newly fallen litter and use N and C flowing between the litter layers, whereas bacteria rely on the flow of substrate into their cells (Laganiere et al., 2010). Certain decomposers have specific relationships with certain plant species and are experts for decomposing their litter (Vivanco and Austin, 2008). Because the decomposer food web, which consists of fauna and microbial communities, differs between forest floors, it affects the rates at which diverse litter fractions mineralize (Laganiere et al., 2010). The quantity and activity of microbial communities, as well as the substrate quality, all contribute to the speed of decomposition of leaf litter (Giebelmann et al., 2011).

6 Conclusion

The SOM is recently an increasingly important role in the study of soil biochemical dynamics of forest ecosystems, as it is the source of the major interactions between the soil, plants, and microorganisms. Definition, classification, and determination of physical, chemical, and biological contents of litter layers can give us information about the biochemical cycles and dynamics of the ecosystem. This study highlighted many interesting potentialities of three species and other environmental effects on biochemical processes of soil and organic matter which are used as indicators of forest ecosystems. Many studies noticed a clear influence between tree species, soil, and its microorganisms when considering the activities of the enzymes responsible for the soil nutrient cycles and C stocks under forest ecosystems. This study recommends that soil and litter layer biochemical properties such as microbial biomass, basal respiration, and enzyme activities at the decomposition stage may be used as the potential indicator of soil quality, nutrient status, and ecosystem nutrient fluxes. Moreover, this study points out the potential of biochemical properties as an expression of biological activity to distinguish different forest species and some other ecological factors. Since soil microbial activity has a direct effect on the balance and productivity of the ecosystem, microbial biomass gives very accurate and fast information about the quality of the soil.

Sustainable land management is essential for the continuation of life on Earth as we are used to and can only be obtained if we understand the interplay between soil physical, chemical, and biological processes. Therefore, the determination of microbial activities in forest soils is very important to better understand the mechanisms behind climate change effects.

The amount of microbial biomass in forest soils is adjusted by the contents and amount of C and N of those soils, and other soil parameters such as temperature, moisture content, pH, clay content which are related to the inherent substrate quality and accessibility. The uptake of C and N into the microbial tissues and the efficient use of C by the microbial biomass primarily depends on the pH value of the environment. Soil biochemical properties and microbial activity strongly related to broad or micro-climatic regions, vegetation types (deciduous, coniferous, and evergreen), and major soil groups indicating that they can be used as an indicator of the balance between substrate availability and decomposition at this broad scale. This chapter highly suggested that soil microbial properties such as MBC, MBN, MR, microbial and metabolic quotient used to be a good indicator of availability or quality of SOM and on the active use and decomposition rates of C and N by microorganisms. Therefore, SOM quality is highly related to the C_{mic}/C_{org} and the N_{mic}/N_t ratio. These soil microbial properties have important implications for the use of soil organic matter, which can be considered as an indicator of the sustainability of forest soil and land use management practices such as clear-felling, burning, fertilization, or tree species change. If deciding on good management practices, we also need to improve our understanding of how management practices, such as fertilization, harvesting, etc., affect soil microbial properties and processes. Recently, although research on the amount and properties of microbial biomass of forest soils has increased, it is still insufficient to assess the amount of nutrients retained in the microbial biomass and its dynamics used to build a robust database. Therefore, there is a significant need for research on the active role of microbial biomass in nutrient cycling in forest soils. This means that the microbial parameters of soils can be diversified in different ways and the development of laboratory protocols has become increasingly important.

References

Adnan, M., Fahad, S., Zamin, M., Shah, S., Mian, I.A., Danish, S., Zafar-ul-Hye, M., Battaglia, M.L., Naz, R.M., Saeed, B., Saud, S., 2020. Coupling phosphate-solubilizing bacteria with phosphorus supplements improve maize phosphorus acquisition and growth under lime induced salinity stress. Plan. Theory 9 (7), 900.

Ajami, M., Heidari, A., Khormali, F., Gorji, M., Ayoubi, S., 2016. Environmental factors controlling soil organic carbon storage in loess soils of a subhumid region, northern Iran. Geoderma 281, 1–10.

Amanuel, W., Yimer, F., Karltun, E., 2018. Soil organic carbon variation in relation to land use changes: the case of birr watershed, upper Blue Nile River Basin, Ethiopia. J. Ecol. Environ. Sci. 42 (1), 1–11.

Anderson, T.H., Domsch, A.K., 1993. The metabolic quotient for CO2 (qCO2) as a specific activity parameter to assess the effects of environmental conditions, such as pH, on the microbial biomass of forest soils. Soil Biol. Biochem. 25 (3), 393–395.

Anonymous, 2021. https://soilquality.org.au/factsheets/microbial-biomass-carbon-nsw (Accessed 10 March 2021).

Arunachalam, K., Arunachalam, A., Melkania, N.P., 1999. Influence of soil properties on microbial populations, activity and biomass in humid subtropical mountainous ecosystems of India. Biol. Fertil. Soils 30 (3), 217–223.

Azam, F., Farooq, S., Lodhi, A., 2003. Microbial biomass in agricultural soils-determination, synthesis, dynamics and role in plant nutrition. Pak. J. Biol. Sci. 6 (7), 629–639.

Babur, E., Dindaroglu, T., 2020. Seasonal changes of soil organic carbon and microbial biomass carbon in different forest ecosystems. In: Environmental Factors Affecting Human Health. IntechOpen, pp. 1–22.

Babur, E., Dindaroğlu, T., Solaiman, Z.M., Battaglia, M.L., 2021a. Microbial respiration, microbial biomass and activity are highly sensitive to forest tree species and seasonal patterns in the eastern Mediterranean Karst Ecosystems. Sci. Total Environ. 775, 145868.

Babur, E., Kara, O., Fathi, R.A., Susam, Y.E., Riaz, M., Arif, M., Akhtar, K., 2021b. Wattle fencing improved soil aggregate stability, organic carbon stocks and biochemical quality by restoring highly eroded mountain region soil. J. Environ. Manag. 288, 112489.

Babur, E., Uslu, Ö.S., Battaglia, M.L., Diatta, A., Fahad, S., Datta, R., Zafar-ul-Hye, M., Hussain, G.S., Danish, S., 2021c. Studying soil erosion by evaluating changes in physico-chemical properties of soils under different land-use types. J. Saudi Soc. Agric. Sci. 20 (3), 190–197.

Basiliko, N., Moore, T.R., Lafleur, P.M., Roulet, N.T., 2005. Seasonal and inter-annual decomposition, microbial biomass, and nitrogen dynamics in a Canadian bog. Soil Sci. 170 (11), 902–912.

Batey, T., 2009. Soil compaction and soil management–a review. Soil Use Manag. 25 (4), 335–345.

Battaglia, M.L., Ketterings, Q.M., Godwin, G., Czymmek, K.J., 2021a. Conservation tillage is compatible with manure injection in corn silage systems. Agron. J. 113 (3), 2900–2912.

Battaglia, M., Thomason, W., Fike, J.H., Evanylo, G.K., von Cossel, M., Babur, E., Iqbal, Y., Diatta, A.A., 2021b. The broad impacts of corn Stover and wheat straw removal for biofuel production on crop productivity, soil health and greenhouse gas emissions: a review. GCB Bioenergy 13 (1), 45–57.

Battle, M., Bender, M.L., Tans, P.P., White, J.W., Ellis, J.T., Conway, T., Francey, R.J., 2000. Global carbon sinks and their variability inferred from atmospheric O2 and δ13C. Science 287 (5462), 2467–2470.

Bauhus, J., Khanna, P.K., 1999. The Significance of Microbial Biomass in Forest Soils. Going Underground-Ecological Studies in Forest Soils. Research Signpost, Trivandrum, India, pp. 77–110.

Bauhus, J., Pare, D., 1998. Effects of tree species, stand age and soil type on soil microbial biomass and its activity in a southern boreal forest. Soil Biol. Biochem. 30 (8–9), 1077–1089.

Belay, B., Pötzelsberger, E., Sisay, K., Assefa, D., Hasenauer, H., 2018. The carbon dynamics of dry tropical Afromontane forest ecosystems in the Amhara region of Ethiopia. Forests 9 (1), 18.

Berg, B., Laskowski, R., 2006. Litter Decomposition: A Guide to Carbon and Nutrient Turnover. Academic Press, Amsterdam.

Cebrian, J., 1999. Patterns in the fate of production in plant communities. Am. Nat. 154 (4), 449–468.

Cebrián, J., Duarte, C.M., 1995. Plant growth-rate dependence of detrital carbon storage in ecosystems. Science 268 (5217), 1606–1608.

Chaplot, V., Bouahom, B., Valentin, C., 2010. Soil organic carbon stocks in Laos: spatial variations and controlling factors. Glob. Chang. Biol. 16 (4), 1380–1393.

Chapman, S.K., Koch, G.W., 2007. What type of diversity yields synergy during mixed litter decomposition in a natural forest ecosystem? Plant Soil 299 (1), 153–162.

Chen, C.R., Condron, L.M., Davis, M.R., Sherlock, R.R., 2000. Effects of afforestation on phosphorus dynamics and biological properties in a New Zealand grassland soil. Plant Soil 220 (1), 151–163.

Clark, D.A., Brown, S., Kicklighter, D.W., Chambers, J.Q., Thomlinson, J.R., Ni, J., 2001. Measuring net primary production in forests: concepts and field methods. Ecol. Appl. 11 (2), 356–370.

Coleman, D.C., Crossley Jr., D.A., 1996. Fundamentals of Soil Ecology. vol. 1 Hardcover.

Conant, R.T., Paustian, K., Del Grosso, S.J., Parton, W.J., 2005. Nitrogen pools and fluxes in grassland soils sequestering carbon. Nutr. Cycl. Agroecosyst. 71 (3), 239–248.

Cookson, W.R., Murphy, D.V., Roper, M.M., 2008. Characterizing the relationships between soil organic matter components and microbial function and composition along a tillage disturbance gradient. Soil Biol. Biochem. 40 (3), 763–777.

Cornwell, W.K., Cornelissen, J.H., Amatangelo, K., Dorrepaal, E., Eviner, V.T., Godoy, O., Hobbie, S.E., Hoorens, B., Kurokawa, H., Pérez-Harguindeguy, N., Quested, H.M., 2008. Plant species traits are the predominant control on litter decomposition rates within biomes worldwide. Ecol. Lett. 11 (10), 1065–1071.

Crawford, D.L., 1988. Biodegradation of agricultural and rural wastes. In: Goodfellow, M., Williams, S.T., Mordaski, M. (Eds.), Actinomycetes in Biotechnology. Academic, London, pp. 433–439.

Diaz-Ravina, M., Carballas, T., Acea, M.J., 1988. Microbial biomass and metabolic activity in four acid soils. Soil Biol. Biochem. 20 (6), 817–823.

Dilly, O., Bloem, J., Vos, A., Munch, J.C., 2004. Bacterial diversity in agricultural soils during litter decomposition. Appl. Environ. Microbiol. 70 (1), 468–474.

Dilly, O., Munch, J.C., 1998. Ratios between estimates of microbial biomass content and microbial activity in soils. Biol. Fertil. Soils 27 (4), 374–379. https://doi.org/10.1007/s003740050446.

Dindaroglu, T., Gundogan, R., Karaoz, M.O., 2019. Determination of spatial distribution of topsoil organic carbon stock using geostatistical technique in a karst ecosystem. Int. J. Glob. Warm. 19 (3), 251–266.

Fontaine, S., Barot, S., Barré, P., Bdioui, N., Mary, B., Rumpel, C., 2007. Stability of organic carbon in deep soil layers controlled by fresh carbon supply. Nature 450 (7167), 277–280.

Gallardo, J.F., Martin, A., Santa, R.I., 1998. Nutrient cycling in deciduous forest ecosystems of the Sierra de Gata mountains: aboveground litter production and potential nutrient return. In: Annales des sciences forestières. vol. 55. EDP Sciences, pp. 749–769. No. 7.

Garcia, F.O., Rice, C.W., 1994. Microbial biomass dynamics in tallgrass prairie. Soil Sci. Soc. Am. J. 58 (3), 816–823.

Gessner, M.O., Swan, C.M., Dang, C.K., McKie, B.G., Bardgett, R.D., Wall, D.H., Hättenschwiler, S., 2010. Diversity meets decomposition. Trends Ecol. Evol. 25 (6), 372–380.

Giebelmann, U.C., Martins, K.G., Brandle, M., et al., 2011. Lack of home-field advantage in the decomposition of leaf litter in the Atlantic rainforest of Brazil. Appl. Soil Ecol. 49, 5–10. https://doi.org/10.1016/j.apsoil.2011.07.010.

Gregorich, E.G., Liang, B.C., Drury, C.F., Mackenzie, A.F., McGill, W.B., 2000. Elucidation of the source and turnover of water soluble and microbial biomass carbon in agricultural soils. Soil Biol. Biochem. 32 (5), 581–587.

Guo, L.B., Gifford, R.M., 2002. Soil carbon stocks and land use change: a meta analysis. Glob. Chang. Biol. 8 (4), 345–360.

Han, D., Wiesmeier, M., Conant, R.T., Kühnel, A., Sun, Z., Kögel-Knabner, I., Hou, R., Cong, P., Liang, R., Ouyang, Z., 2018. Large soil organic carbon increase due to improved agronomic management in the North China Plain from 1980s to 2010s. Glob. Chang. Biol. 24 (3), 987–1000.

Haney, R.L., Franzluebbers, A.J., Hons, F.M., Hossner, L.R., Zuberer, D.A., 2001. Molar concentration of K2SO4 and soil pH affect estimation of extractable C with chloroform fumigation–extraction. Soil Biol. Biochem. 33 (11), 1501–1507.

Hansson, K., 2011. Impact of Tree Species on Carbon in Forest Soils. PhD Thesis, Faculty of Natural Resources and Agricultural Sciences, University of Agricultural Sciences, Uppsala.

Hättenschwiler, S., Jørgensen, H.B., 2010. Carbon quality rather than stoichiometry controls litter decomposition in a tropical rain forest. J. Ecol. 98 (4), 754–763.

Hayes, G.F., Holl, K.D., 2003. Cattle grazing impacts on annual forbs and vegetation composition of mesic grasslands in California. Conserv. Biol. 17 (6), 1694–1702.

Henrot, J., Robertson, G.P., 1994. Vegetation removal in two soils of the humid tropics: effect on microbial biomass. Soil Biol. Biochem. 26 (1), 111–116.

Hooper, D.U., Chapin Iii, F.S., Ewel, J.J., Hector, A., Inchausti, P., Lavorel, S., Lawton, J.H., Lodge, D.M., Loreau, M., Naeem, S., Schmid, B., 2005. Effects of biodiversity on ecosystem functioning: a consensus of current knowledge. Ecol. Monogr. 75 (1), 3–5.

Hoyle, F.C., Murphy, D.V., Fillery, I.R., 2006. Temperature and stubble management influence microbial CO2–C evolution and gross N transformation rates. Soil Biol. Biochem. 38 (1), 71–80.

Jandl, R., Lindner, M., Vesterdal, L., et al., 2007. How strongly can forest management influence soil carbon sequestration? Geoderma 137, 253–268.

Jenkinson, D.S., Ladd, J.N., 1981. Microbial biomass in soil measurement and turnover. In: Paul, E.A., Ladd, J.N. (Eds.), Soil Biochemistry. vol. 5. Marcel Dekker, Inc., New York and Basel, pp. 415–471.

Kanerva, S., Smolander, A., 2007. Microbial activities in forest floor layers under silver birch, Norway spruce and Scots pine. Soil Biol. Biochem. 39 (7), 1459–1467.

Kara, Ö., Bolat, İ., 2008. The effect of different land uses on soil microbial biomass carbon and nitrogen in Bartın province. Turk. J. Agric. For. 32 (4), 281–288.

Kara, O., Babur, E., Altun, L., Seyis, M., 2016 Aug 17. Effects of afforestation on microbial biomass C and respiration in eroded soils of Turkey. J. Sustain. For. 35 (6), 385–396.

Kiikkilä, O., Kanerva, S., Kitunen, V., Smolander, A., 2014. Soil microbial activity in relation to dissolved organic matter properties under different tree species. Plant Soil 377 (1), 169–177.

Koca, D., Smith, B., Sykes, M.T., 2006. Modelling regional climate change effects on potential natural ecosystems in Sweden. Clim. Chang. 78 (2), 381–406.

Krishna, M.P., Mohan, M., 2017 Aug. Litter decomposition in forest ecosystems: a review. Energy Ecol. Environ. 2 (4), 236–249.

Kumar, M., Joshi, M., Todaria, N.P., 2010. Regeneration status of a sub-tropical Anogeissus latifolia forest in Garhwal Himalaya, India. J. For. Res. 21 (4), 439–444.

Kuzyakov, Y., Friedel, J.K., Stahr, K., 2000. Review of mechanisms and quantification of priming effects. Soil Biol. Biochem. 32 (11–12), 1485–1498.

Laganiere, J., Angers, D.A., Pare, D., 2010. Carbon accumulation in agricultural soils after afforestation: a meta-analysis. Glob. Chang. Biol. 16 (1), 439–453.

Laganière, J., Pare, D., Bradley, R.L., 2010. How does a tree species influence litter decomposition? Separating the relative contribution of litter quality, litter mixing, and forest floor conditions. Can. J. For. Res. 40 (3), 465–475.

Lal, R., 2014. Soil carbon management and climate change. Carbon Manage. 4 (4), 439–462.

Lepcha, N.T., Devi, N.B., 2020. Effect of land use, season, and soil depth on soil microbial biomass carbon of eastern Himalayas. Ecol. Process. 9 (1), 1–4.

Lorenz, K., Lal, R., 2005. The depth distribution of soil organic carbon in relation to land use and management and the potential of carbon sequestration in subsoil horizons. Adv. Agron. 88, 35–66. https://doi.org/10.1016/S0065-2113(05)88002-2.

Maková, J., Javoreková, S., Medo, J., Majerčíková, K., 2011. Tree species effect on soil organic matter and soil microorganisms in trace element contaminated soils. J. Cent. Eur. Agric. 12 (4), 745–758.

Manzoni, S., Taylor, P., Richter, A., Porporato, A., Ågren, G.I., 2012. Environmental and stoichiometric controls on microbial carbon-use efficiency in soils. New Phytol. 196 (1), 79–91.

Maraun, M., Scheu, S., 1995. Influence of beech litter fragmentation and glucose concentration on the microbial biomass in three different litter layers of a beechwood. Biol. Fertil. Soils 19, 155–158. https://doi.org/10.1007/BF00336152.

Martikainen, P.J., Palojärvi, A., 1990. Evaluation of the fumigation-extraction method for the determination of microbial C and N in a range of forest soils. Soil Biol. Biochem. 22 (6), 797–802.

Martinez, M.G., Dominguez, M.T., Fernández, C.M.N, 2019. Tree species effect on soil organic matter and soil microorganisms in trace element contaminated soils. In: 8th ISMOM International Symposium on Interactions of Soil Minerals with Organic Components and Microorganisms.

Martín-Lammerding, D., Navas, M., del Mar, A.M., Tenorio, J.L., Walter, I., 2015. LONG term management systems under semiarid conditions: influence on labile organic matter, β-glucosidase activity and microbial efficiency. Appl. Soil Ecol. 96, 296–305.

Matamala, R., Gonzalez-Meler, M.A., Jastrow, J.D., Norby, R.J., Schlesinger, W.H., 2003. Impacts of fine root turnover on forest NPP and soil C sequestration potential. Science 302 (5649), 1385–1387.

McCarthy, A.J., 1987. Lignocellulose-degrading actinomycetes. FEMS Microbiol. Rev. 3 (2), 145–163.

Mélo, M.A., Budke, J.C., Henke-Oliveira, C., 2013. Relationships between structure of the tree component and environmental variables in a subtropical seasonal forest in the upper Uruguay River valley, Brazil. Acta Bot. Bras. 27 (4), 751–760.

Mishra, B.P., Tripathi, O.P., Tripathi, R.S., Pandey, H.N., 2004. Effects of anthropogenic disturbance on plant diversity and community structure of a sacred grove in Meghalaya, Northeast India. Biodivers. Conserv. 13 (2), 421–436.

Murphy, D.V., Sparling, G.P., Fillery, I.R., 1998. Stratification of microbial biomass C and N and gross N mineralisation with soil depth in two contrasting Western Australian agricultural soils. Soil Res. 36 (1), 45–56.

Ozlu, E., 2000. Dynamics of Soil Aggregate Formation in Different Ecosystems. Soil Science (Ph.D. thesis). University of Wisconsin-Madison.

Ozlu, E., Arriaga, F.J., 2021. The role of carbon stabilization and minerals on soil aggregation in different ecosystems. Catena 202, 105303.

Ozlu, E., Kumar, S., 2018. Response of soil organic carbon, pH, electrical conductivity, and water stable aggregates to long-term annual manure and inorganic fertilizer. Soil Sci. Soc. Am. J. 82 (5), 1243–1251.

Pankhurst, C., Kirkby, C., Hawke, B., Harch, B., 2002. Impact of a change in tillage and crop residue management practice on soil chemical and microbiological properties in a cereal-producing red duplex soil in NSW, Australia. Biol. Fertil. Soils 35 (3), 189–196.

Parr, J.F., Papendick, R.I., 1997. Soil quality: relationships and strategies for sustainable dryland farming systems. Ann. Arid Zone 36 (3), 181–191.

Powers, J.S., Corre, M.D., Twine, T.E., Veldkamp, E., 2011. Geographic bias of field observations of soil carbon stocks with tropical land-use changes precludes spatial extrapolation. Proc. Natl. Acad. Sci. 108 (15), 6318–6322.

Prescott, C.E., 2010. Litter decomposition: what controls it and how can we alter it to sequester more carbon in forest soils? Biogeochemistry 101 (1), 133–149.

Priha, O., Smolander, A., 1997. Microbial biomass and activity in soil and litter under Pinus sylvestris, Picea abies and Betula pendula at originally similar field afforestation sites. Biol. Fertil. Soils 24 (1), 45–51.

Ravindran, A., Yang, S.-S., 2014. Effects of vegetation type on microbial biomass carbon and nitrogen in subalpine mountain forest soils. J. Microbiol. Immunol. Infect. 48 (4), 362–369.

Rawat, N., Nautiyal, B.P., Nautiyal, M.C., 2009. Litter production pattern and nutrients discharge from decomposing litter in a Himalayan alpine ecosystem. N. Y. Sci. J. 2 (6), 54–67.

Sarıyıldız, T., Tüfekçioğlu, A., Küçük, M., 2005. Comparison of decomposition rates of beech (Fagus orientalis Lipsky) and spruce (Picea orientalis (L.) Link) litter in pure and mixed stands of both species in Artvin, Turkey. Turk. J. Agric. For. 29, 429–438.

Sayer, E.J., 2006. Using experimental manipulation to assess the roles of leaf litter in the functioning of forest ecosystems. Biol. Rev. 81 (1), 1–31.

Scheer, M.B., 2009. Nutrient flow in rainfall and throughfall in two stretches in an Atlantic Rain Forest in southern Brazil. Floresta 39 (1), 117–130.

Scheu, S., Parkinson, D., 1995. Successional changes in microbial biomass, respiration and nutrient status during litter decomposition in an aspen and pine forest. Biol. Fertil. Soils 19 (4), 327–332.

Schinner, F., Öhlinger, R., Kandeler, E., Margesin, R. (Eds.), 2012. Methods in Soil Biology. Springer Science & Business Media.

Seta, T., Sebsebe, D., Zerihun, W., 2018. Litterfall dynamics in Boter-Becho forest: moist evergreen montane forest of Southwestern Ethiopia. J. Ecol. Nat. Environ. 10 (1), 13–21.

Sharma, P., Rai, S.C., Sharma, R., Sharma, E., 2004. Effects of land-use change on soil microbial C, N and P in a Himalayan watershed. Pedobiologia 48 (1), 83–92.

Sheng, H., Zhou, P., Zhang, Y., Kuzyakov, Y., Zhou, Q., Ge, T., Wang, C., 2015. Loss of labile organic carbon from subsoil due to land-use changes in subtropical China. Soil Biol. Biochem. 88, 148–157.

Shi, J., Cui, L., 2010. Soil carbon change and its affecting factors following afforestation in China. Landsc. Urban Plan. 98 (2), 75–85.

Singh, R.P., Manchanda, G., Singh, R.N., Srivastava, A.K., Dubey, R.C., 2016. Selection of alkalotolerant and symbiotically efficient chickpea nodulating rhizobia from North-West Indo Gangetic Plains. J. Basic Microbiol. 56 (1), 14–25. https://doi.org/10.1002/jobm.201500267.

Sinsabaugh, R.L., Manzoni, S., Moorhead, D.L., Richter, A., 2013. Carbon use efficiency of microbial communities: stoichiometry, methodology and modelling. Ecol. Lett. 16 (7), 930–939.

Smith, P., 2004. Carbon sequestration in croplands: the potential in Europe and the global context. Eur. J. Agron. 20 (3), 229–236.

Smith, J.L., Papendick, R.I., Bezdicek, D.F., Lynch, J.M., 1992. Soil organic matter dynamics and crop residue management. In: Soil Microbial Ecology: Applications in Agricultural and Environmental Management, pp. 65–94.

Sparling, G.P., 1997. Soil microbial biomass, activity and nutrient cycling as indicators of soil health. In: Pankhurst, C., Doube, B.M., Gupta, V.V.S.R. (Eds.), Biological Indicators of Soil Health. CAB International, Wallingford, pp. 97–119.

Spohn, M., 2015. Microbial respiration per unit microbial biomass depends on litter layer carbon-to-nitrogen ratio. Biogeosciences 12 (3), 817–823.

Spohn, M., Chodak, M., 2015. Microbial respiration per unit biomass increases with carbon-to-nutrient ratios in forest soils. Soil Biol. Biochem. 81, 128–133.

Spohn, M., Klaus, K., Wanek, W., Richter, A., 2016. Microbial carbon use efficiency and biomass turnover times depending on soil depth–implications for carbon cycling. Soil Biol. Biochem. 96, 74–81.

Stenberg, B., 1999. Monitoring soil quality of arable land: microbiological indicators. Acta Agric. Scand. B Soil Plant Sci. 49 (1), 1–24.

Strong, W.L., La Roi, G.H., 1985. Root density-soil relationships in selected boreal forests of Central Alberta, Canada. For. Ecol. Manag. 12 (3–4), 233–251.

Swift, M.J., Heal, O.W., Anderson, J.M., Anderson, J.M., 1979. Decomposition in Terrestrial Ecosystems. Univ of California Press.

Tate, K.R., Speir, T.W., Ross, D.J., Parfitt, R.L., Whale, K.N., Cowling, J.C., 1991. Temporal variations in some plant and soil P pools in two pasture soils of widely different P fertility status. Plant Soil 132 (2), 219–232.

Tesfaye, M.A., Bravo, F., Ruiz-Peinado, R., Pando, V., Bravo-Oviedo, A., 2016. Impact of changes in land use, species and elevation on soil organic carbon and total nitrogen in Ethiopian Central Highlands. Geoderma 261, 70–79.

Tolunay, D., Çömez, A., 2008. Türkiye ormanlarinda toprak ve ölü örtüde depolanmiş organik karbon miktarlari. In: Hava Kirliliği ve Kontrolü Ulusal Sempozyumu Bildiri Kitabı, pp. 750–765.

Torsvik, V., Øvreås, T., 2007. Microbial phylogeny and diversity in soil. In: Elsas, J.D., Jansson, J.K., Trevors, J.T. (Eds.), Modern soil microbiology. CRC Press, Boca Raton, FL, pp. 41–47.

Tracy, B.F., Frank, D.A., 1998. Herbivore influence on soil microbial biomass and nitrogen mineralization in a northern grassland ecosystem: Yellowstone National Park. Oecologia 114 (4), 556–562.

Tutua, S.S., Goh, K.M., Daly, M.J., 2002. Decomposition and nitrogen release of understorey plant residues in biological and integrated apple orchards under field conditions in New Zealand. Biol. Fertil. Soils 35 (4), 277–287.

Van Leeuwen, J.P., Djukic, I., Bloem, J., Lehtinen, T., Hemerik, L., De Ruiter, P.C., Lair, G.J., 2017. Effects of land use on soil microbial biomass, activity and community structure at different soil depths in the Danube floodplain. Eur. J. Soil Biol. 79, 14–20.

Vance, E.D., Brookes, P.C., Jenkinson, D.S., 1987. Microbial biomass measurements in forest soils: the use of the chloroform fumigation-incubation method in strongly acid soils. Soil Biol. Biochem. 19 (6), 697–702.

Vásquez-Murrieta, M.S., Govaerts, B., Dendooven, L., 2007. Microbial biomass C measurements in soil of the central highlands of Mexico. Appl. Soil Ecol. 35 (2), 432–440.

Veen, G.F., Sundqvist, M.K., Wardle, D.A., 2015. Environmental factors and traits that drive plant litter decomposition do not determine home-field advantage effects. Funct. Ecol. 29 (7), 981–991.

Vesterdal, L., Raulund-Rasmussen, K., 1998. Forest floor chemistry under seven tree species along a soil fertility gradient. Can. J. For. Res. 28 (11), 1636–1647.

Vesterdal, L., Ritter, E., Gundersen, P., 2002. Change in soil organic carbon following afforestation of former arable land. For. Ecol. Manag. 169 (1–2), 137–147.

Vesterdal, L., Schmidt, I.K., Callesen, I., Nilsson, L.O., Gundersen, P., 2008. Carbon and nitrogen in forest floor and mineral soil under six common European tree species. For. Ecol. Manag. 255 (1), 35–48.

Vitousek, P., 1982. Nutrient cycling and nutrient use efficiency. Am. Nat. 119 (4), 553–572.

Vivanco, L., Austin, A.T., 2008. Tree species identity alters forest litter decomposition through long-term plant and soil interactions in Patagonia, Argentina. J. Ecol. 96 (4), 727–736.

Walters, J., 1999. Environmental Fate of 2, 4-Dichlorophenoxyacetic Acid. Department of Pesticide Regulations, Sacramento, CA, pp. 1–8.

Yang, Y., Guo, J., Chen, G., Yin, Y., Gao, R., Lin, C., 2009. Effects of forest conversion on soil labile organic carbon fractions and aggregate stability in subtropical China. Plant Soil 323 (1), 153–162.

Yang, Y.-J., Lin, W., Singh, R.P., Xu, Q., Chen, Z., Yuan, Y., Zou, P., Li, Y., Zhang, C., 2019. Genomic, transcriptomic and enzymatic insight into lignocellulolytic system of a plant pathogen *Dickeya* sp. WS52 to digest sweet pepper and tomato stalk. Biomolecules 9 (12). https://doi.org/10.3390/biom9120753.

Zhang, M., Zhang, X.K., Liang, W.J., Jiang, Y., Guan-Hua, D.A., Xu-Gao, W.A., Shi-Jie, H.A., 2011. Distribution of soil organic carbon fractions along the altitudinal gradient in Changbai Mountain, China. Pedosphere 21 (5), 615–620.

Zhang, L., Zhuang, Q., Zhao, Q., He, Y., Yu, D., Shi, X., Xing, S., 2016. Uncertainty of organic carbon dynamics in tai-Lake paddy soils of China depends on the scale of soil maps. Agric. Ecosyst. Environ. 222, 13–22.

CHAPTER

Mechanisms of stress adaptation by bacterial communities

10

Saurabh Pandey[a,*], Raunak[b,*], Takshashila Tripathi[c], Masuma Khawary[b], Deeksha Tripathi[b], and Sashi Kant[d]

[a]Department of Biochemistry, School of Chemical and Life Sciences, Jamia Hamdard, New Delhi, India,
[b]Microbial Pathogenesis and Microbiome Lab, Department of Microbiology, Central University of Rajasthan, Ajmer, Rajasthan, India,
[c]Department of Neuroscience, Physiology and Pharmacology, University College London, London, United Kingdom,
[d]Department of Immunology and Microbiology, University of Colorado School of Medicine, Anschutz Medical Campus, Aurora, CO, United States

Abstract

Bacterial communities respond to stimuli in a consorted manner, quite differing from planktonic bacteria. Bacterial growth in biofilm form is the most predominant in all mega and micro-habitats. This matrix stabilized microbial consortia develops in the heterogenic micro-environment creating an interface of any two of solid, liquid or gas where biofilm forms. Biofilm formation offers nutrient acquisition by sorption, synergistic use and recycling. It also benefits by retention of aqueous support and enhanced cell to cell communication. But the most important ones are coordinated behavior and tolerance to stress like biocides and disinfectants.

Another mechanism of coordinated bacterial behavior is quorum sensing. These are the well-orchestrated actions that are directed through differential gene expression and regulated by molecular communication among bacteria. The processes of quorum sensing not only affect the biofilm formation but virulence, pathogenicity, antibiotic release, spore formation, etc.

In a similar context, papulation of bacteria shows the presence of persister cells that differ in physiological status causing enhanced stress tolerance. Very often, persister cells show growth arrest and antibiotic tolerance upon stress stimulation. The name itself justifies the difficult clearance of these cells from the infected host upon routine clinical interventions.

Lastly, the bacterial communities appear as microbiome in normal and disease conditions in humans and other hosts. The composition of the bacterial milieu in the outer and inner surfaces, cavities, and cell linings helps to sustain the host in normal form against continuous environmental stresses. Also, any dysbiosis of the microbiome may have a causal or supportive association with disease conditions.

Thus, the chapter aims to capture the holistic snapshot of the struggle of bacterial communities against chemical, antibiotic and host generated stresses and the underlined mechanisms for their adaptation.

Keywords: Bacteria, Stress, Biofilm, Quorum-sensing, Stress response

*Contributed equally.

Microbial Syntrophy-mediated Eco-enterprising. https://doi.org/10.1016/B978-0-323-99900-7.00013-4

1 Introduction

Efficient and timely sense and response to threatening external stimuli is key to bacterial survival in any environment. If bacteria encounter stress by the appropriate response, they survive and thrive. At any time, bacterial cells are subjected to environmental stresses like heat, pH, salt, detergent, etc. and host generated stress like reactive oxygen and nitrogen species, components of complement system, other humoral and cell-mediated immune responses. Bacteria have evolutionary adaptation to cope up with these stresses and activation of these stress response mechanisms and their cross-talk help the bacterial cells to survive in hostile environments. If failed, the extreme effect will force them to death or dormancy.

As an individualistic approach, every bacterial species studied have a diverse array of stress response mechanisms under the control of multiple regulatory mechanisms. One of the examples is σ factors. The majority of bacterial gene expression is regulated by primary sigma factors example of which is *E. coli* σ^{70}. Whereas expression of alternative sigma factors in multiple numbers, ranging from 2 to 50, equip the bacteria to express the set of genes advantageous to survive in specific environmental stress. Promoter specificity of core RNA polymerase is altered directed to express these genes (Kill et al., 2005). This is an example of an individualistic mechanism of stress adaptation. We will further see many of the population-based stress adaptation mechanisms.

2 Stress, adaptation to stress and survival

Stress is the response to that external pressure that hinders the normal growth and functioning of a bacterial community (Singh et al., 2020). Continuous changes in the surrounding environment parameters like temperature, osmotic balance, substrate availability, etc. make stress response a critical factor for bacterial survival. From bacteria to higher eukaryotes stress response has remained a common factor. To attain stress resistance cell growth stops, acquires a stable state or grows slowly. Maintaining nucleic acid and cell wall integrity with the process of protein folding keeps the cell alive. Small life and auto-destruction on task achievement of mRNA and other macromolecules make them suitable in delaying chances of damage under stress. Speedy molecular processes heal the damages induced due to stress and guard the cells, in case of contact with similar or different kinds of stress. Molecular reactions leading to the synthesis of stress proteins are a well-studied example of stress response.

3 Abiotic factors

Heat stress: Heat shock proteins are ubiquitous in all living organisms including humans, plants, animals, fungi, and bacteria. Heat shock proteins are responsible for the correct folding of newly manufactured polypeptides and the refolding of incorrectly folded or damaged polypeptides (Feder and Hofmann, 1999). The overproduction of Heat shock proteins makes a cell resistant to increased temperature and oxidative stress. The effect of rising temperature on bacteria causes protein denaturing which introduces a stress response including cellular processes.

Cold stress: The effect of cold stress prevents DNA replication, gene translation and transcription due to the stabilization of nucleic acids secondary conformations. The transmembrane influx of the compounds is hindered due to a decrease in enzyme activity and metabolic rate along with the decreased membrane fluidity. Sub-cellular structures are damaged due to the development of intracellular ice crystals, ultimately causing death. Cold shock proteins are synthesized in response to decreased temperature. These proteins are associated with mRNA folding and protein synthesis (Phadtare et al., 1999). Psychrophilic bacteria inhabiting the environment at $-5°$ to $+20\,°C$ have (A) abundant small chain polyunsaturated fatty acids resulting in increased membrane fluidity, (B) enzyme systems are structurally and functionally adaptable (displays plasticity), (C) prevents ice crystal growth due to their ability to produce antifreeze proteins (Chattopadhyay, 2007). DNA supercoiling modifications may have a specific role in cold shock. *E. coli* shows an increase in negative supercoiling of *E. coli* plasmid DNA. The physiological changes in the bacteria due to cold shock increases their rate of survival in frozen food. Psychrophiles show a characteristic feature of a short lag phase and a high rate of growth.

Oxidative stress: It causes disruptions in nucleic acid chains, active site blockage in enzymes, protein molecule cleavage (cross-linking of fragments may be observed later), energy reduction is its effect at the organism level. Bacteria come across reactive oxygen species as both non-radicals like singlet oxygen, H_2O_2 and radicals like OH^-, OH, O_2^-, NO. Singlet oxygen and hydroxyl radicals have a shorter life cycle and higher reactivity which avoids the development of an elimination process by the cell. Peroxidase, Superoxide dismutase and catalase degrade the relatively stable reactive oxygen species H_2O_2 and O_2^- (Cabiscol et al., 2000). Additionally, molecules with antioxidative properties are found in the cell. Carotenoids, fat-soluble vitamins, polyamine are the antioxidative molecules that prevent oxidation of the DNA (Sharma et al., 2012). Anti-stress molecules like methylerythritol cyclopyrophosphate, natural nitroxyls, example a trehalose derivative the trisaccharide lysodektose have been discovered. Under oxidative stress *E. coli* uses two transcription factors SoxR and OxyR. In the presence of O_2^-, SoxR induces antioxidative enzymes. Activating of antioxidative genes due to the presence of H_2O_2 is done by OxyR. OxyR is dual and acts as a receptor for detecting the presence of reactive oxygen species as well as a transcriptor activator: variable conformations are observed in the oxidised and reduced form of this transcription factor. These conformational changes act as cellular emergency signals. OxyR has 6 cysteine amino acid residues 2 of them are Cys199 and Cys208 which play a major role in stress response development. SoxR active site consists of a metal and four residues of amino acid cysteine. SoxR is activated by variability in the metal redox state and this activated protein conveys an emergency signal (Zheng et al., 1999).

Hydrostatic stress: High hydrostatic pressure of 20–130 MPa inhibits microbial growth and a further increase in pressure to 130–800 MPa may cause cell mortality. Membranes and ribosomes are most sensitive to extreme pressure. The growth of *E. coli* cells is inhibited at 55–304 kPa but the synthesis of transcription and translation associated proteins remains unaffected. The cold shock stress response also witnesses this unique stress response of uninterrupted protein synthesis of strictly controlled proteins. Heat shock proteins, cold shock proteins, and other protein synthesis increase in *E. coli* with the increase in pressure (Welch et al., 1993). The introduction of synthesis of stress protein is a comparatively slower process of 60–90 min as compared to the heat shock protein induction which occurs in seconds of temperature increase. Resistance to extreme hydrostatic stress is a major issue of concern as pressurization technology is being widely used to avoid food deterioration by microbes. High pressure along with mild heat leads to the successful inactivation of microbes (Mackey and Mañas, 2014).

Osmotic stress: For surviving extreme salt concentrations microbes adapt a dual mechanism comprising of discharging the excess salt and osmolyte accumulation, substances that should not harm the intracellular structures. A small organic molecule that retains water solubility at higher concentrations are known as osmolytes. Cells exposed to osmotic stress show the presence of high levels of osmolytes in their cytosol. Ectoine, trehalose, glucitin, mannitol, carnitine, proline, sucrose, glycine-betaine, glycerol and small peptides are examples of osmolytes. Intracellularly accumulated Betaine (*N,N,N*-trimethylglycine) is utilized as an osmoregulatory by the bacteria (Sleator and Hill, 2002). Betaine is synthesized by *Staphylococcus*, *E. coli* and *B. subtilis* by affecting transmembrane transportation from medium or by a reaction catalyzed by choline dehydrogenase (Caldas et al., 1999). *Beta vulgaris* and other plant products are a rich source of betaine and their transportation has been well studied in *S. typhimurium* and *E. coli*. The correct folding of partially unfolded polypeptides during osmotic stress is done with the involvement of stress proteins (Sleator and Hill, 2002).

In the battle for survival, bacteria come across a variety of stress. These include multiple abiotic factors which are well known for their contribution to bacterial stress adaptation. Some of the mechanisms of stress adaptation observed in *Lactococcus lactis* (Sanders et al., 1999) are displayed in the flowchart (Fig. 1).

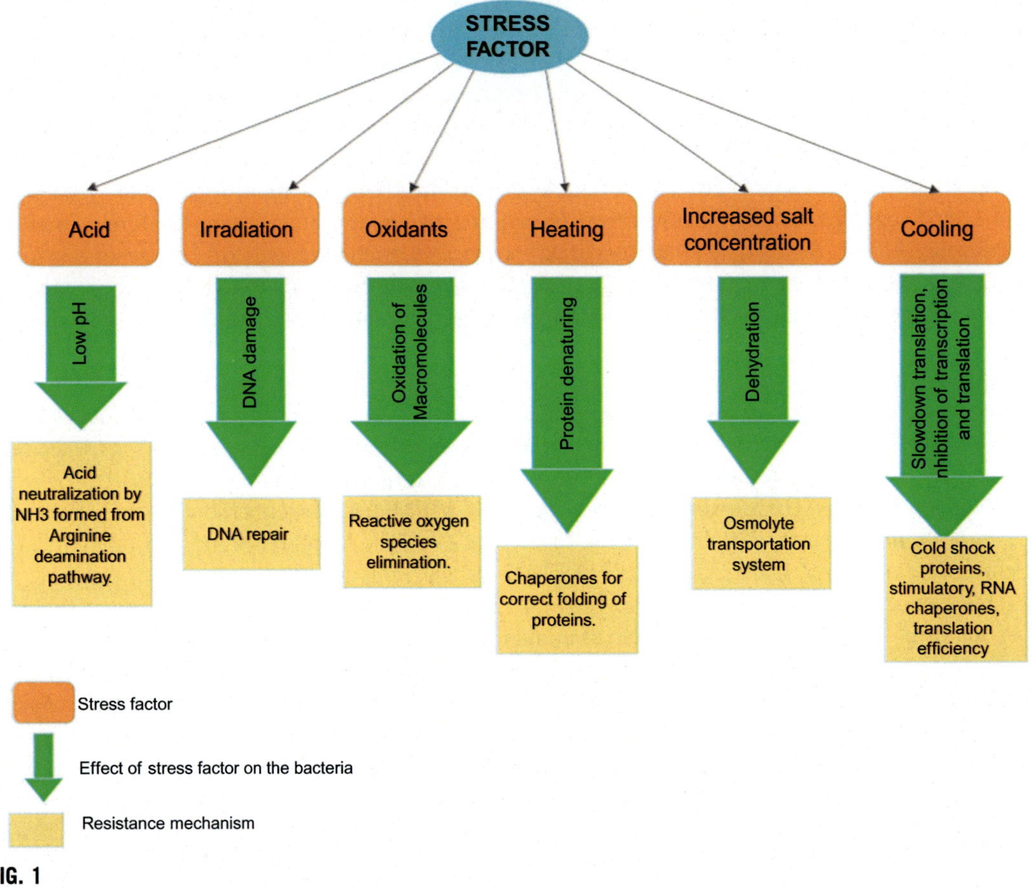

FIG. 1

Schematic view of stress resistance mechanisms in L. *lactis*.

4 Biotic stress factors

Phage, protists, and bacterial host mainly contributes to the biotic factors of stresses. The predator bacteriophages have been found following bacteria in every field from human health to food and biotech industries. The co-evolution of both the bacteria to resist phage infection to phage particles adopting diverse strategies for fighting against these barriers in infecting a bacterial cell a war is on. The bacterial cell adopts a variety of mechanisms to resist phage infection. Avoiding the attachment of phage particles on the surface of bacteria is one such mechanism. Modification of phage receptors present on the bacterial cell surface is achieved with the help of mutation and adding a physical barrier to hide the receptors (Labrie et al., 2010; Dy et al., 2014). In the case of successful attachment, the prevention of DNA entry into the host cell is done with the help of typical, phage-encoded superinfection exclusion. Superinfection exclusion can prevent the entry of phage DNA into the host cells and aid in protecting the lysogenized host from infection by other bacteriophages. Further, abortive infection systems sacrifice the infected bacterial cell to guard the nearby clonal population. Mobile genetic elements such as plasmids and prophages encode for such abortive infection systems (Samson et al., 2013). The mechanical diversity of such systems enables them to act at any stage of phage infection to abolish or decrease the number of phage progeny. Interference in the reproduction of the phage particles in Gram-positive bacteria by phage-inducible chromosomal islands is also a way to resist phage infection (Ram et al., 2012). The family of phage-inducible chromosomal islands is well known to carry and distribute major virulence factors as *Staphylococcus aureus* pathogenicity island (Novick et al., 2010). As compared to other mechanisms of phage resistance *S. aureus* pathogenicity island allow intracellular phage program to produce mature phage particles devoid of phage DNA but consisting of *S. aureus* pathogenicity island DNA (Ram et al., 2014).

Clustered Regularly Inter Spaced Palindromic Repeats (CRISPR) regions make the bacterial cells resistant to viruses by conserving a part of the viral DNA which helps the bacteria to memorize it (Barrangou et al., 2007). CRISPR immunity is acquired in response to viral infection. Generally, immunity is not hereditary since CRISPR immunity is DNA encoded it can be passed onto generations (Barrangou and Marraffini, 2014).

5 Single-cell vs community (biofilm) behavior and advantages

The stress response of a single cell may vary from the stress response of a group of cells arranged as a biofilm. Biofilms are heterogeneous aggregates of microbes that are embedded in the matrix of extracellular polymeric substances, attached to a surface. The microbial consortia create a gradient of nutrient and physical factors of the microenvironment. The cooperative community helps in resource acquisition, synergistic use, and recycling. Any external assaults or exposure to toxins and antibacterial have minimal impact at the core of the biofilm. Biofilms experience various biotic and abiotic stresses and feature an increased tolerance to these stresses. Working as a cooperative community, biofilm effectively overcomes the stresses of antibiotics, disinfectants, predation, ecological competition, and desiccation (Hall-Stoodley et al., 2004; Oliveira et al., 2015). Thus, the formation of biofilm is itself believed to be a response toward stress.

Biofilms display competitive, cooperative, and social behavior (Nadell et al., 2016). Response of individual microbe has a multiplier effect in the presence of biofilm-like community-dwelling. Formation of extracellular matrix, shared secretome, cell to cell communication is beneficial in managing the response to stresses in the biofilm. The distinct phenotypic state of individual cells appears in the biofilm depending upon the cellular microenvironments. Localized stresses can be observed inside a biofilm due to the heterogeneity of microenvironments (Flemming et al., 2016; Stewart and Franklin, 2008). Dental biofilms show internal localized stress due to metabolic processes leading to pH - 4.3, i.e., highly acidic environment at the base of dental biofilm (Bowen et al., 2018). The cells deeper in the biofilm are often starving due to limited diffusion and major nutrient consumption by cells present on the surface. Thus, *Bacillus subtilis* present in inner biofilm produces electrical waves from membrane depolarization. These spatially proliferating membrane depolarisation waves seize the metabolic processes of peripherally located cells in a synchronized manner resulting in the diffusion of nutrients to the deeper cells (Liu et al., 2015; Prindle et al., 2015). Surrounding biofilm colonies also practice this timely nutrient absorption for the welfare of the whole community (Liu et al., 2017). This depicts an example of a localized stress factor introducing a spatially extended, coordinated, multicellular response observed far from the location of stress.

Temperature shifts are also observed within seconds inside the typical biofilm sizes of 10–1000 μm. This is similar to the multicellular behavior of biofilms as observed in the case of responses to temperature variations (Phadtare et al., 1999; Kortmann and Narberhaus, 2012). However, quorum sensing too can play a significant role in increased heat shock tolerance (García-Contreras et al., 2015). However, in the case of *Vibrio cholerae* high intracellular c-di-GMP levels due to the cold shock regulator Csp V and phosphodiesterase RocS, increases the production of extracellular matrix, constituting an abiotic stress response (Townsley et al., 2016). This temperature-driven matrix production of *V. cholerae* is controlled by GTPase BipA which is temperature sensitive and is well maintained in different species (del Peso Santos et al., 2020), indicating that the multicellular responses toward cold shock may be common.

Localized stress may be experienced in a biofilm due to external factors like predation or mechanical shear or because of internal acidification or prophage induction inside a biofilm. Matrix formation and cell signaling in response to localized stress display multicellular behavior. The response to stress away from the location of stress occurs with the help of information transfer, which displays a comparatively high level of organization of community and sociality order (Nadell et al., 2016). Cells that grow on the flow-exposed surfaces experience mechanical shear stress (Pearce et al., 2019; Stewart, 2012). To overcome this stress, increased matrix formation to avoid friction among the cells is observed. Predation stress because of other bacteria, protists, phagocytes, and phages are also experienced by outer surface cells. Stress response due to predation of the neighboring cell is experienced in surrounding cells. In, *Pseudomonas aeruginosa* lysis leads to the instant release of antibacterial compounds by surrounding cells (Le Roux et al., 2015). Biofilm predation by phage is restricted by incorporating selective matrix compounds in biofilm which inhibit the phage entry. Such phage binding matrix compound Curli amyloid fibers are produced in the outer biofilm surface of *E. coli* (Vidakovic et al., 2017). Antibiotics, metabolites, disinfectants are more stress agents involved with the surface cells in a biofilm. Matrix composition, diffusion inhibition are the mechanisms adopted to deal with such stress factors. Metabolic processes and production of electron acceptor gradient in the depth of biofilm results in oxidant limitation. Extracellular cytochromes, nanowires and diffusible active redox metabolites transfer the multiple forms of extracellular electrons to overcome stress caused by oxidant limitation (Costa et al., 2018; Richter et al., 2009; Wang et al., 2019). In *E. coli, Actinomyces odontolyticus, Pseudomonas aeruginosa* cell lysis releases cytosolic proteins and DNA which mainly contributes to biofilm matrix. Biofilm dispersion as single cell in *Pseudomonas aeruginosa* is regulated. While in the case of *Staphylococcus aureus* it is operated by agr quorum-sensing system which synthesizes extracellular protease to dissolve the matrix (Lister and Horswill, 2014). This agr quorum-sensing is a collective cell level response to a global signal. *M. tuberculosis* PPIB protein is localized in secretory and membrane fractions and involved in biofilm formation. It could be targeted for developing an intervention strategy against drug-resistant tuberculosis (González-Zamorano et al., 2009; Pandey et al., 2016; Hasnain et al., 2020).

Behavioral differentiation between the planktonic population of bacteria and multicellular behavior of biofilm is studied using live single-cell microscopy of biofilms (Hartmann et al., 2019) and metabolite imaging techniques (Bellin et al., 2016; Geier et al., 2020). In addition, bacterial single-cell RNA sequencing (Blattman et al., 2020; Kuchina et al., 2021) is the most precise method for behavioral differentiation.

6 Bacterial mechanisms for sensing its population (quorum) and effect of stress

Quorum-sensing is a bacterial cell to cell signaling system which regulates synchronous phenotypic expressions and behavioral coordination after the population density reaches threshold size. Quorum-sensing systems are dependent on the release and detection of extracellular signals. In a fresh culture, the process of signal production begins at a lower concentration, and eventually, the signal is intensified with the rise in population density. At threshold concentrations, this signal peaks and interacts with cognate receptor protein, ultimately leading to a harmonized change in population gene expression. Quorum-sensing guards the cells against a variety of stress factors like thermal, osmotic, heavy metals, etc. It also benefits the cells by producing intracellular enzymes to help the bacterial cells mount a better stress response. Quorum-sensing manages the virulence factor expression and permits the bacteria to fight the immune response by increasing tolerance to oxidative stress produced by the immune cells. The high population size may act as a warning for upcoming stress factors due to the limitation of nutrients, oxygen, and other factors. Biofilm formation in *P. aeruginosa* is controlled by quorum-sensing (Antunes et al., 2010). Relations between quorum-sensing and antibiotic resistance have been observed in nature. Stress tolerance in *P. aeruginosa* is directly associated with quorum-sensing. As the quorum-sensing deficient mutants (*rhlI, lasI and rhlI lasI*) show defected *sodA* and *katA* expressions and concurrently minimized superoxide dismutase and catalase activities. This makes them more sensitive to oxidative stress as compared to the parental strain (Hassett et al., 1999). OxyR, the hydrogen peroxide responsive transactivator, binds with the promoter and impacts the expression of mvfR (pqsR) and rsaL genes which are quorum-sensing transcriptional regulators in *Pseudomonas* (Wei et al., 2012). The starvation of phosphate leading to nutritional stress is linked with quorum-sensing (Lee et al., 2013).

Stress tolerance is related to a major subset of quorum-sensing regulated genes. Increased stress tolerance against abiotic factors due to quorum-sensing in bacteria like *Burkholderia pseudomallei* and *Vibrio* species have been found (Lumijiaktase et al., 2006; McDougald et al., 2003). Quorum-sensing also shows biotic stress tolerance. Resistance from protist predation by biofilm formation in *P. aeruginosa* helps in escaping the biotic stress as compared to quorum-sensing deficient mutants lasR and lasI HSL (Friman et al., 2013).

The quorum-sensing not only regulates the production of bacterial virulence determinants (Cabiscol et al., 2000; Sharma et al., 2012) but also coordinates the production of secondary metabolites (Zheng et al., 1999), initiation of DNA uptake (Welch et al., 1993; Mackey and Mañas, 2014), formation of biofilm (Sleator and Hill, 2002; Caldas et al., 1999) and stress responses (Sharma et al., 2012).

7 Odd man in the population: The persisters

The few bacteria can stochastically exist in the physiologically dormant state, not growing in the population, and thus remain as transiently antibiotic tolerant. The sub-population originated from persisters is not persisters except the few stochastically converted to persisters. Host-pathogen interaction and antibiotic stresses activate certain survival pathways that cause the rare appearance of persisters in the population. This phenotypic heterogeneity increases the chance of survival during stresses. Even with the existence of persisters, the MIC (Minimum Inhibitory Concentration) remains unaltered in generations as they are non-growers. Furthermore, any molecular tool relying on the population of bacterium like bacterial lysate in electrophoresis or western blot is inefficient to capture the persisters and their effects due to their presence in diminutive fractions. But newer methods like fluorescent reporter assays, FACS and transcriptome profiling is used to track the metabolic activity and function of the persisters. The persisters can resume growth by inactivating their pathways of stress adaptation or remain persistent even after the removal of stress (Harms et al., 2016).

The transient antibiotic tolerance is different from the antibiotic resistance generated by a mutation in drug targets or horizontal gene transfer supporting the export of toxins and antibiotics out of the cells. But, persisters are a source of the emergence of drug resistance as they can cause a relapse of infection and require a longer drug treatment regimen. Persisters are a major concern in the treatment of bacterial infection as their low metabolic activities hinder their sensitivity to antibiotics. This makes it difficult for the antibiotics to interfere with the bacteria to kill. An improper dosage or duration of treatment of antibiotic leads to repopulation. However, in resistance, genetic changes inhibit antibiotic activity and cell growth, but in the case of persisters, they stop their growth and metabolism in response to antibiotics. This antibiotic stress tolerance has been well observed in *E. coli*, *P. aeruginosa*, *Lactobacillus acidophilus*, *S. aureus* and *Gardnerella vaginalis* (Singh et al., 2009).

In older inoculum more persisters are found as persister formation is directly proportional to time and the antibiotic used (Luidalepp et al., 2011). Toxin-antitoxin pairs can induce a dormant state ultimately forming persisters (Lewis, 2008; Jayaraman and Wood, 2008). An alarmone, (p)ppGpp guanosine tetraphosphate and pentaphosphate promote persistence by activating the toxin-antitoxin module, decreasing protein synthesis and DNA replication and thus facilitating its entry into the stationary phase (Pacios et al., 2020). A high number of persisters are found in oxygen and nutrient-limited biofilms (Harrison et al., 2005). The center of biofilms has an especially nutrient and oxygen limiting environment, pH variability, presence of reactive oxygen species, starvation, high osmolarity contribute to persister formation. As the cell reaches the stationary phase, persister accumulation begins and the highest persisters accumulate in biofilms (Keren et al., 2004; Spoering and Lewis, 2001). Antibiotics themselves may trigger persister formation due to irregular treatment, exploitation of the MIC or other factors.

References

Antunes, L.C.M., Ferreira, R.B.R., Buckner, M.M.C., Finlay, B.B., 2010. Quorum sensing in bacterial virulence. Microbiology 156, 2271–2282.

Barrangou, R., Marraffini, L.A., 2014. CRISPR-cas systems: prokaryotes upgrade to adaptive immunity. Mol. Cell 54, 234–244.

Barrangou, R., Fremaux, C., Deveau, H., Richards, M., Boyaval, P., Moineau, S., Romero, D.A., Horvath, P., 2007. CRISPR provides acquired resistance against viruses in prokaryotes. Science 315, 1709–1712.

Bellin, D.L., Sakhtah, H., Zhang, Y., Price-Whelan, A., Dietrich, L.E.P., Shepard, K.L., 2016. Electrochemical camera chip for simultaneous imaging of multiple metabolites in biofilms. Nat. Commun. 7, 10535.

Blattman, S.B., Jiang, W., Oikonomou, P., Tavazoie, S., 2020. Prokaryotic single-cell RNA sequencing by in situ combinatorial indexing. Nat. Microbiol. 5, 1192–1201.

Bowen, W.H., Burne, R.A., Wu, H., Koo, H., 2018. Oral biofilms: pathogens, matrix, and polymicrobial interactions in microenvironments. Trends Microbiol. 26, 229–242.

Cabiscol, E., Tamarit, J., Ros, J., 2000. Oxidative stress in bacteria and protein damage by reactive oxygen species. Int. Microbiol. 3, 3–8.

Caldas, T., Demont-Caulet, N., Ghazi, A., Richarme, G., 1999. Thermoprotection by glycine betaine and choline. Microbiology 145, 2543–2548.

Chattopadhyay, M.K., 2007. Antifreeze proteins of bacteria. Resonance 12, 25–30.

Costa, N.L., Clarke, T.A., Philipp, L.A., Gescher, J., Louro, R.O., Paquete, C.M., 2018. Electron transfer process in microbial electrochemical technologies: the role of cell-surface exposed conductive proteins. Bioresour. Technol. 255, 308–317.

del Peso Santos, T., Blake, J., Sit, B., Warr, A., Benes, V., Waldor, M., Cava, F., 2020. A Temperature-Dependent Translational Switch Controls Biofilm Development in Vibrio cholerae. A Temp Transl Switch Control biofilm Dev *Vibrio cholerae*. vol. 2 BioRxiv, pp. 80031–80039.

Dy, R.L., Richter, C., Salmond, G.P.C., Fineran, P.C., 2014. Remarkable mechanisms in microbes to resist phage infections. Annu. Rev. Virol. 1, 307–331.

Feder, M.E., Hofmann, G.E., 1999. Heat-shock proteins, molecular chaperones, and the stress response: evolutionary and ecological physiology. Annu. Rev. Physiol. 61, 243–282.

Flemming, H.C., Wingender, J., Szewzyk, U., Steinberg, P., Rice, S.A., Kjelleberg, S., 2016. Biofilms: an emergent form of bacterial life. Nat. Rev. Microbiol. 14, 563–575.

Friman, V.P., Diggle, S.P., Buckling, A., 2013. Protist predation can favour cooperation within bacterial species. Biol. Lett. 9, 5.

García-Contreras, R., Nuñez-López, L., Jasso-Chávez, R., Kwan, B.W., Belmont, J.A., Rangel-Vega, A., Maeda, T., Wood, T.K., 2015. Quorum sensing enhancement of the stress response promotes resistance to quorum quenching and prevents social cheating. ISME J. 9, 115–125.

Geier, B., Sogin, E.M., Michellod, D., Janda, M., Kompauer, M., Spengler, B., Dubilier, N., Liebeke, M., 2020. Spatial metabolomics of in situ host–microbe interactions at the micrometre scale. Nat. Microbiol. 5, 498–510.

González-Zamorano, M., Mendoza-Hernández, G., Xolalpa, W., Parada, C., Vallecillo, A.J., Bigi, F., Espitia, C., González-Zamorano, M., Hernández, G.M., Xolalpa, W., Parada, C., Vallecillo, A.J., Bigi, F., Espitia, C., 2009. *Mycobacterium tuberculosis* glycoproteomics based on ConA-lectin affinity capture of mannosylated proteins. J. Proteome Res. 8, 721–733.

Hall-Stoodley, L., Costerton, J.W., Stoodley, P., 2004. Bacterial biofilms: from the natural environment to infectious diseases. Nat. Rev. Microbiol. 2, 95–108.

Harms, A., Maisonneuve, E., Gerdes, K., 2016. Mechanisms of bacterial persistence during stress and antibiotic exposure. Science 354 (6318), aaf4268.

Harrison, J.J., Turner, R.J., Ceri, H., 2005. Persister cells, the biofilm matrix and tolerance to metal cations in biofilm and planktonic *Pseudomonas aeruginosa*. Environ. Microbiol. 7, 981–994.

Hartmann, R., Singh, P.K., Pearce, P., Mok, R., Song, B., Díaz-Pascual, F., Dunkel, J., Drescher, K., 2019. Emergence of three-dimensional order and structure in growing biofilms. Nat. Phys. 15, 251–256.

Hasnain, S.E., Ehtesham, N.Z., Tripathi, D., Grover, S., Kumar, A., Alam, A., Pandey, S., 2020. A Medicament for the Treatment of Diseases by Biofilm Forming. US 2020/0188477 A1. USA.

Hassett, D.J., Ma, J.F., Elkins, J.G., McDermott, T.R., Ochsner, U.A., West, S.E.H., Huang, C.T., Fredericks, J., Burnett, S., Stewart, P.S., McFeters, G., Passador, L., Iglewski, B.H., 1999. Quorum sensing in *Pseudomonas aeruginosa* controls expression of catalase and superoxide dismutase genes and mediates biofilm susceptibility to hydrogen peroxide. Mol. Microbiol. 34, 1082–1093.

Jayaraman, A., Wood, T.K., 2008. Bacterial quorum sensing: signals, circuits, and implications for biofilms and disease. Annu. Rev. Biomed. Eng. 10, 145–167.

Keren, I., Kaldalu, N., Spoering, A., Wang, Y., Lewis, K., 2004. Persister cells and tolerance to antimicrobials. FEMS Microbiol. Lett. 230, 13–18.

Kill, K., Binnewies, T.T., Sicheritz-Pontén, T., Willenbrock, H., Hallin, P.F., Wassenaar, T.M., Ussery, D.W., 2005. Genome update: sigma factors in 240 bacterial genomes. Microbiology 151, 3147–3150.

Kortmann, J., Narberhaus, F., 2012. Bacterial RNA thermometers: molecular zippers and switches. Nat. Rev. Microbiol. 10, 255–265.

Kuchina, A., Brettner, L.M., Paleologu, L., Roco, C.M., Rosenberg, A.B., Carignano, A., Kibler, R., Hirano, M., DePaolo, R.W., Seelig, G., 2021. Microbial single-cell RNA sequencing by split-pool barcoding. Science 371 (6531), eaba5257.

Labrie, S.J., Samson, J.E., Moineau, S., 2010. Bacteriophage resistance mechanisms. Nat. Rev. Microbiol. 8, 317–327.

Le Roux, M., Kirkpatrick, R.L., Montauti, E.I., Tran, B.Q., Brook Peterson, S., Harding, B.N., Whitney, J.C., Russell, A.B., Traxler, B., Goo, Y.A., Goodlett, D.R., Wiggins, P.A., Mougous, J.D., 2015. Kin cell lysis is a danger signal that activates antibacterial pathways of *pseudomonas aeruginosa*. Elife 2015, 1–65.

Lee, J., Wu, J., Deng, Y., Wang, J., Wang, C., Wang, J., Chang, C., Dong, Y., Williams, P., Zhang, L.H., 2013. A cell-cell communication signal integrates quorum sensing and stress response. Nat. Chem. Biol. 9, 339–343.

Lewis, K., 2008. Multidrug tolerance of biofilms and persister cells. Curr. Top. Microbiol. Immunol. 322, 107–131.

Lister, J.L., Horswill, A.R., 2014. *Staphylococcus aureus* biofilms: recent developments in biofilm dispersal. Front. Cell. Infect. Microbiol. 4, 178.

Liu, J., Prindle, A., Humphries, J., Gabalda-Sagarra, M., Asally, M., Lee, D.Y.D., Ly, S., Garcia-Ojalvo, J., Süel, G.M., 2015. Metabolic co-dependence gives rise to collective oscillations within biofilms. Nature 523, 7562.

Liu, J., Martinez-Corral, R., Prindle, A., Lee, D.Y.D., Larkin, J., Gabalda-Sagarra, M., Garcia-Ojalvo, J., Süel, G.M., 2017. Coupling between distant biofilms and emergence of nutrient time-sharing. Science 356, 638–642.

Luidalepp, H., Jõers, A., Kaldalu, N., Tenson, T., 2011. Age of inoculum strongly influences persister frequency and can mask effects of mutations implicated in altered persistence. J. Bacteriol. 193, 14.

Lumijiaktase, P., Diggle, S.P., Loprasert, S., Tungpradabkul, S., Daykin, M., Cámara, M., Williams, P., Kunakorn, M., 2006. Quorum sensing regulates dpsA and the oxidative stress response in *Burkholderia pseudomallei*. Microbiology 152, 3651–3659.

Mackey, B.M., Mañas, P., 2014. Inactivation of *Escherichia coli* by high pressure. In: High-Pressure Microbiology. ASM Press, pp. 53–85.

McDougald, D., Srinivasan, S., Rice, S.A., Kjelleberg, S., 2003. Signal-mediated cross-talk regulates stress adaptation in Vibrio species. Microbiology 149, 1923–1933.

Nadell, C.D., Drescher, K., Foster, K.R., 2016. Spatial structure, cooperation and competition in biofilms. Nat. Rev. Microbiol. 14, 589–600.

Novick, R.P., Christie, G.E., Penadés, J.R., 2010. The phage-related chromosomal islands of Gram-positive bacteria. Nat. Rev. Microbiol. 8, 541–551.

Oliveira, N.M., Martinez-Garcia, E., Xavier, J., Durham, W.M., Kolter, R., Kim, W., Foster, K.R., 2015. Biofilm formation as a response to ecological competition. PLoS Biol. 13, 1002191.

Pacios, O., Blasco, L., Bleriot, I., Fernandez-Garcia, L., Ambroa, A., López, M., Bou, G., Cantón, R., Garcia-Contreras, R., Wood, T.K., Tomás, M., 2020. (p)ppGpp and its role in bacterial persistence: new challenges. Antimicrob. Agents Chemother. 64 (10), e01283-20.

Pandey, S., Sharma, A., Tripathi, D., Kumar, A., Khubaib, M., Bhuwan, M., Chaudhuri, T.K., Hasnain, S.E., Ehtesham, N.Z., 2016. *Mycobacterium tuberculosis* peptidyl-prolyl isomerases also exhibit chaperone like activity in-vitro and in-vivo. PLoS One 11, e0150288.

Pearce, P., Song, B., Skinner, D.J., Mok, R., Hartmann, R., Singh, P.K., Jeckel, H., Oishi, J.S., Drescher, K., Dunkel, J., 2019. Flow-induced symmetry breaking in growing bacterial biofilms. Phys. Rev. Lett. 123, 258101.

Phadtare, S., Alsina, J., Inouye, M., 1999. Cold-shock response and cold-shock proteins. Curr. Opin. Microbiol. 2, 2.

Prindle, A., Liu, J., Asally, M., Ly, S., Garcia-Ojalvo, J., Süel, G.M., 2015. Ion channels enable electrical communication in bacterial communities. Nature 527, 59–63.

Ram, G., Chen, J., Kumar, K., Ross, H.F., Ubeda, C., Damle, P.K., Lane, K.D., Penadés, J.R., Christie, G.E., Novick, R.P., 2012. Staphylococcal pathogenicity island interference with helper phage reproduction is a paradigm of molecular parasitism. Proc. Natl. Acad. Sci. U. S. A. 109, 16300–16305.

Ram, G., Chen, J., Ross, H.F., Novick, R.P., Musser, J.M., 2014. Precisely modulated pathogenicity island interference with late phage gene transcription. Proc. Natl. Acad. Sci. U. S. A. 111, 14536–14541.

Richter, H., Nevin, K.P., Jia, H., Lowy, D.A., Lovley, D.R., Tender, L.M., 2009. Cyclic voltammetry of biofilms of wild type and mutant *Geobacter sulfurreducens* on fuel cell anodes indicates possible roles of OmcB, OmcZ, type IV pili, and protons in extracellular electron transfer. Energ. Environ. Sci. 2, 506–516.

Samson, J.E., Magadán, A.H., Sabri, M., Moineau, S., 2013. Revenge of the phages: defeating bacterial defences. Nat. Rev. Microbiol. 11, 675–687.

Sanders, J.W., Venema, G., Kok, J., 1999. Environmental stress responses in *Lactococcus lactis*. FEMS Microbiol. Rev. 23, 483–501.

Sharma, P., Jha, A.B., Dubey, R.S., Pessarakli, M., 2012. Reactive oxygen species, oxidative damage, and antioxidative defense mechanism in plants under stressful conditions. J. Bot. 2012, 1–26.

Singh, R., Ray, P., Das, A., Sharma, M., 2009. Role of persisters and small-colony variants in antibiotic resistance of planktonic and biofilm-associated *Staphylococcus aureus*: an in vitro study. J. Med. Microbiol. 58, 1067–1073.

Singh, R.P., Butter, H.K., Kaur, R., Manchanda, G., 2020. The multifaceted life of microbes: survival in varied environments. In: Singh, R.P., et al. (Eds.), Microbial Versatility in Varied Environments. Springer, Singapore, https://doi.org/10.1007/978-981-15-3028-9_1.

Sleator, R.D., Hill, C., 2002. Bacterial osmoadaptation: the role of osmolytes in bacterial stress and virulence. FEMS Microbiol. Rev. 26, 49–71.

Spoering, A.L., Lewis, K., 2001. Biofilms and planktonic cells of *Pseudomonas aeruginosa* have similar resistance to killing by antimicrobials. J. Bacteriol. 183, 6746–6751.

Stewart, P.S., 2012. Mini-review: convection around biofilms. Biofouling 28, 187–198.

Stewart, P.S., Franklin, M.J., 2008. Physiological heterogeneity in biofilms. Nat. Rev. Microbiol. 6, 199–210.

Townsley, L., Sison Mangus, M.P., Mehic, S., Yildiz, F.H., 2016. Response of *Vibrio cholerae* to low-temperature shifts: CspV regulation of type vi secretion, biofilm formation, and association with zooplankton. Appl. Environ. Microbiol. 82, 4441–4452.

Vidakovic, L., Singh, P.K., Hartmann, R., Nadell, C.D., Drescher, K., 2017. Dynamic biofilm architecture confers individual and collective mechanisms of viral protection. Nat. Microbiol. 3, 26–31.

Wang, F., Gu, Y., O'Brien, J.P., Yi, S.M., Yalcin, S.E., Srikanth, V., Shen, C., Vu, D., Ing, N.L., Hochbaum, A.I., Egelman, E.H., Malvankar, N.S., 2019. Structure of microbial nanowires reveals stacked Hemes that transport electrons over micrometers. Cell 177, 361–369.e10.

Wei, Q., Le Minh, P.N., Dötsch, A., Hildebrand, F., Panmanee, W., Elfarash, A., Schulz, S., Plaisance, S., Charlier, D., Hassett, D., Häussler, S., Cornelis, P., 2012. Global regulation of gene expression by OxyR in an important human opportunistic pathogen. Nucleic Acids Res. 40, 4320–4333.

Welch, T.J., Farewell, A., Neidhardt, F.C., Bartlett, D.H., 1993. Stress response of *Escherichia coli* to elevated hydrostatic pressure. J. Bacteriol. 175, 7170–7177.

Zheng, M., Doan, B., Schneider, T.D., Storz, G., 1999. OxyR and SoxRS regulation of fur. J. Bacteriol. 181, 4639–4643.

CHAPTER

Synergism in microbial communities facilitate the biodegradation of pesticides

11

Yingjie Yang[a], Qianru Chen[a], Naila Ilyas[a], Ping Zou[a], Changliang Jing[a], Bin Li[b], and Yiqiang Li[a]

[a]*Marine Agriculture Research Center, Tobacco Research Institute of Chinese Academy of Agricultural Sciences, Qingdao, China,*

[b]*ChinaTobacco Chengdu Industrial Co., Ltd., Chengdu, China*

Abstract

Pesticide is a well-known material that is used to control the occurrence of numerous pests, weeds, and plant diseases in crops around the world. Despite the benefits of pesticides in agriculture, they are often seen as hazardous for the environment due to their persistence as well as abnormal phenomena. Therefore, the elimination of pesticides from the habitat is an interesting area for investigators. Currently, microbes have become a prevailing tool for the degradation and remediation of pesticides, in-situ. Compared with simple physical and chemical treatment, biological treatment, especially microbial treatment, is an effective and environmentally friendly treatment method. At present, most of the research papers on the microbial treatment of pesticides are presented as single bacteria. The biological disturbance produced by single bacteria added to complex soil and other environments is short-lived and will soon be covered by indigenous microorganisms. The compound microbial agents can not only occupy the soil niche rapidly but also play a key role in the degradation of pesticides in soil and other environments. This review will analyze the molecular mechanism of synergistic degradation of pesticides by mixed microorganisms or consortia (including material-microbial, microbial-microbial, microbial-plant) in the degradation process of various pesticides (including organophosphorus pesticides, organochlorinated pesticides, atrazine pesticides, pyrethroid pesticides, and other insecticides, herbicides, and fungicides) and understand the interaction and symbiosis between the mixed strains. Moreover, the genes/enzymes, pathways, processes and aspects influencing the biodegradation have displayed in this chapter.

Keywords: Synergism effect, Organophosphorus pesticide, Organochlorinated pesticides, Atrazine, Pyrethroid pesticides

1 Introduction

With the sustainable development of the agricultural economy in recent decades, the use of pesticides has been increasing. The quantity of pesticides used in the prevention and control of agricultural diseases and pests is about 6 million tons every year, and the effective utilization rate of pesticides is only 30%, of which the remaining 70% are losses. Pesticide residues in the environment will continue to spread through rain, wind, and other meteorological conditions, resulting in the formation of a huge amount of pesticides residue in the global air, ocean, soil, and organisms. China is a largely agricultural country, and its agricultural production is quite large, which also makes China's pesticide consumption

Microbial Syntrophy-mediated Eco-enterprising. https://doi.org/10.1016/B978-0-323-99900-7.00011-0

occupy the top of the world. It is precise because of the continuous expansion of pesticide use that the problem of pesticide pollution becomes more and more serious, and even pesticide poisoning cases occur frequently.

Biodegradation is the use of microbial metabolic ability to mineralize/transform the organic pollutants into less detrimental, non-hazardous elements, which are then unified into natural biogeochemical cycles (Yang et al., 2020). The biodegradation intensity is affected by several factors, such as oxygen, nutrients, composition, pH value, concentration and bioavailability of the contaminants, etc. It is a non-destructive, cost-effective and sometimes logistically favorable clean-up technology that goes to quicken the naturally occurring biodegradation of pollutants through the optimization of restraining environments (Fuchs et al., 2011).

In modern farming systems, pesticide application partakes as most common practice in agricultural fields in which 2%–3% of it is generally utilized and the rest part is continuing in habitat as a pollutant that is leading to toxicity (WHO, 1990). The microbial mediated biodegradation of pesticides has foremost critical importance for modern agriculture as well as for a safer environment. Microorganisms are versatile in each habitat on the planet, and directed the environmental conditions of their residing niche by their metabolic activities. Indeed, they are profoundly tangled in biogeochemistry, metal precipitation, water purification, and sustenance of plant growth that confirming the elements recycling. In soil, microbes interact with the plant roots, a "hot spot" of microbial bustle and it facilitates the microbial mass enhancement, place for interactions, and genetic exchange (Singh et al., 2019). In plant root, a narrow zone of soil "rhizosphere" is a place or most dynamic interfaces on Earth for the microorganisms as well as invertebrates to perform their activities (Singh et al., 2020). The rhizosphere microbiome is dependent on the plant genotype, root exudates, and the environment (Yang et al., 2020). Hence, the review of expressed microbial communities in pesticide environments are the key of investigation for the various roles played by microorganisms in respective niche and to the identification of the microbial genetic potential for biotechnological application in bioremediation of pesticide, including but not limited to agriculture, industrial, clinical, environmental and renewable of environments.

2 Mechanisms of pesticide biodegradation pathways in microbes

2.1 Organophosphorus pesticide

Organophosphorus pesticide refers to the organic compound pesticide containing phosphorus elements. Most of them are oily liquid, garlic flavor, strong volatility, slightly soluble in water, damaged by alkali. Most of the organophosphorus pesticides are phosphates or thiophosphates and are structured by themethoxy (CH_3O–) or ethoxy (C_2H_5O–) and is substituted with oxygen (O) or sulfur (S) alkoxy, aryloxy, or others.. some organophosphorus pesticides (such as methamidophos) also contain P–N bonds. It can inhibit acetylcholinesterase and accumulate acetylcholine, causing muscarinic, nicotine like and central nervous system symptoms. Most of the organophosphorus pesticides produced in China are insecticides, such as parathion, internal phosphorus, malathion, dimethoate, trichlorfon, and dichlorvos. Due to the low selectivity and high toxicity between target organisms and non-target organisms, organophosphorus pesticides methamidophos, methyl parathion, monocrotophos, parathion, phosphamidon, is banned from 2007 in China. However, this kind of insecticide has the advantages of low cost, high efficiency, and low bioaccumulation. Therefore, some organophosphorus insecticides with good biological selectivity are still widely used. When organophosphorus pesticides enter the soil environment, the microorganisms in the soil produce corresponding enzymes. Under the action of these enzymes, the above bonds are broken, so that the organophosphorus pesticides are degraded.

2.1.1 Organophosphorus pesticide degrading bacteria

Soil microbial system includes bacteria, fungi, actinomycetes, algae, and so on, which can degrade pesticide residues in soil and agricultural products. Bacteria play a dominant role in pesticide degrading microorganisms because of their biochemical adaptability and easy to induce mutation. There are a variety of microorganisms in the soil, and different microorganisms contain different enzymes. Organophosphorus pesticides entering the soil can be degraded by these enzymes. One kind of organophosphorus pesticide might be degraded by diverse groups of microorganisms, and one microorganism can also degrade the several organophosphorus group pesticides. Therefore, the number as well as types of microorganisms in soil display the different-different vital role in the degradation of the organophosphorus pesticide.

2.1.2 Organophosphorus pesticide (OP) degrading genes

Phosphatase belongs to a kind of hydrolase, which is catalyzed the hydrolysis of phosphate bond OP degrading enzyme. Microbiological mediated OP metabolism is majorly governed by two unique families of hydrolases phosphotriesterase (PTE) and the mpd metallo-beta-lactamase. Besides these two main types, there are several types related to the degradation of organphosphorus pesticides, for example, ophc2 gene, opdB gene, three different OP-degrading hydrolases encoding genes from Nocardia species.

2.1.2.1 Opd phosphotriesterase

The first type belonged to opd phosphotriesterase (PTE), and the genes opd is encoded the hydrolytic enzymes functionally and is classified as PTEs. The PTEs belong to the superfamily amidohydrolase. The parathion hydrolase was the first gene product of opd and later it was known as organophosphorus hydrolase. Mulbry et al. (1986) has reported it in the large plasmid which was isolated from *Sphingobium fuliginis* ATCC 27551 (Mulbry and Karns, 1989) and later was in the *Brevundimonas diminuta* GM (Harper et al., 1988; Mulbry et al., 1987). Interestingly, all the reported opd enzymes have the ability to entertain the pH optima, broad temperature and endowed with the capacity to degrade the varied array of OP substrates such as the compound with the bonding of P–CN, P–F, P–O, P–S, etc. (Cycoń et al., 2009; Harper et al., 1988; Kumar et al., 1996; Mulbry et al., 1987).

The *opd* gene and its homologue are widely distributed at species as well as genera level in microbial communities. Such as *opd* gene of *Agrobacterium radiobacter* strain P230, also known as *opd*A is found in *S. fuliginis* ATCC 27551 and *B. diminuta* GM with 88% of identity. Which is revealed that *opd* gene and its homologues are horizontally transferring within and between the genus as well as species. Iyer et al. (2013) have displayed it by maximum likelihood phylogeny within the genus: *Agrobacterium*, *Arthrobacter*, *Clostridium*, *Brevundimonas flavobacterium*, *Sphingobium*, *Mycobacterium*, and *Sulfolobus*, by using the of 12 *opd* and its homologues from representative taxa.

2.1.2.2 Methyl parathion degradation (mpd) gene

mpd genes are a distinct OP degrading pathway that communicates a specialized capability to degrade chlorpyrifos, methyl parathion, and methyl paraoxon. Though, there is no report available for the homology ($>20\%$) of *mpd* genes (Zhongli et al., 2001). The abundance of the *mpd* genes is reported in the bacterial chromosomal region with a single exception of *Pseudomonas putida* WBC-3 plasmid (70kb in size) (Liu et al., 2005). Zhongli et al. (2001) have reported that *mpd* gene encoding plasmid is identical to chromosomal mpd gene of *Plesiomonas* sp. M6 which is strongly revealing the horizontal gene transfer among the genera.

mpd gene is widespread in the isolates belongs to Chian such as *Achromobacter*, *Ochrobactrum*, *Stentrophomonas*, etc. (Yang et al., 2006; Zhang et al., 2005).

The effective chlorpyrifos-degrading *Stenotrophomonas* sp. strain YC-1 was identified (Yang et al., 2006) which is used the chlorpyrifos for its growth, as the solitary source of carbon and phosphorus, and hydrolyzed to 3,5,6-trichloro-2-pyridinol. Fenitrothion, methyl parathion, and parathion are also degraded by the strain YC-1 when these pesticides are provided as the single source of nutrients.

Zhang et al. have isolated the seven methyl parathion-degrading bacteria from a methyl parathion contaminated soil (long-term) and were found similar to the genus *Achromobacter*, *Brucella*, *Ochrobactrum*, and *Pseudaminobacter* and (Zhang et al., 2005). The hydrolase genes reported in these genera are found similar to the *mpd* gene of *Plesiomonas* sp. M6 and all were located in the chromosomal region. Sequence analysis of their mpd genes displayed its conserved nature. The G+C content of the characterized *mpd* genes were markedly dissimilar from the chromosome-located gene and is suggesting that the *mpd* gene might be shifted and expressed among the variety communities of bacteria (Zhang et al., 2005, 2006a).

2.1.2.3 ophc2 gene

OPHC2 in *Pseudomonas pseudoalcaligenes* is a phosphotriesterase, belongs to the superfamily of metallo-β-lactamase (Gotthard et al., 2013). ophc2 was firstly cloned from *Stenotrophomonas* sp. SMSP-1 using the shotgun technique, which was isolated and hydrolyzed the methyl parathion to *p*-nitrophenol (PNP) and dimethyl phosphor-thioate, but unable to degrade it further. *Stenotrophomonas* sp. SMSP-1 is also have the ability to hydrolyzed other OPs, including ethyl parathion, fenthion, fenitrothion, and phoxim (Shen et al., 2010). The ophc2 gene was effectively expressed in *E. coli* and *Pichia pastoris* (Chu et al., 2006; Wu et al., 2004).

2.1.2.4 *opdB* gene

The *opdB* gene was cloned from *Lactobacillus brevis* and is completely diverse from the opd or opdA gene in sequence and structure and also in activity (Islam et al., 2010; Shen et al., 2010; Wu et al., 2004). opdB seems to be correlated to serine-dependent hydrolases which is providing a sturdy sign for it as a key contributor to OP bioremediation efforts in the future (Gotthard et al., 2013; Islam et al., 2010).

Recently, three different OP-degrading ophc2 and opdB hydrolases were isolated a *Nocardia* species which are cytosolic hydrolase ADPase (adpB), aryldialkylphosphatases hydrolase, and unique PTE isolated from *Nocardia* sp. strain B-1, strain SC and strain *Nocardioides simplex* NRRL B-24074, respectively (Mulbry and Karns, 1989; Mulbry, 1992). Finally, a novel and second PTE, *hocA*, was reported from *P. monteilli* (Horne et al., 2002).

2.1.3 Microbial mediated degradation of OPs in soil

Microbial-based degradation of OPs is an important way to degrade and transform organophosphorus pesticides in the soil. The degradation process mainly includes the following processes: one is that the microorganism itself contains the enzyme gene that can degrade the pesticide. When the organophosphorus pesticide enters the soil, the microorganism can immediately produce the degrading enzyme for degrading the organophosphorus pesticide. In this case, the breeding of degrading bacteria is relatively easy. The other is no enzyme system that can degrade the organophosphorus pesticide. When the pesticide enters the environment, the gene of the microorganism will be recombined or changed to produce a new degradation enzyme system. Once the degradation of the organic compound by microorganisms is triggered by the enzymes in cells, the overall mechanisms of degradation are processed into the three steps: firstly, adsorption of compounds on the microbial cell membrane surface, secondly, the adsorbed compounds entering into the cell membrane and finally, interaction and reaction with the degradation enzyme inside the microbial cell.

There are two types of microbial action on pesticides. One is that microorganisms act on pesticides directly, and its essence is an enzymatic reaction. The other is to change the chemical and physical environment through microbial activities and indirectly act on pesticides.

There are three common modes of action. (1) *Mineralization*: it refers to the process that microorganisms decompose organophosphorus pesticides into inorganic substances such as CO_2 and H_2O. (2) *Cometabolism*: it refers to the phenomenon that microorganisms can decompose and metabolize substances that cannot be used when there are available carbon sources. The non-specific enzymes produced in the cometabolism reaction can not only metabolize the growth matrix but also metabolize the target pollutants, which is the key to the microbial cometabolism reaction.

Interspecific cometabolism refers to the joint metabolism of some organophosphorus pesticides by several microorganisms in the same environment. Sometimes a single microorganism cannot complete the degradation of organic phosphorus, and the reasons are various. For example, because the organic matter needs a lot of energy to open the ring, the metabolism of a single microorganism cannot provide enough energy and needs the supplement of external energy; or a certain microorganism only has the function of opening the ring and cannot completely complete the transformation from organic matter to simple inorganic substance, and other bacteria are needed to replace it. The products were further degraded. Therefore, in this case, the cultivation of mixed bacteria is a feasible method to solve the problem.

The degradation of pesticides by symbiotic or single microorganisms is completed by the participation of enzymes, some of which are inherent in microorganisms, and some are produced by variation. The reason why organophosphorus pesticides are easy to decompose in the soil is that microorganisms are secreting phosphoacetate enzyme structures in soil, and the degrading enzymes are more tolerant to abnormal environmental conditions than those producing such enzymes. The degradation consequence of the enzyme is much better than that of the microorganism itself, especially for the low concentration target pesticide. In this case, the degrading bacteria can use other carbon sources and cannot effectively use the target pesticide as the carbon source. The degradation effect is not affected by carbon source.

2.2 Organochlorine pesticides

Organochlorine pesticides can be divided into two categories: benzene and cyclopentadiene. The former includes DDT, BHCs, pentachloronitrobenzene and dicofol; the latter include chlordane, heptachlor, and Aldrin. According to the number and connection mode of benzene ring, they can also be divided into four categories.

The first type of halogenated benzene is a halogenated aromatics class that contains only an aromatic skeleton with substituents like benzene, phenol, aniline but all are halogenated and mostly are its derivatives. The second type of compound that has aromatic rings which are attached to extra moieties such as 2,4-dichlorophenoxyacetic acid (2,4-D), 2,4,5-trichlorophenoxyacetic acid (2,4,5-T) and its analogues as well as OPs herbicides (chlorpyrifos and profenofos). The third type is containing more than one aromatic ring-like polychlorinated biphenyl (PCBs) and polybrominated biphenyl (PBBs). The fourth type is a class of compound Fused Heterocyclic Compound.

2.2.1 Synergistic effect of biodegradation of Endosulfan

Endosulfan, an organochlorine pesticide, contains sulfur, without any aromatic ring, is very common against the insect, pests, and mites. Endosulfan degradation happens by the oxidation to the endosulfan diol in an aquatic system which may further degrade to endosulfan ether/α-hydroxyether/lactone (Walse et al., 2003). It degrades twice in non-sterile conditions compared to sterile conditions signifying its importance in bio-based degradation (Singh et al., 2017). Degradation of Endosulfan in a field clay was found lower at low temperatures as well as lesser water contents (Ghadiri and Rose, 2001). Interestingly, *Bordetella petrii* was found capable of degraded 89% of α and 84% of β isomers of endosulfan, whereas *B. petrii* II degraded 82% of both the isomers (Odukkathil and Vasudevan, 2013). *Achromobacter xylosoxidans* strain C8B was isolated using sulfur-free medium with endosulfan (Singh and Singh, 2011).

A consortium of bacterial and fungal strains was formerly isolated from endosulfan contaminated agricultural land (Abraham and Silambarasan, 2014). *Enterobacter asburiae* JAS5, *Enterobacter cloacae* JAS7, *Klebsiella pneumoniae* JAS8, *Halophilic bacterium* JAS4, and fungi *Aspergillus tamarii* JAS9, *Botryosphaeria laricina* JAS6, and *Lasiodiplodia* sp. JAS12 were isolated and identified. These consortia had ability to degrade $1000\,mg\,L^{-1}$ of endosulfan agriculture soil as well as in water and showed the infra-red (IR) bands at 1400 and 950 cm features of the COOH group and acid dimer band. This has been indicating that endosulfan is hydrolyzed as well as oxidated via TCA cycles.

2.2.2 Synergistic effect of biodegradation of hexachlorocyclohexane

γ-Hexachlorocyclohexane or lindane (γ-HCH, γ-BHC) is an insecticide of the organochlorine group. It has alpha, beta, gamma, and delta predominate stereoisomers in all of eight available forms (Lal et al., 2006). The HCH isomers are crucial for the determination of their persistence as well as toxicology. Though the bio-based transformation and degradation of HCH isomers, under aerobic as well as anaerobic conditions have been described (Deo et al., 1994).

Sphingobium japonicum UT26 mediated degradation of lindane has described the *Lin* gene in this bacterium. In *Pseudomonas paucimobilis* UT26, LinA, LinB, the last step by LinE, meta-cleavage dioxygenase is participated for the conversion of gamma-hexachlorocyclohexane (gamma-HCH) to chlorohydroquinone (CHQ) and then to hydroquinone (HQ) by cleaves aromatic rings with two hydroxyl groups (Miyauchi et al., 1998). Several reports and reviews are published for the biodegradation as well as bioremediation of lindane. For example, microbial system based biodegradation of lindane (γ-Hexachlorocyclohexane) (Zhang et al., 2020), and alpha-, beta-, gamma-, and delta-degradation in the soil-plant system of a contaminated site has been remarkably studied (Calvelo Pereira et al., 2006).

Similarly, the anaerobic sludge detoxification of four isomers was attained and found it degraded. After 20–40 days alpha- and gamma-HCH, while beta- and delta-HCH were only degraded after 102 days (Quintero et al., 2005). Some of the studies indicated that microbial consortium is also the best route to degrade the HCH (Murthy and Manonmani, 2007).

A nine bacterial and a fungal strain-based consortium are tested synergistically as well as individually and found capable to degrade $10\,\mu g\,mL^{-1}$ of γ-HCH (Elcey and Kunhi, 2010). Moreover, the consortia were found latent for bioremediation of HCH polluted soil, waste dump sites, and water system. The microbial consortia was designed by the combination of *Pseudomonas fluorescens* CFR1022, *P. putida* CFR1021, *P. aeruginosa* CFR1023, *P. aeruginosa* CFR1024, *Pseudomonas stutzeri* CFR1027, *Burkholderia cepacia* CFR1026, *B. cepacia* CFR1025, *A. alginolyticus* CFR1028, *Ar. lwoffii* CFR1029, and *Fusarium* sp. CFR225.

Insecticides also disrupt the subtle equilibrium between microflora and their environment. With the increasing concentrations of t-HCH, the microbial populations were declined proportionally and the affected or disappeared populations never been recovered. Bhatt et al. has isolated the latent t-HCH degrading strain and imperiled to further acclimatization to improve their degradation ability (Bhatt et al., 2007).

Lindane biodegradation by fungal isolates suggested there are more possibilities to be able to adequately degrade lindane based on their wide range of enzyme secretions. The comparison between single lindane biodegradation by plant or fungi alone was compared with synergistic lindane degradation and observed that lindane degradation in polluted soil was more effective when the actions of plant's root and fungi were combined (Asemoloye et al., 2017).

2.3 Atrazine pesticides

Atrazine ($C_8H_{14}ClN_5$) or 2-chloro-4-ethylamino-6-isopropylamino-1,3,5-triazine is a type of herbicide and is used for the control of the weed. The bioremediation of atrazine is very significant due to its widely used, stable structure with complexity in degradation, and highly toxic on the organism and human beings. Hence, several types of processing tools and techniques are established and broadly employed for its degradation via different ways like biodegradation, adsorption, photochemical catalysis, etc.

2.3.1 Pathway of biodegradation of atrazine pesticides

Atrazine compound assimilation and degradation were well studied in *Pseudomonas* sp. ADP, and it was reported as the first atrazine-dechlorinating bacteria (Mandelbaum et al., 1995). Enzymatic activities and gene regulation of Atrazine biodegradation were reviewed by the Govantes group (Govantes et al., 2009). Treatment of herbicide atrazine in the contaminated environment was also disclosed and reviewed by He et al. (2019). The enzymatic activity is responsible for atrazine dechlorination by *Pseudomonas* sp. ADP was identified to be atrazine chlorohydrolase (AtzA) to catalyze the reaction of the first step dechlorination (de Souza et al., 1998). The next two enzymes hydroxyatrazine hydrolase (AtzB) and N-isopropylammelide amidohydrolase (AtzC) catalyze the hydroxyatrazine to form cyanuric acid via the Intermediate product N-isopropylammelide (Boundy-Mills et al., 1997; Sadowsky et al., 1998).

Cyanuric acid hydrolase TrzD of *Pseudomonas* sp. strain NRRLB-12227 metabolizes the cyanuric acid to ammonia, biuret, and urea (Karns, 1999), and interestingly the biuret hydrolaseis responsible to produce urea as byproduct (Cook et al., 1985). In *Pseudomonas* sp. ADP, the occurrence of connecting *atz*D, *atz*E, and *atz*F genes were found after plasmid sequencing. These plasmid genes were found to be responsible for metabolism of biuret and cyanuric acid. However, the large plasmid didn't contain the contiguous nature of T*atz*A, *atz*B, and *atz*C genes.

Numerous atrazine-degrading bacteria such as *Alcaligenes* strain SG1, *Rhizobium* strain PATR, *A. radiobacter* J14a, and *Ralstonia picketii* strain D were reported and these are metabolize the hydroxyatrazine to cyanuric acid and then to ammonia and carbon dioxide (Bouquard et al., 1997; Cheng et al., 2005;

Struthers et al., 1998). *Nocariodes* sp. converts atrazine to hydroxyatrazine and it has a wider specificity to substrate than *Pseudomonas* sp. ADP AtzA gene (Topp et al., 2000a). They were isolated with the presence of atrazine as the sole nitrogen source found active metabolize it to the ring carbon atoms such as carbon dioxide (Radosevich et al., 1995; Struthers et al., 1998; Topp et al., 2000b). *Shewanella* sp. YJY4, *Citricoccus* sp. strain TT3, and *Arthrobacter* sp. AD26 were identified to be atrazine-degradating bacterium which confined the atrazine-degrading gene atzA, atzB, and atzC (Li et al., 2008; Yang et al., 2018; Ye et al., 2016).

The presence of dealkylated metabolites in the soil also suggested that there might be diverse pathway of triazine biodegradation, for example, *Rhodococcus* strains. It comprises a cytochrome P450 monooxygenase which is concerned for the oxidative dealkylation of atrazine (Behki et al., 1993; Mulbry, 1994; Nagy et al., 1995).

2.3.2 Synergistic effect of atrazine biodegradation

A synergistic combination is the best way to improve the ability of microorganisms for the degradation of harmful substances. It is exemplified by Jiang et al. as the *Enterobacter* sp. P1 is capable of phosphorus-dissolving ability but lacked of degradation ability for atrazine. But, after mixing it with the *Arthrobacter* sp. strain DNS10, it removes 99.2% of atrazine after 48 h of reaction, while the singularly *Arthrobacter* sp. strain DNS10 mediated atrazine degradation rate was only 38.6% (Jiang et al., 2019). Yu et al. were used the *Aspergillus niger* strain Y3 mycelium pellets for the immobilization of *Arthrobacter* strain ZXY-2 and this synergistic combination were used for atrazine treatment (Yu et al., 2019). Resultantly, the synergistic combination was degraded 57.3 mg/L atrazine completely in 10 h of incubation. Another study for atrazine removal was performed by the mixing of iron-oxidizing bacteria, *Coriolus versicolor*, and white rot fungi and found that a good removal rate of atrazine (98%) by this synergistic mixture (Hai et al., 2012). Jiang et al. (2020) have been exposed to the effect of Zn^{2+} treatments on atrazine removal and were found 94.42% at 48 h, though, in the absence of exogenous Zn^{2+} it was only 66.43%. Microbial mediated rhizoremediation of atrazine is also very effective. As the *Arthrobacter* sp. DNS10 enhances the ability of plant pennisetum (Zhang et al., 2014). This microbial plant interactions degraded 98.1% of the atrazine, while a single bacterium or only plant has degraded 87.4% and 66.7% after a 30-days, respectively. The multibacterial consortium is also reported for the effective degradation of atrazine. A four species combination was studied by Zhang et al. and found it capable to metabolized the atrazine in soil (Zhang et al., 2012).

Studies also conferred that plasmid pDNS10-containing *Arthrobacter* sp. DNS10 was found more efficient in rapid degradation of atrazine when compared it without plasmid pDNS10 (Zhang et al., 2011b).

A eight-membered complex microbial system was also studied for the degradation assay of atrazine (Smith et al., 2005). The bacterial species *Agrobacterium tumefaciens*, *Caulobacter crescentus*, *P. putida*, *S. yaniokuyae*, *Nocardia* sp., *Rhizobium* sp., *Flavimonas oryzihabitans*, and *Variovorax paradoxus* were firstly screened for the presence of atrazine degradation genes (*atz*A,-B,-C,-D,-E,-F and *trz*D,-N) and then all were mixed for the further application.

Overall, the degradation of atrazine is started with the de-chlorination which is controlled by *atz*A, then dealkylation to yield either N-ethylammelide or N-isopropylammelide (Govantes et al., 2009). A study was conducted to proven this hypothesis and 82 strains were isolated from soil by using the agar media with ethylamine as a sole carbon source. Among them, only three were found to be potential to degrade N-ethylammelide, and interestingly all of these three isolates were carrying the *atz*B gene and *atz*C genes. This can be proven that the spreading of these genes among several diverse species in the soil microbiota made it able to degrade the atrazine (Smith and Crowley, 2006).

Four isolates *Alcaligenes faecalis* ND1, *A. tumefaciens* ND4, *Bacillus megaterium* ND3, and *Klebsiella ornithinolytica* ND2, were reported for atrazine degradation (Siripattanakul et al., 2009) and all of them were degrade the atrazine at 33%–51% within 7 days.

2.4 Pyrethroid pesticides

The pyrethroid pesticides are the third most used insecticides class for pests control in crops (Housset and Dickmann, 2009). It accounts for 20% of the world's insecticides. Pyrethroids are classified into the category of Type I and Type II (Proudfoot, 2005; Righi and Palermo-Neto, 2005). Type I is lack of cyano groups and include derivatives such as allethrin, d-phenothrin, permethrin, resmethrin, tetramethrin, etc. Whereas, Type II consists of the cyano group, including derivates, cyfluthrin, cypermethrin, cyhalothrin, delta-methrin, fenpropathrin, fenvalerate, flumethrin, fluvalinate, flucythrinate, and tralomethrin.

2.4.1 Pyrethroid-degrading bacteria

Microbial based degradation of pyrethroids is reported with different genera belonging to *Aspergillus*, *Bacillus*, *Brevibacillus*, *Candida*, *Pseudomonas*, *Micrococcus*, *Raoultella*, *Klebsiella*, and *Trichoderma* (Bhatt et al., 2019; Birolli et al., 2018; Cycoń et al., 2009; Tang et al., 2018; Zhang et al., 2019). These genera produce numerous pyrethroid hydrolases which are highly significant in the bioremediation of pyrethroid. Several fungal species such as *Aspergillus*, *Candida*, and *Trichoderma* are widely reported for the production of pyrethroid hydrolase (Liang et al., 2005; Palmer-Brown et al., 2019a, b).

Esterases, a pyrethroid-degrading hydrolase is widely reported from microbes, plant and animal cells (Bhatt et al., 2020). Carboxylesterase hydrolyzed the ester bond to detoxify the pyrethroids. The biodegradation pathways such as P450 monooxygenases, Aminopeptidase are widely reported for the detoxification of pyrethroid isomers (Chen et al., 2013b; Tang et al., 2017).

Pyrethroid hydrolases reported from microbes, insects, plant, and human cells were phylogenetically analyzed based on the protein structure including *Bacillus*, *Geobacillus*, *Lactobacillus*, *Sphingobium*, *Pseudomonas*, *Salmonella*, *Mesorhizobium*, *Haliangium*, and archaea *Sulfolobus*, mycorrhizal fungi. Bhatt et al. have been recently reviewed the role of esterases in the biodegradation of pyrethroid (Bhatt et al., 2021).

2.4.2 Examples of pyrethroid degrading bacteria and genes

Some bacteria, such as *Sphingobium*, thermophilic crenarchaeota *Sulfolobus*, *Pseudomonas*, Klebsiella, *Bacillus* are often reported from the pyrethroids contaminated soil and are able to degrade it.

Sphingobium sp. strain JZ-2 was isolated from the pesticide-contaminated soil and were found potential to degrade cypermethrin, cyhalothrin, deltamethrin, fenpropathrin, permethrin, fenvalerate, and bifenthrin. Moreover, after harvesting the cell-free extract of *Sphingobium* sp. strain JZ-2 a novel pyrethroid hydrolase was purified that is able to hydrolyze the pyrethroid (Guo et al., 2009). Further, Wei et al. have isolated the novel gene ST2026 from the thermophilic crenarchaeota *Sulfolobus tokodaii* that encodes a putative carboxylesterase (named EstSt7) and it was cloned as well as overexpressed in *E. coli* for functional characterization (Wei et al., 2013). EstPS1 and EstA were reported from *Pseudomonas synxantha* PS1 and *P. fluorescens* A506. Respectively and both were found 88% identical to carboxylesterase family. EstPS1 is well known for the degradation of several pyrethroid, *p*-nitrophenyl esters, etc. (Cai et al., 2017).

The gene encoding pyrethroid-hydrolyzing esterase (EstP) from *Klebsiella* sp. strain ZD112 was cloned. No similarities were found by a database homology search using the nucleotide and deduced amino acid sequences of the esterases and lipases. The purified EstP not only degraded many pyrethroid pesticides and the organophosphorus insecticide malathion but also hydrolyzed rho-nitrophenyl esters of various fatty acids, indicating that EstP is an esterase with broad substrates (Wu et al., 2006).

Some *Bacillus* spp. were often reported to be used to degrade pyrethroid pesticides. *Bacillus subtilis* strain 1D was revealed 95% biodegradation of cypermethrin after the 15 days of incubation. Interestingly, the end products of degraded cypermethrin (aerobically) were acetic acid, cyclopentane palmitoleic acid, cyclododecylamine, phenol, 3-(2,2-dichloroethenyl 2,2-dimethyl) cyclopropane carboxylate,1-decanol, chloroacetic acid and decanoic acid. Most important is the non-toxic end product after Cypermethrin biodegradation and hence it reveals the latent application of this organism in cleaning of pesticide contaminated habitat. The presence of esterase (700 bp) and laccase (1200 bp) genes was also confirmed by the PCR (Gangola et al., 2018).

Analysis of the degradation products by Chen et al. have evaluated the ability of *Bacillus* sp. DG-02 for fenpropathrin and the end product contains the seven metabolites of it. The degradation pattern of fenpropathrin is started with the cleavage of its carboxylester linkage and diaryl bond, followed by aromatic ring degradation and then successive metabolism (Chen et al., 2014). The same biodegradation pattern was studied in *Brevibacterium aureum*, also (Chen et al., 2013a).

2.4.3 Synergistic degradation of pyrethroid pesticides

The hydrolysis step plays a noteworthy role in pyrethroids biodegradation and it could be ascribed to the existence of cyclopropane carboxylic acid moieties, connected to aromatic alcohols via a central ester bond. Ester bond is generally considered susceptible to degrading microbes.

Three bacterial strains from cultivated soil were isolated and identified as *Acinetobacter calcoaceticus* MCm5, *B. parabrevis* FCm9, and *Sphingomonas* sp. RCm6. All were reported to be efficient to degrade cypermethrin as well as other pyrethroids (Akbar et al., 2015). Some of the studies reported the bacterial consortium for the degradation of pyrethroid (±)-lambda-cyhalothrin enantioselective (Birolli et al., 2019). The formulated consortium strains were *Averyella* sp. 4L, *Bacillus* sp. CBMAI 2051, *Bacillus* sp. CBMAI 2065, *Bacillus* sp. CBMAI 1837, *Bacillus* sp. 5G, *Bacillus* sp. 4T, *Bacillus* sp. 6E, *Curtobacterium* sp. CBMAI 1834 and *Pseudomonas* sp. 3F. The three most efficient strains *Bacillus* sp. CBMAI 2065, *Bacillus* sp. CBMAI 2066 and *Bacillus* sp. 2B and are used for the biodegradation of (±)-LC.

2.4.4 Construction of artificial bacteria for pesticide degradation by synthetic biology

Microorganisms are sometimes failed to survive against the different types of pesticides or slow in degradation of it. Presently, synthetic biology offers an influential method to create versatile pesticide degraders. *P. putida* strain KT2440 was engineered for concurrent degradation of carbamates, organophosphates and pyrethroids by enhancing its oxygen-sequestering ability. Also, targeted insertion of pesticide-degrading genes in the chromosome via genome-editing offers the real-time monitoring of the biodegradation process (Gong et al., 2018).

3 Conclusion and future perspectives

Microbial remediation of pollutants by microbial degradation is a safe, effective, and cheap method for the elimination and detoxification of high concentration pesticide residues. It is also an important way to solve the soil pollution caused by pesticides. In the literature reports on microbial remediation of contaminated soil, most of them are about petroleum and heavy metals. The research on pesticides is only about the degradation of single microorganisms in the culture medium or water body, and the research about on synergistic effect of microbial remediation in the soil is less. Therefore, the research on microbial synergistic remediation of pesticide-contaminated soil can protect land resources and promote agriculture sustainable development has very important practical significance.

References

Abraham, J., Silambarasan, S., 2014. Biomineralization and formulation of endosulfan degrading bacterial and fungal consortiums. Pestic. Biochem. Physiol. 116, 24–31.

Akbar, S., Sultan, S., Kertesz, M., 2015. Determination of cypermethrin degradation potential of soil bacteria along with plant growth-promoting characteristics. Curr. Microbiol. 70 (1), 75–84.

Asemoloye, M.D., Ahmad, R., Jonathan, S.G., 2017. Synergistic rhizosphere degradation of gamma-hexachlorocyclohexane (lindane) through the combinatorial plant-fungal action. PLoS One 12 (8), e0183373.

Behki, R., Topp, E., Dick, W., Germon, P., 1993. Metabolism of the herbicide atrazine by Rhodococcus strains. Appl. Environ. Microbiol. 59 (6), 1955–1959.

Bhatt, P., Kumar, M.S., Chakrabarti, T., 2007. Assessment of bioremediation possibilities of technical grade hexachlorocyclohexane (tech-HCH) contaminated soils. J. Hazard. Mater. 143 (1–2), 349–353.

Bhatt, P., Huang, Y., Zhan, H., Chen, S., 2019. Insight into microbial applications for the biodegradation of pyrethroid insecticides. Front. Microbiol. 10, 1778.

Bhatt, P., Bhatt, K., Huang, Y., Lin, Z., Chen, S., 2020. Esterase is a powerful tool for the biodegradation of pyrethroid insecticides. Chemosphere 244, 125507.

Bhatt, P., Zhou, X., Huang, Y., Zhang, W., Chen, S., 2021. Characterization of the role of esterases in the biodegradation of organophosphate, carbamate, and pyrethroid pesticides. J. Hazard. Mater. 411, 125026.

Birolli, W.G., Vacondio, B., Alvarenga, N., Seleghim, M.H.R., Porto, A.L.M., 2018. Enantioselective biodegradation of the pyrethroid (±)-lambda-cyhalothrin by marine-derived fungi. Chemosphere 197, 651–660.

Birolli, W.G., Arai, M.S., Nitschke, M., Porto, A.L.M., 2019. The pyrethroid (±)-lambda-cyhalothrin enantioselective biodegradation by a bacterial consortium. Pestic. Biochem. Physiol. 156, 129–137.

Boundy-Mills, K.L., de Souza, M.L., Mandelbaum, R.T., Wackett, L.P., Sadowsky, M.J., 1997. The atzB gene of Pseudomonas sp. strain ADP encodes the second enzyme of a novel atrazine degradation pathway. Appl. Environ. Microbiol. 63 (3), 916–923.

Bouquard, C., Ouazzani, J., Prome, J., Michel-Briand, Y., Plesiat, P., 1997. Dechlorination of atrazine by a Rhizobium sp. isolate. Appl. Environ. Microbiol. 63 (3), 862–866.

Cai, X., Wang, W., Lin, L., He, D., Huang, G., Shen, Y., Wei, W., Wei, D., 2017. Autotransporter domain-dependent enzymatic analysis of a novel extremely thermostable carboxylesterase with high biodegradability towards pyrethroid pesticides. Sci. Rep. 7 (1), 3461.

Calvelo Pereira, R., Camps-Arbestain, M., Rodríguez Garrido, B., Macías, F., Monterroso, C., 2006. Behaviour of alpha-, beta-, gamma-, and delta-hexachlorocyclohexane in the soil-plant system of a contaminated site. Environ. Pollut. 144 (1), 210–217.

Chen, S., Dong, Y.H., Chang, C., Deng, Y., Zhang, X.F., Zhong, G., Song, H., Hu, M., Zhang, L.H., 2013a. Characterization of a novel cyfluthrin-degrading bacterial strain Brevibacterium aureum and its biochemical degradation pathway. Bioresour. Technol. 132, 16–23.

Chen, S., Lin, Q., Xiao, Y., Deng, Y., Chang, C., Zhong, G., Hu, M., Zhang, L.H., 2013b. Monooxygenase, a novel beta-cypermethrin degrading enzyme from Streptomyces sp. PLoS One 8 (9), e75450.

Chen, S., Chang, C., Deng, Y., An, S., Dong, Y.H., Zhou, J., Hu, M., Zhong, G., Zhang, L.H., 2014. Fenpropathrin biodegradation pathway in Bacillus sp. DG-02 and its potential for bioremediation of pyrethroid-contaminated soils. J. Agric. Food Chem. 62 (10), 2147–2157.

Cheng, G., Shapir, N., Sadowsky, M.J., Wackett, L.P., 2005. Allophanate hydrolase, not urease, functions in bacterial cyanuric acid metabolism. Appl. Environ. Microbiol. 71 (8), 4437–4445.

Chu, X.Y., Wu, N.F., Deng, M.J., Tian, J., Yao, B., Fan, Y.L., 2006. Expression of organophosphorus hydrolase OPHC2 in Pichia pastoris: purification and characterization. Protein Expr. Purif. 49 (1), 9–14.

Cook, A.M., Beilstein, P., Grossenbacher, H., Hütter, R., 1985. Ring cleavage and degradative pathway of cyanuric acid in bacteria. Biochem. J. 231 (1), 25–30.

Cycoń, M., Wójcik, M., Piotrowska-Seget, Z., 2009. Biodegradation of the organophosphorus insecticide diazinon by Serratia sp. and Pseudomonas sp. and their use in bioremediation of contaminated soil. Chemosphere 76 (4), 494–501.

de Souza, M.L., Newcombe, D., Alvey, S., Crowley, D.E., Hay, A., Sadowsky, M.J., Wackett, L.P., 1998. Molecular basis of a bacterial consortium: interspecies catabolism of atrazine. Appl. Environ. Microbiol. 64 (1), 178–184. https://doi.org/10.1128/AEM.64.1.178-184.1998.

Deo, P.G., Karanth, N.G., Karanth, N.G., 1994. Biodegradation of hexachlorocyclohexane isomers in soil and food environment. Crit. Rev. Microbiol. 20 (1), 57–78.

Elcey, C.D., Kunhi, A.A., 2010. Substantially enhanced degradation of hexachlorocyclohexane isomers by a microbial consortium on acclimation. J. Agric. Food Chem. 58 (2), 1046–1054.

Fuchs, G., Boll, M., Heider, J., 2011. Microbial degradation of aromatic compounds—from one strategy to four. Nat. Rev. Microbiol. 9 (11), 803–816. https://doi.org/10.1038/nrmicro2652.

Gangola, S., Sharma, A., Bhatt, P., Khati, P., Chaudhary, P., 2018. Presence of esterase and laccase in Bacillus subtilis facilitates biodegradation and detoxification of cypermethrin. Sci. Rep. 8 (1), 12755.

Ghadiri, H., Rose, C.W., 2001. Degradation of endosulfan in a clay soil from cotton farms of western Queensland. J. Environ. Manag. 62 (2), 155–169.

Gong, T., Xu, X., Dang, Y., Kong, A., Wu, Y., Liang, P., Wang, S., Yu, H., Xu, P., Yang, C., 2018. An engineered Pseudomonas putida can simultaneously degrade organophosphates, pyrethroids and carbamates. Sci. Total Environ. 628–629, 1258–1265.

Gotthard, G., Hiblot, J., Gonzalez, D., Elias, M., Chabriere, E., 2013. Structural and enzymatic characterization of the phosphotriesterase OPHC2 from Pseudomonas pseudoalcaligenes. PLoS One 8 (11), e77995.

Govantes, F., Porrua, O., Garcia-Gonzalez, V., Santero, E., 2009. Atrazine biodegradation in the lab and in the field: enzymatic activities and gene regulation. Microb. Biotechnol. 2 (2), 178–185.

Guo, P., Wang, B., Hang, B., Li, L., Ali, S.W., He, J., Li, S., 2009. Pyrethroid-degrading Sphingobium sp JZ-2 and the purification and characterization of a novel pyrethroid hydrolase. Int. Biodeterior. Biodegrad. 63 (8), 1107–1112.

Hai, F.I., Modin, O., Yamamoto, K., Fukushi, K., Nakajima, F., Nghiem, L.D., 2012. Pesticide removal by a mixed culture of bacteria and white-rot fungi. J. Taiwan Inst. Chem. Eng. 43 (3), 459–462.

Harper, L.L., McDaniel, C.S., Miller, C.E., Wild, J.R., 1988. Dissimilar plasmids isolated from Pseudomonas diminuta MG and a Flavobacterium sp. (ATCC 27551) contain identical opd genes. Appl. Environ. Microbiol. 54 (10), 2586–2589.

He, H., Liu, Y., You, S., Liu, J., Xiao, H., Tu, Z., 2019. A review on recent treatment technology for herbicide atrazine in contaminated environment. Int. J. Environ. Res. Public Health 16 (24).

Horne, I., Sutherland, T.D., Oakeshott, J.G., Russell, R.J., 2002. Cloning and expression of the phosphotriesterase gene hocA from Pseudomonas monteilii C11. Microbiology (Reading) 148 (Pt. 9), 2687–2695.

Housset, P., Dickmann, R., 2009. A promise fulfilled—pyrethroid development and the benefits for agriculture and human health. Bayer CropSci J. 62 (2), 135–144.

Islam, S.M., Math, R.K., Cho, K.M., Lim, W.J., Hong, S.Y., Kim, J.M., Yun, M.G., Cho, J.J., Yun, H.D., 2010. Organophosphorus hydrolase (OpdB) of Lactobacillus brevis WCP902 from kimchi is able to degrade organophosphorus pesticides. J. Agric. Food Chem. 58 (9), 5380–5386.

Iyer, R., Iken, B., Damania, A., 2013. A comparison of organophosphate degradation genes and bioremediation applications. Environ. Microbiol. Rep. 5 (6), 787–798.

Jiang, Z., Li, J., Jiang, D., Gao, Y., Chen, Y., Wang, W., Cao, B., Tao, Y., Wang, L., Zhang, Y., 2020. Removal of atrazine by biochar-supported zero-valent iron catalyzed persulfate oxidation: reactivity, radical production and transformation pathway. Environ. Res. 184, 109260. https://doi.org/10.1016/j.envres.2020.109260.

Jiang, B., Zhang, N., Xing, Y., Lian, L., Chen, Y., Zhang, D., Li, G., Sun, G., Song, Y., 2019. Microbial degradation of organophosphorus pesticides: novel degraders, kinetics, functional genes, and genotoxicity assessment. Environ. Sci. Pollut. Res. Int. 26 (21), 21668–21681.

Karns, J.S., 1999. Gene sequence and properties of an s-triazine ring-cleavage enzyme from Pseudomonas sp. strain NRRLB-12227. Appl. Environ. Microbiol. 65 (8), 3512–3517.

Kumar, S., Mukerji, K.G., Lal, R., 1996. Molecular aspects of pesticide degradation by microorganisms. Crit. Rev. Microbiol. 22 (1), 1–26.

Lal, R., Dogra, C., Malhotra, S., Sharma, P., Pal, R., 2006. Diversity, distribution and divergence of lin genes in hexachlorocyclohexane-degrading sphingomonads. Trends Biotechnol. 24 (3), 121–130.

Li, Q., Li, Y., Zhu, X., Cai, B., 2008. Isolation and characterization of atrazine-degrading Arthrobacter sp. AD26 and use of this strain in bioremediation of contaminated soil. J. Environ. Sci. (China) 20 (10), 1226–1230.

Liang, W.Q., Wang, Z.Y., Li, H., Wu, P.C., Hu, J.M., Luo, N., Cao, L.X., Liu, Y.H., 2005. Purification and characterization of a novel pyrethroid hydrolase from Aspergillus niger ZD11. J. Agric. Food Chem. 53 (19), 7415–7420.

Liu, H., Zhang, J.J., Wang, S.J., Zhang, X.E., Zhou, N.Y., 2005. Plasmid-borne catabolism of methyl parathion and p-nitrophenol in Pseudomonas sp. strain WBC-3. Biochem. Biophys. Res. Commun. 334 (4), 1107–1114.

Mandelbaum, R.T., Allan, D.L., Wackett, L.P., 1995. Isolation and characterization of a Pseudomonas sp. that mineralizes the s-triazine herbicide atrazine. Appl. Environ. Microbiol. 61 (4), 1451–1457.

Miyauchi, K., Suh, S.K., Nagata, Y., Takagi, M., 1998. Cloning and sequencing of a 2,5-dichlorohydroquinone reductive dehalogenase gene whose product is involved in degradation of gamma-hexachlorocyclohexane by Sphingomonas paucimobilis. J. Bacteriol. 180 (6), 1354–1359.

Mulbry, W.W., 1992. The aryldialkylphosphatase-encoding gene adpB from Nocardia sp. strain B-1: cloning, sequencing and expression in Escherichia coli. Gene 121 (1), 149–153.

Mulbry, W.W., 1994. Purification and characterization of an inducible s-triazine hydrolase from Rhodococcus corallinus NRRL B-15444R. Appl. Environ. Microbiol. 60 (2), 613–618.

Mulbry, W.W., Karns, J.S., 1989. Parathion hydrolase specified by the Flavobacterium opd gene: relationship between the gene and protein. J. Bacteriol. 171 (12), 6740–6746.

Mulbry, W.W., Karns, J.S., Kearney, P.C., Nelson, J.O., McDaniel, C.S., Wild, J.R., 1986. Identification of a plasmid-borne parathion hydrolase gene from Flavobacterium sp. by southern hybridization with opd from Pseudomonas diminuta. Appl. Environ. Microbiol. 51 (5), 926–930.

Mulbry, W.W., Kearney, P.C., Nelson, J.O., Karns, J.S., 1987. Physical comparison of parathion hydrolase plasmids from Pseudomonas diminuta and Flavobacterium sp. Plasmid 18 (2), 173–177.

Murthy, H.M., Manonmani, H.K., 2007. Aerobic degradation of technical hexachlorocyclohexane by a defined microbial consortium. J. Hazard. Mater. 149 (1), 18–25.

Nagy, I., Compernolle, F., Ghys, K., Vanderleyden, J., De Mot, R., 1995. A single cytochrome P-450 system is involved in degradation of the herbicides EPTC (S-ethyl dipropylthiocarbamate) and atrazine by Rhodococcus sp. strain NI86/21. Appl. Environ. Microbiol. 61 (5), 2056–2060.

Odukkathil, G., Vasudevan, N., 2013. Enhanced biodegradation of endosulfan and its major metabolite endosulfate by a biosurfactant producing bacterium. J. Environ. Sci. Health B 48 (6), 462–469.

Palmer-Brown, W., de Melo Souza, P.L., Murphy, C.D., 2019a. Cyhalothrin biodegradation in Cunninghamella elegans. Environ. Sci. Pollut. Res. Int. 26 (2), 1414–1421.

Palmer-Brown, W., Miranda-CasoLuengo, R., Wolfe, K.H., Byrne, K.P., Murphy, C.D., 2019b. The CYPome of the model xenobiotic-biotransforming fungus Cunninghamella elegans. Sci. Rep. 9 (1), 9240.

Proudfoot, A.T., 2005. Poisoning due to pyrethrins. Toxicol. Rev. 24 (2), 107–113.

Quintero, J.C., Moreira, M.T., Feijoo, G., Lema, J.M., 2005. Anaerobic degradation of hexachlorocyclohexane isomers in liquid and soil slurry systems. Chemosphere 61 (4), 528–536.

Radosevich, M., Traina, S.J., Hao, Y.L., Tuovinen, O.H., 1995. Degradation and mineralization of atrazine by a soil bacterial isolate. Appl. Environ. Microbiol. 61 (1), 297–302.

Righi, D.A., Palermo-Neto, J., 2005. Effects of type II pyrethroid cyhalothrin on peritoneal macrophage activity in rats. Toxicology 212 (2–3), 98–106.

Sadowsky, M.J., Tong, Z., de Souza, M., Wackett, L.P., 1998. AtzC is a new member of the amidohydrolase protein superfamily and is homologous to other atrazine-metabolizing enzymes. J. Bacteriol. 180 (1), 152–158.

Shen, Y.J., Lu, P., Mei, H., Yu, H.J., Hong, Q., Li, S.P., 2010. Isolation of a methyl parathion-degrading strain Stenotrophomonas sp. SMSP-1 and cloning of the ophc2 gene. Biodegradation 21 (5), 785–792.

Singh, R.P., Buttar, H.K., Kaur, R., Manchanda, G., 2020. The multifaceted life of microbes: survival in varied environments. In: Singh, R., Manchanda, G., Maurya, I., Wei, Y. (Eds.), Microbial Versatility in Varied Environments. Springer, Singapore. https://doi.org/10.1007/978-981-15-3028-9_1.

Singh, R.P., Manchanda, G., Li, Z.F., Rai, A.R., 2017. Insight of proteomics and genomics in environmental bioremediation. In: Handbook of Research on Inventive Bioremediation Techniques. IGI Global, pp. 46–69.

Singh, R.P., Manchanda, G., Maurya, I.K., Maheshwari, N.K., Tiwari, P.K., Rai, A.R., 2019. Streptomyces from rotten wheat straw endowed the high plant growth potential traits and agro-active compounds. Biocatal. Agri. Biotechnol. 17, 507–513.

Singh, N.S., Singh, D.K., 2011. Biodegradation of endosulfan and endosulfan sulfate by Achromobacter xylosoxidans strain C8B in broth medium. Biodegradation 22 (5), 845–857.

Siripattanakul, S., Wirojanagud, W., McEvoy, J., Limpiyakorn, T., Khan, E., 2009. Atrazine degradation by stable mixed cultures enriched from agricultural soil and their characterization. J. Appl. Microbiol. 106 (3), 986–992.

Smith, D., Crowley, D.E., 2006. Contribution of ethylamine degrading bacteria to atrazine degradation in soils. FEMS Microbiol. Ecol. 58 (2), 271–277.

Smith, D., Alvey, S., Crowley, D.E., 2005. Cooperative catabolic pathways within an atrazine-degrading enrichment culture isolated from soil. FEMS Microbiol. Ecol. 53 (2), 265–273.

Struthers, J.K., Jayachandran, K., Moorman, T.B., 1998. Biodegradation of atrazine by Agrobacterium radiobacter J14a and use of this strain in bioremediation of contaminated soil. Appl. Environ. Microbiol. 64 (9), 3368–3375.

Tang, A.X., Liu, H., Liu, Y.Y., Li, Q.Y., Qing, Y.M., 2017. Purification and characterization of a Novel β-cypermethrin-degrading aminopeptidase from Pseudomonas aeruginosa GF31. J. Agric. Food Chem. 65 (43), 9412–9418.

Tang, W., Wang, D., Wang, J., Wu, Z., Li, L., Huang, M., Xu, S., Yan, D., 2018. Pyrethroid pesticide residues in the global environment: an overview. Chemosphere 191, 990–1007.

Topp, E., Mulbry, W.M., Zhu, H., Nour, S.M., Cuppels, D., 2000a. Characterization of S-triazine herbicide metabolism by a Nocardioides sp. isolated from agricultural soils. Appl. Environ. Microbiol. 66 (8), 3134–3141.

Topp, E., Zhu, H., Nour, S.M., Houot, S., Lewis, M., Cuppels, D., 2000b. Characterization of an atrazine-degrading Pseudaminobacter sp. isolated from Canadian and French agricultural soils. Appl. Environ. Microbiol. 66 (7), 2773–2782.

Walse, S.S., Scott, G.I., Ferry, J.L., 2003. Stereoselective degradation of aqueous endosulfan in modular estuarine mesocosms: formation of endosulfan gamma-hydroxycarboxylate. J. Environ. Monit. 5 (3), 373–379.

Wei, T., Feng, S., Shen, Y., He, P., Ma, G., Yu, X., Zhang, F., Mao, D., 2013. Characterization of a novel thermophilic pyrethroid-hydrolyzing carboxylesterase from Sulfolobus tokodaii into a new family. J. Mol. Catal. B Enzym. 97, 225–232.

WHO, 1990. Diet, nutrition and the prevention of chronic diseases. In: WHO Technical Report Series 797. WHO, Geneva.

Wu, N.F., Deng, M.J., Shi, X.Y., Liang, G.Y., Yao, B., Fan, Y.L., 2004. Isolation, purification and characterization of a new organphosphorus hydrolase OPHC2. Chin. Sci. Bull. 49 (3), 268–272.

Wu, P.C., Liu, Y.H., Wang, Z.Y., Zhang, X.Y., Li, H., Liang, W.Q., Luo, N., Hu, J.M., Lu, J.Q., Luan, T.G., Cao, L.X., 2006. Molecular cloning, purification, and biochemical characterization of a novel pyrethroid-hydrolyzing esterase from Klebsiella sp. strain ZD112. J. Agric. Food Chem. 54 (3), 836–842.

Yang, C., Liu, N., Guo, X., Qiao, C., 2006. Cloning of mpd gene from a chlorpyrifos-degrading bacterium and use of this strain in bioremediation of contaminated soil. FEMS Microbiol. Lett. 265 (1), 118–125.

Yang, Y., Liu, L., Singh, R.P., Meng, C., Ma, S., Jing, C., Li, Y., Zhang, C., 2020. Nodule and root zone microbiota of salt-tolerant wild soybean in coastal sand and saline-alkali soil. Front. Microbiol. 11, 2178. https://doi.org/10.3389/fmicb.2020.523142.

Yang, Y., Pratap Singh, R., Song, D., Chen, Q., Zheng, X., Zhang, C., Zhang, M., Li, Y., 2020. Synergistic effect of *Pseudomonas putida* II-2 and *Achromobacter* sp. QC36 for the effective biodegradation of the herbicide quinclorac. Ecotoxicol. Environ. Saf. 188, 109826. https://doi.org/10.1016/j.ecoenv.2019.109826.

Yang, X., Wei, H., Zhu, C., Geng, B., 2018. Biodegradation of atrazine by the novel Citricoccus sp. strain TT3. Ecotoxicol. Environ. Saf. 147, 144–150.

Ye, J.Y., Zhang, J.B., Gao, J.G., Li, H.T., Liang, D., Liu, R.M., 2016. Isolation and characterization of atrazine-degrading strain Shewanella sp. YJY4 from cornfield soil. Lett. Appl. Microbiol. 63 (1), 45–52.

Yu, T., Wang, L., Ma, F., Yang, J., Bai, S., You, J., 2019. Self-immobilized biomixture with pellets of Aspergillus niger Y3 and Arthrobacter. sp ZXY-2 to remove atrazine in water: a bio-functions integration system. Sci. Total Environ. 689, 875–882.

Zhang, R., Cui, Z., Jiang, J., He, J., Gu, X., Li, S., 2005. Diversity of organophosphorus pesticide-degrading bacteria in a polluted soil and conservation of their organophosphorus hydrolase genes. Can. J. Microbiol. 51 (4), 337–343.

Zhang, R., Cui, Z., Zhang, X., Jiang, J., Gu, J.D., Li, S., 2006a. Cloning of the organophosphorus pesticide hydrolase gene clusters of seven degradative bacteria isolated from a methyl parathion contaminated site and evidence of their horizontal gene transfer. Biodegradation 17 (5), 465–472.

Zhang, Y., Jiang, Z., Cao, B., Hu, M., Wang, Z., Dong, X., 2011b. Chemotaxis to atrazine and detection of a xenobiotic catabolic plasmid in Arthrobacter sp. DNS10. Environ. Sci. Pollut. Res. Int. 19 (7), 2951–2958.

Zhang, Y., Cao, B., Jiang, Z., Dong, X., Hu, M., Wang, Z., 2012. Metabolic ability and individual characteristics of an atrazine-degrading consortium DNC5. J. Hazard. Mater. 237–238, 376–381.

Zhang, Y., Ge, S., Jiang, M., Jiang, Z., Wang, Z., Ma, B., 2014. Combined bioremediation of atrazine-contaminated soil by Pennisetum and Arthrobacter sp. strain DNS10. Environ. Sci. Pollut. Res. Int. 21 (9), 6234–6238.

Zhang, X., Hao, X., Huo, S., Lin, W., Xia, X., Liu, K., Duan, B., 2019. Isolation and identification of the Raoultella ornithinolytica-ZK4 degrading pyrethroid pesticides within soil sediment from an abandoned pesticide plant. Arch. Microbiol. 201 (9), 1207–1217.

Zhang, W., Lin, Z., Pang, S., Bhatt, P., Chen, S., 2020. Insights into the biodegradation of lindane (γ-hexachlorocyclohexane) using a microbial system. Front. Microbiol. 11, 522.

Zhongli, C., Shunpeng, L., Guoping, F., 2001. Isolation of methyl parathion-degrading strain M6 and cloning of the methyl parathion hydrolase gene. Appl. Environ. Microbiol. 67 (10), 4922–4925.

CHAPTER

Bioproduction of terpenoid aroma compounds by microbial cell factories

12

Laura Drummond
Microbial Biotechnology, DECHEMA Research Institute, Frankfurt, Germany

Abstract

Terpenoids are important aroma components for food and perfumery products. They belong to a class of chemical compounds structurally defined by their C_5 prenyl diphosphate precursors, which are sequentially assembled by specific enzymes. As the flavor and fragrance industry grows, a consequential increase in demand for aroma chemicals stimulates improvements in their production methods on an industrial scale. Traditionally extracted from plants, terpenoids have recently exhibited a marked tendency toward environmentally-friendly production processes, for which biotechnological production through microbial cell factories configures as the best alternative. Genetic engineering of microorganisms enables the synthesis of terpenoids from simple carbohydrates, using the microbe's metabolism as a factory-like machinery system. Renewable materials can be used as primary carbon sources, qualifying the processes as sustainable and economically competitive, both of which are highly desirable for modern industrial production methods. Moreover, the modular biosynthesis of terpenoids out of C_5 precursors facilitates the metabolic engineering for the formation of larger and more complex compounds. This review provides an overview of the advancements of biotechnological processes toward microbial production of flavor and fragrance terpenoids.

Keywords: Terpenes, Flavor and fragrance, Metabolic engineering, Sustainability, Microbial production

1 Introduction

Terpenes are one of the most abundant and diverse classes of natural products, produced by virtually all types of organisms. Natural products are described as biologically produced chemicals that are useful to humans and are extracted from natural resources for their economic value. Within the organisms, terpene-derived molecules provide several functions in the metabolism, for example in cell structure components, hormones, and intra- and inter-species communication. As examples of terpenes of interest for mankind one can cite beta-carotene, steroids, and rubber; flavors and fragrances like santalol, eucalyptol, limonene, menthol; and nutrients like lycopene, vitamin K and vitamin E. The most valuable examples are medicines, like artemisinin, taxol, and prostratin, which play a central role in the treatments of cancer, malaria, and viral infections (Singh and Sharma, 2015; Vickers et al., 2017). The ones with a low boiling point can be used as biofuels (Tippmann et al., 2013). Volatile terpenes with aroma properties are used in flavor and fragrance industries (Schempp et al., 2018). Terpenes are biologically produced chemicals, traditionally obtained from plant materials, which have in common the fact that they are naturally synthesized out of isoprene-like precursors containing five carbon atoms in their structure. For that reason, they are also called isoprenoids.

Microbial Syntrophy-mediated Eco-enterprising. https://doi.org/10.1016/B978-0-323-99900-7.00004-3

Terpenes have been extracted from natural sources since ancient times, and are still used extensively as aromas, medicines, and commodity chemicals. However, the yield from plants is too low to attend to the ever-rising demand for those compounds. For example, the extraction of artemisinin from *Artemisia annua* generates a quantity equivalent to 1.5% of the plant dry weight in the best cultivars (Weathers et al., 2011), which is not viable for an industrial offer of compounds with lower market value (Pickens et al., 2011). The inherent low yield that natural plant sources provide for the specific chemicals creates limitations to the industrial supply of terpenes. With the current rising demands and urgency for sustainable production chains, microbial production is a much-needed solution to the problem. Humans have used microbes for the production of beverages like beer and wine since immemorial times, and more recently, microbes took a very important role on the industrial production of antibiotics and medicines, among others. The use of engineered strains for the production of insulin and penicillin is the only way to provide these two vital chemicals for the population in the necessary quantity. Thus, the metabolic engineering of known microbes like *Escherichia coli* and *Saccharomyces cerevisiae*, by transferring genes from the terpene original source, is the way to go for reaching industrial scale. The bacteria *E. coli* is the most commonly used prokaryote for the production of recombinant proteins, preferred because of the myriad of tools existing to genetically manipulate it, besides low growing costs and high growth speed (Jia and Jeon, 2016). Another widely used microbe is the yeast *S. cerevisiae*, a highly employed eukaryote for metabolic engineering and commonly more accepted in the food industry, since it is involved with the production of consumer goods and very much involved with our daily routines. High-value terpene aroma compounds are already being produced by microbial cell factories, on an industrial scale by specialized companies. Example cases are beta-farnesene, valencene, nootkatone, santalol, patchoulol, and sclareol (Schempp et al., 2018).

A lot of effort has been invested in the engineering of heterologous metabolic pathways into microbial hosts, mainly through synthetic biology principles and tools. This chapter will explain the concept of using microbial factories for the production of terpenoid aroma compounds, and the main achievements in the area. It will mostly focus on the biosynthesis of terpenes and the use of engineered model organisms like *E. coli* and *S. cerevisiae* for the production of terpene aroma chemicals.

2 Biosynthesis

Even though terpenes are among the most diverse classes of secondary metabolites, all of the existing members of this class are derived from only two basic building blocks of C_5 prenyl diphosphates, the isopentenyl diphosphate (IPP) and dimethylallyl pyrophosphate (DMAPP). These building blocks are condensed to form larger prenyl pyrophosphate precursors which have always a multiple of five carbon atoms in their structures, with a few exceptions that will be discussed later. The prenyl pyrophosphate precursors are then used as substrates by enzymes called terpene synthases to catalyze the formation of final terpene structures. The number of isoprene units and carbon atoms in terpene structures determines its nomenclature, e.g., monoterpenes (C_{10}), sequiterpenes (C_{15}), diterpenes (C_{20}), and triterpenes (C_{30}).

The start of the terpene biosynthetical pathway begins with the universal precursors IPP and DMAPP. These two main precursors, in turn, can be produced out of the organism's natural pathways. Two different pathways are occurring in nature, which are responsible for the production of IPP and DMAPP: the mevalonate pathway (MVA), and the 2-*C*-methyl-D-erythritol 4-phosphate/1-deoxy-D-xylulose 5-phosphate (MEP/DOXP) pathway. The MVA pathway occurs in eukaryotes, some of the bacteria and archaea. It starts with the condensation of two acetyl-CoA molecules and it consists of multi-enzymatic reactions that in the end yield IPP and DMAPP, besides generating NAD(*P*)H and CO_2. The MEP pathway occurs in plant plastids and prokaryotes. It starts from pyruvate and glyceraldehyde-3-phosphate deriving out of the central carbon metabolism. After several enzymatic reaction steps, which require ATP and NAD(*P*)H, it also yields IPP and DMAPP, with the release of CO_2.

After IPP and DMAPP are provided by one of the two pathways (or both, in the case of plants, which possess one in the cytoplasm and another one in the plastid), enzymes called prenyl transferases to catalyze the elongation of chains by condensation of prenyl pyrophosphate units, generating geranyl diphosphate (GPP, C_{10}), farnesyl diphosphate (FPP, C_{15}), or geranylgeranyl diphosphate (GGPP, C_{20}) for the processing by terpene synthases. Terpene synthases then convert the pyrophosphate substrates into various terpene structures, many of which will include a cyclization step (Christianson, 2017). Terpene structures generated by terpene synthases can be further decorated by enzymes like cytochrome P450s and other acyl- and glycosyl-transferases which generate even more diversity to terpene structures (King et al., 2016).

A quite new field of exploration is under development, with the use of additional modules to the terpene biosynthetic pathway. It concentrates on the creation of new molecules with the use of methyltransferases that modify terpene precursors. The first example of such methyltransferases have been described over a decade ago, as a GPP-methyltransferase that forms the unconventional C_{11} terpene 2-methylisoborneol (Dickschat et al., 2007; Komatsu et al., 2008; Wang and Cane, 2008). Since then, other prenyl pyrophosphate methyltransferases have been described, which use IPP and FPP as substrates to form C_6 and C_{16} unconventional terpenes (Radhika et al., 2015; Von Reuss et al., 2018). The use of an additional module has allowed the production of new terpenes with unusual structures (Kschowak et al., 2018; Drummond et al., 2019), a feature that was further enhanced by protein engineering (Ignea et al., 2018; Kschowak et al., 2020).

The engineering of microbes for terpene production must consider all these above mentioned biosynthetical steps. Since there are many tools for genetically engineering those organisms, targets for the creation of a terpene production platform strain must only focus on the supply of precursors from the central carbon metabolism, and the enzymes involved in prenyl diphosphate precursor elongation and final structure decoration.

3 Microbial production

Terpenes are traditionally extracted from plants. The name terpene derives from turpentine, a commodity material rich in terpenes that was extracted from the terebinth tree and used as a solvent. Since terpenes are very useful to mankind, they are under intensive demand, and the extraction of these compounds from natural raw materials leads to resource depletion and even species endangerment. Moreover, exploitation of natural resources for obtaining high-priced refined chemicals leads to other problems such as smuggling and criminality. Besides sustainability, one of the biggest setbacks on the extraction of terpenes from natural sources is the low quantities that are obtained per kilogram of plant. Another problem that arises is the availability of plants. They are subdued to seasonal availability, and with global warming and other natural oscillations from the environment, more barriers are imposed to obtaining terpenes from plant sources. The advance of technological tools to engineer the metabolism of microorganisms, allows us to create strains to produce specific products on larger scales (Fig. 1).

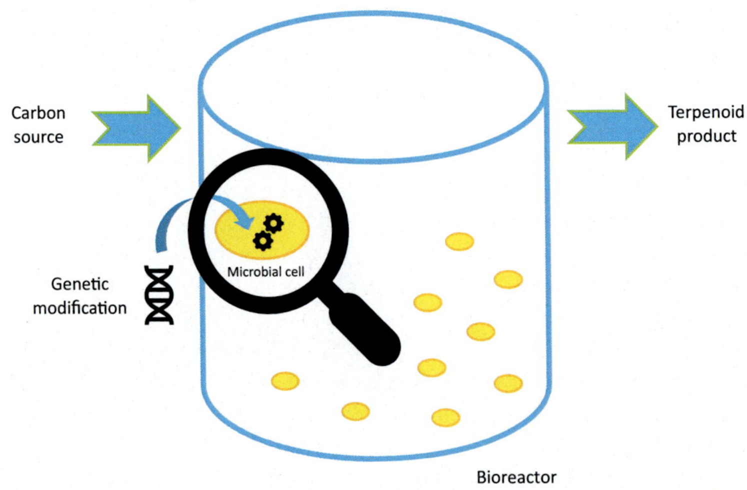

FIG. 1

Microbial production of terpenoids. Microbial hosts are metabolically engineered, through genetic modifications that alter or insert genes from biochemical pathways. The strains generated are cultivated in bioreactors, where they transform carbon sources (enumerated in the text) into terpenoid products.

Credit: Laura Drummond.

One emergent technology that has been applied for years in the pharmaceutical industry, for obtaining for example insulin, is the transformation of microorganisms into small factories for the production of the chemical of interest. In this technique, the microbe is genetically modified, with the insertion of genes responsible for coding the proteins that create the machinery necessary for the production of a determinate compound, out of simple nutrition items present in the culture medium. For example, already in 1978, for the production of insulin by bacteria, the human insulin gene was cloned and expressed in *E. coli*. The same occurs nowadays for the production of different types of terpenes, including aroma compounds. The technique has the advantages of being extremely efficient, reaching over 73 mg/g of cell dry weight (Ma et al., 2019), a great increase when compared with the above-mentioned example from plant extracts, and furthermore the advantage of generating no toxic waste when compared to chemical synthesis. Synthetic flavors and fragrances have been preferred from natural extracts since the second half of the 19th century, due to their physicochemical stability and lower production price; however, chemical synthesis of aromas is nowadays increasingly overshadowed by the sustainability of microbial production. The manipulated microbes, in the end, are not present in the final product, being deactivated, disintegrated, and discarded at the factory, which means that the final product contains no GMOs, an important factor for public acceptance. Microbes also bear the advantageous feature of not relying on water and land resources, besides growing extremely fast. As a comparison, a sandalwood tree will only be ready for harvesting in 4 years, while microbial fermentation for obtaining the same product santalol occurs within a week. Another advantage is the possibility of the use of alternative carbon sources, such as industrial rejects, waste or byproducts of industrial processes.

E. coli naturally produces terpenes, for example, quinones, but only in small quantities. For overproduction of IPP and DMAPP, enzymes from the MEP pathway have been engineered in *E. coli* in other to overcome bottleneck steps. Alternatively, heterologous MVA pathway was newly introduced into bacteria completely, with quite high success. When engineering a strain for the production of the desired compound, there are two main goals for metabolic engineering: engineering the chassis and engineering the terpene downstream pathway. The chassis can be optimized by manipulating the central carbon metabolism, creating shunts and using alternative microbial species with a native expression of desired characteristics, to provide enough substrates for the pathway such as acetyl-CoA and pyruvate/G3P. Pathway engineering unfolds either by heterologously expressing a foreign pathway, or optimizing an existent endogenous pathway. Generally speaking, by inserting a pathway that the organism does not have, it is possible to overcome usual problems with bottlenecks and natural regulation.

Starting from common precursors of the central metabolism, the terpene biosynthesis pathway can be divided into the MVA/MEP pathway and the downstream terpene pathway, which includes prenyl transferases, terpene synthases and terpene decorating enzymes. The pathways belonging to the central carbon metabolism of the organism will provide carbon-containing precursors, such as acetyl-CoA, glyceraldehyde-3-phosphate and pyruvate, and the necessary cofactors, i.e., NAD(*P*)H and ATP. These precursors and cofactors will be formed out of the selected carbon source (e.g., glucose, glycerol, methanol, fatty acids, etc.) employed in the system, depending on the platform organism used. The precursors provided to MVA/MEP pathways by the central carbon metabolism will be converted into IPP and DMAPP. These prenyl pyrophosphate precursors will then be used to form the intermediates GPP, FPP, GGPP which will be catalyzed by enzymes into terpenes. This last step is usually performed by enzymes that are not present in the host organism, so they must necessarily be engineered to heterologously express them.

The terpene biosynthetic pathway has the advantage of being a modular pathway with many steps in common between different terpenes; therefore the strategies used for the metabolic engineering of one product are generally useful for the development of strains for bioproduction of another product. Thus, review articles on the topic are usually well received by the community, and the latest developments of technology are instructive for all researchers working on high yield terpene production.

4 MVA and MEP pathway engineering

The first decision to be made when engineering a microbial strain is to choose between MEP and MVA pathways. Then, the flux through the metabolic pathway has to be adjusted. Both are pathways that involve a series of genes, since they are multi-enzymatic, and the regulation of expression of these genes, as well as the balance between their expressions, have to be considered. Both pathways also include rate-limiting enzymes, which are the bottlenecks of the system, and these enzymes require engineering with the aim of overcoming the flux limitation.

The MVA pathway consumes a total of 1.5 glucose molecules for each IPP molecule and produces four molecules of NAD(*P*)H. With the MEP pathway when one molecule of glucose is consumed, three ATP molecules and two NAD(*P*)H molecules are required. If the cofactor need is converted to glucose need, the total amount of glucose needed for the MEP pathway for each IPP molecule is 1.25 molecule of glucose (Schempp et al., 2018). The MEP pathway is, according to these numbers, more efficient in terms of mass production, i.e., it is more carbon efficient. However, it consumes more cofactor power than the MVA pathway. In summary, the MEP pathway is more carbon-efficient, whereas the MVA pathway is more energy efficient.

The utilization of native MEP in *E. coli* usually yields fewer products than the overexpression of heterologous MVA. By overexpressing enzymes of rate-limiting steps from MEP in *E. coli* lower amounts of isoprene were obtained than when MVA was expressed for the same product (Zhao et al., 2011; Yang et al., 2013). One possible reason for that is the high consumption of cofactor by the MEP pathway. On top of that, the MEP pathway is under heavy control in *E. coli* since it is native to this organism, therefore negative feedback regulation from key step enzymes might occur. Also, some of the intermediates in the MEP pathway, like 1-deoxy-D-xylulose 5-phosphate (DXP), participate in other physiological pathways of the organism (Julliard and Douce, 1991).

In the yeast *S. cerevisiae*, it is possible to engineer the native MVA for terpene production. One of the strategies was to overexpress rate-limiting enzymes (HMG-CoA synthase and reductase) and delete a gene that has the function of downregulating the pathway (Bröker et al., 2018). The expression in yeast of a truncated version of the enzyme, HMGR1 was shown to improve santalene production, because the engineered enzyme did not suffer negative feedback from one of the downstream products FPP (Scalcinati et al., 2012a; Scalcinati et al., 2012b). Another type of yeast, *Yarrowia lipolytica* was engineered for the production of linalool. The strategy used was to overexpress genes that encoded enzymes involved in key steps from the MVA pathway, such as *HMG1* and *IDI1*, together with a mutant version of the gene *ERG20*. The resulting strain was able to produce linalool at almost 7 mg/L (Cao et al., 2017). The heterologous expression of MEP in yeast was not successful for terpene production so far. Interestingly, when MVA and MEP were simultaneously expressed within an organism, terpene production was higher than when each pathway was expressed separately (Yang et al., 2016). This result suggests an interaction between the pathways in the form of synergy, in which each pathway produces intermediates that are used by the other pathway in a complementary way. This could be a useful strategy to use when engineering a host, especially for *E. coli*.

In general, it is complicated to engineer whole multi-step pathways within an organism, because the resulting strains usually show disequilibrium in the metabolism. The imbalance in physiology causes growth problems and ultimately lowers production yields. A possible approach to solve this problem is to engineer the pathways within separate modules. The modules can even be expressed by different species forming a consortium when the species are able to externalize the intermediates to be exchanged. This was shown for co-cultures of *E. coli* and *S. cerevisiae* expressing different modules of the terpene-forming pathways (Zhou et al., 2015). Furthermore, to overcome metabolic imbalance and formulate the best combination, it is possible to modulate the ribosome binding site (RBS) and promoter strength of each gene (Ye et al., 2016).

For gene expression, there is a possibility of plasmid insertion and chromosome engineering. By expressing heterologous enzymes with the help of optimized plasmids in *E. coli*, high levels of cineole and linalool were produced (Mendez-Perez et al., 2017). Nevertheless, the larger the plasmid is, the lower its stability will be, and a compromise must be met. Even though plasmid insertion is the easiest manipulation, the insertion of genes within the genome offers advantages, including the fact that no antibiotic resistance needs to be expressed, which also brings another burden for the cell (George et al., 2015). Moreover, for the cultivation of such strains, antibiotics have to be added to the culture medium, which adds an extra cost to the process. In this specific aspect, the regulation of native pathways offers a more stable and cost-effective alternative to plasmid-based expression. Notwithstanding, genome-based pathway construction encompasses arduous engineering work when compared to the practical and convenient plasmid expression. To tackle these engineering difficulties, the newly introduced CRISPR/Cas-9 methodology of gene modification has been increasingly used, with an application for terpene production in bacteria and yeast (Li et al., 2015; Liu et al., 2018; Schwartz and Wheeldon, 2018). The tools available for strain engineering are several, and there is no single one that outstands from the others alone. The most widely suggested approach for successful engineering of high-titer producing strains is the combination of several techniques in order to achieve an optimum.

A traditional way of increasing productivity is identifying key-step enzymes or bottlenecks and overexpressing them. Another commonly used approach is to search on the database for homologues of an enzyme to find enzymes with the same function but higher affinities. In general, more than one enzyme is engineered, sometimes belonging to different pathways, to obtain the desired effect of increased production. For valencene and amorphadiene production in yeast, one research group engineered HMG1 from the MVA pathway, as well as an FPP synthase and a sesquiterpene synthase. The enzymes, which were expressed in different organelles, were engineered using different strategies, and the combined expression was responsible for reaching an 8–20-fold increase in the yield of desired products (Farhi et al., 2011).

So far, more than 55 terpenes have been produced by engineered microbes (Pang et al., 2019). A large number of successful cases consisted of MVA insertion in bacteria. For the production of monoterpenes, high titers were achieved with *E. coli*, with 2.7 g/L of limonene, 2 g/L of geraniol, and 2.65 g/L of sabinene being produced in fed-batch fermentation (Willrodt et al., 2014; Zhang et al., 2014; Liu et al., 2016). *S. cerevisiae* naturally produces ergosterol in high levels and contains P450 systems, which makes it suitable for engineering its native MVA for the production of complex terpenes. As an example, one can cite the production of sesquiterpenes at 40 g/L that was achieved by Westfall et al. (2012) using yeast as a host. Yeast has also the advantage of being more easily accepted by the general public, especially when producing flavor and aroma compounds for the food industry. Combining the fermentation capabilities of brewer's yeast with acquired characteristics through engineering, Denby et al. (2018) created a strain that added hoppy flavor to the beer, without having to add hop plants to the recipe. Moreover, very high titers were achieved with yeast, for example with the production of beta-farnesene at more than 130 g/L (Meadows et al., 2016).

5 Engineering terpene synthases and prenyl transferases

When engineering a strain, once the host species and pathway is defined, choices have to be made about which genes need to be inserted. For the majority of terpene forming enzymes applied to the production of aroma compounds, these need to be newly introduced into the microorganism, because they are inexistent in this kind of organism. Once the type of enzyme that is going to be expressed is defined, a decision that depends on which product is of interest, a screening is made necessary, to find the one most suitable for production. For most of the aroma compounds produced by microbial cells, the enzymes responsible for terpene formation are already known to science. The variety, or homologue to be expressed, can still be selected from a myriad of available candidates in public databases. For the production of geraniol in *S. cerevisiae*, for example, different plant geraniol synthases were tested, from different species, before one of them was selected (Zhao et al., 2016a). Another option existent is to modify the chosen enzyme through protein engineering, to improve or optimize certain characteristics like affinity, solubility or turnover rate, among others.

Strategies like directed evolution, where mutation and selection processes are performed systematically in the lab, as well as rational protein design, which uses mathematical prediction to change protein structures in a defined way aiming at a specific activity change, can be used, separately or combined, to develop an enzyme with desired characteristics (Chica et al., 2005). Generally, the chosen target for this type of approach is the terpene synthase. Engineering the enzymes from MEP or MVA pathway in such a specific way has shown to be too much time-consuming and result in too low improvement of productivity. Protein engineering can also be performed to modify the specific activity of prenyl transferases. Using a mutant version of an FPP synthase, and a GPP synthase, both originally from plants, monoterpenoid production was improved in *E. coli* and *S. cerevisiae* (Carter et al., 2003; Reiling et al., 2004). When ERG20, which normally catalyzes the formation of GPP and FPP in sequence, was downregulated, together with overexpression of GPP synthase (among other modifications), high titers of geraniol were achieved (Zhao et al., 2017). In another study, the same ERG20 protein had specific residues that are involved with FPP formation but not GPP removed, leading to improvement of the flux of GPP formation to monoterpene production (Ignea et al., 2014). This is greatly facilitated when the protein tertiary structure and function are known, which is not always the case. For many of the terpene synthases, the tertiary structure is not known because insufficient crystal data are provided (or are not provided at all). In these cases, when details of the protein structure are unknown, directed evolution is the most advisable strategy. To select the evolved strains which contain the desired characteristics, good screening methodologies are of key importance. Unfortunately, specific screening methods for terpenes are still rare even though they are in high demand. A very good example of a colorimetric assay for terpene detection is the co-expression with carotenoid-forming enzymes (Furubayashi et al., 2014). The enzymes involved in carotenoid production and target compound production compete for the substrate GPP so that carotenoid formation (with color accumulation) is inversely proportional to the activity of product-forming enzymes. Another useful tool for screening is the use of co-expressed terpene biosensors (Kim et al., 2018b), which can also have a role in regulating gene expression (Dekker and Polizzi, 2017) and effectively tuning a desired biosynthetic pathway. These screening methods specific for terpenes are still rare, and more research needs to be done in this area to develop reliable and practical screening technologies to aid the selection of terpene-producing systems with improved characteristics.

As discussed above, for the directed evolution of proteins, good screening methods are crucial. From those, carotenoid co-expression and colorimetric assay seem to be the best technology developed so far. It is even safe to say those screening methods for terpenes are the main limitation for directed evolution. For rational engineering, on the other hand, the limitation is the poor knowledge about enzyme structures. Future research endeavors in this area should concentrate on providing more information about protein structure and how structure details are responsible for specific activities, and also on developing good screening methods for terpene production so that directed evolution can be an option for enzyme engineering in this area (Li et al., 2020).

Another option for improving enzyme activity is the modification of enzymes through, e.g., fusion enzymes. This is a valid technique that can influence enzyme properties such as activity, expression, and stability. The generation of fusion proteins has been used successfully for the increase in production of sesquiterpenes, epi-cedrol and patchoulol (Albertsen et al., 2011; Navale et al., 2019). By fusing two enzymes, the substrates/products are staying within physical proximity of the active site and do not diffuse easily, improving speed. Such a modification can also improve the solubility of otherwise insoluble proteins like cytochrome P450, as well as co-locate enzymes that perform sequential reactions (Renault et al., 2014). In yeast, an enzyme with multiple functions as GPP, FPP, and geraniol synthase was built for the production of geraniol (Zhao et al., 2016a, b). Sometimes fusion tags can be added to the proteins, which enhance their stability or expression. The fusion of specific scaffolds to proteins can have the function of fixating the protein on membranes or organelles, which can be beneficial because the compartmentalization of pathway modules helps alleviate the toxicity of particular compounds formed (Chessher et al., 2015).

6 Central carbon metabolism

As discussed in the previous sessions, not only the MVA and MEP pathways need to be engineered. Also, the prenyl transferases that provide prenyl diphosphate precursors, and the terpene synthases that provide the final products, are subjected to optimization when engineering a strain for terpene production. The central carbon metabolism, which provides the substrates and cofactors for the MVA and MEP pathways, can also be engineered for higher pools of acetyl-CoA (MVA), or pyruvate and glyceraldehyde-3-phosphate (G3P) (MEP) and cofactors like NADPH and ATP. It is a known fact that the availability of precursors in microbes is usually low, thus limiting the production of terpenes (Vickers and Sabri, 2015). This is a limitation that can be overcome by using a species that naturally has higher pools of the desired precursor, as is the case of *Yarrowia lipolytica* with acetyl-CoA (Zhu and Jackson, 2015) or by engineering. For engineering, two main things have to be taken into consideration. The first one is whether there are pathways native to the strain which compete for the same precursors or somehow inhibit the pathway for their production. Another point is the fact that many of the intermediates can be toxic to the cell when accumulated by the engineered strain.

The availability of precursors can be enhanced by downregulating genes that consume them. In the cell, several other pathways are working at the same time, and the precursors for MVA and MEP pathways can be used by them for other physiological activities, also with the formation of byproducts. By downregulating or even deleting these genes that sequester acetyl-CoA or pyruvate and G3P, higher fluxes into the MVA and MEP pathways can be achieved. Another alternative is to overexpress genes that will do exactly the contrary, which means they will redirect the flux to the production of the precursor. As an example, in yeast, the overexpression of particular genes redirected the flux from ethanol and acetate to acetyl-CoA, which increased the production of santalene (Chen et al., 2013).

The cofactor demand can be quenched by modifications to the central carbon metabolism of the host. This is important especially for the MEP, which consumes both NADPH and ATP to produce IPP out of glucose. Improvements in this area can be done by, for example, overexpressing the enzyme NADH kinase, responsible for converting NADH to NADPH, or by downregulating genes that participate in competing pathways, which also consume the cofactors of interest. Another option is to regulate genes involved in the expression of enzymes related to ATP regeneration. As mentioned before, this kind of strategy focused on cofactor supply is critical for engineering the MEP pathway, because of the necessary reducing power and ATP demanded by this pathway specifically. For the MVA pathway, two NADPH molecules are needed for the production of one DMAPP molecule out of acetyl-CoA. However, the central carbon metabolism provides six molecules of NADH when transforming glucose into acetyl-CoA. For that reason, not so much needs to be done regarding ATP or NADPH production, but rather overexpressing enzymes that convert NADH to NADPH would be a valid approach to improve cofactor supply for the MVA pathway. This was done before with positive results (Zhao et al., 2015). Even so, when comparing the two pathways for terpene precursor formation, the MEP requires much more engineering for cofactor generation than the MVA pathway. Overcoming this problem would require rewriting the central carbon metabolism, which was done through a bypass that avoided CO_2 loss and included the generation of NADH instead of NADPH (Meadows et al., 2016).

The natural production of terpenes and terpene precursors within a host is a major driver of competition between pathways. The microbial hosts inherently use the same precursors for supporting growth metabolism, creating an undesired flux toward the basic functions of the cell. To overcome this problem, the central carbon metabolism has to be redirected to terpene precursor production and uptake by the engineered pathway. One example of this approach is the deletion of genes involved in the expression of competing pathways. This was successfully achieved for geraniol production by *S. cerevisiae*, through suppressing the expression of enzymes responsible for degrading the monoterpene of interest (Zhou et al., 2014; Zhao et al., 2017). However, one must always be careful when deleting genes from the central metabolism of an organism, because this might be lethal for the organism. The modification of genes within an organism can bring advantages in one metabolic pathway but may have negative effects on another level. In that case, downregulation might be a less drastic option for genetic manipulation.

The flux of prenyl pyrophosphate production in a cell is usually directed to FPP formation and then squalene formation, and this particular pathway represents a setback for the production of other terpenes. The gene *ERG9* is responsible for the catalysis of FPP into squalene, and this gene can be downregulated to improve the production of a target terpene. When the expression of *ERG9* was decreased in *S. cerevisiae* (by controlling its expression in the post-translational step), the production of aroma sesquiterpene alcohols was improved 1.85 times (Peng et al., 2017). The enzyme ERG20p which catalyzes the formation of FPP was engineered for the production of GGPP, which in turn potentialized sclareol production (Ignea et al., 2015). On the other hand, engineering the degradation of the protein ERG20, which normally catalyzes the conversion of GPP into FPP, increased the production of monoterpenes in yeast (Peng et al., 2017). This kind of protein degradation strategy is a valid tool for the removal of unwanted activities. Another possible way of dealing with the presence of undesired enzymes in a biological production system is by adding a compound that functions as an inhibitor.

7 Toxicity

A very important aspect that needs to be taken into consideration when engineering strains for terpene production, is the toxicity that these types of compounds pose to a living cell. By overexpressing the enzymes that compose terpene bioproduction pathways, unwanted effects such as negative feedback from intermediates, physiological stress and toxic buildup of metabolites can occur (Sivy et al., 2011; Banerjee and Sharkey, 2014; Tholl, 2015). These effects in turn downgrade the overall efficiency of the system. Even high concentrations of IPP and DMAPP within a cell have a negative influence on growth (George et al., 2018; Miguez et al., 2018). The biological production of terpenes in high concentrations by microbial cells creates an environment that can hinder growth and even bring the cells to death. To deal with this problem, downstream processing techniques like in situ product removal can be applied, which was successfully demonstrated for products like zizaene, nootkatone, and pinene (Kang et al., 2014; Wriessnegger et al., 2014; Aguilar et al., 2019). For microbial strain development, cells have to be engineered for resistance against the toxicity of the product. Some microbial species are naturally more resistant to terpenes and are preferred for strain engineering. It is the case of *Pseudomonas putida* (Inoue and Horikoshi, 1989; Mi et al., 2014; Schempp et al., 2020), as well as some strains of *Rhodococcus* and *Bacillus* (Guan et al., 2015; Kim et al., 2018a), which even though are not considered as model organisms like *E. coli* or *S. cerevisiae*, bring some advantages to the system as platforms.

The cells can be engineered for tolerance to terpenes, for example, with the expression of efflux pumps. For overexpression of efflux pump was demonstrated to be an efficient way of increasing tolerance to high-concentration of terpenes in a strain that produced pinene (Niu et al., 2018). The higher tolerance provided by the efflux pumps also increased the production of pinene by the above-mentioned strain. Several aroma terpenes were shown to have toxic effects on cells and had their production increased when efflux pumps were expressed (Li et al., 2020). However, in a recent study, the higher expression of efflux pumps was shown to confer lower tolerance to geranic acid, even though for other compounds it meant higher tolerance (Schempp et al., 2020).

The toxicity of intermediates can also be tackled with the use of dynamic promoters, which regulate specifically the expression of genes according to the presence of certain compounds in the cell. This type of regulation was used exemplarily for santalene production in yeast, and had stress-dependent regulation (Dahl et al., 2013). In an attempt to find which gene conferred the tolerance to terpenes, an interesting approach was employed for the organism *Marinobacter aquaeolei* where its genomic DNA was first digested, forming several fragments, which were individually placed in vectors and transformed into *E. coli*. The cells were then submitted to the same growth conditions under high pinene concentrations and the one containing the gene that conferred resistance was detected (Tomko and Dunlop, 2015).

When engineering a strain specifically to achieve a goal is not possible or desirable, whole-cell evolution can be an option. For this setting, the growth conditions can be tweaked in a way to promote mutagenesis, and the suitable mutants or mutant populations can then be selected by applying environmental pressure sufficient to kill the ones which are not adapted to the conditions. To speed up the process, mutant inducing techniques can be used, such as UV light or specific chemicals, which knowingly prime the cell for mutations. The only problem with the adaptive evolution approach, besides the fact that it can be very time-consuming, is the lack of suitable screening methods to find the desired mutant with improved characteristics to the goal. For this, in-silico analysis of the genome of the chosen host can help on selecting suitable targets for deletion.

Alternatively, to the use of model organisms, other microbial species or unconventional hosts can also be of interest to engineering for use as terpene producers. Many different *Streptomyces* species are associated with terpene production, and they are good candidates for further strain engineering (Köksal et al., 2012; Phelan et al., 2015, 2017). Further microorganisms that are also studied as potential microbial cell factories include *P. putida*, *Cupriavidus necator*, *Methylobacterium extorquens*, *Corynebacterium glutamicum*, *Yarrowia lipolytica* (Kang et al., 2014; Mi et al., 2014; Sonntag et al., 2015; Krieg et al., 2018; Pang et al., 2019) among many others. The advantages that a non-conventional host can bring are very attractive and comprise the use of different carbon sources other than glucose. Alternative carbon sources can offer the convenience of lower costs for the overall production, but more importantly, they can bring a superior level of sustainability to the technology. Industrial residues, such as industrial biomass waste, methane, cooking oil, wastewater, and even CO_2, can be used as carbon sources by microbial cell factories for terpene production (Trinh and Mendoza, 2016; Krieg et al., 2018; Pang et al., 2019; Kalra et al., 2021). The possibility of using residues from human activity to produce valuable products is a combination of bioremediation with sustainable production of goods, and it is so attractive that the use of microbial cell factories will most probably become a tendency in the upcoming decades.

8 Conclusion

The use of microbes for the industrial production of chemicals is a trend that is not limited to terpenes applied for flavor and fragrance industries. The aroma compound industry nonetheless has benefited from the use of microbes in the last years, due to the flourishing research field and an ever-increasing demand for sustainable products. Flavor and fragrance compounds are now accessible out of sustainable and cost-effective biological factories, which do not depend on raw plant material provision from ecologically endangered sources. The technologies described in this review offer alternatives to the traditional approaches that consume resources in an energy-expensive and environmentally unfriendly way. Besides being energy-efficient and environmentally friendly, microbial cell factories are tunable and very flexible platforms, allowing research and development in industries and universities to repurpose achievements accomplished by groups belonging to different lines of investigation. Moreover, the possibility of the creation of novel non-canonical terpenes, with the addition of recently discovered new methylation modules terpene biosynthesis, adds innovative quality to this field of study. The advancements made in the biotechnological use of microbes have the great advantage of being transferrable technologies. Efforts applied to the development of one microbial platform have the potential of helping the advance of several other research projects, which invests in this research area a rentable and advantageous asset for the establishment of a bioeconomy. The driving force for the advancement in this area, the demand for sustainable ingredients for flavor and fragrance industries, is expected to continue growing; and the use of engineered microbes supplies the demand by offering a suitable option for sustainable bioproduction of chemicals.

References

Aguilar, F., Scheper, T., Beutel, S., 2019. Improved production and in situ recovery of sesquiterpene (+)-zizaene from metabolically-engineered E. coli. Molecules 24 (18). https://doi.org/10.3390/molecules24183356.

Albertsen, L., et al., 2011. Diversion of flux toward sesquiterpene production in Saccharomyces cerevisiae by fusion of host and heterologous enzymes. Appl. Environ. Microbiol. 77 (3), 1033–1040. https://doi.org/10.1128/AEM.01361-10.

Banerjee, A., Sharkey, T.D., 2014. Methylerythritol 4-phosphate (MEP) pathway metabolic regulation. Nat. Prod. Rep. 31 (8), 1043–1055. https://doi.org/10.1039/C3NP70124G. The Royal Society of Chemistry.

Bröker, J.N., et al., 2018. Upregulating the mevalonate pathway and repressing sterol synthesis in Saccharomyces cerevisiae enhances the production of triterpenes. Appl. Microbiol. Biotechnol. 102 (16), 6923–6934. https://doi.org/10.1007/s00253-018-9154-7.

Cao, X., et al., 2017. Enhancing linalool production by engineering oleaginous yeast Yarrowia lipolytica. Bioresour. Technol. 245 (Pt B), 1641–1644. https://doi.org/10.1016/j.biortech.2017.06.105. England.

Carter, O.A., Peters, R.J., Croteau, R., 2003. Monoterpene biosynthesis pathway construction in *Escherichia coli*. Phytochemistry 64 (2), 425–433. https://doi.org/10.1016/s0031-9422(03)00204-8. England.

Chen, Y., et al., 2013. Establishing a platform cell factory through engineering of yeast acetyl-CoA metabolism. Metab. Eng. 15, 48–54. https://doi.org/10.1016/j.ymben.2012.11.002. Belgium.

Chessher, A., Breitling, R., Takano, E., 2015. Bacterial microcompartments: biomaterials for synthetic biology-based compartmentalization strategies. ACS Biomater Sci. Eng. 1 (6), 345–351. https://doi.org/10.1021/acsbiomaterials.5b00059. United States.

Chica, R.A., Doucet, N., Pelletier, J.N., 2005. Semi-rational approaches to engineering enzyme activity: combining the benefits of directed evolution and rational design. Curr. Opin. Biotechnol. 16 (4), 378–384. https://doi.org/10.1016/j.copbio.2005.06.004. England.

Christianson, D.W., 2017. Structural and chemical biology of terpenoid cyclases. Chem. Rev. 117 (17), 11570–11648. https://doi.org/10.1021/acs.chemrev.7b00287.

Dahl, R.H., et al., 2013. Engineering dynamic pathway regulation using stress-response promoters. Nat. Biotechnol. 31 (11), 1039–1046. https://doi.org/10.1038/nbt.2689. United States.

Dekker, L., Polizzi, K.M., 2017. Sense and sensitivity in bioprocessing-detecting cellular metabolites with biosensors. Curr. Opin. Chem. Biol. 40, 31–36. https://doi.org/10.1016/j.cbpa.2017.05.014. England.

Denby, C.M., et al., 2018. Industrial brewing yeast engineered for the production of primary flavor determinants in hopped beer. Nat. Commun. 9 (1), 965. https://doi.org/10.1038/s41467-018-03293-x.

Dickschat, J.S., et al., 2007. Biosynthesis of the off-flavor 2-methylisoborneol by the myxobacterium Nannocystis exedens. Angew. Chem. Int. Ed. 46 (43), 8287–8290. https://doi.org/10.1002/anie.200702496.

Drummond, L., et al., 2019. Expanding the isoprenoid building block repertoire with an IPP methyltransferase from Streptomyces monomycini. ACS Synth. Biol. 8 (6), 1303–1313. https://doi.org/10.1021/acssynbio.8b00525.

Farhi, M., et al., 2011. Harnessing yeast subcellular compartments for the production of plant terpenoids. Metab. Eng. 13 (5), 474–481. https://doi.org/10.1016/j.ymben.2011.05.001. Belgium.

Furubayashi, M., et al., 2014. A high-throughput colorimetric screening assay for terpene synthase activity based on substrate consumption. PLoS One 9 (3). https://doi.org/10.1371/journal.pone.0093317.

George, K.W., et al., 2015. Metabolic engineering for the high-yield production of isoprenoid-based C5 alcohols in *E. coli*. Sci. Rep. 5 (June), 1–12. https://doi.org/10.1038/srep11128. Nature Publishing Group.

George, K.W., et al., 2018. Integrated analysis of isopentenyl pyrophosphate (IPP) toxicity in isoprenoid-producing *Escherichia coli*. Metab. Eng. 47, 60–72. https://doi.org/10.1016/j.ymben.2018.03.004. Belgium.

Guan, Z., et al., 2015. Metabolic engineering of *Bacillus subtilis* for terpenoid production. Appl. Microbiol. Biotechnol. 99 (22), 9395–9406. https://doi.org/10.1007/s00253-015-6950-1.

Ignea, C., et al., 2014. Engineering monoterpene production in yeast using a synthetic dominant negative geranyl diphosphate synthase. ACS Synth. Biol. 3 (5), 298–306. https://doi.org/10.1021/sb400115e. United States.

Ignea, C., et al., 2015. Efficient diterpene production in yeast by engineering Erg20p into a geranylgeranyl diphosphate synthase. Metab. Eng. 27, 65–75. https://doi.org/10.1016/j.ymben.2014.10.008. Belgium.

Ignea, C., et al., 2018. Synthesis of 11-carbon terpenoids in yeast using protein and metabolic engineering. Nat. Chem. Biol. 14 (12), 1090–1098. https://doi.org/10.1038/s41589-018-0166-5.

Inoue, A., Horikoshi, K., 1989. A Pseudomonas thrives in high concentrations of toluene. Nature 338 (6212), 264–266. https://doi.org/10.1038/338264a0.

Jia, B., Jeon, C.O., 2016. High-throughput recombinant protein expression in *Escherichia coli*: current status and future perspectives. Open Biol. 6 (8). https://doi.org/10.1098/rsob.160196.

Julliard, J.H., Douce, R., 1991. Biosynthesis of the thiazole moiety of thiamin (vitamin B1) in higher plant chloroplasts. Proc. Natl. Acad. Sci. U. S. A. 88 (6), 2042–2045. https://doi.org/10.1073/pnas.88.6.2042.

Kalra, R., Gaur, S., Goel, M., 2021. Microalgae bioremediation: a perspective towards wastewater treatment along with industrial carotenoids production. J. Water Process. Eng. 40. https://doi.org/10.1016/j.jwpe.2020.101794, 101794.

Kang, M.-K., et al., 2014. Biosynthesis of pinene from glucose using metabolically-engineered *Corynebacterium glutamicum*. Biotechnol. Lett. 36 (10), 2069–2077. https://doi.org/10.1007/s10529-014-1578-2. Netherlands.

Kim, D., et al., 2018a. Biotechnological potential of rhodococcus biodegradative pathways. J. Microbiol. Biotechnol. 28 (7), 1037–1051. https://doi.org/10.4014/jmb.1712.12017. Korea (South).

Kim, S.K., et al., 2018b. A genetically encoded biosensor for monitoring isoprene production in engineered *Escherichia coli*. ACS Synth. Biol. 7 (10), 2379–2390. https://doi.org/10.1021/acssynbio.8b00164. American Chemical Society.

King, J.R., et al., 2016. Accessing nature's diversity through metabolic engineering and synthetic biology. F1000 Research 5. https://doi.org/10.12688/f1000research.7311.1.

Köksal, M., et al., 2012. Structure of geranyl diphosphate C-methyltransferase from *Streptomyces coelicolor* and implications for the mechanism of isoprenoid modification. Biochemistry 51 (14), 3003–3010. https://doi.org/10.1021/bi300109c.

Komatsu, M., et al., 2008. Identification and functional analysis of genes controlling biosynthesis of 2-methylisoborneol. Proc. Natl. Acad. Sci. U. S. A. 105 (21), 7422–7427. https://doi.org/10.1073/pnas.0802312105.

Krieg, T., et al., 2018. CO2 to terpenes: autotrophic and electroautotrophic α-humulene production with *Cupriavidus necator*. Angew. Chem. Int. Ed. 57 (7), 1879–1882. https://doi.org/10.1002/anie.201711302.

Kschowak, M.J., et al., 2018. Heterologous expression of 2-methylisoborneol/2 methylenebornane biosynthesis genes in *Escherichia coli* yields novel C11-terpenes. PLoS One 13 (4). https://doi.org/10.1371/journal.pone.0196082, e0196082.

Kschowak, M.J., et al., 2020. Analyzing and engineering the product selectivity of a 2-methylenebornane synthase. ACS Synth. Biol. 9 (5), 981–986. https://doi.org/10.1021/acssynbio.9b00432. American Chemical Society.

Li, Y., et al., 2015. Metabolic engineering of *Escherichia coli* using CRISPR-Cas9 meditated genome editing. Metab. Eng. 31, 13–21. https://doi.org/10.1016/j.ymben.2015.06.006. Belgium.

Li, M., et al., 2020. Recent advances of metabolic engineering strategies in natural isoprenoid production using cell factories. Nat. Prod. Rep. https://doi.org/10.1039/c9np00016j.

Liu, W., et al., 2016. Engineering Escherichia coli for high-yield geraniol production with biotransformation of geranyl acetate to geraniol under fed-batch culture. Biotechnol. Biofuels 9 (1), 58. https://doi.org/10.1186/s13068-016-0466-5. BioMed central.

Liu, C.L., et al., 2018. Renewable production of high density jet fuel precursor sesquiterpenes from *Escherichia coli*. Biotechnol. Biofuels 11 (1), 1–15. https://doi.org/10.1186/s13068-018-1272-z. BioMed Central.

Ma, T., et al., 2019. Lipid engineering combined with systematic metabolic engineering of Saccharomyces cerevisiae for high-yield production of lycopene. Metab. Eng. 52, 134–142. https://doi.org/10.1016/j.ymben.2018.11.009.

Meadows, A.L., et al., 2016. Rewriting yeast central carbon metabolism for industrial isoprenoid production. Nature 537 (7622), 694–697. https://doi.org/10.1038/nature19769. Nature Publishing Group.

Mendez-Perez, D., et al., 2017. Production of jet fuel precursor monoterpenoids from engineered Escherichia coli. Biotechnol. Bioeng. 114 (8), 1703–1712. https://doi.org/10.1002/bit.26296.

Mi, J., et al., 2014. De novo production of the monoterpenoid geranic acid by metabolically engineered *Pseudomonas putida*. Microb. Cell Fact. 13 (1), 170. https://doi.org/10.1186/s12934-014-0170-8.

Miguez, A.M., McNerney, M.P., Styczynski, M.P., 2018. Metabolomics analysis of the toxic effects of the production of lycopene and its precursors. Front. Microbiol. 9, 760. https://doi.org/10.3389/fmicb.2018.00760.

Navale, G.R., et al., 2019. Enhancing epi-cedrol production in *Escherichia coli* by fusion expression of farnesyl pyrophosphate synthase and epi-cedrol synthase. Eng. Life Sci. 19 (9), 606–616. https://doi.org/10.1002/elsc.201900103.

Niu, F.-X., et al., 2018. Enhancing production of pinene in *Escherichia coli* by using a combination of tolerance, evolution, and modular co-culture engineering. Front. Microbiol. 9, 1623. https://doi.org/10.3389/fmicb.2018.01623.

Pang, Y., et al., 2019. Biotechnol. Biofuels 12 (1), 1–18. https://doi.org/10.1186/s13068-019-1580-y. BioMed Central. Engineering the oleaginous yeast Yarrowia lipolytica to produce limonene from waste cooking oil.

Peng, B., et al., 2017. A squalene synthase protein degradation method for improved sesquiterpene production in *Saccharomyces cerevisiae*. Metab. Eng. 39, 209–219. https://doi.org/10.1016/j.ymben.2016.12.003. Belgium.

Phelan, R.M., et al., 2015. Engineering terpene biosynthesis in Streptomyces for production of the advanced biofuel precursor bisabolene. ACS Synth. Biol. 4 (4), 393–399. https://doi.org/10.1021/sb5002517. United States.

Phelan, R.M., et al., 2017. Development of next generation synthetic biology tools for use in *Streptomyces venezuelae*. ACS Synth. Biol. 6 (1), 159–166. https://doi.org/10.1021/acssynbio.6b00202. American Chemical Society.

Pickens, L.B., Tang, Y., Chooi, Y.-H., 2011. Metabolic engineering for the production of natural products. Annu. Rev. Chem. Biomol. Eng. 2, 211–236. https://doi.org/10.1146/annurev-chembioeng-061010-114209.

Radhika, V., et al., 2015. Methylated cytokinins from the phytopathogen *Rhodococcus fascians* mimic plant hormone activity. Plant Physiol. 169 (2), 1118–1126. https://doi.org/10.1104/pp.15.00787.

Reiling, K.K., et al., 2004. Mono and diterpene production in *Escherichia coli*. Biotechnol. Bioeng. 87 (2), 200–212. https://doi.org/10.1002/bit.20128.

Renault, H., et al., 2014. Cytochrome P450-mediated metabolic engineering: current progress and future challenges. Curr. Opin. Plant Biol. 19, 27–34. https://doi.org/10.1016/j.pbi.2014.03.004.

Scalcinati, G., Knuf, C., et al., 2012a. Dynamic control of gene expression in Saccharomyces cerevisiae engineered for the production of plant sesquitepene α-santalene in a fed-batch mode. Metab. Eng. 14 (2), 91–103. https://doi.org/10.1016/j.ymben.2012.01.007.

Scalcinati, G., Partow, S., et al., 2012b. Combined metabolic engineering of precursor and co-factor supply to increase α-santalene production by Saccharomyces cerevisiae. Microb. Cell Fact. 11 (1), 117. https://doi.org/10.1186/1475-2859-11-117.

Schempp, F.M., et al., 2018. Microbial cell factories for the production of terpenoid flavor and fragrance compounds. J. Agric. Food Chem., 2247–2258. https://doi.org/10.1021/acs.jafc.7b00473.

Schempp, F.M., et al., 2020. Investigation of monoterpenoid resistance mechanisms in *Pseudomonas putida* and their consequences for biotransformations. Appl. Microbiol. Biotechnol. 104 (12), 5519–5533. https://doi.org/10.1007/s00253-020-10566-3.

Schwartz, C., Wheeldon, I., 2018. CRISPR-Cas9-mediated genome editing and transcriptional control in Yarrowia lipolytica. Methods Mol. Biol. 1772, 327–345. https://doi.org/10.1007/978-1-4939-7795-6_18. United States.

Singh, B., Sharma, R.A., 2015. Plant terpenes: defense responses, phylogenetic analysis, regulation and clinical applications. 3 Biotech 5 (2), 129–151. https://doi.org/10.1007/s13205-014-0220-2.

Sivy, T.L., Fall, R., Rosenstiel, T.N., 2011. Evidence of isoprenoid precursor toxicity in Bacillus subtilis. Biosci. Biotechnol. Biochem. 75 (12), 2376–2383. https://doi.org/10.1271/bbb.110572. Taylor & Francis.

Sonntag, F., et al., 2015. Engineering *Methylobacterium extorquens* for de novo synthesis of the sesquiterpenoid α-humulene from methanol. Metab. Eng. 32, 82–94. https://doi.org/10.1016/j.ymben.2015.09.004.

Tholl, D., 2015. Biosynthesis and biological functions of terpenoids in plants. In: Schrader, J., Bohlmann, J. (Eds.), Biotechnology of Isoprenoids. Springer International Publishing, Cham, pp. 63–106, https://doi.org/10.1007/10_2014_295.

Tippmann, S., et al., 2013. From flavors and pharmaceuticals to advanced biofuels: production of isoprenoids in *Saccharomyces cerevisiae*. Biotechnol. J. 8 (12), 1435–1444. https://doi.org/10.1002/biot.201300028. Germany.

Tomko, T.A., Dunlop, M.J., 2015. Engineering improved bio-jet fuel tolerance in *Escherichia coli* using a transgenic library from the hydrocarbon-degrader *Marinobacter aquaeolei*. Biotechnol. Biofuels 8 (1), 165. https://doi.org/10.1186/s13068-015-0347-3.

Trinh, C.T., Mendoza, B., 2016. Modular cell design for rapid, efficient strain engineering toward industrialization of biology. Curr. Opin. Chem. Eng. 14, 18–25. https://doi.org/10.1016/j.coche.2016.07.005. Elsevier Ltd.

Vickers, C.E., Sabri, S., 2015. Isoprene. Adv. Biochem. Eng. Biotechnol. 148, 289–317. https://doi.org/10.1007/10_2014_303. Germany.

Vickers, C.E., et al., 2017. Recent advances in synthetic biology for engineering isoprenoid production in yeast. Curr. Opin. Chem. Biol. https://doi.org/10.1016/j.cbpa.2017.05.017.

Von Reuss, S., et al., 2018. Sodorifen biosynthesis in the Rhizobacterium Serratia plymuthica involves methylation and cyclization of MEP-derived farnesyl pyrophosphate by a SAM-dependent C-methyltransferase. J. Am. Chem. Soc. 140 (37), 11855–11862. https://doi.org/10.1021/jacs.8b08510.

Wang, C.M., Cane, D.E., 2008. Biochemistry and molecular genetics of the biosynthesis of the earthy odorant methylisoborneol in *Streptomyces coelicolor*. J. Am. Chem. Soc. 130 (28), 8908–8909. https://doi.org/10.1021/ja803639g.

Weathers, P.J., et al., 2011. Artemisinin production in Artemisia annua: studies in planta and results of a novel delivery method for treating malaria and other neglected diseases. Phytochem. Rev. 10 (2), 173–183. https://doi.org/10.1007/s11101-010-9166-0.

Westfall, P.J., et al., 2012. Production of amorphadiene in yeast, and its conversion to dihydroartemisinic acid, precursor to the antimalarial agent artemisinin. Proc. Natl. Acad. Sci. U. S. A. 109 (3), E111–E118. https://doi.org/10.1073/pnas.1110740109.

Willrodt, C., et al., 2014. Engineering the productivity of recombinant *Escherichia coli* for limonene formation from glycerol in minimal media. Biotechnol. J. 9 (8), 1000–1012. https://doi.org/10.1002/biot.201400023. Germany.

Wriessnegger, T., et al., 2014. Production of the sesquiterpenoid (+)-nootkatone by metabolic engineering of Pichia pastoris. Metab. Eng. 24, 18–29. https://doi.org/10.1016/j.ymben.2014.04.001. Belgium.

Yang, J., et al., 2013. Metabolic engineering of Escherichia coli for the biosynthesis of alpha-pinene. Biotechnol. Biofuels 6 (1), 60. https://doi.org/10.1186/1754-6834-6-60.

Yang, X., et al., 2016. Heterologous production of α-farnesene in metabolically engineered strains of Yarrowia lipolytica. Bioresour. Technol. 216, 1040–1048. https://doi.org/10.1016/j.biortech.2016.06.028. Elsevier Ltd.

Ye, L., et al., 2016. Combinatory optimization of chromosomal integrated mevalonate pathway for β-carotene production in *Escherichia coli*. Microb. Cell Fact. 15 (1), 202. https://doi.org/10.1186/s12934-016-0607-3. BioMed central.

Zhang, H., et al., 2014. Microbial production of sabinene—a new terpene-based precursor of advanced biofuel. Microb. Cell Fact. 13 (1), 20. https://doi.org/10.1186/1475-2859-13-20.

Zhao, Y., et al., 2011. Biosynthesis of isoprene in Escherichia coli via methylerythritol phosphate (MEP) pathway. Appl. Microbiol. Biotechnol. 90 (6), 1915–1922. https://doi.org/10.1007/s00253-011-3199-1.

Zhao, X., Shi, F., Zhan, W., 2015. Overexpression of ZWF1 and POS5 improves carotenoid biosynthesis in recombinant Saccharomyces cerevisiae. Lett. Appl. Microbiol. 61 (4), 354–360. https://doi.org/10.1111/lam.12463.

Zhao, D.D., et al., 2016a. Chemical components and pharmacological activities of terpene natural products from the genus Paeonia. Molecules 21 (10). https://doi.org/10.3390/molecules21101362.

Zhao, J., et al., 2016b. Improving monoterpene geraniol production through geranyl diphosphate synthesis regulation in *Saccharomyces cerevisiae*. Appl. Microbiol. Biotechnol. 100 (10), 4561–4571. https://doi.org/10.1007/s00253-016-7375-1. Germany.

Zhao, J., et al., 2017. Dynamic control of ERG20 expression combined with minimized endogenous downstream metabolism contributes to the improvement of geraniol production in Saccharomyces cerevisiae. Microb. Cell Fact. 16 (1), 17. https://doi.org/10.1186/s12934-017-0641-9.

Zhou, J., et al., 2014. Engineering Escherichia coli for selective geraniol production with minimized endogenous dehydrogenation. J. Biotechnol. 169 (1), 42–50. https://doi.org/10.1016/j.jbiotec.2013.11.009. Elsevier B.V.

Zhou, K., et al., 2015. Distributing a metabolic pathway among a microbial consortium enhances production of natural products. Nat. Biotechnol. 33 (4), 377–383. https://doi.org/10.1038/nbt.3095.

Zhu, Q., Jackson, E.N., 2015. Metabolic engineering of Yarrowia lipolytica for industrial applications. Curr. Opin. Biotechnol. 36, 65–72. https://doi.org/10.1016/j.copbio.2015.08.010. England.

CHAPTER

13 Microbial mediated remediation of pesticides: A sustainable tool

Mohit Mishra[a], Siddharth Shankar Bhatt[b], and Mian Nabeel Anwar[c]

[a]*Amity University, Raipur, Chhattisgarh, India,*
[b]*Uttaranchal University, Dehradun, Uttarakhand, India,*
[c]*Department of Civil and Environmental Engineering, University of Alberta, Edmonton, AB, Canada*

Abstract

In modern farming system, pesticide application partakes as most common practice in agricultural fields in which 2%–3% of pesticide is utilized and the rest persists in soil and water causing environmental pollution leading to toxicity (WHO, 1990. Diet, Nutrition, and the Prevention of Chronic Diseases, 797 pp). Pesticide residues remain in surface soil, leading to toxicity in the soil-water environment. A vast majority of the Indian population (56.7%) is engaged in agriculture and is, therefore, exposed to the pesticides used in agriculture. Moreover, microbial biodegradation of pesticides has foremost critical importance for modern agriculture and its environmental impact. Microorganisms occupy virtually every habitat on our planet, and their activities largely determine the environmental conditions of today's world. Indeed, they are heavily involved in biogeochemistry, metal precipitation, water purification and sustenance of plant growth that ensuring the recycling of elements such as carbon and nitrogen. In soil, microbes interact with the plant roots, a "hot spot" of microbial activity, with increased microbial numbers, microbial interactions and genetic exchange. In plant root, a narrow zone of soil that surrounds and is influenced by plant roots, known as rhizosphere, is home to an overwhelming number of microorganisms and invertebrates and is considered to be one of the most dynamic interfaces on Earth. The rhizosphere microbiome is dependent on the plant genotype, root exudates, and the environment. Therefore, the study of expressed microbial communities in pesticide-contaminated and uncontaminated rhizosphere is key to the investigation of the diverse roles played by microorganisms in respective niches and to the identification of the microbial genetic potential for biotechnological application in bioremediation of pesticide, including but not limited to: pharmaceutical, diagnostics, waste treatment, and renewable energy generation.

Keywords: Pesticide, Bioremediation, Rhizosphere, Enzymes, Bioengineering

1 Introduction

Pesticides, as a product of the progress of human civilization, have contributed to solving human food and clothing, enhancing social stability, and promoting social development, and have played a positive role in human health. Especially in the 1930s and 1940s, the successful discovery and production of organic pesticides provided effective means for controlling pest damage (Abubakar et al., 2020; Carvalho, 2017). The broad definition of pesticides refers to chemical synthesis used to prevent, eliminate or control diseases, insects, grasses, and other harmful organisms that harm agriculture and forestry, and to purposefully regulate the growth of plants and insects or derived from biological or other

Microbial Syntrophy-mediated Eco-enterprising. https://doi.org/10.1016/B978-0-323-99900-7.00003-1

natural substances. A substance or a mixture of several substances and their preparations. It refers to a class of drugs used to kill insects, sterilize, and kill harmful animals (or weeds) to protect and promote the growth of plants and crops in agricultural production (Yang et al., 2020). In particular, it is used in agriculture to control pests and regulate plant growth and weeding (Mahmood et al., 2016; Diwakar et al., 2008; Edwards, 1993; Madhun and Freed, 1990) (Table 1).

Depending on the chemical structure the most popular pesticides may be divided into the following groups:

1. Organochlorines (endosulfan, hexachlorobenzene)
2. Organophosphates (diazinon, omethoate, glyphosate)
3. Carbamic and thiocarbamic derivatives
4. Carboxylic acids and their derivatives
5. Urea derivatives
6. Heterocyclic compounds (benzimidazole, triazole derivatives, etc.)
7. Phenol and nitrophenol derivatives
8. Hydrocarbons, ketones, aldehydes, and their derivatives
9. Fluorine-containing compounds
10. Copper-containing compounds

Table 1 List of microbial enzymes involved in pesticide degradation.

Pesticides	Microorganisms	Enzymes	References
1. Organochlorates or chlorinated hydrocarbons Aldrin DDT Alachor Dieldrin 1.4-Dichlorubenzene Endosulfan Heptachlor Methoxychlor Pentachlorontrobenzene BIS Pentachlorophenol (PCP)	*Pseudomonas* sp. *strain ADP* *Ancylobacter* sp. *515* *Agrobacterium* sp. *CZBSAI* *Pseudomonas* sp. *Bacillus* sp. *Micrococcus* sp. *Enterobacter aerogenes* *Enterobacter cloacae* *Klebsiella meumonia* *Bacillus* sp. *Pseudomonas putida* *E. coli* *Hydrogenomonas* sp. *Pseudomonas* sp. *Pseudomonas* sp. *Bacillus* sp. *Flawbacterizane* sp. *Clostridium* sp. *Bosea thucdants* *Sphingomonas paucimobilis* *E. aerogenes* *Cuprirvides* sp. strain	Dehalogenases	Katz et al. (2000) Ewida (2014) Patil et al. (1970) Sharma et al. (2016) Dimitrios et al. (2005) Sethunathan and Yoshida (1973) Fogel et al. (1982) Teng et al. (2017)

Table 1 List of microbial enzymes involved in pesticide degradation—cont'd

Pesticides	Microorganisms	Enzymes	References
2. Organo-phosphate Cadufos Chlorpyrifos Diazinon Dimethoate Ethoprophos Glyphosphate, acephate Malathion Monocrotophos Tetrachlorvinphos	*P. putida* *Flavobacterium* sp. *Achromobacter xylosoxidans (JCp4)* *Ochrobactrum* sp. *(FCpl)* *Pseudomonas capaciea* *Bacillus cereus subtilis*, *Bacillus safensis* *Sphingomonas paucimoblis* *Clostridium* sp. *Arthrobacter* sp. *Pseudomonas aeruginosa AA112* *Rhodococcus* sp. *Stenotrophomonas* Ortiz-Hernández malihophilia, Proteus and *Sánchezvulgaris, Vibrio metschinkouii*, *Serratia ficaria*, *Serratia* sp. *Yersinia enterocolitica*	Organophosphorus hydrolase (OPH), organophosphorus acid anhydrolase (OPAA), Laccase Aspergillus enzyme (A-OPH) Penicillium enzyme (P-OPH)	Dimitrios et al. (2005) Akbar and Sultan (2016) Tiwari et al. (2019) Ishag et al. (2016) Dimitrios et al. (2005) Tiwari et al. (2019) Abo-Amer (2007) Tiwari et al. (2019)
3. Carbamates Aldicarb Carbayl (1-naphthalenyl methyl carbamate) Carbofuran	Arthrobacter sp. *Rhodococcus* sp. *Pseudomonas* sp. *Achromobacter* sp. and *Arthrobacter* sp. *Xanthomonas* sp. and *Pseudomonas cepacia* *Achromobacter* sp. *Pseudomonas* sp. *Flavobacterium* sp. *Pseudomonas* sp. *Flavobacterium* sp. *Achromobacterium*, *Sphingomonas* sp. *Arthrobacter* sp.	Carbofuran hydrolase	Behki and Khan (1994) Chapalamadugu and Chaudhry (1994) Gunasekara et al. (2008) Chaudhry et al. (1988) Sharma et al. (2014a)
Pyrithroid	*Serratia*, *Pseudomonas*, *Aspergillus niger*	Carboxyl esterase, pyrethroid hydrolase phosphotriesterase	Guo-liang et al. (2005)

11. Metal organic and inorganic compounds
12. Natural and synthetic pyrethroids and others.

WHO recommended classification of "Pesticides by Hazard" is shown in Table 2 and revised globally harmonized system (GHS) classification of pesticide is shown in Table 3.

The harm of pesticides to the human body is mainly manifested as acute toxicity and chronic toxicity. Pesticides enter the human body in large quantities through the mouth, inhalation and exhalation

Table 2 WHO recommended classification of pesticides.

WHO class		LD_{50} for rats (mg/kg body wt.) Oral	Dermal	Examples
I_a	Extremely hazardous	<5	<50	Parathion, Dieldrin, Phorate
I_b	Highly hazardous	5–50	50–200	Aldrin, Dichlorvos
II	Moderately hazardous	50–2000	200–2000	DDT, Chlordane
III	Slightly hazardous	Over 2000	Over 2000	Malathion
U	Unlikely to present acute hazard	5000 or higher		Carbetamide, Cycloprothrin

Table 3 GHS Classification of pesticides.

	Classification criteria			
	Oral		Dermal	
GHS category	LD_{50} (mg/kg bw)	Hazard statement	LD_{50} (mg/kg bw)	Hazard statement
Category 1	<5	Fatal if swallowed	<50	Fatal in contact with skin
Category 2	5–50	Fatal if swallowed	50–200	Fatal in contact with skin
Category 3	50–300	Toxic if swallowed	200–1000	Toxic in contact with skin
Category 4	300–2000	Harmful if swallowed	1000–2000	Harmful in contact with skin
Category 5	2000–5000	May be harmful	2000–5000	May be harmful

tract or contact, and the acute pathological reaction shown in a short time is acute poisoning (Singh et al., 2017). Acute poisoning often leads to nerve paralysis and even death, and even causes large-scale deaths, becoming the most obvious pesticide hazard. According to reports from the World Health Organization and the United Nations Environment Program, more than 3 million people worldwide are poisoned by pesticides each year, and 20 million of them die. Today, due to the wide application of pesticides in various aspects, it is impossible for any person living in modern life to avoid daily exposure to various pesticides at very low concentrations, either through food or through drinking water. The resulting possible harm to human health is continuous low-level exposure, which is a potential chronic toxic effect. (Kovach et al., 1992; van der Werf and Hayo, 1996).

However, from the current stage, the use of pesticides is inevitable. For human beings to survive more healthily and safely, to understand, avoid, slow down and solve this increasingly serious problem, it is necessary and necessary to explore and study the environmental pollution mechanism of pesticides.

2 Bioremediation technology of pesticide-contaminated soil

At present, pesticides are seriously polluting the soil all over the world, which not only affects the growth and development of plants but also affects human health through the food chain. It has become a serious problem that restricts agricultural products and food safety and sustainable agricultural development (Singh et al., 2017). Countries around the world have invested a lot of manpower and material resources. Research and develop technologies for remediation of pesticide soil pollution.

At present, there are many methods for remediation of pesticide-contaminated soil, and the establishment of ecological rapid environmental remediation technology is the fundamental solution to this problem. Bioremediation technology has become the most active field of soil environmental protection technology due to its obvious advantages of low consumption, high efficiency, and environmentally safe pure ecological process (Minsheng and Xin, 2004; Gavrilescu, 2005).

The research of bioremediation technology began in the mid-1980s, and there have been successful applications in the 1990s. In a broad sense, contaminated soil bioremediation technology refers to the use of various organisms in the soil (plants, animals, and microorganisms alone or in combination) to absorb, degrade and transform pollutants in the soil to reduce the content of pollutants in the soil to an acceptable level or the process of converting toxic and harmful pollutants into harmless substances. Under this concept, soil bioremediation technology can be divided into three types: phytoremediation, animal remediation, and microbial remediation. Narrowly contaminated soil bioremediation refers to microbial remediation technology, which uses soil microorganisms to use organic pollutants as carbon sources and energy sources to degrade harmful organic pollutants in the soil into harmless inorganic substances (CO and HO) or another harmless material process (Gavrilescu, 2005; Uqab et al., 2016; Niti et al., 2013; Alkorta and Garbisu, 2001).

Microbial remediation technology is the use of microbial life metabolism activities to degrade organic pesticides to restore contaminated soil to a healthy state (Singh et al., 2020). The microorganisms used mainly include indigenous microorganisms, foreign microorganisms, and genetically engineered bacteria. Microbial remediation technology can be divided into in situ remediation, on-site remediation, and ex-situ remediation. In-situ remediation is not only simple in operation and low in cost but also does not damage the soil environment required for plant growth. The pollutant oxidation is safe, and there is no secondary pollution. Good effect, it is an environmentally friendly technology that is efficient, economical and ecologically sustainable (Odukkathil and Vasudevan, 2013).

There are two ways to degrade pesticides by microorganisms: one is that microorganisms directly act on pesticides, using pesticide components as the only carbon source or nitrogen source, and phosphorus source to degrade pesticides through enzymatic reactions. The microorganisms isolated from Pakistani soil such as Fulthorpe can mineralize 2,4-D, and found that adding nitrate, potassium ion, and phosphate can increase the degradation rate (Fulthorpe et al., 1996).

Canada's Stauffer Management company has developed some pesticide-contaminated soil bioremediation technologies for several years. They achieve the purpose of remediation by stimulating the function of degradable indigenous microbial communities in specific environments (Gray et al., 1999). The application of inorganic nitrogen fertilizer and phosphate fertilizer significantly promoted the digestion of atrazine. The digestion rate of atrazine in different treatments was as follows: combined application of nitrogen and phosphorus fertilizers > single application of nitrogen fertilizers > single application of phosphate fertilizers > no fertilizer treatment (Tao et al., 2019). The other is to co-metabolize pesticides with other organic matter. Microbial remediation is different from phytoremediation. Usually, a single microorganism can degrade a variety of pesticides, such as Pseudomonas can degrade DDT, aldrin, toxaphene, and dichlorvos. From 1993 to 1995, Spadaro conducted field trials on the bioremediation of 2,4-D in the soil in Poland. After 7 months of adding anaerobically digested sludge in an anaerobic environment, the 2,4-D in the soil increased from 1 to 100 mg/kg was reduced to 18 mg/kg, and the feasibility of bioremediation was confirmed in large-scale trials (Struthers et al., 1998; Spadaro et al., 1998). In addition, microorganisms can also reduce the effectiveness of pesticides by changing the physical and chemical characteristics of the soil, thereby indirectly playing a role in the remediation of contaminated soil.

Nowadays, bioremediation of pesticide pollution through microbes has entered the genetic level, and the ability of microorganisms to degrade pesticides is improved through genetic recombination and the construction of genetically engineered bacteria. The current research on microbial remediation technology is quite mature. Researchers from all over the world have isolated and screened a large number of degradable microorganisms. (Spadaro et al., 1998; Ortiz-Hernández and Laura, 2013).

3 Metabolic pathways of microbial degradation of pesticides

Microbial pesticide degradation can be divided into enzymatic degradation and non-enzymatic degradation (Struthers et al., 1998; Spadaro et al., 1998). The enzymatic degradation effects are as follows:

(1) Microorganisms use pesticides or certain parts of their molecules as energy and carbon sources, and some microorganisms can use certain pesticides as their sole carbon or nitrogen source. Some can be used immediately by microorganisms, while others cannot be used immediately. Special enzymatic hydrolysis is required before the pesticides can be degraded (Singh, 2008; Fragoeiro and Magan, 2005). (2) Microorganisms degrade pesticides through co-metabolism. Many studies have shown that due to the complex structure of certain chemical pesticides, a single microorganism cannot degrade it, and it needs to be metabolized and degraded by two or more microorganisms (Bollag, 1991; Sharma et al., 2014b; Wang et al., 2010). This field is a hot spot of current research. (3) Detoxification and metabolism. Microorganisms do not obtain nutrients or energy from pesticides but develop detoxification to protect their own survival. Non-enzymatic degradation: Microbial activity changes the pH and causes degradation of pesticides, or produces some auxiliary cofactors or chemical substances participate in the transformation of pesticides, such as dehalogenation, de-hydrocarbonization, hydrolysis of amines and esters, reduction, ring cleavage, etc. (Horvath, 1972; Munnecke et al., 2018; Koushik et al., 2016; Alexander, 1999). The aerobic/anaerobic biodegradation pathways of many refractory pesticides have been clarified. The Biodegradation and Biocatalyst Database of the University of Minnesota in the United States collected 139 metabolic pathways, 910 reactions, and 577 kinds of pesticides and other compounds. Enzymes, 328 microbial entries, 247 biotransformation rules, 50 organic functional groups, which include many pesticides degradation and metabolism pathways and enzymes, such as parathion, atrazine, 2,4-D,4-the metabolic pathways and degradation mechanisms of nitrophenol, tetrahydrofuran, S-triazine, DDT and other pesticides have been listed in detail (Mohn and Tiedje, 1992; Gao et al., 2010).

4 Influencing factors of microbial remediation of pesticide

The nature of the pesticide itself, especially the internal chemical bond, concentration, water solubility, molecular polarity, bioavailability, compound adsorption (Ellis et al., 2001) and environmental factors (temperature, salinity, pH, soil type, redox potential, nutrition substances) (Rieger et al., 2002; Arbeli and Fuentes, 2007) are the main factors affecting the biodegradation and restoration of pesticides. Whether microorganisms can restore environmental pollutants ultimately depends not only on their degradability itself, but also on other factors such as the bioavailability of the pollutants and the ability of bacteria to compete with indigenous microorganisms. Increasing the solubility and bioavailability of pollutants is a necessary condition for successful restoration by biological methods (Aislabie and Lloyd-Jones, 1995; Karthikeyan et al., 2004). The degradation efficiency of pesticides in the soil is also closely related to the activity of microorganisms in the soil, and the activity of microorganisms in the soil is affected by many factors, such as pesticide concentration, soil physical and chemical properties, organic matter types and content, microbial flora composition, etc. (Nam and Kim, 2002; Luthy et al., 1997; Zacharia, 2011).

5 Plant rhizosphere microdomains are an important place to degrade organic pollutants

In 1904, Lorenz Hiltner proposed the concept of the rhizosphere. The rhizosphere is a micro-area of the root-soil interface affected by plant roots, and it is also a place where plants-soil-microbes interact with their environmental conditions (Arias-Estévez et al., 2008). In this micro-region, a large number of plant root secretions are concentrated, including high-molecular-weight secretions and low-molecular-weight secretions. The former mainly includes viscose and extracellular enzymes, and the latter mainly consists of low-molecular organic acids, sugars, phenols, and various amino acids (Singh et al., 2019). The possible mechanism for the rapid degradation of organic pollutants in the rhizosphere microdomains is the catalytic degradation of enzymes released from the roots and the degradation of rhizosphere microorganisms (Kah et al., 2007).

6 Enzymes released from roots can catalyze the degradation of organic pollutants

The enzymes released by plant roots into the soil can directly degrade related compounds, sometimes very fast, making the desorption and mass transfer of organic pollutants from the soil a rate-limiting step. After plant death, enzymes released into the environment can continue to play a decomposing role (Yang et al., 2019). The degradation of organic pollutants by plant-specific enzymes provides strong evidence for the potential of phytoremediation. The EPA laboratory in Athens, Georgia, USA, identified five enzymes from freshwater sediments: dehalogenase, nitric acid Reductase, peroxidase, laccase, and nitrilase, these enzymes all come from plants. Nitrate reductase and laccase can decompose explosive waste (TNT) and combine the broken ring structure into plant material or organic residues to become part of the deposited organic matter. The plant-derived dehalogenase can reduce the chlorine-containing organic solvent trichloroethylene to chlorine ions, carbon dioxide, and water. Although these isolated enzymes can degrade TNT and other organic pollutants, experience has shown that in vitro enzymes have high environmental requirements. Unsuitable acidity, high metal concentrations, or bacterial toxins can all inactivate or destroy enzymes. However, enzymes can be protected in plant tissues or near the root zone, and after being released into the soil, they can maintain their degradation activity for several days (Anderson et al., 1993; Anderson and Coats, 1995). Therefore, phytoremediation depends on the entire plant body to achieve (Fig. 1).

Microbial degradation of pesticides is an important field of environmental restoration science and technology. It is the process by which pesticides is transformed into environmentally compatible substance. The physical and chemical forces are acting upon the pesticides but microorganisms play important role in the degradation of pesticides and convert it into simpler non-toxic compounds. During degradation the carbon dioxide and water are formed by the oxidation of the parent compound and by this process energy is produced which help in the metabolism of the microbes the intracellular and extracellular enzyme of the microbes play a major role in the degradation of the chemical compounds.

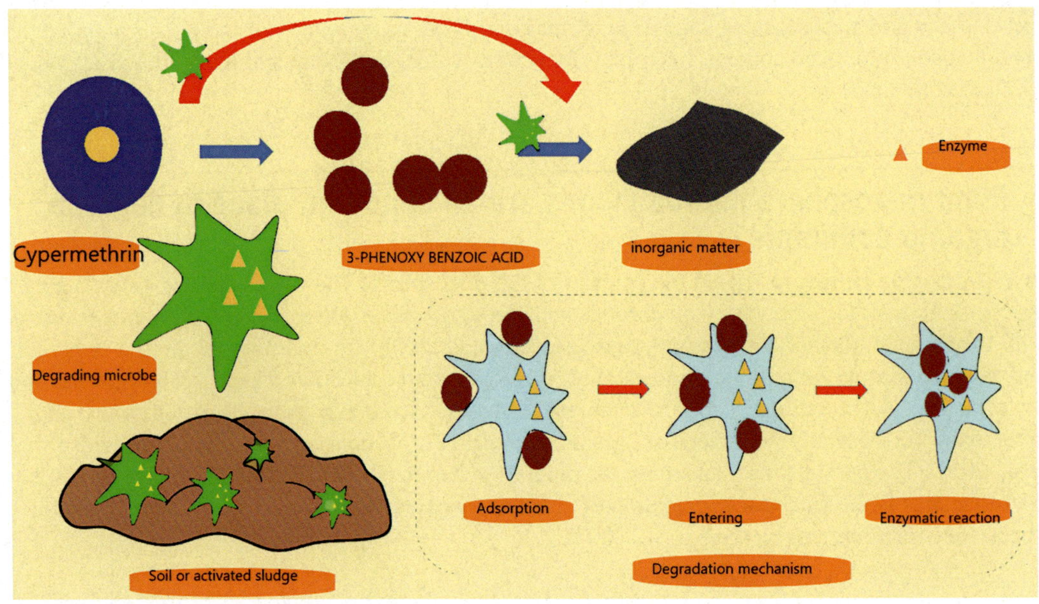

FIG. 1

Pesticide degradation mechanism.

7 Pesticide degradation via microbial associated plant rhizosphere

In the rhizosphere microdomains of plants, root exudates and decomposition products provide nutrients for the reproduction of microorganisms, so that there are a large number of microorganisms near the roots, which promotes the degradation of aliphatic hydrocarbons, polycyclic aromatic hydrocarbons, and pesticides in the rhizosphere microdomains. For example, the research results of Arthur et al. showed that the half-life of atrazine in plant root zone soil was about 75% shorter than that in non-plant control soil, and the number of atrazine degrading bacteria in root zone soil was higher than that in the control soil. The number is nine times more (Esteve-Núñez et al., 2001). Nichols et al. confirmed that the number of microbial communities in the root zone of the same plant is more in contaminated soil than in uncontaminated soil (Schnoor et al., 1995). The mineralization rate of several surfactants in rhizosphere soil is 1.4–1.9 times faster than that of non-rhizosphere soil (Arthur et al., 2000). These studies fully show that the continuous reduction of organic pollutants in the rhizosphere is caused by microbial activities (Nichols et al., 1997; Knaebel et al., 1992; Diez et al., 2017; Nawaz et al., 2011). However, when Fang et al. studied phytoremediation of herbicide pollution, they found that the planting of some plants under herbicide stress did not affect the number of degrading bacteria in the root zone soil, and the mineralization rate of atrazine in the root zone soil was even higher than that of unplanted plants. It may be that the exfoliated roots of the experimental plants do not contain substances that can promote the growth of degrading bacteria, nor can they induce the possible degradation pathways of microorganisms (Sun et al., 2004).

8 Several issues that need further study in bioremediation

The use of organic pollutants caused by microorganisms and phytoremediation pesticides to pollute the soil environment is currently the most promising organism for development and application. Remediation technology, but with the continuous expansion of the scope and connotation of bioremediation, especially for complex contaminated soil ecosystems, each remediation technology must not only overcome its own original shortcomings but also need to further understand and solve the problems in the process of remediation. New phenomena and new problems, such as the discovery of new types of pollutants, the soil remediation process of pollutants and ecological/health risks, the composite mechanism of remediation technologies, and efficient applications.

Therefore, the bioremediation of contaminated soil still faces great challenges and the task is very arduous. Given this, future biological restoration should pay special attention to the following aspects:

1. Remediation process of contaminated soil and ecological risk assessment, strengthen the understanding of the chemical/biological process of microbial metabolism in the remediation of contaminated soil, such as the key links of bioremediation, intermediate processes, enzymology, and co-metabolism mechanisms, rhizosphere effects, etc. Intermediate products, structure, properties and characterization of pollutants in soil, pay close attention to secondary pollution in the process of bioremediation; construct ecotoxicological and biological evaluation methods for the process of bioremediation of contaminated soil, such as ecotoxicological diagnostic indicators and diagnosis mechanism and ecological toxicology standard system; Establish corresponding laws and regulations for bioremediation technology to ensure the public recognition and promotion of remediation technology.
2. Due to the micro-regional, dynamic and complex characteristics of the rhizosphere environment, there are still some difficulties in the research on the rhizosphere stress and rhizosphere repair of organic pollutants. For the dynamics of the rhizosphere under the stress of organic pollutants. The regulation process, especially the root exudates, the leading factor that affects the rhizosphere environment, and the mechanism of the most active biological phase in the soil environment, microbes, in the organic matter rhizosphere pollution ecosystem is currently lacking a systematic understanding. The research work in this area needs to be further strengthened.
3. Carry out research on the screening technology of plant-microorganism complexes to efficiently repair the soil environment contaminated by organic pollutants. This research area includes the screening indicators of plants enriched in specific organic pollutants, the establishment of methods, and the use of molecular biology techniques to study the microbial diversity of plant rhizosphere micro-domains from the genetic level, such as the application of RAPD DNA molecular biology detection technology, as well as ARDRA (amplified rDNA restriction analysis) and PCR-DGGE technology based on 16s rDNA to accurately discover microorganisms that degrade specific organic pollutants.
4. Use molecular biology technology and genetic engineering technology to locate and clone genes that control the secretion of specific secretions from plants to create efficient bioremediation plants. In the long run, it is entirely possible to apply molecular biology and genetic engineering technology to construct efficient and safe genetically engineered bacteria or genetically modified plants and to use plant-microbe joint remediation technology to achieve the goal of remediation of contaminated soil.

References

Abo-Amer, A., 2007. Involvement of chromosomally-encoded genes in malathion utilization by Pseudomonas aeruginosa AA112. Acta Microbiol. Immunol. Hung. 54 (3), 261–277.

Abubakar, Y., Tijjani, H., Egbuna, C., Adetunji, C.O., Kala, S., Kryeziu, T.L., Patrick-Iwuanyanwu, K.C., 2020. Pesticides, history, and classification. In: Natural Remedies for Pest, Disease and Weed Control. Academic Press, pp. 29–42.

Aislabie, J., Lloyd-Jones, G., 1995. A review of bacterial-degradation of pesticides. Soil Res. 33 (6), 925–942.

Akbar, S., Sultan, S., 2016. Soil bacteria showing a potential of chlorpyrifos degradation and plant growth enhancement. Braz. J. Microbiol. 47 (3), 563–570.

Alexander, M., 1999. Biodegradation and Bioremediation. Gulf Professional Publishing.

Alkorta, I., Garbisu, C., 2001. Phytoremediation of organic contaminants in soils. Bioresour. Technol. 79, 273–276.

Anderson, T.A., Coats, J.R., 1995. An overview of microbial degradation in the rhizosphere and its implications for bioremediation. Bioremediation Sci. Appl. 43, 135–143.

Anderson, T.A., Guthrie, E.A., Walton, B.T., 1993. Bioremediation in the rhizosphere. Environ. Sci. Technol. 27 (13), 2630–2636.

Arbeli, Z., Fuentes, C.L., 2007. Accelerated biodegradation of pesticides: an overview of the phenomenon, its basis and possible solutions; and a discussion on the tropical dimension. Crop Prot. 26 (12), 1733–1746.

Arias-Estévez, M., et al., 2008. The mobility and degradation of pesticides in soils and the pollution of groundwater resources. Agric. Ecosyst. Environ. 123 (4), 247–260.

Arthur, E.L., Perkovich, B.S., Anderson, T.A., 2000. Degradation of an atrazine and metolachlor herbicide mixture in pesticide-contaminated soils from two agrochemical dealerships in LOWA. Water Air Soil Pollut. 119, 75–90.

Behki, R.M., Khan, S.U., 1994. Degradation of atrazine, propazine, and simazine by Rhodococcus strain B-30. J. Agric. Food Chem. 42 (5), 1237–1241.

Bollag, J.-M., 1991. Enzymatic Binding of Pesticide Degradation Products to Soil Organic Matter and Their Possible Release. ACS Symposium Series, pp. 122–132.

Carvalho, F.P., 2017. Pesticides, environment, and food safety. Food Energy Secur. 6 (2), 48–60.

Chapalamadugu, S., Chaudhry, G.R., 1994. Hydrolysis of carbaryl by a Pseudomonas sp. and construction of a microbial consortium that completely metabolizes carbaryl. Appl. Environ. Microbiol. 57 (3), 56–60.

Chaudhry, G.R., Ali, A.N., Wheeler, W.B., 1988. Isolation of a methyl parathion-degrading Pseudomonas sp. that possesses DNA homologous to the opd gene from a *Flavobacterium* sp. Appl. Environ. Microbiol. 2, 79–85.

Diez, M.C., et al., 2017. Pesticide dissipation and microbial community changes in a biopurification system: influence of the rhizosphere. Biodegradation 28 (5), 395–412.

Dimitrios, G.K., Fotopoulou, A., Spiroudi, U.M., Singh, B.K., 2005. Non-specific biodegradation of the organophosphorus pesticides, cadusafos and ethoprophos, by two bacterial isolates. FEMS Microbiol. Ecol. 53 (3), 369–378.

Diwakar, J., et al., 2008. Study on major pesticides and fertilizers used in Nepal. Sci. World 6 (6), 76–80.

Edwards, C.A., 1993. The impact of pesticides on the environment. In: The Pesticide Questions. Springer, Boston, MA, pp. 13–46.

Ellis, L.B.M., et al., 2001. The University of Minnesota biocatalysis/biodegradation database: emphasizing enzymes. Nucleic Acids Res. 29 (1), 340–343.

Esteve-Núñez, A., Caballero, A., Ramos, J.L., 2001. Biological degradation of 2, 4, 6-trinitrotoluene. Microbiol. Mol. Biol. Rev. 65 (3), 335–352.

Ewida, 2014. AYI biodegradation of alachlor and endosulphan using environmental bacterial strains. World Applied Science Journal 32 (4), 540–547.

Fogel, S.L., Sewall, R.L., AE., 1982. Enhanced biodegradation of methoxychlor in soil under sequential environmental conditions. Appl. Environ. Microbiol. 44, 113–120.

Fragoeiro, S., Magan, N., 2005. Enzymatic activity, osmotic stress and degradation of pesticide mixtures in soil extract liquid broth inoculated with Phanerochaete chrysosporium and Trametes versicolor. Environ. Microbiol. 7 (3), 348–355.

Fulthorpe, R.R., Rhodes, A.N., Tiedje, J.M., 1996. Pristine soils mineralize 3-chlorbenzoate and 2, 4-dichlorophenoxyacetate via different microbial populations. Appl. Environ. Microbiol. 62 (4), 1159–1166.

Gao, J., Ellis, L.B.M., Wackett, L.P., 2010. The University of Minnesota biocatalysis/biodegradation database: improving public access. Nucleic Acids Res. 38, D488–D491.

Gavrilescu, M., 2005. Fate of pesticides in the environment and its bioremediation. Eng. Life Sci. 5 (6), 497–526.

Gray, N.C.C., Cline, P.R., Moser, G.P., et al., 1999. Full-scale bioremediation of chlorinated pesticides. In: Leeson, A., Alleman, B.C. (Eds.), Fifth International. In Situ and On-Site Bioremediation Symposium, pp. 125–130.

Gunasekara, A.S.R., Goh, A.L., Spurlock, K.S., FC. and Tjeerdema, RS., 2008. Environmental fate and toxicology of carbaryl. Rev. Environ. Contam. Toxicol. 196, 95–121.

Guo-liang, Z., Yue-ting, W., Xin-ping, Q., 2005. Biodegradation of crude oil by *Pseudomonas aeruginosa* in the presence of rhamnolipids. J. Zhejiang Univ. Sci. 6, 725–730.

Horvath, R.S., 1972. Microbial co-metabolism and the degradation of organic compounds in nature. Bacteriol. Rev. 36 (2), 146.

Ishag, A., Abdelbagi, E.O., Hammad, A.M., Elsheikh, E.A., Elsaid, O.E., Hur, J.H., Laing, M.D., 2016. Biodegradation of chlorpyrifos, malathion, and dimethoate by three strains of bacteria isolated from pesticide-polluted soils in Sudan. J. Agric. Food Chem. 64 (45), 8491–8498.

Kah, M., Beulke, S., Brown, C.D., 2007. Factors influencing degradation of pesticides in soil. J. Agric. Food Chem. 55 (11), 4487–4492.

Karthikeyan, R., Davis, L.C., Erickson, L.E., Al-Khatib, K., Kulakow, P.A., Barnes, P.L., Hutchinson, S.L., Nurzhanova, A.A., 2004. Potential for plant-based remediation of pesticide-contaminated soil and water using nontarget plants such as trees, shrubs, and grasses. Crit. Rev. Plant Sci. 23 (1), 91–101.

Katz, I., Green, M., Dosoretz, C., 2000. Characterization of atrazine degradation and nitrate reduction by Pseudomonas sp. strain ADP. Adv. Environ. Res. 4, 219–224.

Knaebel, D.B., Vestal, J.R., Can, J., 1992. Surfactants degradation in rhizosphere soil. Microbiology 38, 643–653.

Koushik, D., et al., 2016. Rapid dehalogenation of pesticides and organics at the interface of reduced graphene oxide–silver nanocomposite. J. Hazard. Mater. 308, 192–198.

Kovach, J., Petzoldt, C., Degni, J., Tette, J., 1992. A method to measure the environmental impact of pesticides. N. Y. Food Life Sci. Bull. 139, 1–8.

Luthy, R.G., Aiken, G.R., Brusseau, M.L., et al., 1997. Sequestration of hydrophobic organi-contaminants by geosorbents. Environ. Sci. Technol. 31, 3341–3347.

Madhun, Y.A., Freed, V.H., 1990. Impact of pesticides on the environment. In: Pesticides in the Soil Environment: Processes Impacts and Modeling. vol. 2. SSSA Book Series, pp. 429–466.

Mahmood, I., Imadi, S.R., Shazadi, K., Gul, A., Hakeem, K.R., 2016. Effects of pesticides on environment. In: Plant, Soil and Microbes. Springer, Cham, pp. 253–269.

Minsheng, Y., Xin, L., 2004. Biodegradation and bioremediation of pesticide pollution. J. Ecol. 23 (1), 73–77.

Mohn, W.W., Tiedje, J.M., 1992. Microbial reductive dehalogenation. Microbiol. Rev. 56 (3), 482–507.

Munnecke, D.M., et al., 2018. Microbial metabolism and enzymology of selected pesticides. In: Biodegradation and Detoxification of Environmental Pollutants. CRC Press, pp. 1–32.

Nam, K., Kim, J.Y., 2002. Persistence and bioavailability of hydrophobic organic compounds in the environment. Geosci. J. 6 (1), 13–21.

Nawaz, K., et al., 2011. Eco-friendly role of biodegradation against agricultural pesticides hazards. Afr. J. Microbiol. Res. 5 (3), 177–183.

Nichols, T.D., Wolf, D.C., Rogers, H.B., et al., 1997. Rhizosphere microbial populations in contaminated soils. Water Air Soil Pollut. 95, 165–178.

Niti, C., Sunita, S., Kamlesh, K., Rakesh, K., 2013. Bioremediation: an emerging technology for remediation of pesticides. Res. J. Chem. Environ 17, 4.

Odukkathil, G., Vasudevan, N., 2013. Toxicity and bioremediation of pesticides in agricultural soil. Rev. Environ. Sci. Biotechnol. 12 (4), 421–444.

Ortiz-Hernández, Laura, M., et al., 2013. Pesticide biodegradation: mechanisms, genetics and strategies to enhance the process. Biodegrad. Life Sci. 1, 251–287.

Patil, K.C., Matsumura, F., Boush, G.M., 1970. Degradation of endrin, aldrin, and DDT by soil microorganisms. Appl. Microbiol. 19 (5), 879–881.

Rieger, P.G., Meier, H.M., Gerle, M., et al., 2002. Xenobiotics in the environment: present and future strategies to obviate the problem of biological persistence. J. Biotechnol. 94, 101–123.

Schnoor, J.L., Licht, L.A., Mccultcheon, S.C., 1995. Phytoremediation of organic and nutrient contaminants. Environ. Sci. Technol. 29 (7), 318–323.

Sethunathan, N., Yoshida, T., 1973. A *Flavobacterium* sp. that degrades diazinon and parathion. Can. J. Microbiol. 19, 873–875.

Sharma, S., Verma, P.P., Kaur, M., 2014a. Isolation, purification and estimation of IAA from Pseudomonas sp. using high-performance liquid chromatography. J. Pure Appl. Sci. 8 (4), 1–7.

Sharma, T., Rajor, A., Toor, A.P., 2014b. Degradation of imidacloprid in liquid by Enterobacter sp. strain ATA1 using co-metabolism. Biorem. J. 18 (3), 227–235.

Sharma, A., Pankaj, K.P., Gangola, S., Kumar, G., 2016. Chapter 6: Microbial degradation of pesticides for environmental cleanup. In: Bioremediation for Industrial Pollutants. GenNext Publications, Lanham, pp. 179–205.

Singh, D.K., 2008b. Biodegradation and bioremediation of pesticide in soil: concept, method and recent developments. Indian J. Microbiol. 48 (1), 35–40.

Singh, R.P., Anwar, M.N., Singh, D., Bahuguna, V., Manchanda, G., Yang, Y., 2020. Deciphering the key factors for heavy metal resistance in gram-negative bacteria. In: Singh, R., Manchanda, G., Maurya, I., Wei, Y. (Eds.), Microbial Versatility in Varied Environments. Springer, Singapore. https://doi.org/10.1007/978-981-15-3028-9_7.

Singh, R.P., Manchanda, G., Li, Z., Rai, A.R., 2017. Insight of proteomics and genomics in environmental bioremediation. In: Bhakta, J. (Ed.), Handbook of Research on Inventive Bioremediation Techniques. IGI Global, pp. 46–69. https://doi.org/10.4018/978-1-5225-2325-3.ch003.

Singh, R.P., Manchanda, G., Maurya, I.K., Maheshwari, N.K., Tiwari, P.K., Rai, A.R., 2019. Streptomyces from rotten wheat straw endowed the high plant growth potential traits and agro-active compounds. Biocatal. Agric. Biotechnol. 17, 507–513.

Spadaro, J.T., Webb, E.L., Schmid, H., et al., 1998. Bioremediation of soil containing 2,4-D, 2,4,5-T, dichlorprop, and silvex. In: Wickramanayake, G.D., Hinchee, R.E. (Eds.), Designing and Applying Treatment Technologies, Remediation of Chlorinated and Recalcitrant Compounds, pp. 183–188. [C]. The First International Conference on Remediation of Chlorinated and Recalcitrant Compounds.

Struthers, J.K., Jayachandran, K., Moorman, T.B., 1998. Biodegradation of atrazine by agrobacterium radiobacter J14a and use of this strain in bioremediation of contaminated soil. Appl. Environ. Microbiol. 64 (9), 3368–3375.

Sun, H., et al., 2004. Plant uptake of aldicarb from contaminated soil and its enhanced degradation in the rhizosphere. Chemosphere 54 (4), 569–574.

Tao, Y., et al., 2019. Efficient removal of atrazine by iron-modified biochar loaded Acinetobacter l woffii DNS32. Sci. Total Environ. 682, 59–69.

Teng, Y., Wang, X., Zhu, Y., Chen, W., Christie, P., Li, Z., Luo, Y., 2017. Biodegradation of pentachloronitrobenzene by Cupriavidus sp. YNS-85 and its potential for remediation of contaminated soils. Environ. Sci. Pollut. Res. 24 (10), 9538–9547.

Tiwari, B., Kharwar, S., Tiwari, D.N., 2019. Pesticides and rice agriculture. Cyanobacteria, 303–325.

Uqab, B., Mudasir, S., Nazir, R., 2016. Review on bioremediation of pesticides. J. Biomed. Bioeng. 7, 343.

van der Werf, Hayo, M.G., 1996. Assessing the impact of pesticides on the environment. Agric. Ecosyst. Environ. 60 (2–3), 81–96.

Wang, G., et al., 2010. Co-metabolism of DDT by the newly isolated bacterium, Pseudo xanthomonas sp. wax. Braz. J. Microbiol. 41 (2), 431–438.

Yang, Y.-J., Lin, W., Singh, R.P., Xu, Q., Chen, Z., Yuan, Y., Zou, P., Li, Y., Zhang, C., 2019. Genomic, transcriptomic and enzymatic insight into lignocellulolytic system of a plant pathogen *Dickeya* sp. WS52 to digest sweet pepper and tomato stalk. Biomolecules 9 (12). https://doi.org/10.3390/biom9120753.

Yang, Y., Pratap Singh, R., Song, D., Chen, Q., Zheng, X., Zhang, C., Zhang, M., Li, Y., 2020. Synergistic effect of Pseudomonas putida II-2 and *Achromobacter* sp. QC36 for the effective biodegradation of the herbicide quinclorac. Ecotoxicol. Environ. Saf. 188, 109826. https://doi.org/10.1016/j.ecoenv.2019.109826.

Zacharia, J.T., 2011. Identity, physical and chemical properties of pesticides. In: Pesticides in the Modern World-Trends in Pesticides Analysis. Intechopen, pp. 1–18.

CHAPTER

14

Applying enzymatic biomarkers of the in situ microbial community to assess the risk of coastal sediment

Elisamara Sabadini-Santos[a], Vanessa de Almeida Moreira[a], Angelo Cezar Borges de Carvalho[a], Juliana Ribeiro Nascimento[a], Jose V. Lopez[b], Luiz Francisco Fontana[c], Ana Elisa Fonseca Silveira[a], and Edison Dausacker Bidone[a]

[a]*Geoscience (Geochemistry) Pos-Graduation Program, Chemistry Institute, Fluminense Federal University—UFF, Niteroi, RJ, Brazil,*

[b]*Halmos College of Natural Sciences and Oceanography, Nova Southeastern University, Dania Beach, FL, United States,*

[c]*Laboratory of Micropaleontology, Rio de Janeiro State Federal University—UNIRIO, Rio de Janeiro, RJ, Brazil*

Abstract

This study applied the Quality Ratio (QR) index to integrate geochemical (TOC, fine grain content, and metal concentrations) and microbiological (Esterases (EST) and Dehydrogenase (DHA) activities of the in situ microbial community) parameters in order to classify the potential ecological risk of coastal sediments in dredging activities. Total concentrations (C) of Hg, Cd, As, Pb, Cr, Cu, and Zn (indicators of the complex mixture of contaminants in sediments) were determined in sediments inside Guanabara and Sepetiba bays (Rio de Janeiro, Brazil) and in oceanic dump sites outside the bays (C_0) to calculate the contamination factor ($CF = C/C_0$) and the degree of contamination (ΣCF). Likewise, DHA and EST activities were determined—respectively, biomarkers of energy production in the cell and hydrolase of organic matter outside the cell—which are altered under adverse conditions (e.g., contamination). The QR, a function of the microbial term DHA/EST and the geochemical term ($TOC \times \Sigma CF$)/fine-grained content, was able to classify the sediments into three classes of risk: low ($QR \geq 10^{-1}$), moderate ($10^{-2} \leq QR < 10^{-1}$), and high ($QR \geq 10^{-3}$). The QR was able to segregate the hot spots of contamination of the bays. The QR was also applied to an acute assay and successfully identified the microbial community shift under a contamination gradient when mixing with dredged sediments. Thus QR provides an accessible (low cost and fast) and efficient alternative for assessing both the quality of coastal sediments and the ex situ bioassays, as required by Brazilian legislation for dredging sediments, as well as for other developing countries.

Keywords: Trace metals, Esterases, Dehydrogenase, Bacteria, Dredging

Microbial Syntrophy-mediated Eco-enterprising. https://doi.org/10.1016/B978-0-323-99900-7.00008-0

1 Introduction

Life has probably been present on Earth since the Earlier Archean (Nisbet and Sleep, 2001), a common prokaryotic ancestor of eubacteria, archaea, and eukaryotes (López-García and Moreira, 1999). Widely accepted evidence of this are stromatolites, the principal features of platform and shelf carbonates during the first 85% of Earth's history and described as laminated organo-sedimentary structures (Grotzinger and Knoll, 1999). Although some of these features can be interpreted equally as nonbiological in origin (Lowe, 1994), they were diagnosed as fossil biogenic structures of a microbial mat by others (e.g., Burne and Moore, 1987; Visscher and Stolz, 2005; Dupraz et al., 2009; Decho and Gutierrez, 2017).

Dupraz et al. (2009) define microbial mats as "ecosystems that arguably greatly affected the conditions of the biosphere on Earth through geological time." Even if microorganisms occur as a pure culture of free-living cells in nature, much more abundant are complex microbial communities attached to surfaces, forming aggregates such as sludge, mat, or biofilm (Knopka, 2006; Flemming and Wingender, 2010; Decho and Gutierrez, 2017). This lifestyle is possible because microbial cells are embedded in a self-produced matrix of hydrated biopolymers, known as extracellular polymeric substances (EPS). The potential stabilizing effects of EPS provide numerous advantages, creating their own microhabitats and allowing resource capture, strongly enhancing the survival, metabolic efficiency, and adaptation of cells (Meyer-Reil and Köster, 2000; Sobolev and Begonia, 2008; Flemming and Wingender, 2010). Those advantages explain not only the microbial distribution and abundance on Earth but also its activities, affecting the environment and providing broader processes on a macroscopic and even global scale (Knopka, 2006; Jião et al., 2010; Decho and Gutierrez, 2017).

Carbonate processes resulting in the formation of lithified mats (or microbialites) are possibly due to the microbial lifestyle within the ESP matrix, which may provide a template for mineral trapping, binding, and precipitation (Dupraz et al., 2009; Decho and Gutierrez, 2017). Environmental conditions impact the calcium carbonate saturation index, but the microbial metabolisms are also a key component to organomineralization of the "alkalinit" engine. This is not only true for carbonate mineralization but also for the precipitation or dissolution/weathering of several minerals while providing microorganisms with nutrients- and surface-associated living habitats (Ehrlich and Newman, 2009; Dong, 2010; Gadd, 2010). Indeed, microorganisms can metabolize broad ranges of substrates, effectively changing element speciation, toxicity, and mobility during interactions of these minerals. The great metabolic versatility allows microorganisms to use essentially all chemicals on Earth as an energy source and allow the oxidation of an element coupled to reduction of a terminal electron acceptor in a thermodynamically favorable reaction (Knopka, 2006; Meyer-Reil and Köster, 2000; Flemming and Wingender, 2010).

Microbial communities are essential for numerous geochemical processes (Decho, 2000; Ehrlich and Newman, 2009; Dong, 2010; Gadd, 2010). These issues are investigated by geomicrobiology, while microbial reactions of geochemical processes and their kinetics are the objects of study of microbial biogeochemistry. Interrelationships among microorganisms, plants, and animals are issues of microbial ecology. Although, these subjects are often studied in the context of the cycle of elements in the environment and they have overlapped to some degree (Ehrlich and Newman, 2009).

The establishment of a microbial community within a biofilm reduces changes in environmental perturbations, provides resources, and leads to various levels of microbial interaction as symbiosis—syntrophy among others (Flemming and Wingender, 2010; Meyer-Reil and Köster, 2000; Knopka, 2006). Syntrophy is microbial metabolic cooperation in a sequence of processes of mutual benefit, whereby products of one metabolic group of microorganisms form the substrate for others, enabling a microbial community to survive with minimal energy resources (Morris et al., 2013). Syntrophy embraces the evolution pathway of early life. Both the "hyperthermophile Eden" and the "hyperthermophile Noah" hypotheses agree that the common ancestor lived in a chemotrophic community, in which syntrophic metabolisms were originally sulfate reducers, fermenters, and methanogens. The development of anoxygenic and oxygenic photosynthetic probably came later (Nisbet and Sleep, 2001).

Predictably, life within the biofilm evolves into a functional structure of niches and complex metabolic interactions of microbial populations providing resilience to the whole community (Knopka, 2006). Moreover, the genes available for exchanging inevitably dying cells within the biofilm generate mutation and horizontal gene transfers (Meyer-Reil and Köster, 2000; Knopka, 2006; Allison and Martiny, 2008; Flemming and Wingender, 2010; Nogales et al., 2011). As a result, the microbial community rapidly adapts to environmental disturbance from the Archean to the present, reaching functional redundancy and stability, which ensured a ubiquitous spreading (Flemming and Wingender, 2010) and an equivalent trophic functioning for animals (Steffan et al., 2015).

Even if the disturbances caused to the microbial community have been greatly neglected, probably due to the difficulties in the analysis, their high diversity, and unknown definition of patterns (Allison and Martiny, 2008; Nogales et al., 2011), microorganisms are sensitive to disturbances and can be suitable bioindicators for environmental risk assessment (Busch et al., 2015; Cornall et al., 2016; Saxena et al., 2015). Recently, microbial assemblage usefulness in the European Marine Strategy Framework Directive was shown to be highly effective for environmental quality assessment (Caruso et al., 2016). Great progress determining the microbial diversity in the environment was achieved by the development of genomic technologies over the past 20 years, however without simultaneously analyzing physiological and ecological processes (Knopka, 2006). Nevertheless, these tools are useful to detect the effects of anthropogenic pressures on the microbial community structure and function, assessing changes in the environmental health status (Caporaso et al., 2011; Caruso et al., 2016; Mumtaz Moiz et al., 2010).

Besides community structure changes, the contamination effects disrupting homeostasis of the microbial community can be measured in various approaches described in the literature, such as enzymatic activity changes (Nascimento et al., 2019; Obbard et al., 1994; Sabadini-Santos et al., 2014a; Said and Lewis, 1991; da Waite et al., 2016). The Quality Ratio index (QR) presented in this chapter is based on this approach and it considers the ecotoxicological and human health risks through exposure to the complex contaminating mixture of coastal sediments which are due to processes directly linked to the interaction between the sedimentary substrate and microbial activity. As will be seen, the QR has the geomicrobiological characteristics appropriate for the problem of dredging in the Brazilian coastal system and elsewhere. The arguments for this are presented and discussed using the case studies of the Guanabara (GB) and Sepetiba (SB) bays located in the State of Rio de Janeiro, Brazil, which contain important harbor structures.

2 Sediment dredging in coastal systems and study areas

Dredging in the coastal zone is done for the establishment, expansion of capacity, operation and maintenance of ports and navigation security, as well as other hydraulic works. The increase in port dredging is mainly due to the growth of globalized trade and the increase in the number and size of ships.

Frequent dredging has become more necessary and consists of the removal of sediment that originates from continental erosion, having silted the estuaries' navigation channels (Manap and Voulvoulis, 2015).

In the specific case of developing countries, including Brazil, contamination control policies were ineffective to deal with the expansion of the population and urbanization, as well as economic activities in the coastal region, as contaminants continue to be released into the environment and thereby causing toxicity. Therefore, from a human and ecological health risk perspective, it is necessary to assess sediment contamination and its proper management in dredging work and final disposal of dredged sediments (Silveira et al., 2017).

The Brazilian port system is composed of 37 public ports and more than 40 private terminals and is responsible for more than 90% of the cargo handling carried out as a part of the international market, which demonstrates its strategic importance. As of 2007, in phase 1 of the National Dredging Program (PND), approximately 73 million m^3 were dredged in 16 ports (Secretaria Especial De Portos—SEP, 2007); in phase 2, which started in 2012, for maintenance of dredging for the next 10 years, dredging of almost 45 million m^3 of sediment was planned (Secretaria Especial De Portos—SEP, 2012).

The two studied bays, Guanabara Bay (GB) and Sepetiba Bay (SB), present the challenge of dredging 12 million m^3 of sediment in GB alone (Secretaria Especial De Portos—SEP, 2012); while SB from a single industrial source close to the port facilities (Companhia Siderúrgica Mercantil Ingá SA) suffered the discharge of more than 2 million tons of tailings and 50 million liters of effluents contaminated by heavy metals (Molisani et al., 2004).

2.1 Guanabara Bay

Guanabara Bay is a semi-enclosed, low-energy microtidal estuarine environment that communicates with the Atlantic Ocean, located at 22°40′00″–23°05′00″ S and 43°00′00″–43°20′00″ W, at Rio de Janeiro city and its metropolitan region (Fig. 1). The Guanabara Bay basin is surrounded by rainforests-covered mountains at a distance ranging from 1000 to 1500 m. The annual average rainfall is 1170 mm, and the mean annual temperature is 23.7°C, with rainy summers (December to April) and dry winters (June to August), representative of a tropical climate with a strong marine influence (JICA, 1994).

Its area and volume are 384 km^2 and 2.2×10^9 m^3. The bathymetry has a range between 84% < 10 m and 46% < 5 m, with an average of 7.7 m, with a central channel of 30–40 m deep delimited by 10 m isobathymetry. The average water temperature is 24.2 ± 2.6°C and salinity is 29.5 ± 4.8. As the bay widens, the system changes from being moderately stratified to weakly stratified, resulting in a 50% renewal of the bay's water volume in 11.4 days (an average estimate for the entire bay). The times of renewal in the innermost areas are significantly longer (Kjerfve et al., 1997).

The GB basin (around 4080 km^2) includes 15 municipalities, more than 11 million inhabitants, and around 14,000 industries. GB also hosts two oil refineries along its shores in addition to the port of Rio de Janeiro (the second biggest port in Brazil), two naval bases, shipyards, ferries, fishing activities, marinas, sewage outfall, and so on (Meniconi et al., 2012).

There are 55 rivers that discharge into the bay. The average discharge value ranges from 0.1 m^3/s to 55 m^3/s, with a total average value of 260 m^3/year (Amador, 1997). Intensive inputs of largely untreated domestic sewage, constant runoff, and industrial effluents contribute to the condition of chronic pollution of the bay (INEA, 2016). Approximately, 500 tBOD/day is discharged by the rivers into the bay.

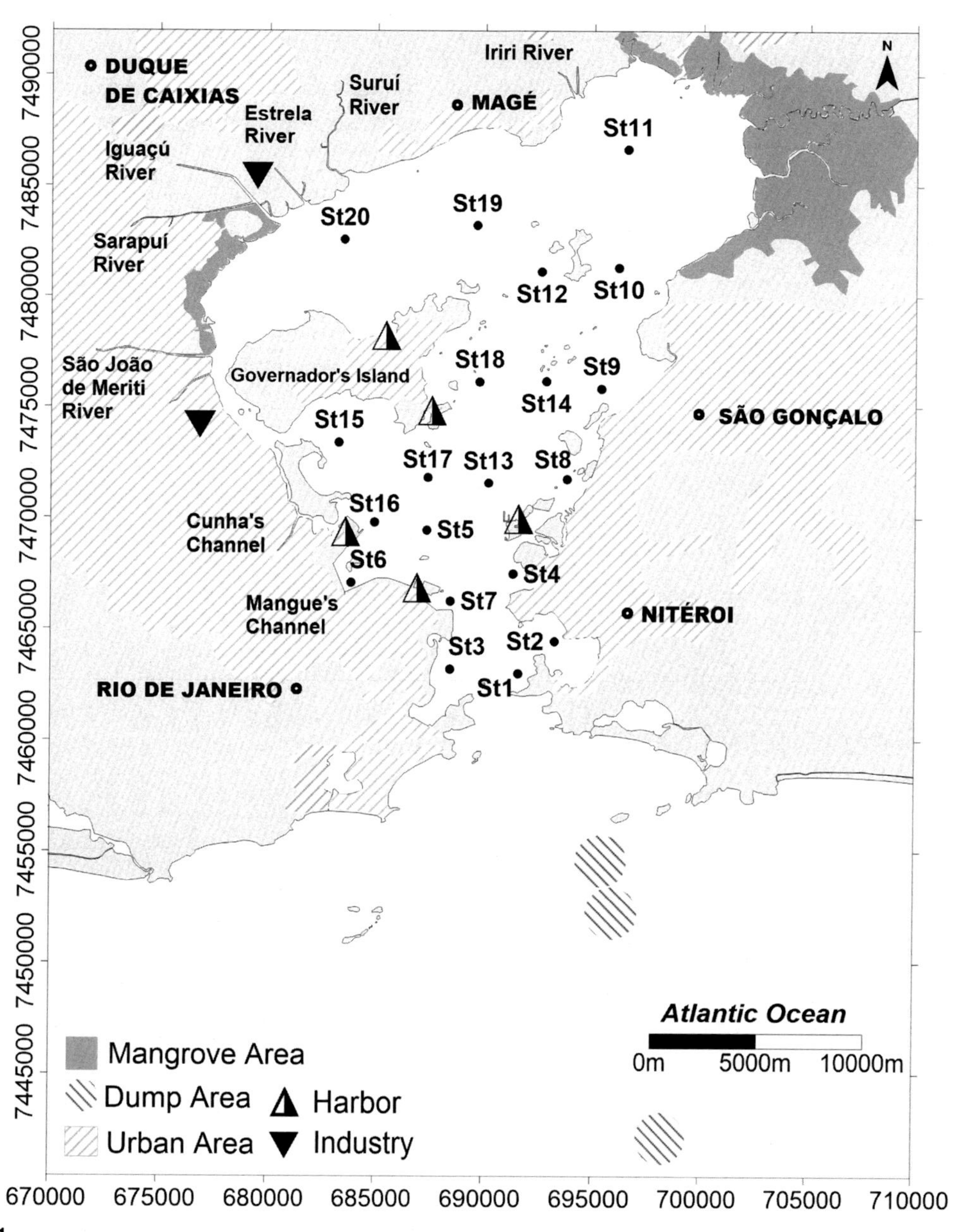

FIG. 1

Sampling stations, dump area, mainland uses, and covers at Guanabara Bay.

Untreated sewage is responsible for 80% of BOD discharge, which includes nutrients and a significant amount of toxic chemicals (metals and organic micropollutants). The remaining 20% is industrial waste that accounts for most of the chemical pollution that accumulates in the sediments (Silveira et al., 2017).

The GB basin area has been significantly modified by anthropogenic activities (e.g., destruction of peripheral bay ecosystems such as wetlands, mangroves, and soils), exacerbating siltation (over the last 50 years, sedimentation rates have been estimated to range from 0.65 to 2.2 cm $year^{-1}$), contamination, and eutrophication in GB as the population grew without appropriate environmental management or sanitation structures (Bidone and Lacerda, 2004). A high degree of eutrophication is spreading from these highly urbanized sectors to other sectors, threatening the bay's water quality (Carreira et al., 2002; Sabadini-Santos et al., 2014b).

The sediments of GB show a similar distribution in terms of grain size, organic matter (OM), and toxic contaminants. A lower hydrodynamic capacity and the proximity of sources of toxic contaminants contribute to this distribution pattern. The western and northwestern sectors also present the highest pollution levels in the sediment in addition to some other hot spots in other sectors (e.g., harbors) (Cordeiro et al., 2015; Baptista-Neto et al., 2016). Therefore, GB is characterized by the widespread presence of complex mixtures of contaminants in the sediments.

2.2 Sepetiba Bay

Sepetiba bay is a semi-enclosed water body that communicates with the Atlantic Ocean, located at 22°55′00″S and 23°03′60″S and 43°56′30″W and 43°36′20″W, 60 km from Rio de Janeiro city (Fig. 2). The bay's climate is typically hot-humid tropical, with a mean annual precipitation of 1400 mm and mean evaporation of 960 mm. Its area and volume vary according to the tidal cycle between 419 and 447 km^2 and $3.06.10^9 m^3$–$2.38.10^9 m^3$. The bathymetry ranges between 2 and 12 m with an average of 6 m in depth, except in the two channels (24 m and 31 m) which allow access to the Sepetiba Port. The circulation pattern of the SB—wind and tide drive the clockwise superficial currents adding seawater and carrying fresh water and fluvial sediments for the south of the bay—results in a relatively low estimated turnover time (around 6 days) of the water mass (Kjerfve, 2001), a great mixture of the water column, and the absence of stratification. In general, salinity is between 20% and 34%, with the bottom of the bay and coastal areas presenting a salinity below 30%. In the central part, and close to the entrance of the bay, the salinity varies between 30% and 34% (Molisani et al., 2004; INEA—Instituto Estadual do Ambiente, 2014).

The SB basin (around 2000 km^2) includes 10 municipalities (totally or partially inserted), around 7,610,000 inhabitants, an industrial complex—composed of more than 400 industries of textile, metallurgy, smelting, and petrochemical sectors—tourism, fisheries, military, and university research facilities (Censo, 2010; Trevisan et al., 2020). Apart from the marine ecosystem, the bay supports 40 km^2 of mangrove forest, developed in the most eastern part and has an important role for biodiversity and the supply of organic matter for the bay (Carreira et al., 2009; Carvalho et al., 2020).

The rocky northern shore borders the Serra do Mar Mountain chain, covered by tropical rain forests. Agriculture (fruits and vegetables) also transform tropical rain forest slopes, beach ridges, and wetlands. Besides industrialization and urbanization—hydraulic works opening and rectifying river channels, highways, roads, etc.—in continental areas in the SB basin, all generate more polluted sediments (Molisani et al., 2004).

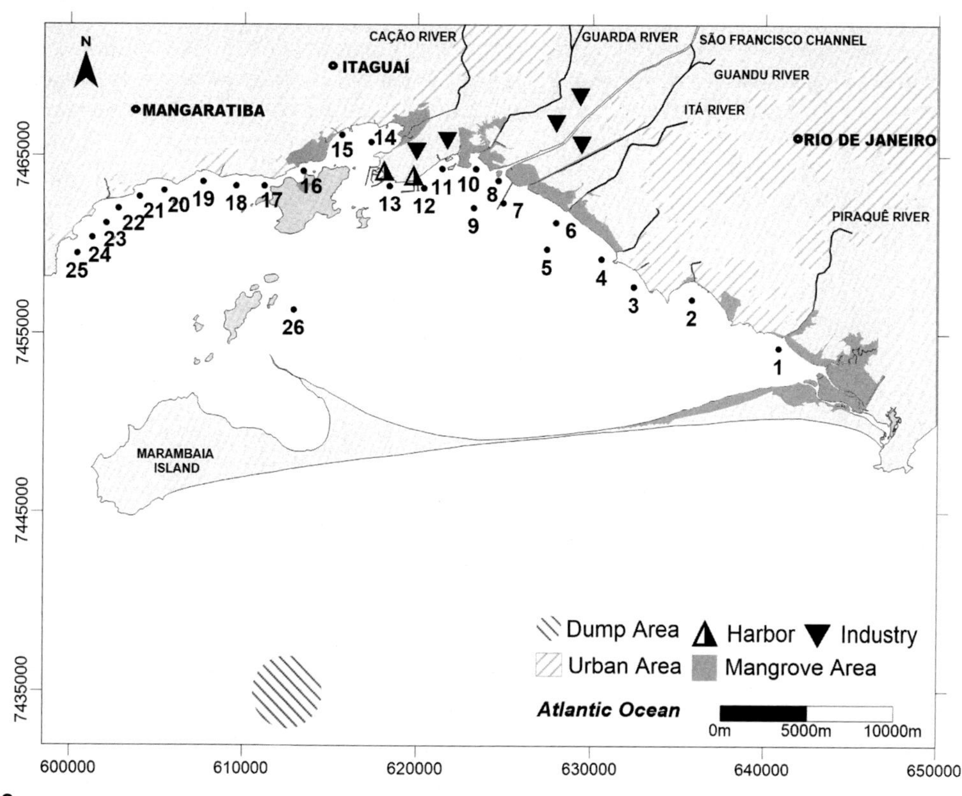

FIG. 2

Sampling stations, dump area, mainland uses, and covers at Sepetiba Bay.

The São Francisco River Channel is the largest tributary, contributing 86% of the river's water supply (Smoak et al., 2012). The wet season (January–March) shows that around 700–800 t/day is discharged into Sepetiba Bay through the São Francisco Canal, representing 20–30% of total suspended matter entering the bay (Molisani et al., 2002). The sedimentation rate in SB ranges from 0.3–0.8 cm/year (Gomes et al., 2009) to 2.4 cm/year (Borges and Nittrouer, 2015). The average annual discharge of particulate matter in suspension estimated for SB is over 2 million t/year with more than 20,000 tDBO/year (CENTRAN, 2009). The dominant sediments are represented by sand-silty clastics. It should be noted that the finer the sediments, the greater their ability to retain contaminants.

3 Sediment dredging legislation: In situ microbial bioindicators/biomarkers replacing bioassays and toxicity tests ex situ

More than 90 countries including Brazil became signatories to the Convention on the Prevention of Marine Pollution by Dumping of Wastes and Other Matter (LC 72; London Convention, 1972)—revised in 1996 by the London Protocol (London Protocol, 1996) that came into force in 2006. It is committed to monitor, predict, and mitigate the impacts from the dredged coastal sediment on the biota. This also affects human health since they act as a deposit and source of contaminants, originated by the input of industrial and domestic effluents in nature or by residuals of treated ones in the water of the drainage basin of the coastal body.

There is a great concern among managers and stakeholders in determining the risk that sediments can pose to biota and human consumers of marine food. Numerical criteria for sediment quality guidelines (SQGs) have been developed in several countries as application tools for the design of monitoring programs to assess the need for remediation or to assess the quality of the material to be dredged (MacDonald et al., 2003).

Aiming to comply with this perspective, the Brazilian legislation by ways of the National Environment Council Resolution n°454/2012 (CONAMA, 2012) establishes the criteria and methodologies to be followed from the dredging planning phase to the final disposal of the contaminated sediments.

Threshold concentrations of contaminants—each contaminant concentration presented in dry weight sediment—with the lowest (L1) and highest (L2) probability of causing adverse effects to biota are established, and ecotoxicological assays ex situ must be carried out when the concentration exceeds L1. In practice, these values guide the three categories of waste classification regarding disposal at sea, as is recommended in LP96. The resolution adopted guiding values for Europe, the USA, and Canada (CCME—Canadian Council of Ministers of the Environment, 1999; Environment Canada and Ministère du Développement durable, de l'Environnement et des Parcs du Québec, 2007; Hamburg Port Authority, 2011; Long and Chapman, 1985; MacDonald, 1994), which incorporate reference values established by the concentration of contaminants based on background values, ecotoxicological studies, bioassays from other countries, and such (Hamburg Port Authority, 2011).

The adverse effects on biota are obtained from bioassays contained in an extensive database of studies performed in developed temperate countries, e.g., the USA BEDS—Biological Effect Database for Sediments (Macdonald et al., 1996). However, indicated organisms are not always representative of the Brazilian study areas and others in the world's tropical zones. The comparison with the SQGs revealed that the Brazilian transcribed standards have low sensitivity for predicting impacts on biota when compared to the specific values of the site (Buruaem et al., 2013).

Also, ecotoxicological tests ex situ do not accurately mimic natural environmental conditions (Håkanson, 1980; Long and Chapman, 1985; Rosado et al., 2015), since the tested organisms are kept under constant physical-chemical conditions, which does not occur in natural environment.

As an alternative to the bioassays, the use of bioindicators and biomarkers in situ are representative tools of the study site being useful for ecological risk assessment (Adams et al., 2001; Bartell, 2006; Kienzl et al., 2003; Lorenz, 2003; Market et al., 2003) and also holds for sustainable use of biodiversity (UNCED—United Nations Conference on Environment and Development, 1992). In this case, the measured biological response represents a native population/community/ecosystem level and not only individual-/species-specific responses; in addition, the in situ approach considers the vast genetic diversity of organisms while laboratory tests use selected, genetically homogeneous organisms (Market et al., 2003). What is more, the conservative character is respected regarding the preservation of the environmental conditions to which the organisms are exposed in their natural environment. The use of organisms in situ also allows for the measurement of a biological response to exposure to multiple contaminants concurrently, while laboratory bioassays usually address the effect of exposure to contaminants in isolation (Adams et al., 2001; Bartell, 2006).

In pollution scenarios, bioindicators can be used to provide information on the variation of pollutants over time and space (Hamza-Chaffai, 2014). To be considered the perfect bioindicator, the organism must have the following characteristics: (i) be sedentary or make a small displacement during all phases of its life cycle; (ii) have abundant biomass and a wide geographic distribution that make it possible to repeat sampling and compare results; (iii) be easy to collect and identify; (iv) to be able to accumulate high concentrations of contaminants without leading to death and without completely regulating its transformation/excretion; (v) have a long life so that you can compare the effect of contaminants on your body during the different stages of its life cycle; and (vi) be resistant to physical-chemical and seasonal variations in the environment (Hamza-Chaffai, 2014; Rainbow and Phillips, 1993; Zhou et al., 2008).

Different from the bioindicators, biomarkers are biological exposure responses at the sub-organism level (i.e., at the molecular, cellular, biochemical, structural, or physiological level) and they can occur from levels of low biological classification to the highest (Adams et al., 2001; Peakall, 1994). Responses that occur at the level of individual, population, and ecosystem are generally accepted to have ecological relevance and tend to be less reversible and cause more damage than effects at lower levels (Hamza-Chaffai, 2014).

According to Hamza-Chaffai (2014), biomarkers are generally classified into exposure and effect markers. The first is the result of the interaction between a xenobiotic and a target molecule or cell; it is used to predict the dose received by an individual, which can be related to changes in the organism resulting in disease. The second is defined as a measure of the biochemical, physiological, behavioral, or other alterations within an organism that, according to their magnitude, can be recognized as established or potential health impairment.

Microorganisms fulfill all the requirements as a good bioindicator, being widely abundant and diverse in every kind of environment. Moreover, their metabolic and physiological diversity makes them very important in the nutrients cycle and for their subsequent availability to organisms of different trophic levels (Azam, 1998; Caruso et al., 2016; Decho, 2000; Prosser et al., 2007). Its high contact surface has an affinity for trace concentrations of substances present in the environment, and because of their sensitivity they respond quickly in different situations of environmental stress (Nogales et al., 2011), as represented in studies reporting the response of bacterial enzymatic biomarkers in contamination scenarios (Acosta-Martínez and Tabatabai, 2001; Li et al., 2018; Liu et al., 2019; Nascimento et al., 2019; Tyler, 1976; da Waite et al., 2016). Thus changes are quickly reflected in microbial organisms making them great indicators of the early stages of contamination (Bloem and Breure, 2003). Despite the fact that prokaryotic organisms dominate the seafloor (Danovaro et al., 2010; Whitman et al., 1998) and that they are important in the structuring and functioning of the ecosystem, they continue to be neglected as their incorporation in methods aim to assess the status of marine environments, and consequently they are not considered in the coastal strategy framework planning (Caruso et al., 2016).

An alternative to bioassays and toxicity tests for establishing the risk assessment for toxic mixtures and acceptable levels in sediments is to analyze the composition of the mixture and apply an algorithm relating the concentration of individual contaminants to the total risk of the mixtures (Ragas et al., 2010). This allows for a simple, fast, efficient, and technically justified approach to the information needed to meet the requirements, aimed at decisions concerning dredging plans and management (Linkov et al., 2006), especially in the phase prior to the preparation of the dredging plan itself. In this strategy, microorganisms can play an essential role. Therefore, it is necessary to integrate microbial parameters (biological dimension) with indicators of the main processes acting in coastal sediments (geochemical dimension), which would meet the two main dimensions/requirements included in the dredging legislation. The QR described in the next section aims at this index alternative.

4 Quality ratio (QR) index

The quality ratio (QR) is a new index proposed by Nascimento et al., 2019 that simplifies the complex relationship between geochemical and microbial parameters for the risk assessment of coastal sediments chronically contaminated by multiple contaminants. The QR was developed by combining the concepts of two models: stressed ecosystem development (Odum, 1969, 1985) and the time-and-dose-dependent model of biofilm multimetal resistance and tolerance (Harrison et al., 2007), emphasizing the effectiveness of the enzymatic biomarkers in evaluating the hypothesis of homeostatic disturbance under multiple environmental stress (e.g., eutrophication and metal contamination) (Nascimento et al., 2019). Therefore, the QR presents an innovative approach, especially in regard to the relationships with the microbiological parameters that are far from evident in the routine of geochemical monitoring.

The QR integrates a microbial term (developed from esterase and dehydrogenase enzymatic activities) and a geochemical term (constituting fine grain, total organic carbon, and metal concentrations) to rank sediment sampling into three risk classes (Fig. 3). The parameters used in the index are interconnected and integrated environmental processes, involving different natures (physical, geochemical, and biological) such as silting, eutrophication, and anoxia. This interconnection between the parameters is supported by quantitative information and significant correlations between geochemical and microbial processes (see Nascimento et al., 2019).

The microbial term interprets disrupted homeostasis manifested by increased energetic demand (Odum, 1985), here expressed as increased EST activities, and reducing energy production—ATP generation compromised by metals (Van Beelen and Doelman, 1997)—here represented by lower DHA

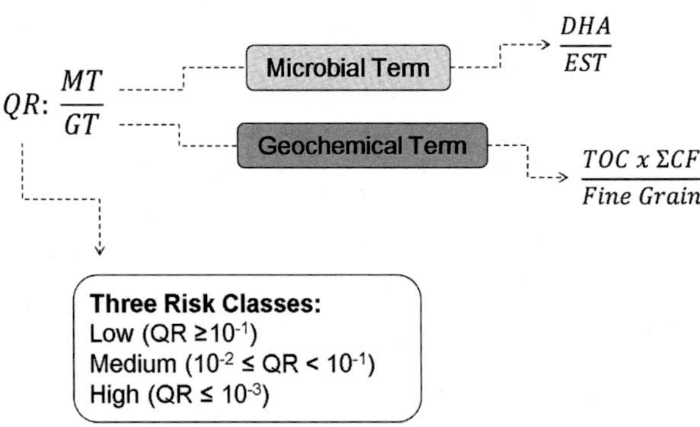

FIG. 3

QR equation. *DHA*, dehydrogenase (μg INT-F g^{-1}); *EST*, esterase (μg FDA $h^{-1} g^{-1}$); *TOC*, total organic carbon (% w w^{-1}); Fine grain: silt plus clay content (%); $\sum$CF: sum of contamination factors—relationship between actual and background metal concentrations that comprise As, Cd, Cr, Cu, Hg, Pb, and Zn (dimensionless).

activity. DHA acts on ATP production in the electron transport chain, whereas ESTs hydrolyze organic matter outside the cell membrane, so both are related to energy metabolism and viability. These enzymes proposed as biomarkers are interdependent and can be affected in situations of stress generated by environmental contamination and considering their involvement in energy metabolism, the use of the DHA/EST ratio is justified.

Although MT efficiently interprets these alterations in homeostasis, it is necessary to link it to sediment parameters, here represented by the geochemical term (GT), to effectively quantify the impact of contamination. The GT aggregates geochemical matrix processes such as complexation and adsorption processes of organic matter (TOC) and metals (ΣCF) onto fine-grained sediments, essentially deposited in areas of low hydrodynamic energy. Therefore, TOC and ΣCF values were considered dependent on the proportion of fine-grained sediments, with the product of TOC and ΣCF reflecting synergistic contamination effects. Aggregating metallic contaminants (As, Cd, Cr, Cu, Hg, Pb, and Zn) in the form of a contamination factor (CF) is a coherent strategy because—in addition to forming one of the most dangerous groups of recognized contaminants—they are indicators of the complex mixture of pollutants associated with fine sediments in the coastal system (including organic pollutants). It is pointless to segregate or individualize the effect of a single contaminant. Thus, in the QR approach, metals (ΣCF) are indicators of this multiple environmental stress for the microbial community. It is important to highlight that during the development of the index risk, the first approach proposed for the geochemical term (GT) was the PERI—Potential Ecological Risk Index (Håkanson, 1984). However, PERI did not show any significant correlation with the microbiological term (MT) or with the TOC. This result may be related to the discrepancy between the toxicity factor given to mercury (90%) and or other metals in the formulation of PERI (see Silveira et al., 2017). In addition, as organic matter is considered in both the PERI formulation and the QR formulation, organic matter is overestimated. All of this distorts the correlations with PERI, damaging both the classification of sediments in risk classes and their prediction.

The QR was calibrated using a sensitivity analysis based on Brazilian legislation to assess the quality of sediments (CONAMA, 2012). The sensitivity analysis investigated how the risk index is changed according to the variations in its constituent terms, in an approach widely used for model calibration (Pianosi et al., 2016). The results indicate that the QR is a feasible tool for measuring the potential risk of any coastal sediment, especially in developing countries with serious technical limitations, since its evaluated parameters are cheap, fast, and easy to obtain (see Nascimento et al., 2019).

The resources proposed by Rosado et al. (2015) for the selection of parameters to compose an algorithm were followed. The biomarkers used in the QR to assess the suitability of the microbial community are appropriate because (i) they are applicable to all types of sediments; (ii) are comparable across locations and over time; (iii) considering organisms in situ; and (iv) they are inexpensive to obtain and easy to understand. Thus the QR fulfills the function of an effective environmental index because expressing the risk associated with physical and chemical parameters, it integrates the effects of organisms in situ through biomarkers, and its mathematical formulation allows disaggregating the index into its constituents/individual indicators, which is a property for a good index. The indicators showed significant statistical correlations among themselves and with the terms MT and GT (c.f. item 5 below), corroborating the relationships between processes formulated in the QR. Following the premises of the preparation of the QR, the index was successful when applied to the sediments of the Guanabara Bay where it was proposed, in the sediments of the Sepetiba Bay and in contaminated sediments and soils around a gold mine.

5 Case studies of the quality ratio index

The Quality Ratio index (QR) was applied to two coastal bays of relevant economic and urban importance in Brazil, located in the state of Rio de Janeiro: Guanabara Bay—GB (Nascimento et al., 2019) and Sepetiba Bay—SB (Moreira et al., 2021). QR was also applied in continental areas to soils and river sediments under the direct influence of effluents and tailings from the largest gold mine in Brazil, Morro do Ouro, located in the municipality of Paracatu in the State of Minas Gerais. Bearing in mind that this chapter is dedicated to the application of QR only to dredging of coastal sediments, the case of Paracatu can be found in Sabadini-Santos et al. (2020).

5.1 Metal sediment contamination in GB and SB bays

High organic matter loads from continental drainage, increased runoff, and anthropogenic contaminants that bioaccumulate in the trophic web cause coastal sediments to represent a risk to biota and human health in GB and SB, demanding attention regarding the monitoring and managing of the sedimentary contamination (c.f. item 2). Table 1 shows the ranges of metal concentrations in the GB (Nascimento et al., 2019) and SB (Moreira et al., 2021) bays, at the disposal sites of the dredged material, and the Brazilian legal thresholds for dredging sediments, L1 and L2 (CONAMA, 2012) c.f. item 3.

In the two bays, the concentrations on the disposal sites are between 1 and 2 orders of magnitude below the legal values. Considering the maximum concentration values of metals:

(a) At GB only As is below L1, i.e., with the lowest probability of causing adverse effects; Cd, Cr, Cu, and Pb are between L1 and L2 and Hg and Zn above L2. The stations belonging to N-NW sectors of the Guanabara Bay present metal concentrations higher than L1, reflecting the input of the most polluted rivers, the concentrated industrial activities (including an oil refinery), and the large population contingent (Sabadini-Santos et al., 2014a). Other contamination hot spots are associated

Table 1 The ranges of metal concentrations in the Guanabara and Sepetiba bays, at the disposal sites of the dredged material, and the Brazilian legal thresholds for dredging sediments, L1 and L2.

	As (mg kg^{-1})	Cd	Cr	Cu	Hg	Pb	Zn
GB (n=20)[a]	0.2–4.1	0.02–1.77	0.02–88	0.02–108	0.16–4.91	3–105	1–545
Median GB[a]	1.6	0.82	35	26	0.55	30	216
GB dump site[a]	1.3	0.10	10	1.1	0.02	2	7
SB (n=24)[b]	0.6–10.8	0.01–4.70	4–64	0.6–34	0.01–0.25	2–42	37–891
SB - St 11	16.2	46.8	39	44.4	0.13	81	7660
SB - St 12	12.7	16.8	46	30.1	0.12	44	3310
Median SB[b]	5.0	0.40	43	9.2	0.03	11	167
SB dump site[b]	4.35	0.02	1.69	1.81	0.02	1.85	12.04
L1[c]	19	1.2	81	34	0.3	46.7	150
L2[c]	70	7.2	370	270	1.0	218	410

[a] *Nascimento et al. (2019).*
[b] Moreira et al. in process of publication.
[c] CONAMA 454.

with sewage outfall: Rio de Janeiro and Niterói harbors and the São Gonçalo region (the second largest city in the state of Rio de Janeiro with poor sanitation conditions at sector E of GB).

(b) At SB the values of As, Cr, and Hg are close to L1; Cu and Pb are between L1 and L2, and Cd and Zn are far above L2. The latter corroborate the contamination by metals (mainly Zn and Cd) since the 1960s, as a result of the activity of the former metallurgy Mercantil Ingá S.A., deactivated in 1997, persisting to the present (Molisani et al., 2006). The highest concentrations are at St 11 and St 12, near the Mercantil Ingá S.A. and in the internal sector of the bay and under the influence of the main rivers responsible for the untreated domestic and industrial effluent loads (see item 2). GB shows higher concentrations of Cu, Hg, Cr, and Pb, and SB of As, Cd, and Zn.

Recall that the legislation states that concentration values above L1 indicate that ecotoxicological assays ex situ must be carried out. The question is whether the QR index is capable of replacing these bioassays, especially in the phase that precedes the elaboration of the dredging executive plan itself?

Three premises to answer the question (c.f. item 4): (i) an algorithm is better suited to address the potential risk of sediments contaminated by complex mixtures; (ii) a good index must allow the breakdown of its algorithm into its individual constituents; and (iii) it is necessary to evaluate the relationships between the constituents and the terms geo and bio, which would allow synthesizing the synergy acting between processes that occur in the sedimentary substrate, allowing a measure of the ecotoxicological risk of the sediments.

5.2 Geochemical parameter behavior—GT—In GB and SB sediments

Table 2 shows the ranges of values obtained in the two bays for the GT constituent parameters.

(a) Fiori et al. (2013) studied the sediments of five bays in the State of Rio de Janeiro, establishing a correlation between TOC and organic matter (O.M.) $r=0.86$ and $P<.01$. In GB in general, the TOC values correspond to levels of organic matter (OM) varying between 8% and 22% with an important capacity to retain contaminants—corroborated by Sabadini-Santos et al. (2014a,b)—which is reflected in the concentration of metals above L1 and the average $\sum$CF of the bay is 100-fold higher than the dump site $\sum$CF. The CF Zn is responsible for 31% of the $\sum$CF, followed by the CF Hg and CF Cu, accounting for 28% and 24%, respectively. These parameters associated with high levels of fine grain result in a relatively high GT average (maximum of 2668)—remembering that GT=(TOC × Sum CF)/Fine Grain.

(b) In SB, the TOC values correspond to O.M. between 9% and 17% and are able to retain metals in the sediments of the bay—corroborating (Rodrigues et al. (2017))—of an average $\sum$CF 60-fold higher than the dump site $\sum$CF, when not considering the St 11 and St 12, sampled in the area of the sedimentary hot spot of the former metallurgic company Ingá. The CF of the Cr is responsible for 32% of the $\sum$CF, followed by the CF of Cd and Zn, 28% and 19% of the $\sum$CF, respectively. These parameters associated with the relatively low fine grain content—even in the river mouths—result in a GT average fivefold higher than GB (maximum of 11,772, excluding St 11 and St 12).

The distribution pattern of the GT values follows that of its constituent parameters, presenting the highest values in the areas with the lowest hydrodynamic energy and proximity to the continental inputs of contaminants, especially in the N-NW and E sectors and in the harbor areas in GB and especially close to the metallurgic company Ingá, near the area of port activities and spreading to the internal sector of SB (Fig. 4A and D).

Table 2 The ranges of values obtained in the two bays for the GT constituent parameters.

	Fine Grain (%)	TOC ($\% w w^{-1}$)	ΣCF	GT	DHA (μg INT-F g^{-1})	EST (μg FDA $h^{-1} g^{-1}$)	MT
GB ($n = 20$)[a]	4–100	0.03–5.15	31–491	21–2668	1.1–108.6	0.3–18.9	1–18
Median GB	92	3.55	108	450	36.3	8.8	4
SB ($n = 24$)[b]	0.9–26	0.20–3.45	10–3078	90–41,728	7.1–67.5	1.6–19.2	0.4–26
median SB	10	1.20	77	815	28.7	10.3	2.8
SB - St 11[b]	24.8	3.4	3078	41,728	23.4	4.9	4.7
SB - St 12[b]	13.7	2.7	1191	23,338	15.4	12.1	1.3

[a] *Nascimento et al. (2019)*
[b] Moreira et al. (2021).

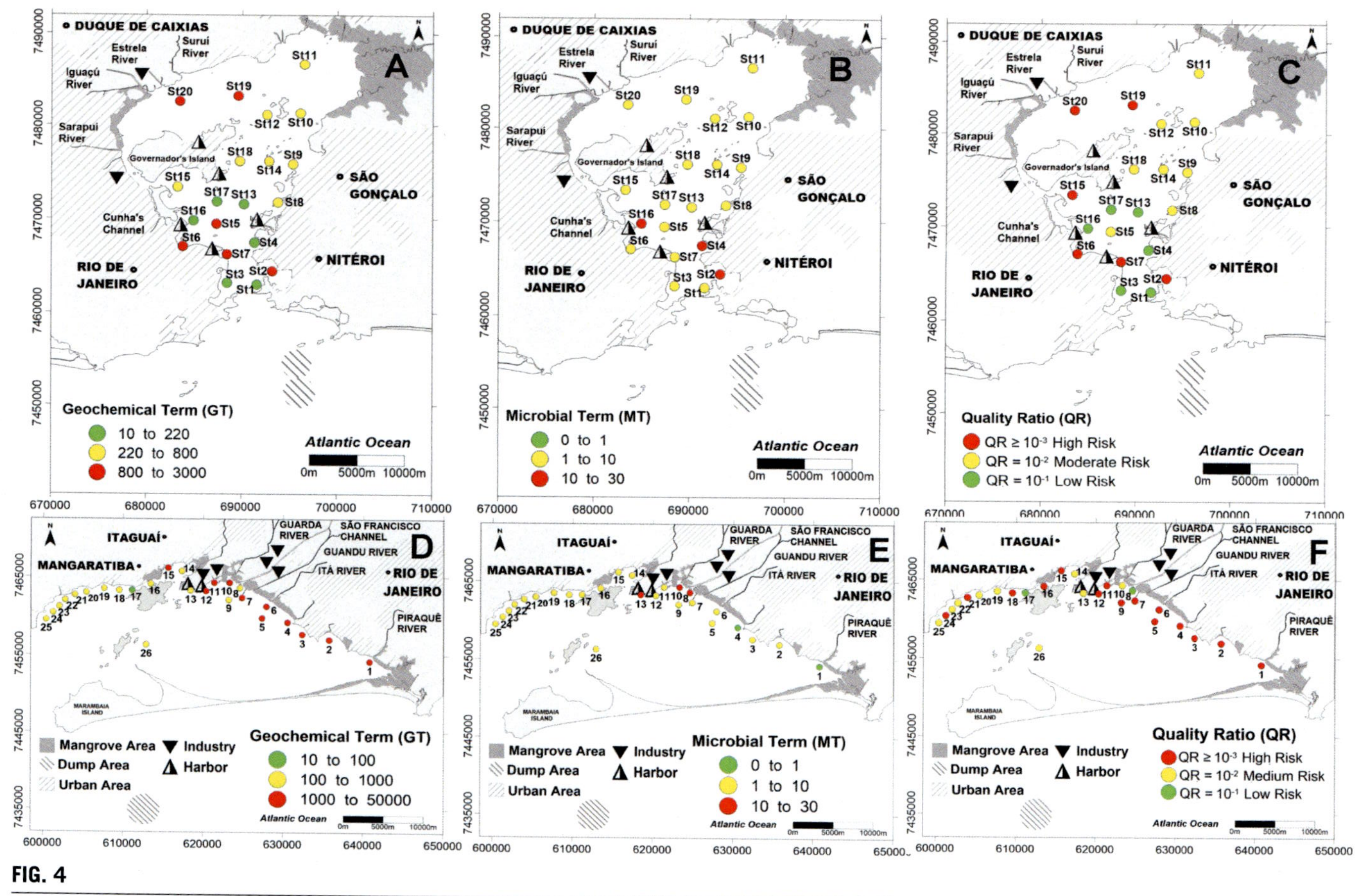

FIG. 4

Case studies of the Quality Ratio index.

Therefore, GT is able to indicate and classify the areas with the most contaminated sediments and the causes of this contamination. As explained in item 4, well-known geochemical parameters allow for the justification and forecasting of contaminants' bioavailability for the biota (Förstner and Schoer, 1984) once fine grain, organic matter, and total organic carbon represent some of the main metal-binding geochemical matrices (Förstner and Wittmann, 1981). The spatial pattern and the low hydrodynamic capacity as well as the proximity to the river's mouth of contaminated loads to the bays play a key role in the distribution of coastal sediments, accumulating fine grains, organic matter, and contaminants. Thus accumulation generates an anoxic environment, favoring geochemical processes of sorption and complexations (e.g., the formation of metallic sulfides, methyl-mercury, and others) and the release of contaminants when exposed to oxidation due to dredging (Silveira et al., 2017). In addition, metals (mainly Cr, Cd, and Zn) appear as major contaminants in SB—corroborating the literature (e.g., Molisani et al., 2006; Nascimento et al., 2016, 2017)—while in GB other contaminants such as oil (in agreement with, e.g., Mauad et al., 2015, Meniconi et al., 2002) are relevant in addition to metals and contribute to the toxicity of their sediments.

5.3 Microbial parameter behavior and their integration with GT in GB and SB sediments

The number of bacterial cells counted under the microscope was 1–2 orders of magnitude higher in SB (10^8–10^9cell g^{-1}) than in GB (10^7 cell g^{-1}), however both bays had similar enzyme activity averages and ranges (Table 2). Although both bays are chronically contaminated by metals, metals are being used as indicators of multiple contaminants present in coastal bays, particularly of those whose drainage basins are densely populated and industrialized. Organic pollution is another concern in GB (e.g., Mauad et al., 2015; Meniconi et al., 2012) and thus exacerbating pollution pressures—despite water renewal, stratification, and anoxia (see item 2)—may be an explanation for the biomass restriction and perhaps the process of phenotypic resistance/tolerance diversification (Harrison et al., 2007; Su et al., 2014; Wang and Feng, 2007), which might be clarified by further genomic investigations.

Indeed, decreased microbial biomass under adverse conditions represents disrupted homeostasis (Harrison et al., 2007), which is also manifested by (i) increased energetic demand (Odum, 1985) and (ii) reduced energy production through compromising the electron transport chain (Van Beelen and Doelman, 1997). Because organic matter is one of the most important contaminant-binding geochemical matrices (explained in item 4 and here assigned by positive and significant correlations between TOC and $\sum$CF, Table 3), it represents a route of contaminant exposure for the microbial community.

Table 3 Polynomial and exponential correlations between the MT and the constituents of GT.

		Fine grain	TOC	ΣCF
GB (n=20)	TOC	Exp*. r=0.83 P<.05		
	ΣCF	Exp. r=0.52 P<.05	Exp. r=0.61 P<.05	
	MT	Pol**. r=0.60 P<.01	Pol. r=−0.50 P<.05	Pol. r=−0.48 P<.05
SB (n=26)	TOC	Exp. r=0.43 P<.05		
	ΣCF	Exp. r=0.53 P<.01	Exp. r=0.72 P<.01	
	MT	Pol. r=0.39 P<.05	Pol. r=−0.37 P<.1	Pol. r=−0.41 P<.05

**Exp. = exponential, **Pol. = polynomial.*

When degrading the organic matter (intense EST activities), contaminants become bioavailable for the microbial community compromising DHA activity. Thus the QR approach uses the ratio between DHA and EST enzymatic activities (MT) to estimate these biological responses to the exposure to contaminants. It is worth to highlight that EST and DHA are present in both aerobic and anaerobic metabolisms (Stubberfield and Shaw, 1990; Trevors et al., 1982), hence being representative of the entire microbial community.

To meet the premise that the QR is capable of replacing bioassays in dredging planning, it is necessary to consider two points. First, that the term GT is linked to the notion of potential risk, observing what is established by the legislation as well as its capacity to identify the action of major geochemical processes on the behavior of metals. Second, that GT does not measure the ecotoxicological risk of sediments, a central issue in the legislation on dredged sediments. Table 3 shows significant polynomial correlations between the MT and the constituents of GT. Positive correlations between fine grain, TOC, CF, and MT corroborate the microbial lifestyle (item 1), allowing resource capture and creating their own microhabitats, e.g., anaerobic metabolisms in a coastal shallow and eutrophicated water column. Correlations between MT and TOC, and between MT and ΣCF are negative, due to increased toxicity with increasing levels of TOC since organic matter accumulates contamination, implying a stronger biological response (reducing MT) under conditions of high eutrophication and contamination (Fig. 4 B and E).

The ecological potential risk of sediments from GB and SB were classified as high ($QR \leq 10^{-3}$), moderate ($10^{-2} \leq QR < 10^{-1}$), and low risks ($QR \geq 10^{-1}$) from applying the QR index. The distribution pattern of the risk follows that of its GT, presenting the highest risk in the areas with the lowest hydrodynamic energy and proximity to the continental inputs of contaminants, especially, N-NW and E sectors and harbor areas in GB and especially close to the metallurgic company Ingá, near the area with port activities and spreading to the internal sector of SB (Fig. 4C and F).

6 Quality ratio performance assessment: Taxonomic approach

As discussed earlier, microbial communities are ubiquitous and are also submitted to changes under environmental disturbance. We had hypothesized that the metabolic viability of the microbial community responds to environmental disturbances, in a chronically and/or acute way, and can be used to assess ecological risk in sediments. That is why the Quality Ratio index (QR), as previously described in this chapter, was developed by Nascimento et al. (2019). However, some questions still remain: (i) Which taxonomy groups are representative in these sediments? (ii) Do these groups change as a result of prolonged contamination? and (iii) Can QR infer these changes on microbial community structure?

To answer these questions, two different dredging disposal scenarios (onshore and offshore) were simulated, on a laboratory scale. Microbial biomarkers (EST, DHA, and CELL) and geochemical parameters (grain size, $\sum$CF, and TOC) were monitored under contamination gradient and timescale, and QR was also applied (Nascimento et al., 2020, 2019). Microbial community composition was addressed by amplifying the 16S rRNA gene and sequencing was done using MiSeq platform (Illumina), following the standard procedures of the Earth Microbiome Project (http://www.earthmicrobiome.org/emp-standard-protocols/16s/ (Bokulich et al., 2012; Caporaso et al., 2011; Quast et al., 2013; Thompson et al., 2017).

The assay simulating offshore dumping was developed and fully described by Nascimento et al. (2020). Dredged sediments from GB were mixed with sediments from outside of the bay area (D) in three different proportions (25%, 50%, and 75% m:m) which we labeled GBD25, GBD50, and GBD75. These mixtures were kept at 20 °C and all parameters were measured after 24 h of incubation. Sequencing was performed for every sample of this assay. Onshore disposal simulations were also conducted by mixing (in three different proportions of 25%, 50%, and 75% m:m) dredged sediments from GB with ferralsol (S) which we labeled GBS25, GBS50, and GBS75. They were also kept at 20°C and all parameters were measured weekly, over a period of 28 days (T0–24 h, T1–7 days, T2–14 days, T3–21 days, T4–28 days). In this case, sequencing was performed for samples at T0 and T4 (Nascimento, 2018).

To answer our first question of which taxonomy groups are representative in these sediments, we obtained a total of 29,179 OTUs by means of high-throughput 16S rRNA sequencing. Fig. 5 highlights the most representative OTUS (> 3% of relative abundance in at least one replicate) and the respective taxonomy classification is listed in Table 4.

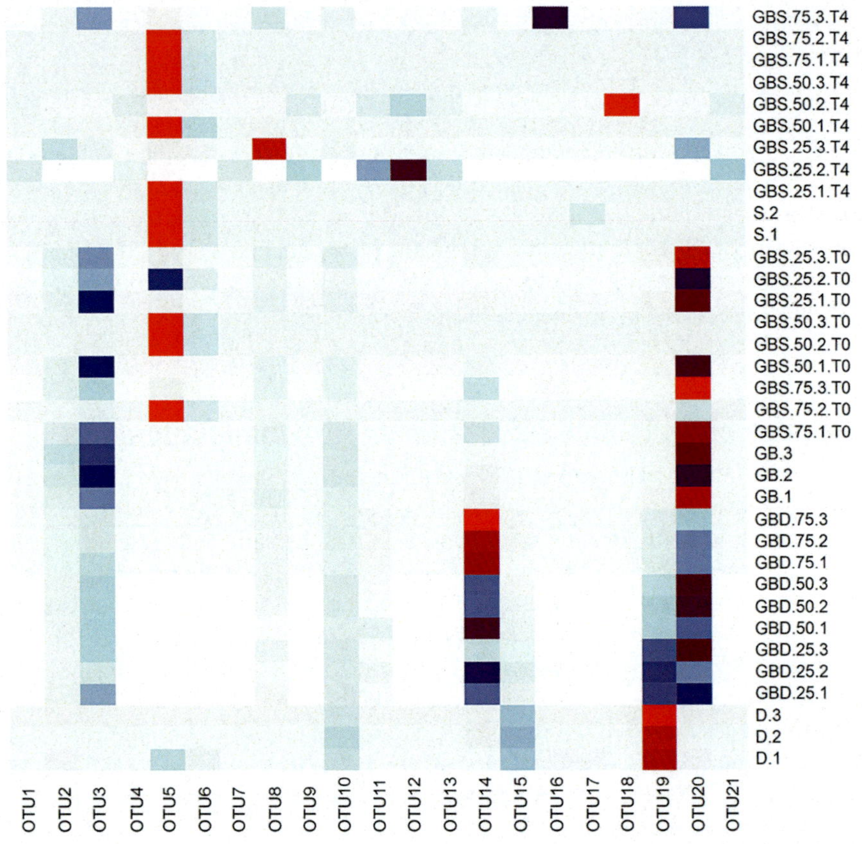

FIG. 5

Heatmap highlighting OTUs with relative abundance > 3%. White and light-blue shades represent low abundance; dark-blue and red shades represent greater relative abundance. The respective taxonomy identifications for each OTU are shown in Table 4.

Table 4 Taxonomy classification of OTUs identified in Fig. 5.

OTU number	OTU ID	Kingdom	Phylum	Class	Order	Family	Genus	Species
OTU1	New.Reference OTU287	Bacteria	Bacteroidetes	Bacteroidia	Bacteroidales	Prolixibacteraceae	Prolixibacter	
OTU2	GQ249534.1.1200			Flavobacteriia	Flavobacteriales	Flavobacteriaceae	Actibacter	
OTU3	EF125407.1.1477		Chloroflexi	Anaerolineae	Anaerolineales	Anaerolineaceae		
OTU4	JF345523.1.1302		Firmicutes	Bacilli	Bacillales	Family.XII	Exiguobacterium	
OTU5	JQ712536.1.1265					Paenibacillaceae	Paenibacillus	
OTU6	KM260652.1.1476							
OTU7	HG798413.1.1376				Lactobacillales	Streptococcaceae	Lactococcus	*Lactococcus lactis*
OTU8	AM176883.1.1405		Proteobacteria	Alphaproteobacteria	Rhizobiales	Rhodobiaceae	Anderseniella	
OTU9	DQ071270.1.1513			Deltaproteobacteria	Desulfovibrionales	Desulfovibrionaceae	Desulfovibrio	
OTU10	JQ580211.1.1503				Myxococcales	Sandaracinaceae		
OTU11	JN118552.1.1427			Epsilonproteobacteria	Campylobacterales	Campylobacteraceae	Arcobacter	
OTU12	HG932573.1.1402							
OTU13	New.Reference OTU619							
OTU14	FJ901685.1.1333					Helicobacteraceae	Sulfurovum	
OTU15	HQ153954.1.1438			Gammaproteobacteria	BD7-8.marine.group			
OTU16	GU120553.1.1467				Chromatiales	Ectothiorhodospiraceae	Thioalkalispira	
OTU17	JX522461.1.1370				Pseudomonadales	Pseudomonadaceae	Pseudomonas	
OTU18	FJ900839.1.1396							
OTU19	JQ825030.1.1500				Xanthomonadales	JTB255. marine.benthic.group		
OTU20	AB530194.1.1501			Milano-WF1B-44				
OTU21	AF509468.1.1359		Thermotogae	Thermotogae	Petrotogales	Petrotogaceae		

Sample D was dominated by OTU 19 (JTB255 marine benthic group), OTU 15 (BD7–8 marine group), and OTU 10 (Sandaracinaceae) representing around 13%, 5%, and 3% of relative abundance in the three replicates, respectively (Fig. 5). The JTB255 marine benthic group has been reported in many marine studies and expresses a great range of metabolisms which explains the ubiquity and high abundance in sediments (Bienhold et al., 2016; Corinaldesi et al., 2018; Mußmann et al., 2017; Probandt et al., 2018). In addition, Aylagas et al. (2017) found these three taxa related to good environmental conditions while analyzing the microbial composition at locations subjected to different anthropogenic impacts in Spain.

In the GB sample, the greatest relative abundance corresponds to OTU20 (Milano-WF1B-44), followed by OTU3 (Anaerolineaceae), comprising about 2 to 3% of relative abundance in each replicate (Fig. 5). These groups are often found in anaerobic and sulfur-enriched environments (Guo et al., 2019; Müller et al., 2018; Trembath-Reichert et al., 2016; Ugarelli et al., 2018). OTUs related to Anaerolineaceae were found to be associated with anthropogenically impacted areas along the northern coast of Spain (Aylagas et al., 2017). Based on what is available in the literature, it is noteworthy that these groups reflect the environmental damage caused by the lack of sanitary conditions and other anthropogenic impacts in the study area.

In the GBD mixtures, the OTU14 (Sulfurovum) is the most representative, representing between 3% and 7% of the relative abundance. Although it is very expressive in the mixtures, it has a relative abundance of only 0.08% (GB) and 0.01% (D) in sediments (Fig. 5). Microbes from genus Sulfurovum are denitrifying sulfur-oxidizing bacteria, often found in suboxic or anoxic niches, associated with anaerobic methane-oxidizing archaea and sulfate-reducing bacteria (Lipsewers et al., 2017; Pjevac et al., 2014; Probandt et al., 2018; Trembath-Reichert et al., 2016). The presence of this type of organism also is congruent with the eutrophic characteristics of Guanabara Bay, which assembles factors for sulfate reduction: C:S ratio expressing the reduced trend, labile TOC, and available reactive sulfur (Sabadini-Santos et al., 2014a,b).

OTU 5 and OTU6 (Paenibacillus) are highly expressed in S and most of GBS replicates (T0 and T4) and reach more than 80% of relative abundance in some replicates (Fig. 5). The majority of Paenibacillus species are found in soils, associated with plant roots. Indeed, some species from this genus can also produce antibiotics and may be utilized in the removal or degradation of pollutants (as petroleum components), through bioflocculation or enzymatic activities (Grady et al., 2016). In this experiment, these OTUs are resistant, since they remain highly abundant during the experimental timescale (Allison and Martiny, 2008; Harrison et al., 2007).

The most abundant OTUs for GBS-T0 are also representative of GB sediments, such as OTU20 (Milano-WF1B-44) and OTU3 (Anaerolineaceae) and are less represented in GBS-T4. The laboratory conditions might support the increase of the OTU4 (*Exiguobacterium*) abundance since it is absent in every replicate of S and GB samples; however, it is observed in every GBS replicate, especially in GBS-T4 samples. Microbes from the *Exiguobacterium* genus are found in a great range of environments, such as permafrost, soil, freshwater mirabilites, and root rhizosphere. Hyson et al. (2015) identified some tolerance genes, including genes linked to tolerance for trace metals while sequencing an *Exiguobacterium* lineage isolated in Pennsylvania (USA). Based on this finding, we suggest that the increase of OTU4 in GBS-T4 samples represents the adaptation to contamination effects.

The taxonomic profile could also answer our second question “Do these groups change as a result of prolonged contamination?” since the most representative OTUs of each sediment and mixture reflect the respective environmental characteristics. But elaborating on this investigation, Fig. 6 exhibits the β

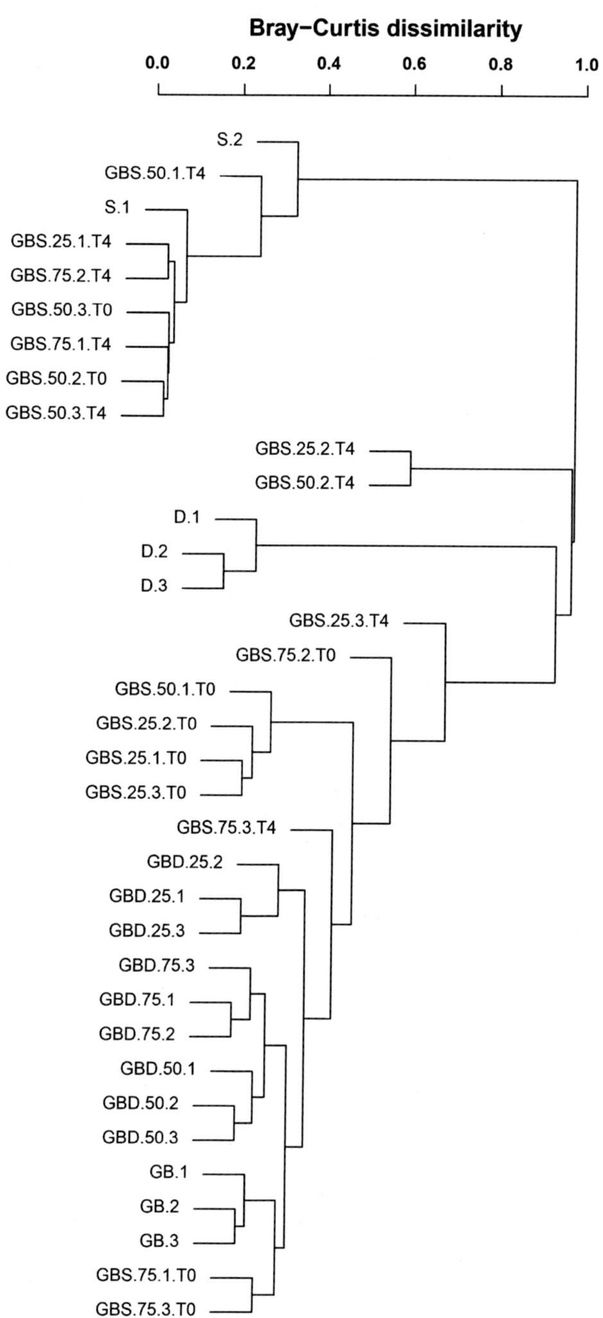

FIG. 6

Cluster dendrogram calculated by Bray-Curtis dissimilarity, obtained through sequencing of GB and D sediments; S sample; the mixtures between GB and D sediments (GBD); and the mixtures between GB and S in the first and fourth weeks (GBS-T0 and GBS-T4, respectively).

diversity cluster dendrogram targeted by Bray-Curtis dissimilarity (Bray and Curtis, 1957). It is possible to distinguish three main groups, where: (1) most of GBD and GBS-T0 replicates are closely related to GB replicates; (2) the three replicates of D sample; and (3) most of GBS-T4 replicates related to S samples. Throughout the mixtures, it is notable that the microbial communities shifted from an aerobic (D and S samples) to an anaerobic profile consistent with the eutrophic and reducing characteristics of the GB sample. The sensitivity of the community present in the D and S samples is noticeable, since the composition of the community changed in 24 h. However, after 28 days of incubation, microbial communities from GBS mixtures seem to resemble the original composition of S sample, suggesting the community's resilience.

In general, microbial communities from GBS-T4 mixtures are resilient since the relative abundance of the original microbes from soil decreases and then revert to a similar pattern from the beginning of the experiment. Allison and Martiny (2008) highlighted the rapid evolutionary processes related to microbial communities such as mutations and lateral gene transfer. These mechanisms lead to adaptation to changes in environmental conditions and consequent recolonization after a disturbance scenario—here represented by the addition of sediment from GB.

Before answering the third question "Can QR infers these changes on microbial community structure?" it is necessary to understand the changes in the environmental parameters evaluated and Table 4 summarizes the results. The quality ratio (QR) estimated for the present data indicates a high risk for GB sediments, all GBD mixtures, and GBS75-T0 ($QR \leq 10^{-3)}$; medium risk for S, GBS25-T0, GBS50-T0, and GBS25-T4 samples ($QR = 10^{-2}$); low risk was observed for samples D, GBS50-T4, and GBS75-T4 ($QR = 10^{-1}$). DHA activities could not be detected at GBD50 and GBD75, but to calculate the microbial term (MT) and respective QR, the detection limit of 0.02 μg INTF g^{-1} was applied instead. We cannot affirm the exact MT value, but it certainly is sufficient to consider $QR < 10^{-3}$.

Finally, in response to the third question, we applied the Non-metric multidimensional scaling (NMDS) to ordinate the samples. For a better interpretation, the environmental variables were fitted as vectors by overlaying information onto ordination diagrams (Fig. 7). The arrows point to the direction of the gradient where the most rapid change occurs in the variable. The length of the arrow is the strength of the gradient, and is proportional to the correlation between ordination and variable (Oksanen et al., 2015).

Grain size (Sand and Mud), Esterases enzymes (EST), Quality Ratio—and its terms MT and GT—and $\sum$CF are the most significant vectors ($P < .05$) to explain the gradient of contamination and, consequently, shift in microbial community profile. It is possible to observe that vectors representing QR, MT, and Sand are pointed to D replicates, suggesting great environmental conditions and also a better microbial community availability. It is important to keep in mind that the higher the QR values, the greater the environmental conditions. By contrast, vectors associated with contamination processes, such as $\sum$CF and GT, are directed to GB, GBD, and GBS-T0 replicates, which are the most contaminated samples (Fig. 7).

Our findings emphasize the importance of considering the effects of risk assessment on the community level and underscore the susceptibility of microbial communities under a contamination gradient. Some laboratory assays such as Microtox, Mutatox, and ToxScreen, and others to assess LC50, for example (Janssen and Persoone, 1993; Johnson and Long, 1998; Ulitzur et al., 2002), are widely used in monitoring studies to evaluate the toxicity in a short time period. The Brazilian law for dredging management (CONAMA, 2012) also requires laboratory tests to verify toxicity and the potential risk in sediments. Unlike we are discussing in this chapter, these available standardized tests are not

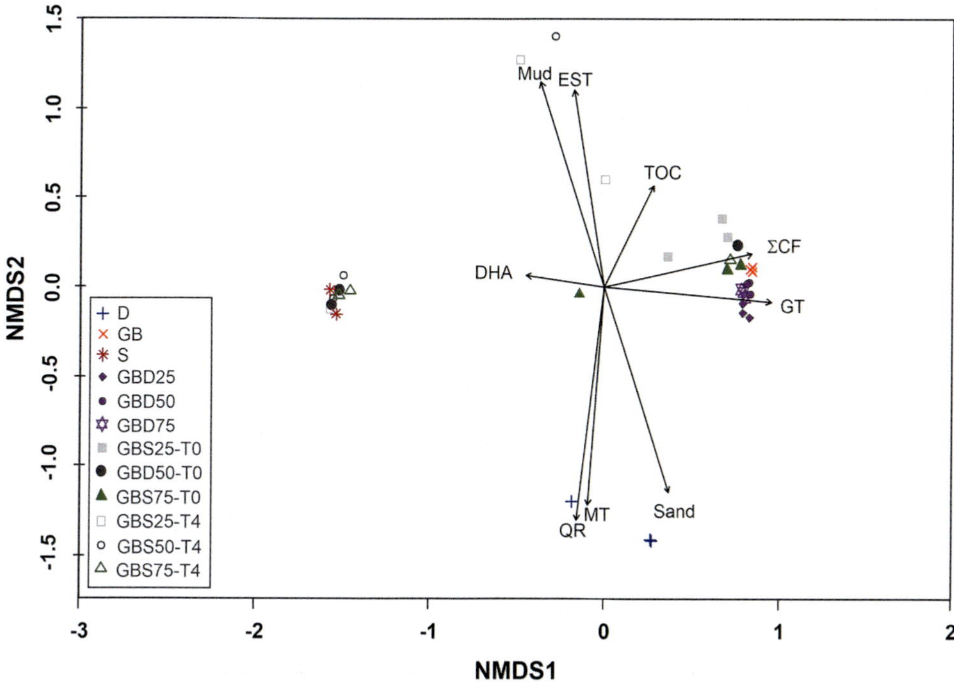

FIG. 7

NMDS representing the ordination of the samples according to discrepancies on community composition. The vectors represent the geochemical parameters evaluated, and the length of the arrow is proportional to the correlation between ordination and variable.

representative of the natural system. Thus the QR (Nascimento et al., 2019) reinforces the use of in situ biota to better represent the local conditions and contribute to improve environmental policies.

7 Conclusion

The microbial community has evolved in a syntrophic way within a biofilm, in which complementary metabolic populations share information and better adapt to the environmental disturbances from the Archean to the present day. Thus ubiquitous spreading ensures that it is suitable as bioindicators for environmental risk assessment and the Quality Ratio (QR) index makes use of it.

The reduction of hydrodynamic capacity from the entrance to the bottom of coastal bays and the contaminated loads of a river's input plays a key role in the distribution of coastal sediments, accumulating fine grains, organic matter, and contaminants under anoxic conditions—in QR expressed, respectively, as total organic carbon (TOC) and the sum of the contamination factor of metals (ΣCF), as indicators of multiple contaminants. Thus coastal sediments release contaminants when exposed to oxidation caused by dredging, which is a matter of concern due to the increasing dredging demand worldwide.

The QR index approach was able to integrate the environmental variables in a geochemical term (GT = TOC × ΣCF/Fine-Grain) and the in situ microbial response to contamination exposure in a microbial term (MT). The MT is the ration between the dehydrogenase activity of the electron transfer system (DHA) and the esterases activities (EST) of the organic matter hydrolysis. The exposure to contaminants causes homeostasis disturbance, increasing energy demand (EST), and decreasing intracellular ATP generation (DHA), reflected in a reduced MT.

The QR index was applied to coastal sediments for measuring the potential risk and was able to segregate the hot spots of contamination. The QR was also applied to an acute assay and successfully identified the microbial community shift under a contamination gradient when mixing with dredged sediments. Thus QR proved to be robust, fast, economical, and representative of the microbial exposure to multiple contaminants in situ. Its conservative character regarding the preservation of the environmental parameters to which the microbiota are exposed in a natural environment, once the enzyme activity analysis was performed right after the sampling, makes the QR a great alternative to ex situ bioassays established in legislations for the monitoring of contaminated sediments and soils.

Acknowledgments

This study was financed by the Coordenação de Aperfeiçoamento de Pessoal de Nível Superior—Brasil (CAPES/ Finance Code 001), PRINT-UFF-FEEDBACKS Project (CAPES/ Finance Code 88887.310301/2018-00) and Programa de Doutorado Sanduíche no Exterior (CAPES/ Finance Code 88881.135581/2016-01) and by the Conselho Nacional de Desenvolvimento Científico e Tecnológico (CNPq/Universal Finance Code 449631/2014-1 and CNPq/PDJ Finance Code 155406/2018-3). The authors are also thankful to professors Mirian Crapez, Emmanoel Silva-Filho, Renato Cordeiro, and Micheli Ferreira from Universidade Federal Fluminense (UFF) for lab facilities.

References

Acosta-Martínez, V., Tabatabai, M.A., 2001. Arylamidase activity in soils: effect of trace elements and relationships to soil properties and activities of amidohydrolases. Soil Biol. Biochem. 33 (1), 17–23. https://doi.org/10.1016/S0038-0717(00)00109-7.

Adams, S.M., Giesy, J.P., Tremblay, L.A., Eason, C.T., 2001. The use of biomarkers in ecological risk assessment: recommendations from the Christchurch conference on biomarkers in ecotoxicology. Biomarkers 6 (1), 1–6. https://doi.org/10.1080/135475001452724.

Allison, S.D., Martiny, J.B.H., 2008. Resistance, resilience, and redundancy in microbial communities. Proc. Natl. Acad. Sci. U. S. A. 105 (Suppl. 1), 11512–11519. https://doi.org/10.1073/pnas.0801925105.

Amador, E.D.S., 1997. Baía de Guanabara e ecossistemas periféricos: homem e natureza, first ed. Rio de Janeiro.

Aylagas, E., Borja, Á., Tangherlini, M., Dell'Anno, A., Corinaldesi, C., Michell, C.T., Irigoien, X., Danovaro, R., Rodríguez-Ezpeleta, N., 2017. A bacterial community-based index to assess the ecological status of estuarine and coastal environments. Mar. Pollut. Bull. 114 (2), 679–688. https://doi.org/10.1016/j.marpolbul.2016.10.050.

Azam, F., 1998. Microbial control of oceanic carbon flux: the plot thickens. Science 280 (5364), 694–696. https://doi.org/10.1126/science.280.5364.694.

Baptista-Neto, J.A., Barreto, C.F., Vilela, C.G., da Fonseca, E.M., Melo, G.V., Barth, O.M., 2016. Environmental change in Guanabara Bay, SE Brazil, based in microfaunal, pollen and geochemical proxies in sedimentary cores. Ocean Coast. Manag. 143, 4–15. https://doi.org/10.1016/j.ocecoaman.2016.04.010.

Bartell, S.M., 2006. Biomarkers, bioindicators, and ecological risk assessment—a brief review and evaluation. Environ. Bioindic. 1 (1), 60–73. https://doi.org/10.1080/15555270591004920.

Bidone, E.D., Lacerda, L.D., 2004. The use of DPSIR framework to evaluate sustainability in coastal areas. Case study: Gunabara Bay Basin, Rio de Janeiro, Brazil. Reg. Environ. Change 4 (1), 5–16.

Bienhold, C., Zinger, L., Boetius, A., Ramette, A., 2016. Diversity and biogeography of bathyal and abyssal seafloor bacteria. PLoS One 11 (1), 1–20. https://doi.org/10.1371/journal.pone.0148016.

Bloem, J., Breure, A.M., 2003. Microbial indicators. In: Market, B.A., Breure, A.M., Zechmeister, H.G. (Eds.), Bioindicators and Biomonitors: Principles, Concepts and Applications, first ed. Elsevier Science Ltd, Amsterdam.

Bokulich, N.A., Subramanian, S., Faith, J.J., Gevers, D., Gordon, J.I., Knight, R., Mills, D.A., Caporaso, J.G., 2012. Quality-filtering vastly improves diversity estimates from Illumina amplicon sequencing. Nat. Methods 10 (1), 57–59. https://doi.org/10.1038/nmeth.2276.

Borges, H.V., Nittrouer, C.A., 2015. Coastal sedimentation in a tropical Barrier-Island system during the past century in Sepetiba Bay, Brazil Sedimentação Costeira em um Sistema de Ilha-de-Barreira Tropical Durante o Último Século na Baía de Sepetiba. Anu. Inst. Geocienc. 39 (2), 5–14. https://doi.org/10.11137/2016_2_05_14.

Bray, J.R., Curtis, J.T., 1957. An ordination of the upland forest communities of southern Wisconsin. Ecol. Monogr. 27 (4), 325–349. https://doi.org/10.2307/1942268.

Burne, R.V., Moore, L.S., 1987. Microbialites: organosedimentary deposits of benthic microbial communities. Palaios 2, 241–254.

Buruaem, L.M., de Castro, Í.B., Hortellani, M.A., Taniguchi, S., Fillmann, G., Sasaki, S.T., et al., 2013. Integrated quality assessment of sediments from harbour areas in Santos-São Vicente estuarine system, southern Brazil. Estuar. Coast. Shelf Sci. 130, 179–189. https://doi.org/10.1016/j.ecss.2013.06.006.

Busch, J., Nascimento, J.R., Magalhães, A.C.R., Dutilh, B.E., Dinsdale, E., 2015. Copper tolerance and distribution of epibiotic bacteria associated with giant kelp Macrocystis pyrifera in southern California. Ecotoxicology 24, 1131–1140. https://doi.org/10.1007/s10646-015-1460-6.

Caporaso, J.G., Lauber, C.L., Walters, W.A., Berg-Lyons, D., Lozupone, C.A., Turnbaugh, P.J., Fierer, N., Knight, R., 2011. Global patterns of 16S rRNA diversity at a depth of millions of sequences per sample. Proc. Natl. Acad. Sci. U. S. A. 108, 4516–4522. https://doi.org/10.1073/pnas.1000080107.

Carreira, R.S., Wagener, A.L., Readman, J.W., Fileman, T.W., Macko, S.A., Veiga, A., 2002. Changes in the sedimentary organic carbon pool of a fertilized tropical estuary, Guanabara Bay, Brazil: an elemental, isotopic and molecular marker approach. Mar. Chem. 79, 207–227. https://doi.org/10.1016/S0304-4203(02)00065-8.

Carreira, R.S., Ribeiro, P.V., Silva, C.E.M., Farias, C.O., 2009. Hidrocarbonetos e Esterois como indicadores de fontes e destino de matéria orgânica em sedimentos da Baía de Sepetiba, Rio De Janeiro. Quim. Nova 32, 1805–1811. https://doi.org/10.1590/S0100-40422009000700023.

Caruso, G., La Ferla, R., Azzaro, M., et al., 2016. Microbial assemblages for environmental quality assessment: knowledge, gaps and usefulness in the European marine strategy framework directive. Crit. Rev. Microbiol. 42 (6), 883–904. https://doi.org/10.3109/1040841X.2015.1087380.

Carvalho, A.C.B., Silva, L.J., Dick, D.P., Moreira, V.D.A., Vicente, M.D.C., 2020. Análise espectroscópica da matéria orgânica no sedimento superficial da Baía de Sepetiba, Rio De Janeiro, Brasil. Quim. Nova 42, 552–557. https://doi.org/10.21577/0100-4042.20170518.

CCME—Canadian Council of Ministers of the Environment, 1999. Canadian Sediment Quality Guidelines for the Protection of Aquatic Life. Canadian Environmental Quality Guidelines—Summary Tables. https://www.elaw.org/system/files/sediment_summary_table.pdf. (Accessed 13 April 2020).

Censo, 2010. Censo Demográfico—Características Gerais da População. Resultados da Amostra. Instituto Brasileiro de Geografia e Estatística. http://censo2010.ibge.gov.br/resultados.html.

CENTRAN, 2009. Infra-estrutura Portuária Nacional de Apoio ao Comércio Exterior—Relatório Final—Centro de Excelência em Engenharia de Transportes (CENTRAN), Secretaria de Planejamento e Investimentos Estratégicos do Ministério do Planejamento, Orçamento e Gestão (SPI/MP) e Secretaria Especial de Portos da Presidência da República (SEP/PR) (impublished report).

CONAMA, 2012. Resolução no 454, de 01 de novembro de 2012. Diário Oficial da União, p. 17. http://www2.mma.gov.br/port/conama/legiabre.cfm?codlegi=693.

Cordeiro, R.C., Machado, W., Santelli, R.E., Figueiredo, A.G., Seoane, J.C.S., Oliveira, E.P., et al., 2015. Geochemical fractionation of metals and semimetals in surface sediments from tropical impacted estuary (Guanabara Bay, Brazil). Environ. Earth Sci. 74 (2), 1363–1378. https://doi.org/10.1007/s12665-015-4127-y.

Corinaldesi, C., Tangherlini, M., Manea, E., Dell'Anno, A., 2018. Extracellular DNA as a genetic recorder of microbial diversity in benthic deep-sea ecosystems. Sci. Rep. 8 (1), 1–9. https://doi.org/10.1038/s41598-018-20302-7.

Cornall, A., Rose, A., Streten, C., McGuinness, K., Parry, D., Gibb, K., 2016. Molecular screening of microbial communities for candidate indicators of multiple metal impacts in marine sediments from northern Australia. Environ. Toxicol. Chem. 35, 468–484. https://doi.org/10.1002/etc.3205.

da Waite, C.C.C., da Silva, G.O.A., Bitencourt, J.A.P., Sabadini-Santos, E., Crapez, M.A.C., 2016. Copper and lead removal from aqueous solutions by bacterial consortia acting as biosorbents. Mar. Pollut. Bull. 109 (1), 386–392. https://doi.org/10.1016/j.marpolbul.2016.05.044.

Danovaro, R., Company, J.B., Corinaldesi, C., D'Onghia, G., Galil, B., Gambi, C., et al., 2010. Deep-sea biodiversity in the Mediterranean Sea: the known, the unknown, and the unknowable. PLoS One 5 (8), 25. https://doi.org/10.1371/journal.pone.0011832.

Decho, A.W., 2000. Microbial biofilms in intertidal systems: an overview. Cont. Shelf Res. 20, 1257–1273.

Decho, A.W., Gutierrez, T., 2017. Microbial extracellular polymeric substances (EPSs) in ocean systems. Front. Microbiol. 8, 922. https://doi.org/10.3389/fmicb.2017.00922.

Dong, H., 2010. Mineral-microbe interactions: a review. Front. Earth Sci. 4 (2), 127–147.

Dupraz, C., Reid, R.P., Braissant, O., Decho, A.W., Norman, R.S., Visscher, P.T., 2009. Processes of carbonate precipitation in modern microbial mats. Earth Sci. Rev. 96, 141–162. https://doi.org/10.1016/j.earscirev.2008.10.005.

Ehrlich, H.L., Newman, D.K., 2009. Geomicrobiology, fifth ed. CRC Press, New York, ISBN: 978-0-8493-7906-2.

Environment Canada and Ministère du Développement durable, de l'Environnement et des Parcs du Québec, 2007. Criteria for the Assessment of Sediment Quality in Quebec and Application Frameworks: Prevention, Dredging and Remediation. http://planstlaurent.qc.ca/fileadmin/publications/diverses/Registre_de_dragage/Criteria_sediment_2007E.pdf. (Accessed 13 April 2020).

Fiori, C.da S.F., Rodrigues, A.P. de C., Santelli, R.E., Cordeiro, R.C., Carvalheira, R.G., Araújo, P.C., Castilhos, Z.C., Bidone, E.D., 2013. Ecological risk index for aquatic pollution control: a case study of coastal water bodies from the Rio de Janeiro State, southeastern Brazil. Geochim. Bras. 2 (5), 24–36. https://doi.org/10.5327/Z0102 9800201300010003.

Flemming, H.C., Wingender, J., 2010. The biofilm matrix. Nat. Rev. Microbiol. 19, 139–150. https://doi.org/10.1038/nrmicro2415.

Förstner, U., Schoer, J., 1984. Some Typical Examples of the Importance of the Role of Sediments in the Propagation and Accumulation of Pollutants. Viena. http://inis.iaea.org/search/search.aspx?orig_q=RN:15040908.

Förstner, U., Wittmann, G.T.W., 1981. Metal Pollution in the Aquatic Environment. Metal Pollution in the Aquatic Environment, second ed. Springer Verlag, https://doi.org/10.1007/978-3-642-69385-4_1.

Gadd, G.M., 2010. Metals, minerals and microbes: geomicrobiology and bioremediation. Microbiology 156, 609–643.

Gomes, F.deC., Godoy, J.M., Godoy, M.L.D.P., Carvalho, Z.L., Lopes, R.T., Sanchez-Cabeza, J.A., et al., 2009. Metal concentrations, fluxes, inventories and chronologies in sediments from Sepetiba and Ribeira Bays: a comparative study. Mar. Pollut. Bull. 59 (4–7), 123–133. https://doi.org/10.1016/j.marpolbul.2009.03.015.

Grady, E.N., MacDonald, J., Liu, L., Richman, A., Yuan, Z.C., 2016. Current knowledge and perspectives of Paenibacillus: a review. Microb. Cell Fact. 15 (1), 1–18. https://doi.org/10.1186/s12934-016-0603-7.

Grotzinger, J.P., Knoll, A.H., 1999. Stromatolites in Precambrian carbonates: evolutionary mileposts or environmental dipsticks? Annu. Rev. Earth Planet. Sci. 27, 313–358. https://doi.org/10.1146/annurev.earth.27.1.313.

Guo, G., Ekama, G.A., Wang, Y., Dai, J., Biswal, B.K., Chen, G., Wu, D., 2019. Advances in sulfur conversion-associated enhanced biological phosphorus removal in sulfate-rich wastewater treatment: a review. Bioresour. Technol. https://doi.org/10.1016/j.biortech.2019.03.142.

Håkanson, L., 1980. An ecological risk index for aquatic pollution control. A sedimentological approach. Water Res. 14, 975–1001.

Håkanson, L., 1984. Aquatic contamination and ecological risk. An attempt to a conceptual framework. Water Res. 18 (9), 1107–1118. https://doi.org/10.1016/0043-1354(84)90225-2.

Hamburg Port Authority, 2011. Assessment Criteria for Dredged Material with special focus on the North Sea Region Prepared by Henrich Röper/Axel Netzband With support from DGE/Dredging in Europe. http://www.ospar.org/v_measures/get_page.asp?v0=09-04e_Guidelines. (Accessed 13 April 2020).

Hamza-Chaffai, A., 2014. Usefulness of bioindicators and biomarkers in pollution biomonitoring. Int. J. Biotechnol. Wellness Ind. 3 (1), 19–26. https://doi.org/10.6000/1927-3037.2014.03.01.4.

Harrison, J.J., Ceri, H., Turner, R.J., 2007. Multimetal resistance and tolerance in microbial biofilms. Nat. Rev. Microbiol. 5 (12), 928–938. https://doi.org/10.1038/nrmicro1774.

Hyson, P., Shapiro, J.A., Wien, M.W., 2015. Draft genome sequence of Exiguobacterium sp. strain BMC-KP, an environmental isolate from Bryn Mawr, Pennsylvania. Genome Announc. 3 (5). https://doi.org/10.1128/genomeA.01164-15, 01164-15.

INEA, 2016. Gestão das águas> Instrumentos de Gestão de Recursos Hídricos> Planos de Bacias Hidrográficas> Baía de Guanabara. www.inea.rj.gov.br. (Accessed 1 November 2016).

INEA—Instituto Estadual do Ambiente, 2014. Gerenciamento costeiro. http://www.inea.rj.gov.br/Portal/Agendas/GESTAODEAGUAS/Gerenciamentocosteiro/index.htm&lang. (Accessed 12 February 2014).

Janssen, C.R., Persoone, G., 1993. Rapid toxicity screening tests for aquatic biota 1. Methodology and experiments with Daphnia magna. Environ. Toxicol. Chem. 12 (4), 711–717. https://doi.org/10.1002/etc.5620120413.

Jião, N., et al., 2010. Microbial production of recalcitrant dissolved organic matter: long-term carbon storage in the global ocean. Nat. Rev. Microbiol. 8, 593–599.

JICA, 1994. The Study on Recuperation of the Guanabara Bay Ecosystem. Japan International Cooperation Agency: Kokusai Kogyo Co., Ltd, Tokyo. https://openjicareport.jica.go.jp/619/619/619_703_11198470.html. (Accessed 21 October 2020).

Johnson, B.T., Long, E.R., 1998. Rapid toxicity assessment of sediments from estuarine ecosystems: a new tandem in vitro testing approach. Environ. Toxicol. Chem. 17 (6), 1099. https://doi.org/10.1897/1551-5028(1998)017.

Kienzl, K., Riss, A., Vogel, W., Hackl, J., Götz, B., Breure, A.M., 2003. Bioindicators and biomonitors for policy, legislation and administration. In: Market, B.A., Zechmeister, H.G. (Eds.), Bioindicators and Biomonitors: Principles, Concepts and Applications, first ed. Elsevier Science Ltd., Amsterdam, https://doi.org/10.1016/S0927-5215(03)80133-9.

Kjerfve, B., 2001. Circulation and Transport in Baía de Sepetiba, Rio de Janeiro. Niterói, Brazil: CNPq-UFF PRONEX Project Report.

Kjerfve, B., Ribeiro, C.H., Dias, G.T., Filippo, A.M., Quaresma, V.D.S., 1997. Oceanographic characteristics of an impacted coastal bay: Baía de Guanabara, Rio de Janeiro, Brazil. Cont. Shelf Res. 17, 1609–1643. https://doi.org/10.1016/S0278-4343(97)00028-9.

Knopka, A., 2006. Microbial ecology: searching for principles. Microbe 1 (4), 175–179.

Li, H., Yao, J., Gu, J., Duran, R., Roha, B., Jordan, G., et al., 2018. Microcalorimetry and enzyme activity to determine the effect of nickel and sodium butyl xanthate on soil microbial community. Ecotoxicol. Environ. Saf. 163, 577–584. https://doi.org/10.1016/j.ecoenv.2018.07.108.

Linkov, I., Satterstrom, F.K., Kiker, G., Batchelor, C., Bridges, T., Ferguson, E., 2006. From comparative risk assessment to multi-criteria decision analysis and adaptive management: recent developments and applications. Environ. Int. 32 (8), 1072–1093. https://doi.org/10.1016/j.envint.2006.06.013.

Lipsewers, Y.A., Vasquez-Cardenas, D., Seitaj, D., Schauer, R., Hidalgo-Martinez, S., Damsté, J.S.S., Meysman, F.J.R., Villanueva, L., Boschker, H.T.S., 2017. Impact of seasonal hypoxia on activity and community structure of chemolithoautotrophic bacteria in a coastal sediment. Appl. Environ. Microbiol. 83 (10). https://doi.org/10.1128/AEM.03517-16.

Liu, J.L., Yao, J., Lu, C., Li, H., Li, Z.F., Duran, R., et al., 2019. Microbial activity and biodiversity responding to contamination of metal(loid) in heterogeneous nonferrous mining and smelting areas. Chemosphere 226, 659–667. https://doi.org/10.1016/j.chemosphere.2019.03.051.

London Convention, 1972. Convention on the Prevention of Marine Pollution by Dumping of Wastes and Other Matter. London Convention, London.

London Protocol, 1996. Protocol to the Convention on the Prevention of Marine Pollution by Dumping of Wastes and Other Matter. London Convention, London.

Long, E.R., Chapman, P.M., 1985. A sediment quality triad: measures of sediment contamination, toxicity and infaunal community composition. Mar. Pollut. Bull. 16 (10), 405–415.

López-García, P., Moreira, D., 1999. Metabolic symbiosis at the origin of eukaryotes. Trends Biochem. Sci. 24 (3), 88–93. https://doi.org/10.1016/S0968-0004(98)01342-5.

Lorenz, C.M., 2003. Bioindicators for ecosystem management, with special reference to freshwater systems. In: Market, B.A., Breure, A.M., Zechmeister, H.G. (Eds.), Bioindicators and Biomonitors: Principles, Concepts and Applications, first ed. Elsevier Science Ltd., Amisterdam.

Lowe, D.R., 1994. A biological origin of described stromatolites older than 3.2Ga. Geology 22, 387–390.

Macdonald, D.D., Carr, R.S., Calder, F.D., Long, E.R., Ingersoll, C.G., 1996. Development and evaluation of sediment quality guidelines for Florida coastal waters. Ecotoxicology 5 (4), 253–278. https://doi.org/10.1007/BF00118995.

MacDonald, D.D., 1994. Approach to the Assessment of Sediment Quality in Florida Coastal Waters. Volume 1-Development and Evaluation of Sediment Quality Assessment Guidelines. Florida. https://www.waterboards.ca.gov/water_issues/programs/tmdl/docs/303d_policydocs/239.pdf. (Accessed 13 April 2020).

MacDonald, D.D., Ingersoll, C.G., Smorong, D.E., Lindskoog, R.A., 2003. In: Development and Applications of Sediment Quality Criteria for Managing Contaminated Sediment in British Columbia, British Columbia Ministry of Water. Land and Air Protection Environmental Management Branch, British Columbia.

Manap, N., Voulvoulis, N., 2015. Environmental management for dredging sediments–the requirement of developing nations. J. Environ. Manage. 147, 338–348.

Market, B.A., Breure, A.M., Zechmeister, H.G., 2003. Definitions, strategies and principles for bioindication/biomonitoring of the environment. In: Market, B.A., Breure, A.M., Zechmeister, H.G. (Eds.), Bioindicators and Biomonitors: Principles, Concepts and Applications, first ed. Elsevier Science Ltd., Amisterdam.

Mauad, C.R., de Wagener, A.L.R., Massone, C.G., da Aniceto, M.S., Lazzari, L., Carreira, R.S., de Farias, C.O., 2015. Urban rivers as conveyors of hydrocarbons to sediments of estuarine areas: source characterization, flow rates and mass accumulation. Sci. Total Environ. 506–507, 656–666. https://doi.org/10.1016/j.scitotenv.2014.11.033.

Meniconi, M.D.F.G., Terezinha Gabardo, I., Rocha Carneiro, M.E., Maria Barbanti, S., Cruz Da Silva, G., German Massone, C., 2002. Brazilian oil spills chemical characterization—case studies. Environ. Forensic 3, 303–321. https://doi.org/10.1006/enfo.2002.0101.

Meniconi, M.D.F.G., et al., 2012. Baía de Guanabara: síntese do conhecimento ambiental. Petrobrás, Rio de Janeiro, p. 337.

Meyer-Reil, L.A., Köster, M., 2000. Eutrophication of marine waters: effects on benthic microbial communities. Mar. Pollut. Bull. 41, 255–263.

Molisani, M.M., Marins, R.V., Machado, W.T., Paraquetti, H.H.M., Lacerda, L.D., 2002. Some implications of inter basin water transfers mercury emission to Sepetiba Bay from Paraı'ba do Sul River, Brazil. In: Lacerda, L.D., Kremer, H., Kjerfve, B., Salomons, W., Marshall, C. (Eds.), South AMERICAN BASINS LOICZ Global

Change Assessment and Synthesis of River Catchment—Coastal Sea Interaction and Human Dimensions. LOICZ Reports and Studies no. 21, Texel, pp. 113–117.

Molisani, M.M., Marins, R.V., Machado, W., Paraquetti, H.H.M., Bidone, E.D., Lacerda, L.D., 2004. Environmental changes in Sepetiba Bay, SE Brazil. Reg. Environ. Chang. 4 (1), 17–27. https://doi.org/10.1007/s10113-003-0060-9.

Molisani, M.M., Kjerfve, B., Silva, A.P., Lacerda, L.D., 2006. Water discharge and sediment load to Sepetiba Bay from an anthropogenically-altered drainage basin, SE Brazil. J. Hydrol. 331 (3–4), 425–433. https://doi.org/10.1016/j.jhydrol.2006.05.038.

Moreira, V.A., Carvalho, A.C.B., Fontana, L.F., Bidone, E.D., Sabadini-Santos, E., 2021. Applying enzymatic biomarkers of the in situ microbial community to assess the sediment risk from Sepetiba Bay (Brazil). Mar. Pollut. Bull. 169. https://doi.org/10.1016/j.marpolbul.2021.112547, 112547.

Morris, B.E.L., Henneberger, R., Huber, H., Moissl-Eichinger, C., 2013. Microbial syntrophy: interaction for the common good. FEMS Microbiol. Rev. 37, 384–406.

Müller, A.L., Pelikan, C., de Rezende, J.R., Wasmund, K., Putz, M., Glombitza, C., Kjeldsen, K.U., Jørgensen, B.B., Loy, A., 2018. Bacterial interactions during sequential degradation of cyanobacterial necromass in a sulfidic arctic marine sediment. Environ. Microbiol. 20 (8), 2927–2940. https://doi.org/10.1111/1462-2920.14297.

Mumtaz Moiz, M., Suk, W.A., Yang, R.S.H., 2010. Introduction to mixtures toxicology and risk assessment. In: Principles and Practice of Mixtures Toxicology, pp. 1–25, https://doi.org/10.1002/9783527630196.ch1.

Mußmann, M., Pjevac, P., Krüger, K., Dyksma, S., 2017. Genomic repertoire of the Woeseiaceae/JTB255, cosmopolitan and abundant core members of microbial communities in marine sediments. ISME J. 11 (5), 1276–1281. https://doi.org/10.1038/ismej.2016.185.

Nascimento, J.R., 2018. Integração De Indicadores Geoquímicos e Microbiológicos na Avaliação da Disposição Oceânica e Continental de Sedimentos Dragados da Baía de Guanabara – RJ, BRASIL. Tese de Doutorado. Univerisade Federal Fluminense, Niterói, RJ, https://doi.org/10.22409/PPG-Geo.2018.d.12065992786.

Nascimento, J.R., Bidone, E.D., Rolão-Araripe, D., Keunecke, K.A., Sabadini-Santos, E., 2016. Trace metal distribution in white shrimp (*Litopenaeus schmitti*) tissues from a Brazilian coastal area. Environ. Earth Sci. 75, 990. https://doi.org/10.1007/s12665-016-5798-8.

Nascimento, J.R., Sabadini-Santos, E., Carvalho, C., Keunecke, K.A., Cesar, R., Bidone, E.D., 2017. Bioaccumulation of heavy metals by shrimp (*Litopenaeus schmitti*): a dose–response approach for coastal resources management. Mar. Pollut. Bull. 114, 1007–1013. https://doi.org/10.1016/j.marpolbul.2016.11.013.

Nascimento, J.R., Silveira, A.E.F., Bidone, E.D., Sabadini-Santos, E., 2019. Microbial community activity in response to multiple contaminant exposure: a feasible tool for sediment quality assessment. Environ. Monit. Assess. 191 (6), 392. https://doi.org/10.1007/s10661-019-7532-y.

Nascimento, J.R., Easson, C.G., de Jurelevicius, D.A., Lopez, J.V., Bidone, E.D., Sabadini-Santos, E., 2020. Microbial community shift under exposure of dredged sediments from a eutrophic bay. Environ. Monit. Assess. 192 (8). https://doi.org/10.1007/s10661-020-08507-8.

Nisbet, E.G., Sleep, N.H., 2001. The habitat and nature of early life. Nature 409, 1083–1091.

Nogales, B., Lanfranconi, M.P., Piña-Villalonga, J.M., Bosch, R., 2011. Anthropogenic perturbations in marine microbial communities. FEMS Microbiol. Rev. 35 (2), 275–298. https://doi.org/10.1111/j.1574-6976.2010.00248.x.

Obbard, J.P., Sauerbeck, D., Jones, K.C., 1994. Dehydrogenase activity of the microbial biomass in soils from a field experiment amended with heavy metal contaminated sewage sludges. Sci. Total Environ. 142, 157–162. https://doi.org/10.1016/0048-9697(94)90323-9.

Odum, E.P., 1969. The strategy of ecosystem development. Science 164 (3877), 262–270.

Odum, E.P., 1985. Trends expected in stressed ecosystems. Bioscience 35 (7), 419–422. https://doi.org/10.2307/1310021.

Oksanen, A.J., Blanchet, F.G., Friendly, M., Kindt, R., Legendre, P., Mcglinn, D., Minchin, P.R., Hara, R.B.O., Simpson, G.L., Solymos, P., Stevens, M.H.H., Szoecs, E., 2015. Multivariate Analysis of Ecological Communities in R: Vegan Tutorial. R Package Version 1.7.

Peakall, D.B., 1994. The role of biomarkers in environmental assessment (1). Introduction. Ecotoxicology 3, 157–160. https://doi.org/10.1007/BF00117082.

Pianosi, F., Beven, K., Freer, J., Hall, J.W., Rougier, J., Stephenson, D.B., Wagener, T., 2016. Sensitivity analysis of environmental models: a systematic review with practical workflow. Environ Model Softw. 79, 214–232. https://doi.org/10.1016/j.envsoft.2016.02.008.

Pjevac, P., Kamyshny, A., Dyksma, S., Mußmann, M., 2014. Microbial consumption of zero-valence sulfur in marine benthic habitats. Environ. Microbiol. 16 (11), 3416–3430. https://doi.org/10.1111/1462-2920.12410.

Probandt, D., Eickhorst, T., Ellrott, A., Amann, R., Knittel, K., 2018. Microbial life on a sand grain: from bulk sediment to single grains. ISME J. 12 (2), 623–633. https://doi.org/10.1038/ismej.2017.197.

Prosser, J.I., Bohannan, B.J.M., Curtis, T.P., Ellis, R.J., Firestone, M.K., Freckleton, R.P., et al., 2007. The role of ecological theory in microbial ecology. Nat. Rev. Microbiol. 5 (5), 384–392. https://doi.org/10.1038/nrmicro1643.

Quast, C., Pruesse, E., Yilmaz, P., Gerken, J., Schweer, T., Yarza, P., Peplies, J., Glöckner, F.O., 2013. The SILVA ribosomal RNA gene database project: improved data processing and web-based tools. Nucleic Acids Res. 41 (D1), 590–596. https://doi.org/10.1093/nar/gks1219.

Ragas, A.M., Teuschler, L.K., Posthuma, L., Cowan, C.E., 2010. Human and ecological risk assessment of chemical mixtures. In: Spurgeon, D.J., Pohl, H.R., Loureiro, S., Lokke, H., Gestel, C.A.M.V. (Eds.), Mixture Toxicity: Linking Approaches From Ecological and Human Toxicology. CRC Press, Boca Raton.

Rainbow, P.S., Phillips, D.J.H., 1993. Cosmopolitan biomonitors of trace metals. Mar. Pollut. 26 (11), 593–601.

Rodrigues, S.K., Abessa, D.M.S., de C Rodrigues, A.P., Soares-Gomes, A., Freitas, C.B., Santelli, R.E., Freire, A.S., Machado, W., 2017. Sediment quality in a metalcontaminated tropical bay assessed with a multiple lines of evidence approach. Environ. Pollut. 228, 265–276. https://doi.org/10.1016/j.envpol.2017.05.045.

Rosado, D., Usero, J., Morillo, J., 2015. Application of a new integrated sediment quality assessment method to Huelva estuary and its littoral of influence (Southwestern Spain). Mar. Pollut. Bull. 98 (1–2), 106–114. https://doi.org/10.1016/j.marpolbul.2015.07.008.

Sabadini-Santos, E., Silva, T.S., Lopes-Rosa, T.D., et al., 2014a. Microbial activities and bioavailable concentrations of Cu, Zn, and Pb in sediments from a tropic and eutrothicated bay. Water Air Soil Pollut. 225. https://doi.org/10.1007/s11270-014-1949-2, 1949.

Sabadini-Santos, E., Senez, T.M., Silva, T.S., Moreira, M.R., Mendonça-Filho, J.G., Santelli, R.E., Crapez, M.A.C., 2014b. Organic matter and pyritization relationship in recent sediments from a tropical and eutrophic bay. Mar. Pollut. Bull. 89 (1–2), 220–228. https://doi.org/10.1016/j.marpolbul.2014.09.055.

Sabadini-Santos, E., Castilhos, Z.C., Bidone, E.D., 2020. Microbial activities response to contamination in soil and sediments rich in as surrounding an industrial gold mine. Water Air Soil Pollut. 231 (7), 1–9. https://doi.org/10.1007/s11270-020-04734-4.

Said, W., Lewis, D.L., 1991. Quantitative assessment of the effects of metals on microbial degradation of organic chemicals. Appl. Environ. Microbiol. 57 (5), 1498–1503. Retrieved from http://aem.asm.org/content/57/5/1498.short.

Saxena, G., Marzinelli, E.M., Naing, N.N., He, Z., Liang, Y., Tom, L., et al., 2015. Ecogenomics reveals metals and land-use pressures on microbial communities in the waterways of a megacity. Environ. Sci. Technol. 49 (3), 1462–1471. https://doi.org/10.1021/es504531s.

Secretaria Especial De Portos—SEP, 2007. Programa Nacional de Dragagem – PND1/PAC1. Brasília, DF. Disponível em: Acesso: out 2020 http://www.portosdobrasil.gov.br/assuntos-1/pnd/arquivos/programa-nacional-de-dragagem-pnd1-pac-1.pdf.

Secretaria Especial De Portos—SEP, 2012. Programa Nacional de Dragagem – PND2/PAC2. Brasília, DF. Disponível em: Acesso: out 2020 http://www.portosdobrasil.gov.br/assuntos-1/pnd/arquivos/cronograma-de-execucao-das-obras-de-dragagem.pdf.

Silveira, A.E.F., Nascimento, J.R., Sabadini-Santos, E., Bidone, E.D., 2017. Screening-level risk assessment applied to dredging of polluted sediments from Guanabara Bay, Rio de Janeiro, Brazil. Mar. Pollut. Bull. 118 (1–2), 368–375. https://doi.org/10.1016/j.marpolbul.2017.03.016.

Smoak, J.M., Sanders, C.J., Patchineelam, S.R., Moore, W.S., 2012. Radium mass balance and submarine groundwater discharge in Sepetiba Bay, Rio de Janeiro State, Brazil. J. South Am. Earth Sci. 39, 44–51.

Sobolev, D., Begonia, M.F.T., 2008. Effects of heavy metal contamination upon soil microbes: Lead-induced changes in general and denitrifying microbial communities as evidenced by molecular markers. Int. J. Environ. Res. Public Health 5, 450–456. https://doi.org/10.3390/ijerph5050450.

Steffan, S.A., Chikaraishi, Y., Currie, C.R., Horn, H., Gaines-Day, H.R., Pauli, J.N., et al., 2015. Microbes are trophic analogs of animals. Proc. Natl. Acad. Sci. 112 (49), 15119–15124. https://doi.org/10.1073/pnas.1508782112.

Stubberfield, L.C.E., Shaw, P.J.A., 1990. A comparison of tetrazolium reduction and FDA\hydrolysis with other measures of microbial\activity. J. Microbiol. Methods 12, 151–162.

Su, H., Pan, C., Ying, G., et al., 2014. Contamination profiles of antibiotic resistance genes in the sediments at a catchment scale. Sci. Total Environ. 490, 708–714. https://doi.org/10.1016/j.scitotenv.2014.05.060.

Thompson, L.R., Sanders, J.G., McDonald, D., Amir, A., Ladau, J., Locey, K.J., Prill, R.J., Tripathi, A., Gibbons, S.M., Ackermann, G., Navas-Molina, J.A., Janssen, S., Kopylova, E., Vázquez-Baeza, Y., González, A., Morton, J.T., Mirarab, S., Zech Xu, Z., Jiang, L., et al., 2017. A communal catalogue reveals Earth's multiscale microbial diversity. Nature 551 (7681), 457–463. https://doi.org/10.1038/nature24621.

Trembath-Reichert, E., Case, D.H., Orphan, V.J., 2016. Characterization of microbial associations with methanotrophic archaea and sulfate-reducing bacteria through statistical comparison of nested magneto-FISH enrichments. PeerJ 4, 1913. https://doi.org/10.7717/peerj.1913.

Trevisan, C.L., et al., 2020. Development of a dredging sensitivity index, applied to an industrialized coastal environment in Brazil. Sci. Total Environ. 748, 141294. https://doi.org/10.1016/j.scitotenv.2020.141294.

Trevors, J.T., Mayfield, C.I., Inniss, W.E., 1982. Measurement of electron transport system (ETS) activity in soil. Microb. Ecol. 8, 163–168. https://doi.org/10.1007/BF02010449.

Tyler, G., 1976. Influence of vanadium on soil phosphatase activity. J. Environ. Qual. 5 (2), 216–217. https://doi.org/10.2134/jeq1976.00472425000500020023x.

Ugarelli, K., Laas, P., Stingl, U., 2018. The microbial communities of leaves and roots associated with turtle grass (*Thalassia testudinum*) and manatee grass (*Syringodium filliforme*) are distinct from seawater and sediment communities, but are similar between species and sampling sites. Microorganisms 7 (1), 4. https://doi.org/10.3390/microorganisms7010004.

Ulitzur, S., Lahav, T., Ulitzur, N., 2002. A novel and sensitive test for rapid determination of water toxicity. Environ. Toxicol. 17 (3), 291–296.

UNCED—United Nations Conference on Environment and Development, 1992. Agenda 21. Rio de Janeiro. https://www.un.org/esa/dsd/agenda21/Agenda21.pdf.

Van Beelen, P., Doelman, P., 1997. Significance and application of microbial toxicity tests in assessing ecotoxicological risks of contaminants in soil and sediment. Chemosphere 34 (3), 455–499. https://doi.org/10.1016/S0045-6535(96)00388-8.

Visscher, P.T., Stolz, F.J., 2005. Microbial mats as bioreactors: populations, processes and products. Palaeogeogr. Palaeoclimatol. Palaeoecol. 219, 87–100. https://doi.org/10.1016/j.palaeo.2004.10.016.

Wang, X.Y., Feng, J., 2007. Assessment of the effectiveness of environmental dredging in South Lake, China. Environ. Manag. 40 (2), 314–322. https://doi.org/10.1007/s00267-006-0132-y.

Whitman, W.B., Coleman, D.C., Wiebe, W.J., 1998. Prokaryotes: the unseen majority. Proc. Natl. Acad. Sci. U. S. A. 95 (12), 6578–6583. https://doi.org/10.1073/pnas.95.12.6578.

Zhou, Q., Zhang, J., Fu, J., Shi, J., Jiang, G., 2008. Biomonitoring: an appealing tool for assessment of metal pollution in the aquatic ecosystem. Anal. Chim. Acta 606 (2), 135–150. https://doi.org/10.1016/j.aca.2007.11.018.

Index

Note: Page numbers followed by *f* indicate figures and *t* indicate tables.

F

G

M

N

O

P

Q

T

U

V

W

X

Y